AF847307

Springer Proceedings in Physics

Volume 169

The series Springer Proceedings in Physics, founded in 1984, is devoted to timely reports of state-of-the-art developments in physics and related sciences. Typically based on material presented at conferences, workshops and similar scientific meetings, volumes published in this series will constitute a comprehensive up-to-date source of reference on a field or subfield of relevance in contemporary physics. Proposals must include the following:—name, place and date of the scientific meeting—a link to the committees (local organization, international advisors etc.)—scientific description of the meeting—list of invited/plenary speakers—an estimate of the planned proceedings book parameters (number of pages/articles, requested number of bulk copies, submission deadline).

More information about this series at http://www.springer.com/series/361

Jorge Rocca • Carmen Menoni • Mario Marconi
Editors

X-Ray Lasers 2014

Proceedings of the 14th International Conference on X-Ray Lasers

Springer

Editors
Jorge Rocca
Colorado State University
Fort Collins
Colorado
USA

Mario Marconi
Colorado State University
Fort Collins
Colorado
USA

Carmen Menoni
Colorado State University
Fort Collins
Colorado
USA

ISSN 0930-8989 ISSN 1867-4941 (electronic)
Springer Proceedings in Physics
ISBN 978-3-319-19520-9 ISBN 978-3-319-19521-6 (eBook)
DOI 10.1007/978-3-319-19521-6

Library of Congress Control Number: 2015949815

Springer Cham Heidelberg New York Dordrecht London
© Springer International Publishing Switzerland 2016
This work is subject to copyright. All rights are reserved by the Publisher, whether the whole or part of the material is concerned, specifically the rights of translation, reprinting, reuse of illustrations, recitation, broadcasting, reproduction on microfilms or in any other physical way, and transmission or information storage and retrieval, electronic adaptation, computer software, or by similar or dissimilar methodology now known or hereafter developed.
The use of general descriptive names, registered names, trademarks, service marks, etc. in this publication does not imply, even in the absence of a specific statement, that such names are exempt from the relevant protective laws and regulations and therefore free for general use.
The publisher, the authors and the editors are safe to assume that the advice and information in this book are believed to be true and accurate at the date of publication. Neither the publisher nor the authors or the editors give a warranty, express or implied, with respect to the material contained herein or for any errors or omissions that may have been made.

Printed on acid-free paper

Springer International Publishing AG Switzerland is part of Springer Science+Business Media (www.springer.com)

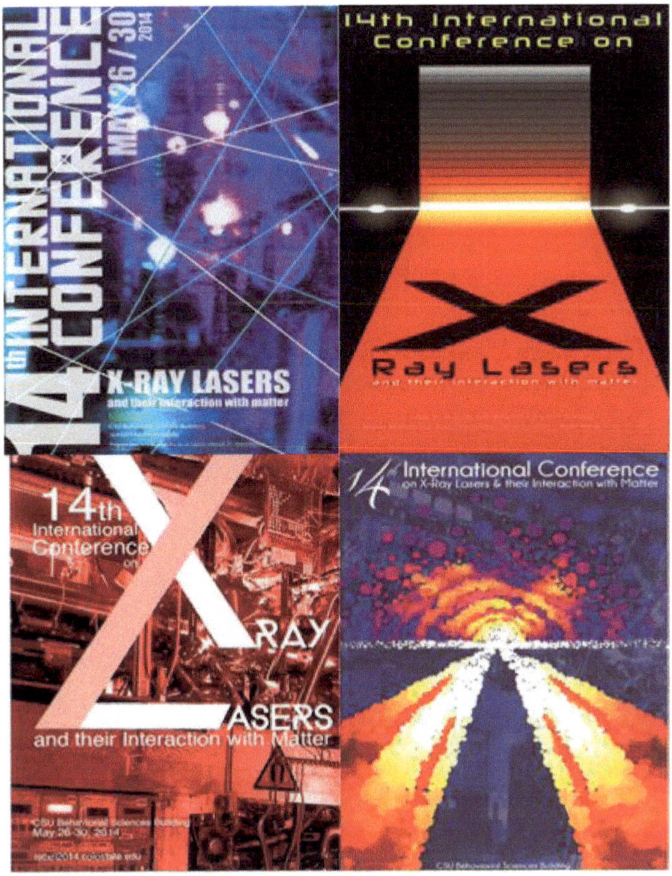

4 1

3 2

Conference posters by: Aubrey Michelle Aronson (1), Audrey Yue Yamamoto (2, 3), and Hannah Lynn Hibbs (4), students of Liberal Arts at Colorado State University. These posters are among 54 submissions to a Poster Exhibition and Competition The poster competition was under the direction of Professor Phil Risbeck of the Colorado State University Art Department.

To Jim Dunn
We would like to dedicate this conference proceedings to Dr. Jim Dunn. Jim, an outstanding scientist and a dear friend, is recognized for major contributions to the field of X-ray lasers, in which he was active for most of his career. He also made significant contributions to the study of high energy density plasmas. Jim passed away on March 31st, 2014. He co-chaired the 2002 ICXRL, in Aspen, Colorado, and he was engaged in the organization of the 2014 ICXRL until the last moment. He was hoping to be able to attend. This was not to happen, and many of us miss him. However, years from now, the 31st March and 26th May 2014 will intermingle and chronology will be lost. And then, in some way, it will be right to claim that Jim was with us, participating in this conference.*

Jorge Rocca
Carmen Menoni
Mario Marconi

*The last two sentences are adapted from an idea expressed in the prologue of "El Hacedor" ("The Maker"), by Jorge Luis Borges.

Preface

This volume includes papers presented at the 14th International Conference on X-Ray Lasers that took place in the campus of Colorado State University in Fort Collins, located in the foothills of the Colorado Rocky Mountains during May 26–30, 2014. This conference is the 14th in a series of bi-annual meetings that were held in locations all over the world, starting with a meeting in Aussois, France, in 1986. Through the past 28 years this, conference has been the primary international event in which the latest advances in the development of X-ray laser sources and applications is reported. The conference also covers advances in supporting technology and instrumentation, and alternative methods to generate coherent x-ray radiation.

The Fort Collins meeting was attended by more than 80 participants from 18 countries. It comprised invited and contributed talks, and a poster session. The technical content of these proceedings highlights recent achievements in the development of plasma-based X-ray lasers, the generation of coherent X-ray radiation by the generation high harmonic and free-electron lasers, and proposed new schemes for the generation of ultrashort wavelength coherent radiation. It also covers applications of compact soft X-ray sources to nano-scale imaging, lithography, photochemistry, and laser-material interactions.

A highlight of this conference was the display of 54 artistic posters illustrating the generation and application of X-ray Laser beams, that are the creative realization of students from the Arts Department at Colorado State University. Four of these posters are shown in these proceedings. Following the tradition of keeping this conference as a truly international event, the International Advisory Board of the Conference choose Japan as the location of the next conference, schedule to take place in 2016.

We would like to thank the International Advisory Board of the Conference for its role in organizing the scientific program. We also would like to thank the valuable assistance of Bob Bower, and Kaarin Goncz in organizing and helping to run the meeting smoothly, Prof. Phil Risbeck and his students for the creation of the

54 artistic posters that illustrated the Conference, to Bart Hoff for enlightening an evening pool party with his musical talent, and to Colorado State University Provost Rick Miranda for securing the modern auditorium in which the oral presentations took place.

Fort Collins, Colorado

Jorge Rocca
Carmen Menoni
Mario Marconi

Contents

Contributors

D. E. Adams JILA, University of Colorado at Boulder, Boulder, CO, USA

M. Agåker Department of Physics and Astronomy, Uppsala University, Uppsala, Sweden

I. U. Ahad Institute of Optoelectronics, Military University of Technology, Warsaw, Poland

M. Albrecht IoP ASCR, Na Slovance 2, Prague 8, Czech Republic

FNSPE, CTU, Prague, Czech Republic

C. N. Anderson Center for X-Ray Optics, Berkeley Lab, Berkeley, CA, USA

E. H. Anderson National Science Foundation Engineering Research Center for Extreme Ultraviolet Science and Technology, Colorado State University, Fort Collins, USA

Center for X-Ray Optics, Lawrence Berkeley National Laboratory, Berkeley, CA, USA

A. Apolonski Max-Planck-Institut für Quantenoptik, Garching, Germany

Ludwig-Maximilians-Universität München, Garching, Germany

I. A. Artyukov Lebedev Physical Institute, Moscow, Russia

B. Aurand Helmholtz Institute Jena, Jena, Germany

GSI Helmholtz Centre for Heavy Ion Research, Darmstadt, Germany

Department of Physics, Lund University, Lund, Sweden

G. Avaria Colorado State University, Fort Collins, CO, USA

National Science Foundation ERC for Extreme Ultraviolet Science and Technology, Fort Collins, CO, USA

Department of Electrical and Computing Engineering, Colorado State University, Fort Collins, CO, USA

Comisión Chilena de Energía Nuclear, Santiago, Chile

J. Balmer Institute of Applied Physics, University of Bern, Bern, Switzerland

R. A. Banici National Institute for Lasers, Plasma and Radiation Physics (INFLPR), Magurele, Romania

J. Bao SRI International, Menlo Park, CA, USA

C. Bargsten Colorado State University, Fort Collins, CO, USA

A. Bartnik Institute of Optoelectronics, Military University of Technology, Warsaw, Poland

C. Baumgarten NSF ERC for Extreme Ultraviolet Science and Technology, Fort Collins, CO, USA

Department of Physics, Colorado State University, Fort Collins, CO, USA

E. Baynard LUMAT, CNRS, Universite Paris-Sud, Orsay Cedex, France

LASERIX, Centre Laser Universite Paris Sud, Sud, Paris

K. Bergmann Fraunhofer Institute for Laser Technology, Aachen, Germany

E. R. Bernstein Engineering Research Center for Extreme Ultraviolet Science and Technology, Fort Collins, USA

Department of Chemistry and NSF ERCEUVST, Colorado State University, Fort Collins, CO, USA

M. Berrill Oak Ridge National Laboratory, OAK Ridge, USA

NSF ERC for Extreme Ultraviolet Science and Technology, Fort Collins, CO, USA

Department of Electrical and Computer Engineering, Colorado State University, Fort Collins, CO, USA

J. Bierbach Helmholtz-Institut Jena, Jena, Germany

Friedrich-Schiller-Universität Jena, Jena, Germany

P. Bolton Quantum Beam Science Center, Japan Atomic Energy Agency (JAEA), Kyoto, Japan

C. Bostedt Linac Coherent Light Source, SLAC National Accelerator Laboratory, Menlo Park, CA, USA

J. D. Bozek Linac Coherent Light Source, SLAC National Accelerator Laboratory, Menlo Park, CA, USA

N. F. Brejnholt Lawrence Livermore National Laboratory, Livermore, CA, USA

R. Bresenitz Experimental Physics of EUV, RWTH Aachen University, JARA-FIT, Aachen, Germany

Peter Grünberg Institut (PGI-9), Forschungszentrum Jülich GmbH, JARA-FIT, Jülich, Germany

W. S. Brocklesby Optoelectronics Research Centre, University of Southampton, Southampton, UK

S. Brose Chair for the Technology of Optical Systems, RWTH Aachen University and JARA-FIT, Aachen, Germany

J. Bußmann Experimental Physics of EUV, RWTH Aachen University, JARA-FIT, Aachen, Germany

Peter Grünberg Institut (PGI-9), Forschungszentrum Jülich GmbH, JARA-FIT, Jülich, Germany

S. V. Bulanov Quantum Beam Science Center, Japan Atomic Energy Agency (JAEA), Kyoto, Japan

M. Busquet APTEC Inc, Muskegon, USA

Ellicott City, ARTEP Inc, MD, USA

A. Calisti Aix Marseille Université, CNRS, PIIM UMR 7345, Marseille, France

D. Carlton National Science Foundation Engineering Research Center for Extreme Ultraviolet Science and Technology, Colorado State University, Fort Collins, USA

Center for X-Ray Optics, Lawrence Berkeley National Laboratory, Berkeley, CA, USA

D. B. Carlton Engineering Research Center for Extreme Ultraviolet Science and Technology, Fort Collins, USA

Center for X-Ray Optics, Lawrence Berkeley National Laboratory, Berkeley, CA, USA

S. Carron Monterro Linac Coherent Light Source, SLAC National Accelerator Laboratory, Menlo Park, CA, USA

H. Carstens Max-Planck-Institut für Quantenoptik, Garching, Germany

Ludwig-Maximilians-Universität München, Garching, Germany

K. Cassou LPGP, CNRS, Universite Paris Sud, Orsay Cedex, France

Campus de lENSTA, Chemin de la Huniere, LASERIX, Universite Paris-Sud, Palaiseau Cedex, France

W. Chao Engineering Research Center for Extreme Ultraviolet Science and Technology, Fort Collins, USA

Center for X-Ray Optics, Lawrence Berkeley National Laboratory, Berkeley, CA, USA

National Science Foundation Engineering Research Center for Extreme Ultraviolet Science and Technology, Colorado State University, Fort Collins, USA

I. U. Chaudhary University of Engineering and Technology, Lahore, Pakistan

L. M. Chen Beijing National Laboratory of Condensed Matter Physics, Institute of Physics, CAS, Beijing, China

R. N. Coffee Linac Coherent Light Source, SLAC National Accelerator Laboratory, Menlo Park, CA, USA

Stanford PULSE Institute, SLAC National Accelerator Laboratory, Menlo Park, California

O. Cohen Solid State Institute and Physics Department, Technion, Haifa, Israel

G. V. Cojocaru National Institute for Lasers, Plasma and Radiation Physics (INFLPR), Magurele, Romania

Faculty of Physics, University of Bucharest, Magurele, Romania

Laboratoire de Physique des Gaz et des Plasmas, Universite Paris-Sud (UMR-CNRS 8578), Cedex, France

M. Coughlan Department of Physics and Astronomy, Queen's University Belfast, Belfast, UK

S. Cousens Department of Physics and Astronomy, Queen's University Belfast, Belfast, UK

D. C. Crick Department of Microbiology, Immunology, and Pathology, Colorado State University, Fort Collins, USA

O. Culfa York Plasma Institute, The University of York, York, UK

S. Daboussi Laboratoire d'Optique Appliquée, ENSTA ParisTech - CNRS UMR 7639 - École Polytechnique, Chemin de la Hunière, Palaiseau, France

R. Dabu National Institute for Lasers, Plasma and Radiation Physics (INFLPR), Magurele, Romania

H. Daido Tsuruga Head Office, Applied Laser Technology Institute, Japan Atomic Energy Agency, Tsuruga, Japan

S. Danylyuk Chair for the Technology of Optical Systems, RWTH Aachen University and JARA-FIT, Aachen, Germany

A. Dehlinger Berlin Laboratory for innovative X-ray Technologies (BLiX), Berlin, Germany

Max-Born Institut für Nichtlineare Optik und Kurzzeitspektroskopie (MBI), Berlin, Germany

O. Delmas Campus de lENSTA, Chemin de la Huniere, LASERIX, Universite Paris-Sud, Palaiseau Cedex, France

Laboratoire de Physique des Gaz et des Plasmas, Universite Paris-Sud (UMR-CNRS 8578), Cedex, France

J. Demailly Campus de lENSTA, Chemin de la Huniere, LASERIX, Universite Paris-Sud, Palaiseau Cedex, France

Laboratoire de Physique des Gaz et des Plasmas, Universite Paris-Sud (UMR-CNRS 8578), Cedex, France

M.-A. Descalle Lawrence Livermore National Laboratory, Livermore, CA, USA

F. Dong Department of Chemistry and NSF ERCEUVST, Colorado State University, Fort Collins, CO, USA

B. Dromey Department of Physics and Astronomy, Queen's University Belfast, Belfast, UK

J. Dunn Lawrence Livermore National Laboratory, Livermore, CA, USA

B. Ecker Helmholtz Institute Jena, Jena, Germany

GSI Helmholtz Centre for Heavy Ion Research, Darmstadt, Germany

E. Eckner Helmholtz-Institut Jena, Jena, Germany

Friedrich-Schiller-Universität Jena, Jena, Germany

B. Erk DESY, Hamburg, Germany

M. Förster Friedrich-Schiller-Universität Jena, Jena, Germany

A. Y. Faenov Joint Institute for High Temperatures, Russian Academy of Science, Moscow, Russia

Institute for Academic Initiatives, Osaka University, Osaka, Japan

M. Farjardo Centro de Fisica de Plasmas, Instituto Superior Técnico, Lisbon, Portugal

R. Feifel Department of Physics, University of Gothenburg, Gothenburg, Sweden

M. Fernández-Perea Lawrence Livermore National Laboratory, Livermore, CA, USA

R. M. Feshchenko P.N. Lebedev Physical Institute,Russian Academy of Sciences, Moscow, Russia

H. Fiedorowicz Institute of Optoelectronics, Military University of Technology, Warsaw, Poland

J. Filevich Engineering Research Center for Extreme Ultraviolet Science and Technology, Fort Collins, USA

Electrical and Computer Engineering, Colorado State University, Fort Collins, USA

E. Fill Max-Planck-Institut für Quantenoptik, Garching, Germany

Ludwig-Maximilians-Universität München, Garching, Germany

P. Fischer Center for X-Ray Optics, Berkeley Lab, Berkeley, CA, USA

A. Fleischer Solid State Institute and Physics Department, Technion, Haifa, Israel

Department of Physics and Optical Engineering, Ort Braude College, Karmiel, Israel

T. Fok Institute of Optoelectronics, Military University of Technology, Warsaw, Poland

O. Frolov Academy of Sciences of the Czech Republic, Institute of Plasma Physics, v.v.i., Prague, Czech Republic

D. F. Gardner JILA, University of Colorado at Boulder, Boulder, CO, USA

J. Gautier IoP ASCR, Na Slovance 2, Prague 8, Czech Republic

LOA, ENSTA, Chemin de la Hunière, Palaiseau cedex, France

K. A. Goldberg Center for X-Ray Optics, Berkeley Lab, Berkeley, CA, USA

D. Gordon Naval Research Lab, Washington DC, USA

M. Grisham Colorado State University, Fort Collins, CO, USA

National Science Foundation ERC for Extreme Ultraviolet Science and Technology, Fort Collins, CO, USA

Department of Electrical and Computing Engineering, Colorado State University, Fort Collins, CO, USA

P. Grychtol JILA, University of Colorado, Boulder, CO, USA

O. Guilbaud Laboratoire de Physique des Gaz et des Plasmas, Universite Paris-Sud (UMR-CNRS 8578), Cedex, France

LASERIX, Centre Laser Universite Paris Sud, Sud, Paris

S. Guilbaud Campus de lENSTA, Chemin de la Huniere, LASERIX, Universite Paris-Sud, Palaiseau Cedex, France

E. M. Gullikson Lawrence Berkeley National Laboratory, Berkeley, CA, USA

Center for X-Ray Optics, Berkeley Lab, Berkeley, CA, USA

Y. Q. Guo Department of Chemistry and NSF ERCEUVST, Colorado State University, Fort Collins, CO, USA

P. L. Hagelstein Massachusetts Institute of Technology, Cambridge, MA, USA

N. Hasegawa Quantum Beam Science Center, Japan Atomic Energy Agency (JAEA), Kyoto, Japan

S. Heinbuch Department of Electrical and Computer Engineering, NSF ERCEUVST, Colorado State University, Fort Collins, CO, USA

A. Hoffmann IOQ, Abbe Center of Photonics, Friedrich Schiller University, Jena, Germany

R. Hollinger Colorado State University, Fort Collins, CO, USA

S. Holzberger Max-Planck-Institut für Quantenoptik, Garching, Germany

Ludwig-Maximilians-Universität München, Garching, Germany

I. D. Howlett National Science Foundation Engineering Research Center for Extreme Ultraviolet Science and Technology, Colorado State University, Fort Collins, USA

Electrical and Computer Engineering, Colorado State University, Fort Collins, USA

K. Huang Beijing National Laboratory of Condensed Matter Physics, Institute of Physics, CAS, Beijing, China

M. Ilchen Linac Coherent Light Source, SLAC National Accelerator Laboratory, Menlo Park, CA, USA

European XFEL GmbH, Hamburg, Germany

T. Imazono Quantum Beam Science Center, Japan Atomic Energy Agency (JAEA), Kyoto, Japan

L. Ionel National Institute for Lasers, Plasma and Radiation Physics (INFLPR), Magurele, Romania

M. Ishino Quantum Beam Science Center, Japan Atomic Energy Agency (JAEA), Kyoto, Japan

K. A. Janulewicz Department of Physics and Photon Science, Gwangju Institute of Science and Technology, Gwangju, Republic of Korea

R. Jarocki Institute of Optoelectronics, Military University of Technology, Warsaw, Poland

T. M. Jeong Center for Relativistic Laser Science, Institute for Basic Science (IBS), Gwangju, Korea

Advanced Photonics Research Institute, Gwangju Institute of Science and Technology (GIST), Gwangju, Korea

F. Jia Institute of Applied Physics, University of Bern, Bern, Switzerland

R. Jung Max-Born Institut für Nichtlineare Optik und Kurzzeitspektroskopie (MBI), Berlin, Germany

Berlin Laboratory for innovative X-ray Technologies (BLiX), Berlin, Germany

L. Juschkin Experimental Physics of EUV, RWTH Aachen University, JARA-FIT, Aachen, Germany

Peter Grünberg Institut (PGI-9), Forschungszentrum Jülich GmbH, JARA-FIT, Jülich, Germany

Chair for the Experimental Physics of EUV, RWTH Aachen University and JARA-FIT, Aachen, Germany

J. Küpper Center for Free-Electron Laser Science, Hamburg, Germany

University of Hamburg, Hamburg, Germany

DESY, Hamburg, Germany

M. Kando Quantum Beam Science Center, Japan Atomic Energy Agency (JAEA), Kyoto, Japan

H. C. Kapteyn JILA, University of Colorado at Boulder, Boulder, CO, USA

Y. Kato Graduate School for Photonics and Industries (GPI), Shizuoka, Japan

T. Kawachi Quantum Beam Science Center, Japan Atomic Energy Agency (JAEA), Kyoto, Japan

O. Kazamias Campus de lENSTA, Chemin de la Huniere, LASERIX, Universite Paris-Sud, Palaiseau Cedex, France

S. Kazamias Laboratoire de Physique des Gaz et des Plasmas, Universite Paris-Sud (UMR-CNRS 8578), Cedex, France

LASERIX, Centre Laser Universite Paris Sud, Sud, Paris

D. Keiss Colorado State University, Fort Collins, CO, USA

O. Kfir Solid State Institute and Physics Department, Technion, Haifa, Israel

T. Kierspel Center for Free-Electron Laser Science, Hamburg, Germany

University of Hamburg, Hamburg, Germany

DESY, Hamburg, Germany

C. M. Kim Institute of Basic Research and Advanced Photonics Research Institute, Gwangju Institute of Science and Technology, Gwangju, Republic of Korea

Center for Relativistic Laser Science, Institute for Basic Science (IBS), Gwangju, Korea

Advanced Photonics Research Institute, Gwangju Institute of Science and Technology (GIST), Gwangju, Korea

H. Kim Chair for the Experimental Physics of EUV, RWTH Aachen University and JARA-FIT, Aachen, Germany

Optoelectronics Research Center, University of Southampton, Southampton, UK

H. T. Kim Center for Relativistic Laser Science, Institute for Basic Science (IBS), Gwangju, Korea

Advanced Photonics Research Institute, Gwangju Institute of Science and Technology (GIST), Gwangju, Korea

I. J. Kim Center for Relativistic Laser Science, Institute for Basic Science (IBS), Gwangju, Korea

Advanced Photonics Research Institute, Gwangju Institute of Science and Technology (GIST), Gwangju, Korea

S.-S. Kim Samsung Electronics Co., Ltd., Hwasung, Gyeonggi, Korea

V. Kimberg Max Planck Institute for the Physics of Complex Systems, Dresden, Germany

Center for Free-Electron Laser Science, Hamburg, Germany

H. Kiriyama Quantum Beam Science Center, Japan Atomic Energy Agency (JAEA), Kyoto, Japan

M. Kishimoto Quantum Beam Science Center, Japan Atomic Energy Agency, Kyoto, Japan

A. Klisnick Institute of Fluid Physics, China Acdemy of Engineering Physics, Mianyang, China

Univ Paris 11, CNRS, ISMO, Orsay, France

Institut des Sciences Moleculaires d'Orsay, CNRS, Universite Paris-Sud, Orsay Cedex, France

R. Knut JILA, University of Colorado, Boulder, CO, USA

Electromagnetics Division, NIST, Boulder, CO, USA

K. Kolacek Academy of Sciences of the Czech Republic, Institute of Plasma Physics, v.v.i., Prague, Czech Republic

K. Kondo Quantum Beam Science Center, Japan Atomic Energy Agency (JAEA), Kyoto, Japan

J. Kostecki Institute of Optoelectronics, Military University of Technology, Warsaw, Poland

M. Kozlova IoP ASCR, Na Slovance 2, Prague 8, Czech Republic

IPP ASCR, Za Slovankou 3, Prague 8, Czech Republic

IoP ASCR, ELI Beamlines Project, Prague, Czech Republic

F. Krausz Max-Planck-Institut für Quantenoptik, Garching, Germany

Ludwig-Maximilians-Universität München, Garching, Germany

J. Krzywinski Linac Coherent Light Source, SLAC National Accelerator Laboratory, Menlo Park, CA, USA

T. Kuehl Helmholtz Institute Jena, Jena, Germany

GSI Helmholtz Centre for Heavy Ion Research, Darmstadt, Germany

S. Kuschel Helmholtz-Institut Jena, Jena, Germany

Friedrich-Schiller-Universität Jena, Jena, Germany

I. Kuznetsov Engineering Research Center for Extreme Ultraviolet Science and Technology, Fort Collins, USA

Electrical and Computer Engineering, Colorado State University, Fort Collins, USA

B. Landgraf IOQ, Abbe Center of Photonics, Friedrich Schiller University, Jena, Germany

Helmholtz Institute Jena, Jena, Germany

O. Larroche CEA DIF, Bruyères le Châtel, Arpajon Cedex, France

T. T. Thuy Le Laboratoire de Physique des Gaz et des Plasmas, Universite Paris XI, Orsay cedex, France

Loratoire d'Optique Appliquée (LOA), ENSTA-École Polytechnique, Palaiseau, France

D. Lee Samsung Electronics Co., Ltd., Hwasung, Gyeonggi, Korea

K. H. Lee Center for Relativistic Laser Science, Institute for Basic Science (IBS), Gwangju, Korea

S. K. Lee Center for Relativistic Laser Science, Institute for Basic Science (IBS), Gwangju, Korea

Advanced Photonics Research Institute, Gwangju Institute of Science and Technology (GIST), Gwangju, Korea

H. Legall Berlin Laboratory for innovative X-ray Technologies (BLiX), Berlin, Germany

Max-Born Institut für Nichtlineare Optik und Kurzzeitspektroskopie (MBI), Berlin, Germany

C. L. S. Lewis Department of Physics, Queens University Belfast, Belfast, UK

D. Z. Li Institute of High Energy Physics, CAS, Beijing, China

J. Li Colorado State University, Fort Collins, CO, USA

National Science Foundation ERC for Extreme Ultraviolet Science and Technology, Fort Collins, CO, USA

Department of Electrical and Computing Engineering, Colorado State University, Fort Collins, CO, USA

L. Li Loratoire d'Optique Appliquée (LOA), ENSTA-École Polytechnique, Palaiseau, France

Laboratoire d'Optique Appliquee, UMR7639, ENSTA ParisTech, CNRS, Ecole Polytechnique, Palaiseau, France

School of Nuclear Science and Technology, Lanzhou University, Lanzhou, China

W. Li NSF Engineering Research Center for Extreme Ultraviolet Science and Technology and Electrical and Computer Engineering Department, Colorado State University, Fort Collins, USA

Y. T. Li Beijing National Laboratory of Condensed Matter Physics, Institute of Physics, CAS, Beijing, China

N. Lilienfein Max-Planck-Institut für Quantenoptik, Garching, Germany

Ludwig-Maximilians-Universität München, Garching, Germany

A. O. Lindahl Department of Physics, University of Gothenburg, Gothenburg, Sweden

Stanford PULSE Institute, SLAC National Accelerator Laboratory, Menlo Park, California

R. A. London Lawrence Livermore National Laboratory, Livermore, CA, USA

P. Loosen Chair for the Technology of Optical Systems, RWTH Aachen University and JARA-FIT, Aachen, Germany

Y. Luo Princeton University, Princeton, NJ, USA

B. Luther National Science Foundation ERC for Extreme Ultraviolet Science and Technology and Colorado State University, Fort Collins, USA

Oak Ridge National Laboratory, Oak Ridge, USA

A. Lutman Linac Coherent Light Source, SLAC National Accelerator Laboratory, Menlo Park, CA, USA

O. D. Mücke Center for Free-Electron Laser Science, Hamburg, Germany

DESY, Hamburg, Germany

Y. Ma Beijing National Laboratory of Condensed Matter Physics, Institute of Physics, CAS, Beijing, China

M. Magnitskiy M.V. Lomonosov Moscow State University, Moscow, Russia

E. Malm NSF Engineering Research Center for Extreme Ultraviolet Science and Technology and Electrical and Computer Engineering Department, Colorado State University, Fort Collins, USA

Center for X-Ray Optics, Lawrence Berkeley National Laboratory, Berkeley, CA, USA

M. C. Marconi NSF Engineering Research Center for Extreme Ultraviolet Science and Technology and Electrical and Computer Engineering Department, Colorado State University, Fort Collins, USA

A. Le Marec Institute of Fluid Physics, China Acdemy of Engineering Physics, Mianyang, China

Institut des Sciences Moleculaires d'Orsay, CNRS, Universite Paris-Sud, Orsay Cedex, France

ISMO, CNRS, Université Paris-Sud, Orsay Cedex, France

Univ Paris 11, CNRS, ISMO, Orsay, France

M. Maruyama Quantum Beam Science Center, Japan Atomic Energy Agency (JAEA), Kyoto, Japan

B. Matouš Department of Physics and Photon Science, Gwangju Institute of Science and Technology, Gwangju, Republic of Korea

T. Maxwell Linac Coherent Light Source, SLAC National Accelerator Laboratory, Menlo Park, CA, USA

G. Maynard Laboratoire de Physique des Gaz et des Plasmas, Universite Paris XI, Orsay cedex, France

M. C. H. McKubre SRI International, Menlo Park, CA, USA

L. Meng Institute of Fluid Physics, China Acdemy of Engineering Physics, Mianyang, China

C. S. Menoni NSF Engineering Research Center for Extreme Ultraviolet Science and Technology and Electrical and Computer Engineering Department, Colorado State University, Fort Collins, USA

A. Mestrallain Campus de lENSTA, Chemin de la Huniere, LASERIX, Universite Paris-Sud, Palaiseau Cedex, France

J. Miao Department of Physics and Astronomy, and California Nano Systems Institute, University of California, Los Angeles, CA, USA

R. Miyakawa Center for X-ray Optics, Lawrence Berkeley National Lab, Berkeley, CA, USA

N. Monserud NSF Engineering Research Center for Extreme Ultraviolet Science and Technology and Electrical and Computer Engineering Department, Colorado State University, Fort Collins, USA

A. Morozov Princeton University, Princeton, NJ, USA

M. Mucke Department of Physics and Astronomy, Uppsala University, Uppsala, Sweden

T. Mullins Center for Free-Electron Laser Science, Hamburg, Germany

University of Hamburg, Hamburg, Germany

DESY, Hamburg, Germany

M. M. Murnane JILA, University of Colorado at Boulder, Boulder, CO, USA

C. H. Nam Center for Relativistic Laser Science, Institute for Basic Science (IBS), Gwangju, Korea

Dept. of Physics and Photon Science, GIST, Gwangju, Korea

S. Namba Graduate School of Engineering, Hiroshima University, Higashihiroshima, Japan

P. P. Naulleau Center for X-ray Optics, Lawrence Berkeley National Lab, Berkeley, CA, USA

J. Nejdl IoP ASCR, Na Slovance 2, Prague 8, Czech Republic

IPP ASCR, Za Slovankou 3, Prague 8, Czech Republic

National Science Foundation Engineering Research Center for Extreme Ultraviolet Science and Technology, Colorado State University, Fort Collins, USA

ELI Beamlines project, Institute of Physics of AS CR, Prague, Czech Republic

H. Nembach Electromagnetics Division, NIST, Boulder, CO, USA

O. Neveu Laboratoire de Physique des Gaz et des Plasmas, Universite Paris-Sud (UMR-CNRS 8578), Cedex, France

J. Nilsen Lawrence Livermore National Laboratory, Livermore, CA, USA

M. Nishikino Quantum Beam Science Center, Japan Atomic Energy Agency (JAEA), Kyoto, Japan

J. Nordgren Department of Physics and Astronomy, Uppsala University, Uppsala, Sweden

M. Odstrčil Experimental Physics of EUV, RWTH Aachen University, JARA-FIT, Aachen, Germany

Optoelectronics Research Centre, University of Southampton, Southampton, UK

E. Oliva Loratoire d'Optique Appliquée (LOA), ENSTA-École Polytechnique, Palaiseau, France

Laboratoire de Physique des Gaz et des Plasmas, Universite Paris XI, Orsay cedex, France

K. H. Pae Advanced Photonics Research Institute, Gwangju Institute of Science and Technology (GIST), Gwangju, Korea

J. Park Samsung Electronics Co., Ltd., Hwasung, Gyeonggi, Korea

D. Patel NSF Engineering Research Center for Extreme Ultraviolet Science and Technology and Electrical and Computer Engineering Department, Colorado State University, Fort Collins, USA

G. Paulus Helmholtz-Institut Jena, Jena, Germany

Friedrich-Schiller-Universität Jena, Jena, Germany

W. Pengfei State Key Laboratory of High Field Laser Physics, Shanghai Institute of Optics and Fine Mechanics, Chinese Academy of Sciences, Shanghai, China

College of Physics and Electronic Information Engineering, Wenzhou University, Wenzhou, Zhejiang Province, China

T. A. Pikuz Quantum Beam Science Center, Japan Atomic Energy Agency (JAEA), Kyoto, Japan

Joint Institute of High Temperature, Russian Academy of Science, Moscow, Russia

A. S. Pirozhkov Quantum Beam Science Center, Japan Atomic Energy Agency (JAEA), Kyoto, Japan

M. Pittman LUMAT, CNRS, Universite Paris-Sud, Orsay Cedex, France

Campus de lENSTA, Chemin de la Huniere, LASERIX, Universite Paris-Sud, Palaiseau Cedex, France

Laboratoire de Physique des Gaz et des Plasmas, Universite Paris-Sud (UMR-CNRS 8578), Cedex, France

M. J. Pivovaroff Lawrence Livermore National Laboratory, Livermore, CA, USA

I. V. Pogorelsky Accelerator Test Facility, Collider Accelerator Dept., Brookhaven National Laboratory, Upton, NY, US

D. Popmintchev JILA, University of Colorado, Boulder, CO, USA

T. Popmintchev JILA, University of Colorado, Boulder, CO, USA

N. L. Popov Lebedev Physical Institute, Moscow, Russia

C. Pratsch Helmholtz-Zentrum Berlin (HZB), Berlin, Germany

A. Prieto Colorado State University, Fort Collins, CO, USA

C. Proux Loratoire d'Optique Appliquée (LOA), ENSTA-École Polytechnique, Palaiseau, France

A. Pukhov Institut für Theoretische Physik, Heinrich-Heine-Universität Düsseldorf, Düsseldorf, Germany

I. Pupeza Max-Planck-Institut für Quantenoptik, Garching, Germany

Ludwig-Maximilians-Universität München, Garching, Germany

M. A. Purvis Colorado State University, Fort Collins, CO, USA

C. Rödel Helmholtz-Institut Jena, Jena, Germany

Friedrich-Schiller-Universität Jena, Jena, Germany

B. A. Reagan NSF ERC for Extreme Ultraviolet Science and Technology, Fort Collins, CO, USA

Department of Electrical and Computer Engineering, Colorado State University, Fort Collins, CO, USA

J. C. Robinson Lawrence Livermore National Laboratory, Livermore, CA, USA

J. Rocca Colorado State University, Fort Collins, CO, USA

Colorado State University, National Science Foundation ERC for Extreme Ultraviolet Science and Technology, Fort Collins, CO, USA

Department of Electrical and Computing Engineering, Colorado State University, Fort Collins, CO, USA

Department of Physics, Colorado State University, Fort Collins, CO, USA

Center for X-Ray Optics, Lawrence Berkeley National Laboratory, Berkeley, CA, USA

N. Rohringer Max Planck Institute for the Physics of Complex Systems, Dresden, Germany

Center for Free-Electron Laser Science, Hamburg, Germany

D. Rolles DESY, Hamburg, Germany

D. Ros Laboratoire de Physique des Gaz et des Plasmas, Universite Paris-Sud (UMR-CNRS 8578), Cedex, France

Loratoire d'Optique Appliquée (LOA), ENSTA-École Polytechnique, Palaiseau, France

Campus de lENSTA, Chemin de la Huniere, LASERIX, Universite Paris-Sud, Palaiseau Cedex, France

Laboratoire de Physique des Gaz et des Plasmas, Universite Paris XI, Orsay cedex, France

S. J. Rose South Kensington Campus, Imperial College London, London, UK

A. Rossal Department of Physics, University of York,, York YO10 5DD, UK

A. K. Rossall York Plasma Institute, Department of Physics, University of York, Heslington, York, UK

J. E. Rubensson Department of Physics and Astronomy, Uppsala University, Uppsala, Sweden

D. Rudolf Experimental Physics of EUV, RWTH Aachen University, JARA-FIT, Aachen, Germany

Peter Grünberg Institut (PGI-9), Forschungszentrum Jülich GmbH, JARA-FIT, Jülich, Germany

L. Ruxin State Key Laboratory of High Field Laser Physics, Shanghai Institute of Optics and Fine Mechanics, Chinese Academy of Sciences, Shanghai, China

School of Physical Science and Technology, Shanghai Tech University, Shanghai, China

D. Ryan Colorado State University, Fort Collins, CO, USA

S. G. Rykovanov Helmholtz-Institut Jena, Jena, Germany

C. Såthe Department of Physics and Astronomy, Uppsala University, Uppsala, Sweden

A. Sanchez-Gonzalez Imperial College, London, United Kingdom

A. Sasaki Quantum Beam Science Center, Japan Atomic Energy Agency (JAEA), Kyoto, Japan

J. Schmidt Academy of Sciences of the Czech Republic, Institute of Plasma Physics, v.v.i., Prague, Czech Republic

M. Scully Princeton University, Princeton, NJ, USA

Texas A&M University, College Station, TX, USA

M. Seaberg JILA, University of Colorado at Boulder, Boulder, CO, USA

S. Sebban IoP ASCR, Na Slovance 2, Prague 8, Czech Republic

Laboratoire d'Optique Appliquee, UMR7639, ENSTA ParisTech, CNRS, Ecole Polytechnique, Palaiseau, France

C. Seim Berlin Laboratory for innovative X-ray Technologies (BLiX), Berlin, Germany

Max-Born Institut für Nichtlineare Optik und Kurzzeitspektroskopie (MBI), Berlin, Germany

E. Seres IOQ, Abbe Center of Photonics, Friedrich Schiller University, Jena, Germany

Atominstitut (ATI), Vienna University of Technology, Vienna, Austria

Helmholtz Institute Jena, Jena, Germany

J. Seres IOQ, Abbe Center of Photonics, Friedrich Schiller University, Jena, Germany

Atominstitut (ATI), Vienna University of Technology, Vienna, Austria

A. Sergeevich Pirozhkov Quantum Beam Science Directorate, Japan Atomic Energy Agency, Kyoto, Japan

M. Shahzad York Plasma Institute, The University of York, York, UK

E. Shanblatt JILA, University of Colorado at Boulder, Boulder, CO, USA

J. Shaw Electromagnetics Division, NIST, Boulder, CO, USA

M. Shazad Department of Physics, University of York,, York YO10 5DD, UK

Z. M. Sheng Laboratory for Laser Plasmas (Ministry of Education) and Department of Physics, Shanghai Jiao Tong University, Shanghai, China

J. W. Shin Department of Chemistry and NSF ERCEUVST, Colorado State University, Fort Collins, CO, USA

V. N. Shlyaptsev National Science Foundation ERC for Extreme Ultraviolet Science and Technology, Fort Collins, CO, USA

Department of Electrical and Computing Engineering, Colorado State University, Fort Collins, CO, USA

M. Siegrist Institute of Applied Physics, University of Bern, Bern, Switzerland

S. Simion National Institute for Lasers, Plasma and Radiation Physics (INFLPR), Magurele, Romania

I. Skobelev Joint Institute for High Temperatures, Russian Academy of Science, Moscow, Russia

R. Soufli Lawrence Livermore National Laboratory, Livermore, CA, USA

C. Spielmann IOQ, Abbe Center of Photonics, Friedrich Schiller University, Jena, Germany

Helmholtz Institute Jena, Jena, Germany

E. Spiller Spiller X-ray Optics, Livermore, CA, USA

P. Sprangle Naval Research Lab, Washington DC, USA

R. Squibb Department of Physics, University of Gothenburg, Gothenburg, Sweden

F. Staub Institute of Applied Physics, University of Bern, Bern, Switzerland

H. Stiel Max-Born Institut für Nichtlineare Optik und Kurzzeitspektroskopie (MBI), Berlin, Germany

Berlin Laboratory for innovative X-ray Technologies (BLiX), Berlin, Germany

J. Straus Academy of Sciences of the Czech Republic, Institute of Plasma Physics, v.v.i., Prague, Czech Republic

S. Suckewer Princeton University, Princeton, NJ, USA

T. Suemoto Institute of Solid State Physics (ISSP), Univeristy of Tokyo, Kashiwa, Japan

J. H. Sung Center for Relativistic Laser Science, Institute for Basic Science (IBS), Gwangju, Korea

Advanced Photonics Research Institute, Gwangju Institute of Science and Technology (GIST), Gwangju, Korea

Y. Suzuki Tsuruga Head Office, Applied Laser Technology Institute, Japan Atomic Energy Agency, Tsuruga, Japan

A. Svidzinsky Princeton University, Princeton, NJ, USA

Texas A&M University, College Station, TX, USA

A. Szczurek Institute of Optoelectronics, Military University of Technology, Warsaw, Poland

M. Szczurek Institute of Optoelectronics, Military University of Technology, Warsaw, Poland

J. Tümmler Max-Born Institut für Nichtlineare Optik und Kurzzeitspektroskopie (MBI), Berlin, Germany

K. Ta Phuoc IoP ASCR, Na Slovance 2, Prague 8, Czech Republic

LOA, ENSTA, Chemin de la Hunière, Palaiseau cedex, France

G. J. Tallents Department of Physics, University of York,, York YO10 5DD, UK

University of Engineering and Technology, Lahore, Pakistan

York Plasma Institute, The University of York, York, UK

F. L. Tanzella SRI International, Menlo Park, CA, USA

F. Tissandier LOA, ENSTA, Chemin de la Hunière, Palaiseau cedex, France

LPGP, CNRS, LOA, CNRS, ENSTAe, Palaiseau, France

F. G. Tomasel Department of Electrical and Computing Engineering, Colorado State University, Fort Collins, CO, USA

Advanced Energy Industries, Fort Collis, CO, USA

T. Tomita Faculty of Engineering, University of Tokushima, Tokushima, Japan

A. Towsend Colorado State University, Fort Collins, CO, USA

E. Turgut JILA, University of Colorado, Boulder, CO, USA

Faculty of Physics, University of Bucharest, Magurele, Romania

R. G. Ungureanu National Institute for Lasers, Plasma and Radiation Physics (INFLPR), Magurele, Romania

L. Urbanski NSF ERC for Extreme Ultraviolet Science and Technology, Fort Collins, CO, USA

Department of Electrical and Computer Engineering, Colorado State University, Fort Collins, CO, USA

D. Ursescu ELI-NP Department, Horia Hulubei National Institute for Physics and Nuclear Engineering, (IFINHH), Magurele, Romania

National Institute for Lasers, Plasma and Radiation Physics (INFLPR), Magurele, Romania

A. V. Vinogradov Lebedev Physical Institute, Moscow, Russia

Ł. Węgrzyński Institute of Optoelectronics, Military University of Technology, Warsaw, Poland

P. W. Wachulak Institute of Optoelectronics, Military University of Technology, Warsaw, Poland

S. Wang Colorado State University, National Science Foundation Engineering Research Center for Extreme Ultraviolet Science and Technology, Fort Collins, CO, USA

W. M. Wang Beijing National Laboratory of Condensed Matter Physics, Institute of Physics, CAS, Beijing, China

Y. Wang Colorado State University, National Science Foundation Engineering Research Center for Extreme Ultraviolet Science and Technology, Fort Collins, CO, USA

C. Weninger Max Planck Institute for the Physics of Complex Systems, Dresden, Germany

Center for Free-Electron Laser Science, Hamburg, Germany

K. A. Wernsing NSF ERC for Extreme Ultraviolet Science and Technology, Fort Collins, CO, USA

Department of Electrical and Computer Engineering, Colorado State University, Fort Collins, CO, USA

L. A. Wilson Central Laser Facility, STFC Rutherford Appleton Laboratory, Oxfordshire, UK

G. Winkler Atominstitut (ATI), Vienna University of Technology, Vienna, Austria

M. Woolston National Science Foundation ERC for Extreme Ultraviolet Science and Technology, Fort Collins, CO, USA

Department of Electrical and Computing Engineering, Colorado State University, Fort Collins, CO, USA

H. Xia Princeton University, Princeton, NJ, USA

Texas A&M University, College Station, TX, USA

M. Yamagiwa Quantum Beam Science Center, Japan Atomic Energy Agency (JAEA), Kyoto, Japan

W. C. Yan Beijing National Laboratory of Condensed Matter Physics, Institute of Physics, CAS, Beijing, China

M. Yeung Department of Physics and Astronomy, Queen's University Belfast, Belfast, UK

Helmholtz-Institut Jena, Jena, Germany

L. Yin Colorado State University, National Science Foundation Engineering Research Center for Extreme Ultraviolet Science and Technology, Fort Collins, CO, USA

H. Yun Center for Relativistic Laser Science, Institute for Basic Science (IBS), Gwangju, Korea

P. Zeitoun Laboratoire d'Optique Appliquee, UMR7639, ENSTA ParisTech, CNRS, Ecole Polytechnique, Palaiseau, France

M. Zepf Helmholtz-Institut Jena, Jena, Germany

Department of Physics and Astronomy, Queen's University Belfast, Belfast, UK

B. Zhang JILA, University of Colorado at Boulder, Boulder, CO, USA

J. Zhang Laboratory for Laser Plasmas (Ministry of Education) and Department of Physics, Shanghai Jiao Tong University, Shanghai, China

V. Zhaunerchyk Department of Physics, University of Gothenburg, Gothenburg, Sweden

Z. Zhinan State Key Laboratory of High Field Laser Physics, Shanghai Institute of Optics and Fine Mechanics, Chinese Academy of Sciences, Shanghai, China

X. Zhizhan State Key Laboratory of High Field Laser Physics, Shanghai Institute of Optics and Fine Mechanics, Chinese Academy of Sciences, Shanghai, China

School of Physical Science and Technology, Shanghai Tech University, Shanghai, China

D. Zusin JILA, University of Colorado, Boulder, CO, USA

Part I
X-Ray Lasers

Progress and Prospects of Coherent X-Ray Research Using High Power Lasers in Japan Atomic Energy Agency

T. Kawachi, M. Nishikino, A. Sasaki, M. Ishino, N. Hasegawa, T. Imazono, T. A. Pikuz, A. Y. Faenov, M. Magnitskiy, M. Maruyama, T. Tomita, T. Suemoto, A. S. Pirozhkov, S. V. Bulanov, H. Kiriyama, M. Kando, M. Yamagiwa, K. Kondo, P. Bolton and Y. Kato

Abstract Recent progress of the coherent x-ray research using the transient collisional excitation (TCE) laser and higher-order harmonics from relativistic plasmas is reviewed. We also describe the future prospects of this research field in JAEA which involve short wavelength coherent x-ray generation using a petawatt class driver laser and application studies using high average power TCE laser.

1 Introduction

Short pulse x-ray sources become indispensable diagnostic tools in modern science and technology and are widely used in probing substances for new material development, protein crystallography in innovative drug development, and non-destructive x-ray imaging etc. The improvement of the sources is also important subject, and in particular source development of coherent x-rays is intensively studied. Since the laser-driven sources have potentials to downsizing and ultra-short duration toward atto-sec region, we carry out the development of soft x-ray lasers (SXRLs) and higher-order harmonics generation (HHG) and their applications.

T. Kawachi (✉) · M. Nishikino · A. Sasaki · M. Ishino · N. Hasegawa · T. Imazono · T. A. Pikuz· M. Maruyama · A. S. Pirozhkov · S. V. Bulanov · H. Kiriyama · M. Kando · M. Yamagiwa· K. Kondo · P. Bolton
Quantum Beam Science Center,
Japan Atomic Energy Agency (JAEA), Kyoto, Japan
e-mail: kawachi.tetsuya@jaea.go.jp

T. A. Pikuz · A. Y. Faenov
Joint Institute of High Temperature, Russian Academy of Science, Moscow, Russia

M. Magnitskiy
M.V. Lomonosov Moscow State University, Moscow, Russia

T. Tomita
Faculty of Engineering, University of Tokushima, Tokushima, Japan

T. Suemoto
Institute of Solid State Physics (ISSP), Univeristy of Tokyo, Kashiwa, Japan

Y. Kato
Graduate School for Photonics and Industries (GPI), Shizuoka, Japan

© Springer International Publishing Switzerland 2016
J. Rocca et al. (eds.), *X-Ray Lasers 2014*, Springer Proceedings in Physics 169,
DOI 10.1007/978-3-319-19521-6_1

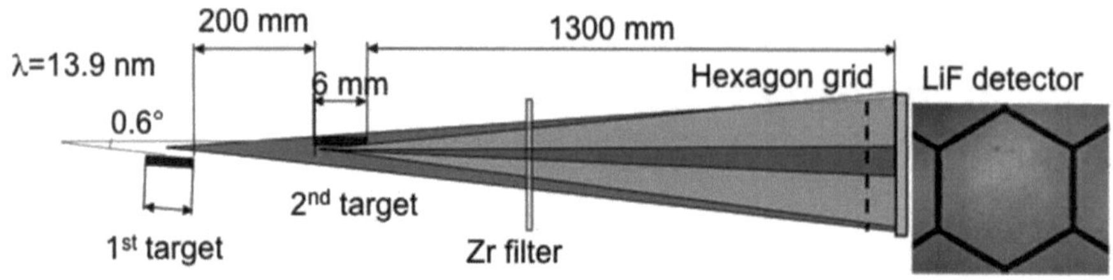

Fig. 1 The experimental set up for the double target geometry. We could observe "extraordinary" coaxial fringe patterns under the slightly misaligned condition of the traget position

In this paper, we describe several new results in the coherent x-ray researches and our future plan. In the source development, the observation of x-ray mirage in the gain medium plasma of the 13.9 nm SXRL [1], which is strongly related to the temporal evolution of electron density profile of gain region, and recent progress of HHG from relativistic plasma are presented. As the application study, we show the surface dynamics of femto-second laser ablation of metal observed by the SXRL interferometer and reflectometer [2]. Finally we mention the prospects for the future: By the financial support from government, we have started upgrade of the ultra-intense Ti:Sapphire laser, "J-KAREN", toward PW level in order to develop coherent x-rays in keV region and high energy particle beams. By use of this "J-KAREN-P" together with high average power pumping source developed separately in JAEA, we will explore the possibility of ultrashort-pulse short-wavelength lasers and open up new applications of high average power coherent x-rays.

2 Observation of X-Ray Mirage Using the 13.9 nm Laser

Beginning was the observation of extraordinary fringes in the alignment procedure of the 13.9 nm SXRL. The set up is shown in Fig. 1. The two targets were used for the SXRL generation, and the first one worked as x-ray seeder and the second one worked as x-ray amplifier [3]. The distance between these two targets was 200 mm. Under the slightly misaligned condition, we could find coaxial fringe patterns on the LiF crystal detector [4]. The fringe patterns obtained in the detector indicated that as the distance from the center of the arc became large, the fringe interval became smaller. This characteristics was the typical in the interference patterns generated by two coherent sources located separately on the observation axis.

In order to reproduce the experimental coaxial fringe patterns, a model calculation was performed, and under the assumption of that the two sources were separated by 203 mm the experiment and the calculation results became perfectly consistent. The origin of the first source was obvious, *i.e.*, the exit position of the first target. Therefore we could conclude that the second source was located in the x-ray amplifier. In principle, since the phase of the SXRL beam from the second target was independent from that of the first one, in order to explain the present interference

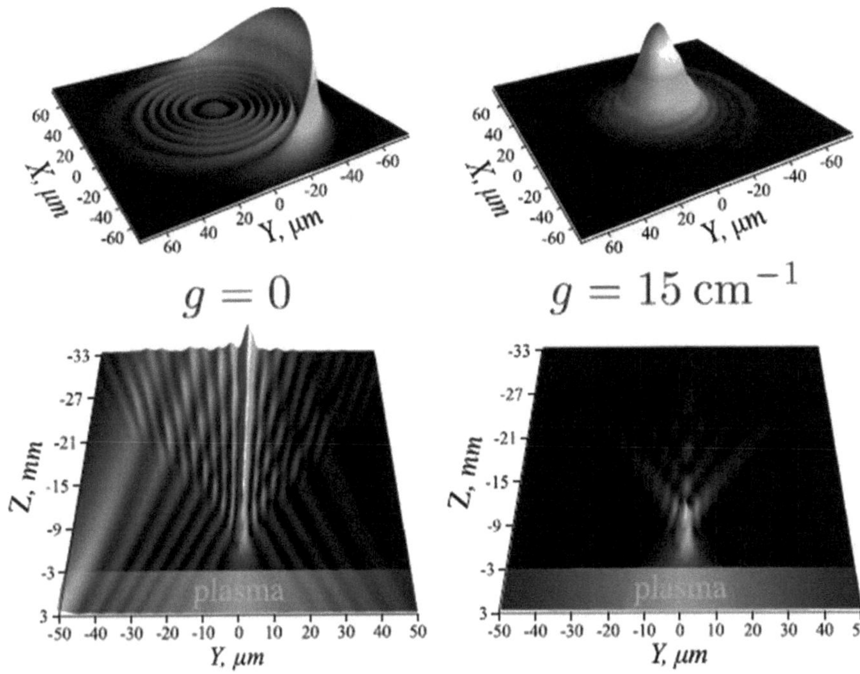

Fig. 2 Calculated spatial profiles of the x-ray mirage. The *left-hand side* is the case without amplification gain (g), and the *right-hand side* is with $g = 15\,\mathrm{cm}^{-1}$

fringes, a small portion of SXRL beam from the first plasma changed the direction in the x-ray amplifier and formed "virtual source" in the x-ray amplifier, *i.e.*, x-ray mirage [1]. The x-ray beam from the virtual source together with the rest of the x-ray beam from the first plasma might form the interference fringes at the detector position. Indeed the hydrodynamics simulation suggested that the density hump could be formed in the x-ray amplifier, and it worked as a concave lens for the x-ray.

In order to reproduce the present fringes quantitatively, theoretical investigation was performed in which the wave-equations of SXRL beam from virtual source and the rest of the SXRL beam are treated, and the spatial profiles of the electron density and amplification gain are taken into account in the dielectric permittivity of plasma. For given 2D gain profile and 2D electron density profile [3], we derived the intensity distribution of virtual source of x-ray mirage, Fig. 2 shows the intensity profiles of the virtual source in the x-y plane and in the y-z plane for the case of no amplification gain (left-hand side) and the case of amplification gain of $15\,\mathrm{cm}^{-1}$ (right-hand-side). In the absence of the amplification gain, shape of the source does not corresponds to the point source. However, under the condition with certain amplification gain, the virtual source becomes perfect point source with the spatial profile close to Gaussian.

Present experimental result indicates the potential of the plasma as an x-ray optics. The obtained perfect interference fringes means that this x-ray optics can

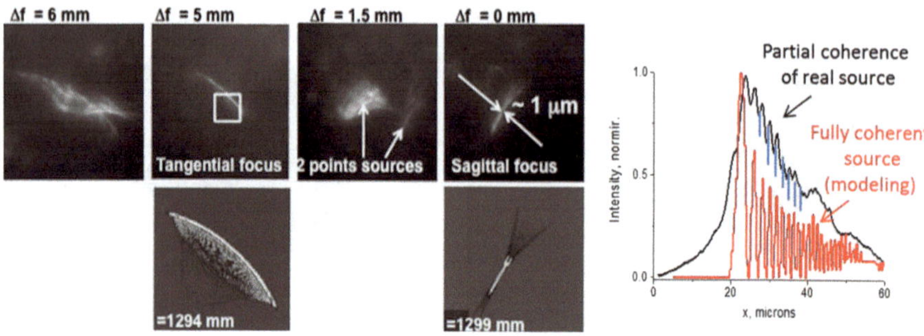

Fig. 3 The images of the source of the HH signal at a wavelength of 18 nm from the relativistic plasma (*left-hand-side*). From the tangential focus and sagittal focus, we could obtain the source size. From the fringe pattern of the tangential focus, we found the HH signal is partially coherent

keep the wave-front of x-ray with other advantages of plasma optics such as highly permitted intensity for the input beam. We will continue to survey this possibility as the next research theme.

3 Beam Characterization of High-Order Harmonics Emission from a Relativistic Plasma

A couple of years ago, we have demonstrated the ganeration of intense higher-order harmonics (HH) from He gas plasma irradiated by a laser pulse with the intensity of more than 10^{18} W/cm^2 [5]. The wavelength of HH signal reached to the water-window region, and regardless of the polarization of the pumping laser, both the odd- and even-order HH signals were obtained. Our theoretical investigation implied that localized high density regions could be generated due to the laser induced modulation in the bow wave, and strong oscillation of these high density region by laser electric field induced the HH emission. In this scheme, we could increace the incident laser intensity up to wave-breaking region, therefore the upper-limit of the laser was estimated to be 10^{23} W/cm^2, under which the expected wavelength of the HH reached in several 10 keV region.

Recently we have conducted an experiment to measure the source size. We have prepared off-axis diagnostic tool which is based on a spherical mirror imaging system with slightly off-axis angle. We choose the HH at 18 nm, and uses the LiF crystal detector. By changing the detector position, the images of the source are obtained experimentally as shown in Fig. 3. The magnification factor is around 5.5. We can obtain both the tangential focus and sagittal focus, and the corresponding theoretical images are attached, where Δf in Fig. 3 indicates the distance of the LiF crystal detector measured from the sagittal focus. The best focus, or width of the tangential and sagittal focus, is aound 5 μ, which leads to the source size of 0.9 μ.

The graph in the right-hand side of Fig. 3 shows the cross section of the tangential focus (See the square in the image of the focus in the left-hand side) with corresponding area of the theoretical model. The theoretical model assume fully spatial coherent beam, and the sharpe fringe patterns are obtained. On the other hand, we could obtain the experimental fringe pattern although they are not sharpe; this means that the HH signal is partial coherent. Since the shape of the source plasma is vertically long with the Rayleigh length, and multiple sources can be generated in this plasma column. The present low spatial coherence may be due to the superposition of these multiple sources in plasma. For further information, we need further experimental investigation.

4 Application of XRL for Probing Surface Dynamics of Femto-Second Laser Ablation

Understanding of optical laser ablation process is important not only as the fundamental process of laser fabrication but also as the physics of laser-matter interaction. Several years ago, Germany group observed metal surface irrardiated by a femtosecond laser pulse, and they found Newton's ring (NR) formation in the ablation area after 500 ps from the pump by use of 400 nm laser probe [6]. They concluded that the considered ablation area consists of outer-shell structure called as ablation front and distorted surface. Between the outer-shell and the distorted surface was considered to be transparent fluid, and the NR was originated from the interference of the reflection from the ablation front and that from the distorted surface. However, due to the limitaion of the wavelength of the probe, the detailed dynamics especially in the early time region has not been understood yet.

Soft x-ray laser interferometer and reflectometer developed in our institute enables us to observe the surface deformation with the depth resolution of 1 nm with pico-second accuracy [7], and we apply it to probe the surface dynamics in early time region of femto-second laser ablation.

A gold sample plate was irradiated by a 40 fs-duration Ti:Sapphire laser at a wavelength of 800 nm. The spatial intensity profile was Gaussain with the FWHM of 60 μm with the energy fluence of 1.3 J/cm^2 at the peak of the Gaussian. The details of the soft x-ray laser interferometer and reflectometer was described in ref [2].

Figure 4 shows the temporal evolution of the distorted surface observed by the interferometer; (a) and (b), and the reflectometer; (c) and (d). Apparent NRs was obtained in the reflective images for the early time region ($t \sim 1$ ns). From the reflectivity and visibility of the fringes, it could be said that the ablation front was something like a thin film, and between the ablation front and the distorted surface was transparent for SXRL wavelength region, which was consistent with the previous results.

The interferometer provided the information of the distorted surface, although slight mixing of NR could be seen in Fig. 4b (narrow fringes). The contribution from NR was removed in the analysis, and we derived the temporal evolution of the

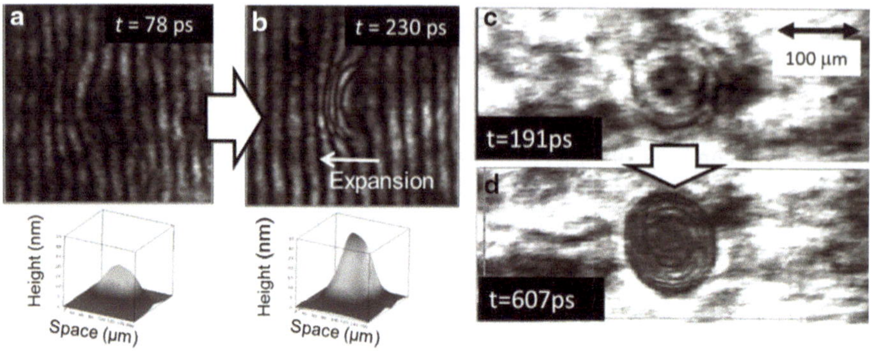

Fig. 4 Temporal evolution of the surface dynamics observed by the interferometer; (**a**) and (**b**), and the reflectometer; (**c**) and (**d**)

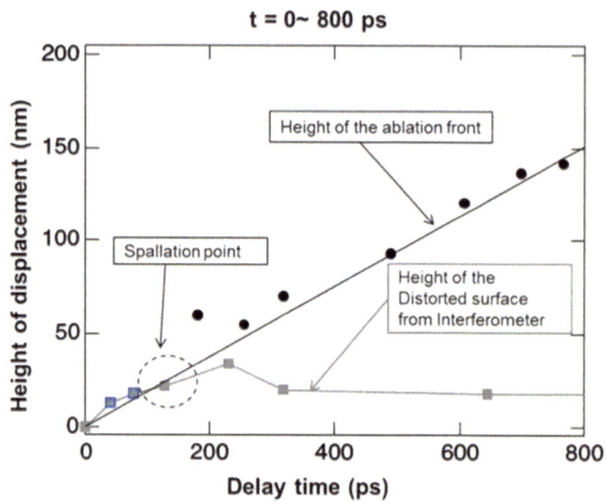

Fig. 5 Height of the outer-shell and the distorted surface as the function of the time measured from the pumping timing

height of the distorted surface. The height of the ablation front was derived from the data of NR with the result of the distorted surface. Figure 5 shows the height of the distorted surface and the ablation front as the function of the time measured from the pumping timing.

The displacement of the ablation front is in accordance with uniform motion, and the interpolation in the time region of $t = 0 \sim 150$ ps indicates that the ablation front and the distortion surface is identical in this time region. At $t = 150$ ps, separation of the ablation front from the distroted surface, *i.e.*, spallative process, occurs. This is, to author's knowledge, the first observation of the spallation of the materials in the laser ablation processes and becomes a good benchmark of the theoretical modeling such as molecular dynamics (MD) simulation.

The height of the distorted surface reaches the maximum at $t = 230$ ps, and start to decrease. It should be noted that at around $t = 2 \sim 3$ ns, the height of the surface becomes negative, and the crater formation starts. This implies that after reaching the peak of the height, the evaporation of the surface becomes substantial and the surface is scraped off.

Present experimental results involve a plenty of new things relevant to the foundamental processes in femto-second laser ablation, and detailed comparison with theoretical model will give us the deep understanding of the laser-matter interactions.

5 Future Porspects of High Power Laser Science in JAEA

In this section we describe the future prospects of coherent x-ray research using high power lasers in JAEA.

Based upon the high peak power and high average power laser development, we will continuously make effort in the radiation source development such as intense coherent x-ray, high energy charged particle beam and intense THz radiation. These new radiation sources are applicable in the wide variety of research field as an ultra-fast probe, compact accelerator and chemical reaction control, etc.

Recently we obtained $ 12 M supplementary budget from government to promote laser-driven ion acceleration research with the energy of more than 100 MeV and coherent x-ray source development in keV region. By use of this budget, we have started the upgrade of 200 TW Ti:Sapphire laser, J-KAREN laser, up to 1 PW level. The repetition-rate will be also improved from each 30 min to 10 s. In April 2014, the installation has been almost finished, and we are now preparing the set-up of the total system toward the first laser shot which is expected in the end of the year.

Using this J-KAREN-P laser, we will conduct the experiment of HH generation from relativistic plasma toward shorter wavelength region and preliminary experiment of gamma-flash [8]. Theory suggests that the demonstration of gamma-flash in the MeV region requires the laser intensity of more than 10^{23} W/cm^2, but even in 10^{22} W/cm^2, we can expect the x-ray in 10–100 keV energy range, although the conversion efficiency is not so high. This may become a principle proof of this scheme.

The research using the transient collsional excitation (TCE) XRLs will be shifted to industrial applications such as nano-fabrication and the EUV mask inspection and so on. Therefore we will replace the CPA Nd:glass driver laser by Ti:Sapphire based system in a couple of years. Pumping scheme will changed from normal incidence pumping to GRIP scheme. Repetition rate will start from 10 Hz, and at the same time we will start the development of the 100 Hz YAG pumping laser with 5 J output energy with a few ns duration. By use of this new pumping laser, we will realize high average power SXRL system.

6 Summary

We review the recent progress and future prospects of coherent x-ray research using high power lasers in Japan Atomic Energy Agency. Research highlights by use of the 13.9 nm laser are the first demonstration of x-ray mirage and the first observation of spallation dynamics in the femto-second laser ablation. In the higher-order harmonics (HH) from the relativistic plasma, the source size of the HH emissions is investigated, and we have assured that the source is partially coherent. In the future prospects, we will explore the possibility of ultrashort-pulse short-wavelength lasers and open up new applications of high average power coherent x-rays by use of upgraded J-KAREN-P and high average power TCE laser in GRIP scheme.

References

1. Magnitskiy, S., Nagorskiy, N., Faenov, A., Pikuz, T., Tanaka, M., Nishikino, M., Fukuda, Y., Kando, M., Kawachi, T., Kato, Y.: Observation and theory of x-ray mirages. Nat. Commun. **4**, 1–7 (2013). (ncomms2923)
2. Tomita, T., Yamamoto, M., Hasegawa, N., Terakawa, K., Minami, Y., Nishikino, M., Ishino, M., Kaihori, T., Ochi, Y., Kawachi, T., Yamagiwa, M., Suemoto, T.: Experimental verification of femto-second laser ablation schemes by time-resolved soft x-ray reflective imaging. Opt. Express **20**, 29329 (2012)
3. Nishikino, M., Hasegawa, N., Kawachi, T., Yamatani, H., Sukegawa, K., Nagashima, K.: Characterization of a high-brilliance soft x-ray laser at 13.9 nm using an oscillator-amplifier configuration. Appl. Opt. **47**, 1129–1133 (2008)
4. Faenov, A., Kato, Y., Tanaka, M., Pikuz, T., Kishimoto, M., Ishino, M., Nishikino, M., Fukuda, Y., Bulanov, S., Kawachi, T.: Submicrometer-resolution in situ imaging of the focus pattern of soft x-ray laser by color center formation in LiF crystal. Opt. Lett. **34**, 941–943 (2009)
5. Pirozhkov, A., et al.: Soft x-ray harmonics comb from relativistic electron spike. Phys. Rev. Lett. **108**, 135004 (2012)
6. Sokolowski-Tinten, K., Bialkowski, J., Cavalleri, A., von der Linde, D., Oparin, A., Meyer-ter-Vehn, J., Anisimov, S.: Transient states of matter during short pulse laser ablation. Phys. Rev. Lett. **81**, 224–227 (1998)
7. Suemoto, T., Terakawa, K., Ochi, Y., Tomita, T., Yamamoto, M., Hasegawa, N., Deki, M., Minami, Y., Kawachi, T.: Single-shot picosecond interferometry with one-nanometer resolution for dynamical surface morphology using a soft x-ray laser. Opt. Express **18**, 14114–14122 (2010)
8. Nakamura, T., Koga, J., Esirkepov, T., Kando, M., Korn, G., Bulanov, S.: High-power g-ray generation in ultraintense laser-plasma interaction. Phys. Rev. Lett. **108**, 195001 (2012)

Advances in High Average Power, 100 Hz Repetition Rate Table-Top Soft X-Ray Lasers

Brendan A. Reagan, Cory Baumgarten, Mark Berrill, Keith A. Wernsing, Mark Woolston, Lukasz Urbanski, Wei Li, Mario C. Marconi, Vyacheslav N. Slyaptsev, Carmen S. Menoni and Jorge J. Rocca

Abstract We discuss new results of the demonstration of 100 Hz table-top soft x-ray lasers operating at wavelengths from 10.9 nm to 18.9 nm and report the highest soft x-ray laser average power to date at sub-20 nm wavelength. The results include the generation of average powers up to 0.2 mW at $\lambda = 18.9$ nm, 0.1 mW at $\lambda = 13.9$ nm, and 20 μW at $\lambda = 11.9$ nm. These lasers are driven by a compact chirped pulse amplification laser system featuring diode-pumped, cryogenically-cooled Yb:YAG power amplifiers that produces 1 J pulses of 5 ps FWHM duration at 100 Hz repetition rate. The driver laser pulse shape was tailored to more efficiently pump soft x-ray plasma amplifiers operating at sub-15 nm wavelengths leading to a threefold increase in the $\lambda = 13.9$ nm laser pulse energy and lasing down to 10.9 nm. The pump pulse profile consisting of a nanosecond ramp followed by two closely spaced peaks of picosecond duration, is shown to create a plasma with an increased density of Ni-like lasing ions at the time of peak temperature, resulting in a larger gain coefficient over a temporally and spatially enlarged space. The development of rotating solid targets with high shot capacity has allowed the uninterrupted operation of the $\lambda = 18.9$ nm soft x-ray laser for hundreds of thousands of consecutive shots making it suitable for applications in nanoscience and nanotechnology that require high photon flux at short wavelengths. As a proof-of-principle demonstration of the utility of this laser in such applications we have lithographically printed an array of nanometer-scale features through coherent Talbot self imaging. These results open the path to milliwatt average power table-top soft x-ray lasers.

B. A. Reagan (✉) · C. Baumgarten · M. Berrill · K. A. Wernsing · M. Woolston · L. Urbanski · W. Li · M. C. Marconi · C. S. Menoni · J. J. Rocca
NSF ERC for Extreme Ultraviolet Science and Technology, Fort Collins, CO, USA

B. A. Reagan · M. Berrill · K. A. Wernsing · M. Woolston · L. Urbanski · W. Li · M. C. Marconi · V. N. Slyaptsev · C. S. Menoni · J. J. Rocca
Department of Electrical and Computer Engineering, Colorado State University, Fort Collins, CO, USA

C. Baumgarten · J. J. Rocca
Department of Physics, Colorado State University, Fort Collins, CO, USA

© Springer International Publishing Switzerland 2016

J. Rocca et al. (eds.), *X-Ray Lasers 2014*, Springer Proceedings in Physics 169, DOI 10.1007/978-3-319-19521-6_2

1 Introduction

There is great interest in the development of bright, coherent soft x-ray sources for a large number of applications. Despite the significant progress recently made in plasma-based soft x-ray lasers (SXRLs) [1–11], their average power is at present much lower than that produced by soft x-ray FELs [12, 13]. In the case of SXRLs, their average power has been limited by the relatively low repetition rate of the high energy optical wavelength pump lasers used to drive them and by low pumping efficiency. Capillary discharge soft x-ray lasers can produce milliwatt average power [14, 15], but are presently limited to longer wavelengths [16]. Laser-pumped $\lambda = 10$–20 nm soft x-ray lasers have been limited repetition rates of 10 Hz or less, resulting in average powers ranging from 1 to 20 μW [4, 6]. We previously reported the first table-top SXRL capable of 100 Hz repetition rate operation, generating 0.15 mW average power on the $\lambda = 18.9$ nm line of Ni-like Mo [17]. Recently the diffraction grating used to conduct the measurement was calibrated at a synchrotron facility showing that the average power we had reported was underestimated and was the corrected value is ~ 0.2 mW. This achievement was made possible by the development of a diode-pumped, picosecond Yb:YAG driver laser. Here, we report the demonstration of a 0.1 mW average power, gain-saturated, $\lambda = 13.9$ nm soft x-ray laser operating at 100 Hz repetition rate. The infrared driver laser pulse was temporally tailored to more efficiently pumps the plasma amplifier, resulting in a threefold enhancement of the $\lambda = 13.9$ nm laser energy [18]. Additionally, we demonstrated high repetition rate, gain saturated operation of a $\lambda = 11.9$ nm laser from a Sn plasma. We also report the continuous, hour-long operation of the $\lambda = 18.9$ nm Ni-like Mo laser at high repetition rates [19]. Finally, using this robust laser we have made a proof-of-principle demonstration of the printing of nano-scale features through coherent Talbot lithography.

2 Demonstration of 50–100 Hz Operation of Sub-15 nm Wavelength Lasers

The plasma amplifiers were excited by a single shaped pump pulse. Although the use of a single pulse has been used to drive table-top soft x-ray lasers before [5,9,17,20], the pulse employed here has different characteristics. The advantage of shaping the temporal sequence of pulses used to efficiently create and heat soft x-ray plasma amplifiers has been recognized [3, 9]. In particular, it was recently shown that the addition of a short duration (2 ps FWHM) pre-pulse with a fraction of the intensity of the main pulse can significantly increase the soft x-ray laser pulse energy produced by a $\lambda = 13.9$ nm Ni-like Ag laser when pumping with relatively low driving laser pulse energies [3]. In another recent paper, this scheme was expanded upon to allow the generation of 4.7 μJ pulses at $\lambda = 13.9$ nm using less than 2 J of total pump laser energy [21]. In recent previous work, we have used a single shaped pump laser pulse of < 1 J energy to drive lasing on the $\lambda = 18.9$ nm transition of Ni-like Mo at 100 Hz repetition. rate that resulted in an average power of 0.15 mW [17] (which we subsequently realized was ~ 0.2 mW after calibration of the spectrometer diffraction grating). Here, in order to more efficiently pump the soft x-ray amplifier plasma and allow generation at shorter wavelengths, we made use of a tailored temporal pulse profile that consists of a ~ 2 ns long, low amplitude ramp that creates and ionizes the plasma and a short duration intense pulse that precedes the main peak of the pulse. Hydrodynamic/atomic physics plasma simulations show that the pump pulse profile, consisting of a nanosecond ramp followed by two peaks of picosecond duration, creates a plasma with an increased density of Ni-like ions at the time of maximum temperature that results in a larger gain coefficient over a temporally and spatially enlarged region leading to a three-fold increase in the soft x-ray laser output pulse energy. These simulations are presented in detail in [18]. This increase in pumping efficiency combined with the increased repetition rate results in record-high average power at these wavelengths from a table-top device. Using this approach, bright lasing was also demonstrated at a number of wavelengths between 10.9 nm and 14.7 nm, including the high repetition rate, gain-saturated operation of a $\lambda = 11.9$ nm laser from Ni-like Sn. These results will enable new technologic applications and basic research experiments requiring high average photon flux, and opens a path to further scale compact SXRLs to milliwatt average power in this wavelength range and possibly at shorter wavelengths.

The high repetition rate diode pumped $\lambda = 1.03$ μm CPA laser system based on cryogenic Yb:YAG amplifiers was used to obtain the results reported here is described in references [17, 22, 23]. This laser produces 1 J pulses with durations as short as 5 ps FWHM at 100 Hz repetition rates. The laser pulses are focused using cylindrical optics onto a solid target at a grazing incidence [11, 24] angle of 32° to form a 30 μm FWHM wide by 5 mm long line focus as shown in Fig. 1a. For prolonged high repetition rate operation we made use of a circular target consisting of a 100 mm diameter copper disk (Fig. 1a) with a foil of the desired lasing material (in this case Mo, Ag, and Sn) bonded to its face. The target material is mechanically polished to a smooth flat face suitable to produce an axially uniform plasma column when irradiated with the driving laser. The target is rotated using a motorized stage to present a fresh surface for each consecutive laser shot allowing for prolonged laser operation. Single shot, on-axis EUV plasma emission spectra were obtained using a 1200 lines/mm, flat-field, gold-coated diffraction grating positioned at grazing incidence and an x-ray sensitive CCD. For the high repetition rate soft x-ray laser measurements, an EUV-sensitive 10×1 mm silicon photodiode was placed in the imaging plane of the spectrometer at the location corresponding to the laser wavelength. Thin foil filters were used to reject visible plasma emission and also to prevent detector saturation. Absolute energy measurements were estimated using the efficiency of the diffraction grating, the reported responsivities of the Si photodiode and CCD, and the transmission efficiency of the thin filters measured *in situ*.

The pump pulse profile was measured via second harmonic generation (SHG) cross-correlation of the driving laser pulse with 100 μJ, 700 fs FWHM pulses produced by a Yb:KYW regenerative amplifier. The presence of any non-negligible

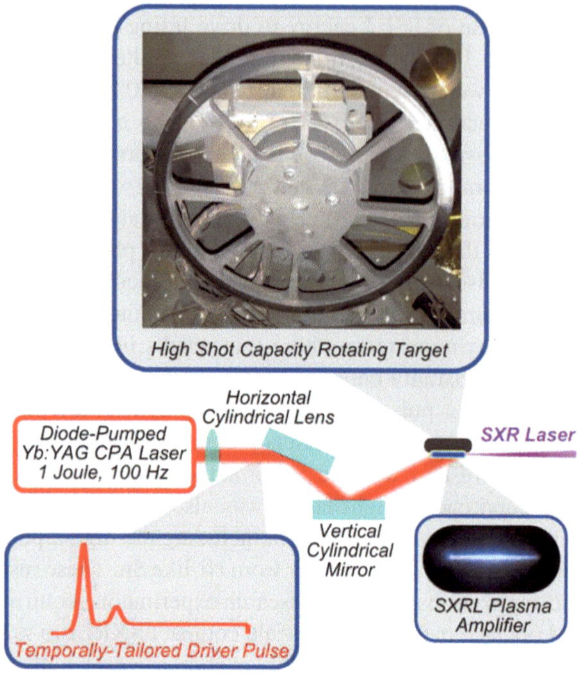

Fig. 1 Set up and target configuration for high repetition rate SXR laser operation. A high shot capacity, rotating target is irradiated by ~ 1 J pulses produced by diode-pumped CPA laser focused into a high aspect ratio *line* focus by cylindrical optics. The pump pulse temporal profile consists of a long ns ramp ending in a sequence of *two* closely spaced picosecond duration peaks

pulses before the 3.5 ns temporal window shown was ruled out by measurement with a fast photodiode. The main peak of the driving laser is preceded by an intentionally added pedestal that ramps from an intensity of 10^{-4} relative to the main peak at 2.7 ns before the peak, to about 10^{-3} before its onset. Additionally, a short picosecond peak preceding the main peak is generated by splitting the seed pulse beam from the pulse stretcher using a polarizing beamsplitter and a $\lambda/2$ waveplate, and subsequently recombining the beams into a single beam with an adjustable delay before seeding the regenerative amplifier. This feature precedes the main peak by 15 ps with a relative intensity of 12 %. Because all of the temporal features originate from within the regenerative amplifier cavity, they are collinear and share the same spatial mode profile making this truly a single pulse and avoiding the temporal and spatial overlapping that is otherwise required when using multiple pulses to drive soft x-ray lasers. Delays of 10–70 ps, and relative intensities within 10–25 % of the main pump peak were experimentally observed to produce similar $\lambda = 13.9$ nm laser pulse energies, which makes this a robust system.

Figure 2 shows the measured laser pulse energy of 2000 consecutive pulses of the $\lambda = 13.9$ nm laser operating at 100 Hz repetition rate on the $4d^1S_0 \rightarrow 4p^1p_1$ transition of Ni-like Ag obtained using the tailored pump pulse profile with the

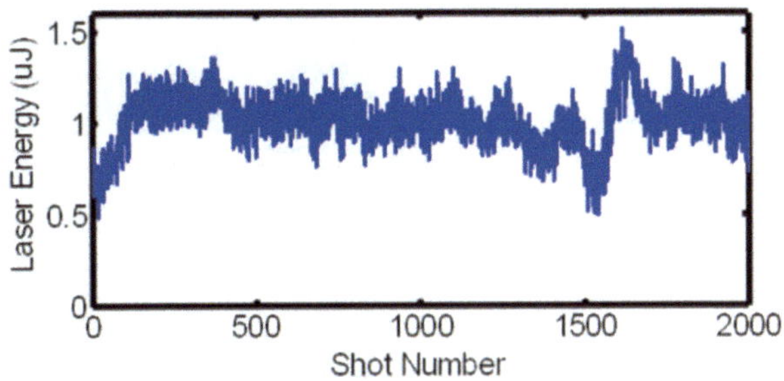

Fig. 2 Sequence of pulses from a 13.9 nm Ni-like Ag laser operating at 100 Hz repetition rate. The average pulse energy corresponds to an average power of 0.1 mW. The variation of the output pulse energy around shot 1500 is the result of shooting over a target region that was already irradiated

dual peak shown in Fig. 1b. The total pump pulse energy on target was 900 mJ. The circular target described earlier, which has a width of 1 cm in the direction of the line focus axis and an outer diameter of 100 mm, was rotated at 10 deg/s resulting in a distance between successive shots of ~90 μm at 100 Hz repetition rate. The mean SXRL pulse energy was measured to be 1.0 μJ with a shot-to-shot standard deviation of 14 %, resulting in an average power of 0.1 mW. This is to our knowledge the highest average power reported for a compact, coherent source at this wavelength.

We also used the combination of the compact diode-pumped driver laser and the specially tailored driving laser pulse profile technique to demonstrate bright SXRLs at shorter wavelengths down to $\lambda = 10.9$ nm, as summarized by the single-shot on-axis spectra of Fig. 3. The same pumping geometry and detection configuration described above was employed. Rectangular slab targets were used for the case of Te, In, Cd, and Pd, and a circular target of the form described above was used for Sn. The $\lambda = 11.9$ nm Sn laser was operated at 50 Hz repetition rate and produced an average power of 20 μW [18]. For the shortest wavelength investigated, $\lambda = 10.9$ nm from Ni-like Te, the total driver laser pump energy of 0.9 J used to obtain lasing is significantly lower than that reported in previous experiments [6]. No lasing was observed when the single peak temporal pumping profile was used with Te or Sn targets, and lower laser energy was observed for Pd, Ag, Cd and In targets.

3 Hour-long Operation of a High Repetition Rate $\lambda = 18.9$ nm Soft X-Ray Laser and Demonstration of Defect-Free Nanoprinting Through Talbot Lithography Using a $\lambda = 18.9$ nm Soft X-Ray Laser

Figure 4 shows data corresponding to 1 h of continuous operation of the Ni-like Mo $\lambda = 18.9$ nm SXRL at 50 Hz repetition rate. The laser output pulse energy is quite stable with very few low intensity shots. This continuous sequence of 1.8×10^5

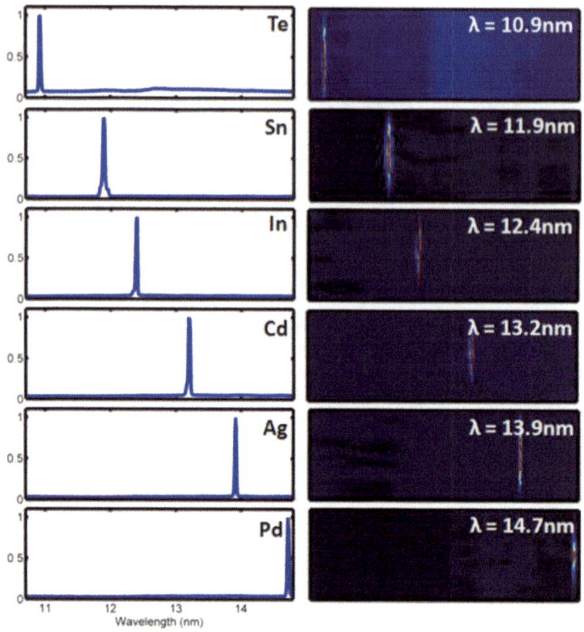

Fig. 3 Single shot, on-axis plasma emission spectra of $4d^1S_0 \rightarrow 4p^1p_1$ Ni-like ion soft x-ray lasers produced by irradiating solid targets of Te, Sn, In, Cd, Ag, and Pd with a single tailored pulse of 0.9 J energy produced by a diode-pumped Yb:YAG laser. The *right* column shows the raw CCD image of each spectrograph. All lasers down to 11.9 nm were operated in gain saturation

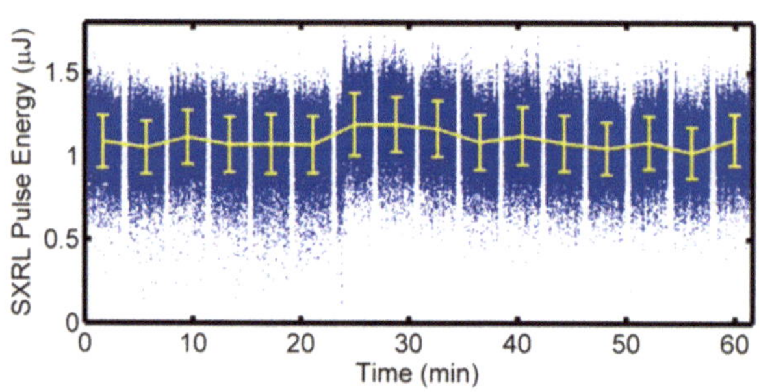

Fig. 4 Measured SXRL pulse energy at 50 Hz repetition rate for just over 1 h of operation. 180,000 consecutive shots were recorded. The *line* shows the running average of the pulse energy with the error bars representing the shot to shot standard deviation of each section of pulse sequence. The periodic gaps in the data correspond to transfers of the data from a digitizing oscilloscope during which the laser continued to operate

shots was acquired with one rotation of the target of Fig. 1a. We have also continuously operated this laser with similar performance at 100 Hz repetition rate for 30 min [19].

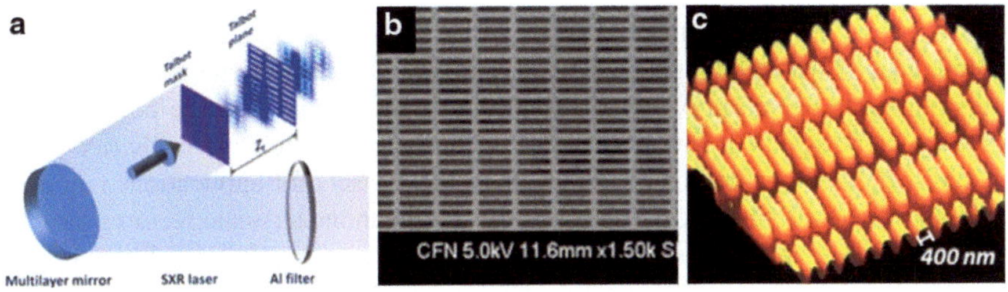

Fig. 5 **a** Talbot lithography experimental setup. A *two* dimensional periodic mask generates a Talbot image at the sample plane that is registered in photoresist. **b** Scanning electron microscope image of the Talbot mask. **c** Atomic force microscope image of the nanometer-scale features printed using the 18.9 nm laser

As a first demonstration of these high average power power soft x-ray lasers in applications requiring high photon flux we printed an array of nanometer-scale features using coherent Talbot lithography. The photo-lithographic approach used in this experiment is based on the self-imaging produced when a periodic transmission mask is illuminated with a coherent light beam. A semitransparent mask composed of an array of tiles each having an arbitrary design produces self images that replicate the mask on the surface of a photoresist [25]. When illuminated with coherent light, the tiled diffractive mask produces images which are 1 × replicas at certain locations (Talbot planes). Because the Talbot images are generated by the diffraction of the thousands of cells in the mask, a defect in any of the unitary cells is averaged over a very large numbers of tiles consequently rendering a virtually defect-free image [25]. Figure 5a shows the scheme of the experimental setup used to demonstrate Talbot printing. A scanning electron microscope (SEM) image of the mask is shown in Fig. 5b. The overall size of the mask is $0.5 \times 0.5\,\text{mm}^2$, with a calculated first Talbot distance $Z_T = 2.65$ mm. However, the print was made at half this distance where a phase-shifted Talbot image is produced, yielding a numerical aperture (NA) of 0.186. Figure 5c is an atomic force microscope image of the print in the photoresist showing reproduced lines is ~400 nm FWHM width in this initial proof-of-principle demonstration. The high flux of the soft x-ray laser allowed for efficient printing with an exposure of 30 s operating the laser at 50 Hz repetition rate. Ultimately, the short wavelength of this laser in combination with masks of larger NA will lead to defect-free printing of patterns with sub-20 nm resolutions.

4 Summary

In summary, we have demonstrated the high repetition rate, high average power, operation of soft x-ray lasers at wavelengths between 11.9 and 18.9 nm. Results include the continuous operation of a compact $\lambda = 18.9$ nm SXRL at 100 Hz repetition rate for extended periods of time with average powers up to 0.2 mW. Combining a soft x-ray plasma amplifier heated by a diode-pumped Yb:YAG laser with a tailored pulse shape and a high shot capacity rotating target we demonstrated 100 Hz repetition rate operation of $\lambda = 13.9$ nm laser with 0.1 mW average power. These results can be scaled to several hours of continuous operation and to shorter wavelengths. As a demonstration of the utility of these lasers for applications requiring a high average photon flux we printed an array of nanometer-scale features through coherent lithography. These compact high average power soft x-ray lasers can be expected to open the door to numerous photon flux intensive applications on a table-top. The results also open the path to milliwatt average power table-top soft x-ray lasers.

Acknowledgements The work supported AMOS program of the Office of Basic Energy Sciences, U.S. Department of Energy and was conducted using equipment funded by NSF MRI grant 0521649 and the NSF Center for Extreme ultraviolet Science and Technology.

References

1. Suckewer, S., Jaeglé, P.: X-ray laser: past, present, and future. Laser Phys. Lett. **6**, 411–436 (2009)
2. Alessi, D., Wang, Y., Luther, B.M., Yin, L., Martz, D.H., Woolston, M.R., Liu, Y., Berrill, M., Rocca, J.J.: Efficient excitation of gain-saturated sub-9-nm-wavelength tabletop soft-x-ray lasers and lasing down to 7.36 nm. Phys. Rev. **X 1**, 021023 (2011)
3. Banici, R.A., Cojocaru, G.V., Ungureanu, R.G., Dabu, R., Ursescu, D., Stiel, H.: Pump energy reduction for a high gain Ag x-ray laser using one long and two short pump pulses. Opt. Lett. **37**, 5130–5132 (2012)
4. Martz, D.H., Alessi, D., Luther, B.M., Wang, Y., Kemp, D., Berrill, M., Rocca, J.J.: High-energy 13.9 nm table-top soft-x-ray laser at 2.5 Hz repetition rate excited by a slab-pumped Ti: sapphire laser. Opt. Lett. **35**, 1632–1634 (2010)
5. Zimmer, D., Zielbauer, B., Pittman, M., Guilbaud, O., Habib, J., Kazamias, S., Ros, D., Bagnoud, V., Kuehl, T.: Optimization of a tabletop high-repetition-rate soft x-ray laser pumped in double-pulse single-beam grazing incidence. Opt. Lett. **35**, 450–452 (2010)
6. Alessi, D., Martz, D.H., Wang, Y., Berrill, M., Luther, B.M., Rocca, J.J.: Gain-saturated 10.9 nm tabletop laser operating at 1 Hz repetition rate. Opt. Lett. **35**, 414–416 (2010)
7. Grünig, M., Imesch, C., Staub, F., Balmer, J.E.: Saturated x-ray lasing in Ni-like Sn at 11.9 nm using the GRIP scheme. Opt. Commun. **282**, 267–271 (2009)
8. Goddet, J., Sebban, S., Gautier, J., Zeitoun, P., Valentin, C., Tissandier, F., Marchenko, T., Lambert, G., Ribières, M., Douillet, D., Lefrou, T., Iaquaniello, G., Burgy, F., Maynard, G., Cros, B., Robillard, B., Mocek, T., Nejdl, J., Kozlova, M., Jakubczak, K.: Aberration-free laser beam in the soft x-ray range. Opt. Lett. **34**, 2438–2440 (2009)
9. Kim, H.T., Choi, I.W., Hafz, N., Sung, J.H., Yu, T.J., Hong, K.-H., Jeong, T.M., Noh, Y.-C., Ko, D.-K., Janulewicz, K.A.: Demonstration of a saturated Ni-like Ag x-ray laser pumped by a single profiled laser pulse from a 10-Hz Ti: sapphire laser system. Phys. Rev. A **77**, 023807 (2008)
10. Ochi, Y., Hasegawa, N., Kawachi, T., Nagashima, K.: Development of a chirped pulse amplification laser with zigzag slab Nd:glass amplifiers dedicated to x-ray laser research. Appl. Opt. **46**, 1500–1506 (2007)
11. Wang, Y., Larotonda, M.A., Luther, B.M., Alessi, D., Berrill, M., Shlyaptsev, V.N., Rocca, J.J.: Demonstration of high-repetition-rate tabletop soft-x-ray lasers with saturated output at wavelengths down to 13.9 nm and gain down to 10.9 nm. Phys. Rev. A **72**, 053807 (2005)

12. Ackermann, W., et al.: Operation of a free-electron laser from the extreme ultraviolet to the water window. Nat. Photonics **1**, 336–342 (2007)
13. Allaria, E., et al.: Highly coherent and stable pulses from the FERMI seeded free-electron laser in the extreme ultraviolet. Nat. Photonics **6**, 699–704 (2012)
14. Benware, B.R., Macchietto, C.D., Moreno, C.H., Rocca, J.J.: Demonstration of a high average power tabletop soft x-ray laser. Phys. Rev. Lett. **81**, 5804 (1998)
15. Macchietto, C.D., Benware, B.R., Rocca, J.J.: Generation of millijoule-level soft-x-ray laser pulses at a 4-Hz repetition rate in a highly saturated tabletop capillary discharge amplifier. Opt. Lett. **24**, 1115–1117 (1999)
16. Frati, M., Seminario, M., Rocca, J.J.: Demonstration of a 10-μJ tabletop laser at 52.9 nm in neonlike chlorine. Opt. Lett. **25**, 1022–1024 (2000)
17. Reagan, B.A., Wernsing, K.A., Curtis, A.H., Furch, F.J., Luther, B.M., Patel, D., Menoni, C.S., Rocca, J.J.: Demonstration of a 100 Hz repetition rate gain-saturated diode-pumped table-top soft x-ray laser. Opt. Lett. **37**, 3624–3626 (2012)
18. Reagan, B.A., Berrill, M., Wernsing, K.A., Baumgarten, C., Woolston, M., Rocca, J.J.: High-average-power, 100-Hz-repetition-rate, tabletop soft-x-ray lasers at sub-15-nm wavelengths. Phys. Rev. A **89**, 053820 (2014)
19. Reagan, B.A., Li, W., Urbanski, L., Wernsing, K.A., Salsbury, C., Baumgarten, C., Marconi, M.C., Menoni, C.S., Rocca, J.J.: Hour-long continuous operation of a tabletop soft x-ray laser at 50–100 Hz repetition rate. Opt. Express **21**, 28380–28386 (2013)
20. Janulewicz, K.A.: Single-pulse low-energy-driven transient inversion x-ray lasers. IEEE J. Sel. Top. Quantum Electron. **10**, 1368–1372 (2004)
21. Ursescu, D., Cojocaru, G., Ungureanu, R., Banici, R., Delmas, O., Pittman, M., Guilbaud, O., Kazamias, S., Cassou, K., Demailly, J., Neveu, O., Elsa, B., Ros, D.: Thin film beam splitter multiple short pulses generation for enhanced Ni-like Ag x-ray laser emission. Opt. Lett. **39**, 2246–2249 (2014)
22. Curtis, A.H., Reagan, B.A., Wernsing, K.A., Furch, F.J., Luther, B.M., Rocca, J.J.: Demonstration of a compact 100 Hz, 0.1 J, diode-pumped picosecond laser. Opt. Lett. **36**, 2164–2166 (2011)
23. Reagan, B.A., Curtis, A.H., Wernsing, K.A., Furch, F.J., Luther, B.M., Rocca, J.J.: Development of high energy diode-pumped thick-disk Yb: YAG chirped-pulse-amplification lasers. IEEE J. Quantum Electron. **48**, 827–835 (2012)
24. Keenan, R., Dunn, J., Patel, P.K., Price, D.F., Smith, R.F., Shlyaptsev, V.N.: High-repetition-rate grazing-incidence pumped x-ray laser operating at 18.9 nm. Phys. Rev. Lett. **94**, 103901 (2005)
25. Isoyan, A., Jiang, F., Cheng, Y., Cerrina, F., Wachulak, P., Urbanski, L., Rocca, J., Menoni, C., Marconi, M.: Talbot lithography: self-imaging of complex structures. J. Vaccum Sci. Technol. B **27**, 2931–2937 (2009)

Possibility of Recombination Gain Increase in CV Ions at 4.0 nm Via Coherence

Y. Luo, A. Morozov, D. Gordon, P. Sprangle, A. Svidzinsky, H. Xia, M. Scully and S. Suckewer

Abstract This paper is about the recent experimental results on amplification of the CV line in the "water window" at 4.03 nm from resonance transition to the ground level of He-like ions in recombination scheme. The indication of the amplification of the CV line has been observed when an elongated narrow plasma channel was created, where high intensity 100 fs beams, optimal for creating CV ions in high density plasma, was propagated up to 0.5–0.6 mm. Without channeling the effective plasma length was much shorter and there was no indication of amplification.The large interest in gain generation in He-like ions in the transition to ground state is due to the possibility of applying a recently developed theory of Lasing Without Inversion (LWI) in XUV and X-ray regions to largely increase the gain for such transitions. The presented results of the indication of CV line amplifications are being discussed from the point of view of using LWI as a superradiance gain increase, hence to construct a very compact soft X-ray laser in the "water window".The last part of the paper is related to the application of the ultra-intensive fs plasma laser, which is currently in the process of development by using stimulated Raman backscattering (SRBS) to create a plasma amplifier and compressor, as the pump for compact laser operating in the "water window" and also at shorter wavelengths.

1 Introduction

Tremendous progress has been made in beam quality, compactness, and especially in the application of soft X-ray lasers (SXLs), since the first large gain-length products (gl) were demonstrated in the soft X-ray region nearly three decades ago. The qualities of SXL beams, reaching 10 nm and slightly below, have been improved in their coherence, divergence and stability (i.e. very good reproducibility from shot-to-shot) is presented in a number of recent and past reviews [1–9], including presentations at this conference (for example [10–12]).

S. Suckewer (✉) · Y. Luo · A. Morozov · A. Svidzinsky · H. Xia · M. Scully
Princeton University, Princeton, NJ 08544, USA
e-mail: suckewer@princeton.edu

A. Svidzinsky · H. Xia · M. Scully
Texas A&M University, College Station, TX, USA

D. Gordon · P. Sprangle
Naval Research Lab, Washington DC, USA

© Springer International Publishing Switzerland 2016
J. Rocca et al. (eds.), *X-Ray Lasers 2014*, Springer Proceedings in Physics 169,
DOI 10.1007/978-3-319-19521-6_3

Special attention has been directed toward SXL compactness. The development of the table top SXLs, which will have a number of practical applications in science and technology, has become a high priority for many researchers in the field. Of course, the compactness of SXL depends on the efficiency of the lasing system (especially quantum efficiency) and the pumping efficiency. For the maximum quantum efficiency, systems in which lasing is in transition to ground state are very attractive. As for pumping efficiency, longitudinal pumping usually is more efficient than transverse pumping.

To realize lasing to the ground states of ions in the soft– and x-ray regions, ultrashort laser pump pulses have to be used in order to provide population inversion between the excited and ground states in ions. With the development of very high intensity fsec laser pulses it becomes possible to realize recombination lasing to the ground states in ions, which are first stripped of electrons by optical field ionization, OFI.

Achieving gain using recombination scheme is highly desirable in the pursuit of x-ray lasers due to requiring relatively low pumping pulse energy. This, combined with the high quantum efficiency achieved by using the transition to the ground state, makes the creation of a practical tabletop x-ray laser (XRL) quite feasible. Furthermore, the highly favorable scaling of the required pumping energy with decreasing the wavelengths may enable reaching the "water window" (the wavelength range 2.3–4.4 nm, for which absorption in water is low) with a tabletop x-ray laser system. Hence, such XRL would be an excellent source for high resolution X-ray microscopy of biological cells in their natural environment, placed in a relatively small biological laboratory.

The highly efficient recombination system utilizing transition to ground states of ions was theoretically predicted [13, 14], and experimentally high gl was demonstrated in Hydrogen-like Li III in a transition to the ground state ($2 \rightarrow 1$ transition) at 13.5 nm [15,16] and explained theoretically [17] with added computer simulation of experimental results. This computer simulation takes into account the uniqueness of the OFI processes such as the non-Maxwellian nature of created plasma and its relatively low temperature, which is a very important feature for generating high gain in fast recombining ions in plasma. For example, the experiment on obtaining high gain in the $2 \rightarrow 1$ transition in LiIII ions has been an important benchmark for computer simulations. Better efficiency of the gain generation using $\sim 0.25\ \mu$m pumping wavelength in [15] in comparison to $\sim 1.0\ \mu$m in [16], both experiments use preplasma in microcapillaries for waveguiding pumping beams at $\sim 3 \times 10^{17}$ W/cm^2 intensities, was verified by computer modeling. Furthermore, by investigating the effects of different laser and plasma parameters on the gain generation at various lasing wavelengths, the optimal conditions required to achieve high gain in CVI ions in $2 \rightarrow 1$ transition at 3.4 nm were predicted [18], providing a "prescription" for designing and conducting experiments.

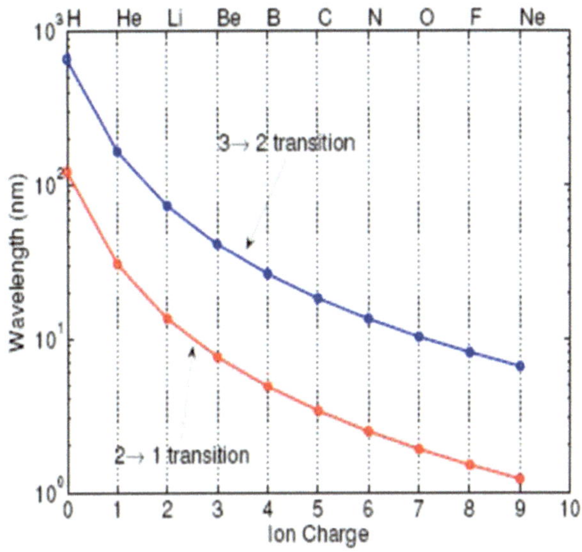

Fig. 1 Wavelength scaling of recombination schemes for $2 \rightarrow 1$(in *red*) and $3 \rightarrow 2$ transitions for H-like ions

2 Remarks About Recombination XRL: Rapid Convergence to Short Wavelengths

The recombination scheme is attractive due to its fast convergence to short wavelength of $3 \rightarrow 2$ and $2 \rightarrow 1$ transitions in H-like ions as illustrated by Fig. 1. Especially attractive are the transitions to ground state ($2 \rightarrow 1$ transitions) with their high quantum efficiency and for which the "water window" is attainable for CV and CVI ions using pumping lasers with relatively low energy, but at high intensity and repetition rate. Our extensive computer calculations have shown the feasibility of achieving high gain with the transient recombination scheme in the "water window" with peak pump intensity in the order of $I_p \sim 10^{18}$–10^{19} W/cm^2 and duration of pulses less than 100 fs for He-like ions CV (lower intensity range) and H-like ions CVI (higher intensity range). The crucial aspect of the generation of high gains for these transitions is a strong non-Maxwellian electron energy distribution function. Such distribution function, obtained by computer simulation of Optical Field Ionization, OFI, in the process of creating a fully ionized carbon, is shown in Fig. 2 from [18]. These computer calculations indicate that most of the electrons from the fully stripped carbon, have very low kinetic energies in the order of 10–20 eV, in comparison to the ~400–500 eV ionization energies of the CV and CVI ions.

Electrons in the high-energy tail of the distribution function, due to their high velocities and low cross section in interaction with ions at such velocities, can quickly escape the narrow plasma channel without participating in the recombination processes. After a short time, in the order of a few picoseconds, these electrons

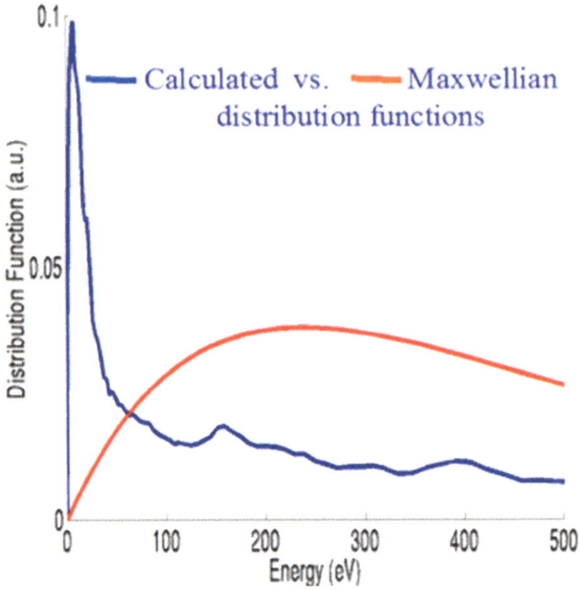

Fig. 2 Calculated highly non-Maxwellian electron distribution (in *blue*) after fully stripped carbon ions off electrons by means of OFI; corresponding Maxwellian distribution is shown (in *red*)

proceed to thermal equilibrium, which reach Maxwellian electron distribution in about 15–20 ps with it maximum energy at about 250–300 eV, much too high for effective three-body recombination process.

Therefore, the very short and very intensive pulses can create fully ionized, high density carbon plasma with very non-Maxwellian electron energy distribution, leading to a 3.4 nm gain generation in CVI followed by 4.03 nm gain in CV. The gain generation in CVI takes place in a very short time, hence potentially leading to a very short lasing time, during which the ground state of CVI would rapidly populate, followed by its fast recombination to highly excited states of CV. Due to this process gain in the singlet states of the CV ions is expected to be obtained, while the electrons are still highly non-Maxwellian.

3 Predicted Gain for 2 → 1 Transition in CVI Ions and Experimental Setup

Gain generation in H-like ions, as we have already indicated, requires the atoms to be totally stripped of electrons, as the first step. Next, these atomic nucleus in the plasma have to recombine very fast by three-body, non-radiative process (nucleus + 2 electrons), providing H-like ions in highly excited states (three-body recombination is approximately proportional to main quantum number n to

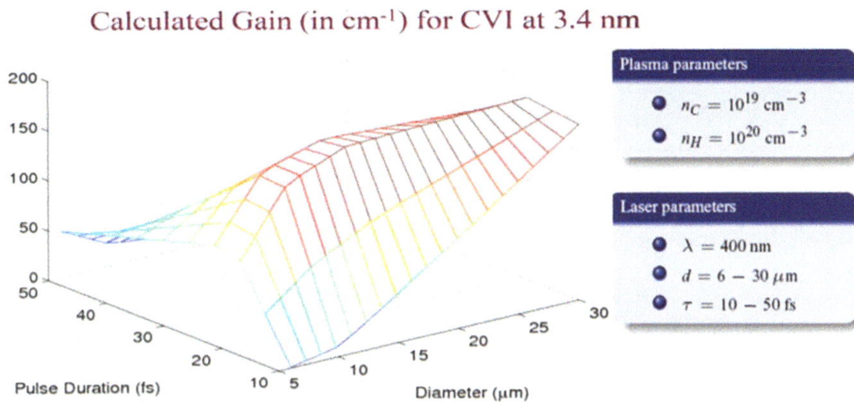

Fig. 3 Predicted gain (from [18]) in H-like CVI at 3.4 nm as a function of pumping beam diameter (in μm) and its pulse duration (in fs) for initial carbon and hydrogen densities $n_C = 10^{19}$ cm^{-3} and $n_H = 10^{20}$ cm

power 4). Moreover, the electron densities have to be high (in range 10^{17}–10^{20} cm^{-3}, depending on the nucleus charge Z), and in order to create a $2 \rightarrow 1$ population inversion, the ionizing pulses have to be of very high intensity and very short duration. Short pumping pulse duration prevents significant plasma heating, which the main source for OFI process is the residual heating, or the above threshold ionization (ATI) heating. This heating arises from the variation in the oscillation phase f_d between the free electrons from the ionization of atoms and ions and the phase of the laser electric field. The average residual energy is proportional to the quiver energy of the electrons in the laser field, $\varepsilon_q = e^2E^2/4m_e\nu^2$, where e is the electron charge, E is the laser peak electric field, m_e is the electron mass and $\nu$is pumping laser frequency.

Because three-body recombination rate is proportional to the square of the electron density, N_e^2, and quantum number n to power 4, the transition from fully ionized carbon, C^{6+}, to H-like CVI occurs primarily to the states with high n, while collisional and radiative transitions to level $n = 2$ are faster than to ground level $n = 1$, creating a population inversion between $n = 2$ and $n = 1$. In addition to minimizing plasma heating, the ultrashort pumping pulses are also necessary due to the very short radiative lifetime, τ, of the first excited levels of ions and decreasing this time as Z^{-4} (Z is atomic number of ions). In fact, this time has to be much shorter than the radiative decay time in order to be shorter than the collisional decay time at high N_e. It was also shown that the gain can be enhanced and become less stringently—dependent on exactly matching the required experimental parameters, if hydrogen is added into the carbon plasma.

The gain (in cm^{-1}), which was obtained by computer calculations [18], is shown in Fig. 3 for $2 \rightarrow 1$ transition at 3.4 nm in CVI for the range of pumping beam intensity $I \approx 7 \times 10^{18}$–$1 \times 10^{19}$ W/cm^2, where the lower intensity was for the 50 fs pumping pulses and the higher intensity was for the 10 fs pulses. The initial carbon

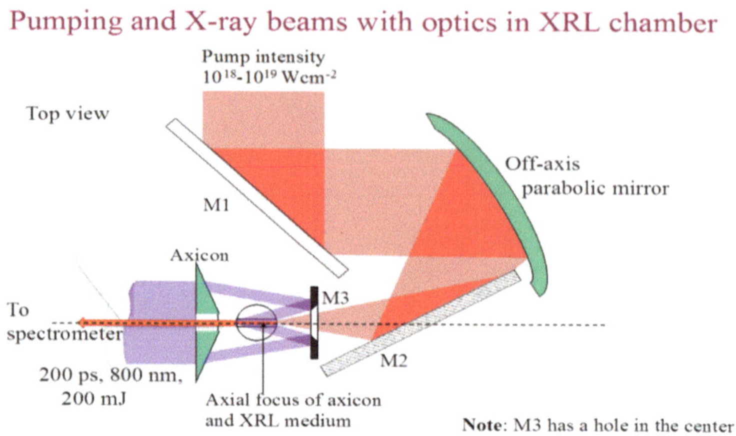

Fig. 4 Experimental setup

density was $n_C = 10^{19}$ cm^{-3} with added hydrogen density of $n_H = 10^{20}$ cm^{-3}. The gain is presented as a function of the pumping beam diameter and its pulse duration. It is apparent that the maximum gain, $g \approx 150$–$180\,\text{cm}^{-1}$, is obtained for pulse duration of 20–30 fs, and this gain is also less dependent on the pumping beam diameter. For pumping pulses ~ 50 fs and up to 100 fs the predicted gain drops to $50\,\text{cm}^{-1}$ and below, but still is quite high.

Experimental setup, which is shown schematically in Fig. 4, was designed following computer gain predictions for optimal laser pulses and plasma medium parameters as well as gained experience with development high gain in $2 \rightarrow 1$ transition in H-like LiIII ions [15]. The multipass Ti:Sa amplifier provided 2 J, 800 nm 200 ps pulses at 5 Hz repetition rate. The pulse was optically compressed down to 100 fs and was focused by an off-axis parabolic mirror ($f = 150$ mm) into the preformed plasma in the gas jet. The plasma was initiated by an optical breakdown using part of a 200 ps pulse, splitted before the compressor. It was focused by an axicon lens in order to create a pre-formed plasma waveguide, whose purpose was to enable the propagation of the ultra-intense pump pulses over a much longer distance than its Rayleigh length, pioneered by Milchberg and his group [19, 20] and provided computer modeling by Sprangel et al. [21]. In these experiments, the near diffraction-limited focus of the pulse was achieved and enabled reaching 2×10^{19} W/cm^2 intensities with 300 mJ of pulse energy. The extensive diagnostics were implemented to monitor absolute intensity distribution of the 100 fs pulses in the plasma channel, the electron density distributions as well as the neutral gas density in the gas jet was monitored using a shear-type of interferometer with time and spatial resolutions of 200 fs and ~ 5 μm, respectively. The spectra in "water window" region in direction of fs beam propagation in plasma were recorded using a grazing incidence flat-field spectrometer having a diffraction grating of 1200 groves/mm with a variable density of grooves. The spectra were registered using a 1024×1024 back illuminated CCD camera. Two 150 μm stainless steel pinholes were placed

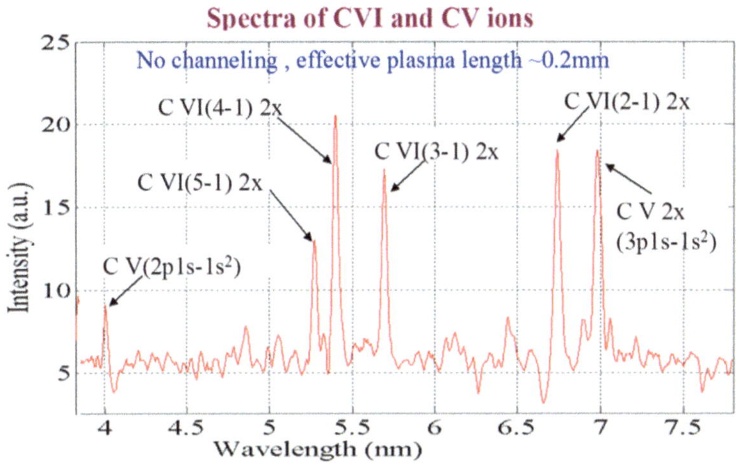

Fig. 5 Spectra of CV and CVI in "water window" wavelength range created with 100 fs beam without channeling and with pulse energy/intensity: 200 mJ/$\sim$4–5 $\times$ 10^{18} W/cm^2, effective plasma length was: l_{eff} $\sim$0.2 mm, gas jet mixture: 20 % of C_2H_6 and 80 % of H_2 note: notation 2x, for example in CVI(2 – 1) 2x, indicates *second* order of CVI(2 – 1) line

between the spectrometer slit and the plasma to reduce the amount of visible light from the plasma and stray light from the laser from reaching the detector, very significantly improving signal-to-noise ratio in the recorded spectra.

4 Recorded CV and CVI Spectra in "Water Window"

The spectra were recorded in the wavelength range of 2.3–4.4 nm, covering the "water window", in search for conditions to generate gain in transitions 2 $\rightarrow$ 1 at 3.4 and 4.0 nm in CVI and CV ions, respectively. For this purpose the spectra were taken along the propagation path of the laser pulses, perpendicularly to the gas jet's nozzle slit and both the neutral gas density and the plasma density were monitored simultaneously in a single shot. The example of CV and CVI line emissions in "water window" wavelength range is shown in Fig. 5 for tightly focused 100 fs beam onto thin plasma layer of the effective depth $l_{eff} \approx 0.2$ mm without channeling. The plasma was created in a gas jet mixture of 20 % C_2H_6 and 80 % H_2. The CVI lines dominate the spectra for these pumping laser pulses of intensities in the range of $\sim$4–5 $\times$ 10^{18} W/cm^2 for 200 mJ pulse energy, although the CV (3 p – 1 s^2) in second order is intensive. Moreover, the intensity ratio of CVI (4 – 1)2x to CVI (3 – 1)2x indicates quite strong population inversion between levels four and three when taking into account $A_{41} g_4/A_3\ g_3 \approx 0.2$, where A_{nm} are spontaneous emission coefficients from upper level n to lower level m, and g_n are statistical weights of levels n. Similar population inversion was observed for levels three and two in CVI

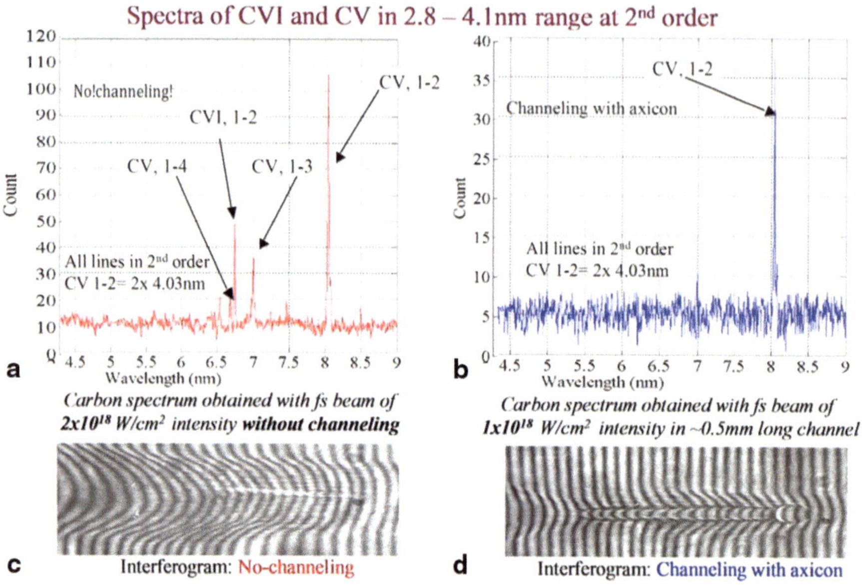

Fig. 6 **a** Carbon spectrum using only a fs beam (no channeling), the pulse energy: 150 mJ, gas jet mixture: 20 % of C_2H_6 and 80 % of H_2; **b** Carbon spectrum with channeling (waveguiding with axicon), energies of channeling and fs beams were 150 mJ each, gas jet mixture as in (**a**); **c** Interferogram of plasma created by 100 fs laser pulses with 150 mJ pulse energy; **d** Interferogram of plasma created by the 100 fs laser pulses propagated in pre-formed plasma waveguide, which was created by axicon *line*-focused of 200 ps pulses with 150 mJ pulse energy; the delay between fs and ps pulses was 1.2 ns, gas jet mixture was as in (**a**).The laser pulses propagate from *right* to *left*

and CV, when longer wavelengths were monitored to see simultaneously all these lines in second order.

For lower pumping laser intensities the He-like CV lines became dominant. For example, for pumping intensity of $\sim 2 \times 10^{18}$ W/cm^2 with pulse energy $\sim$ 150 mJ and the same gas jet mixture as in Fig. 5, the spectrum in Fig. 6a is dominated by CV lines, which is reasonable considering that the peak intensity is sufficient to create predominately CV ions but much less CVI ions. Such dominance of CV ions was even more pronounced when the axicon created plasma channel has been used to enable propagation of the fs pulses in longer plasma, as shown in Fig. 6b, while the intensities of fs pulses were twice lower than in Fig. 6a. Under this condition, only the line from the CV 2$\rightarrow$1 transition could be observed. This indicates the encouraging possibility of achieving lasing in 2$\rightarrow$1 transition in CV ions at 4.03 nm, which is within the "water window". It is very interesting and provides additional encouragement for reaching our goal that the spectrum in Fig. 6b is very similar to the spectrum of H-like LiIII in Fig. 1 in [16], which was recorded in a series of experiments leading to high gl for 2$\rightarrow$1 transition at 13.5 nm.

The recording spectra shown in Fig. 6a, b and the neutral particles and plasma densities were simultaneously monitored with an interferometry. The interferogram of the plasma created by a single ultra-intense 100 fs pulses with 150 mJ pulse energy taken immediately after it passed through the jet is shown in Fig. 6c. Although the focal spot of the fs pulse had a diameter of ~5 μm, the resulted plasma was over 200 μm wide presumably due to the ionization-induced refraction and other nonlinear effects during the propagation of the fs pulse with effective plasma length $l_{eff} \approx 0.2$ mm.

The interferogram of the plasma created by propagating the same 100 fs pulse, but through the preformed plasma waveguide is shown in Fig. 6d. The 200 ps pulse had 150 mJ energy and arrived at the gas target 1.2 ns prior to the arrival of the fs pulse. The gas target used in this set of data was the same as in Fig. 6a. Diluting the C_2H_6 gas with H_2 lead to two advantages: (1) to suppress parametric instabilities during the ionization by the Bessel beam and (2) to provide larger density of low energy electrons due to much lower ionization potential of hydrogen atoms than that of CVI and CV ions, which is favorable for fast three-body recombination of C^{6+} to CVI (C^{5+}). The pre-formed plasma waveguide effectively confined the propagating the fs pulse maintaining its high intensity over distance $l_{eff} \approx 0.5$ mm, apparently already having a very important effect on the intensity of the CV(2 − 1) line in Fig. 6b spectrum.

5 Possibility of Gain Increase in CV (2 → 1) Transition at 4.0 nm Via Quantum Coherence

Recent work on quantum coherence control of transitions in He atoms and He-like ions has predicted the possibility of generating lasing action in the XUV and X-ray regions via superradiance radiation enhancement and even generation gain without population inversion [22–24]. In general, the quantum coherence phenomenon in atomic and radiation physics has led to many interesting and unexpected results. For example, by preparing an atomic system in a coherent superposition of states, under certain conditions, it is possible for atomic coherence to cancel absorption but not emission, which is the basis for lasing without inversion, LWI. Frequently this is accomplished in three- or four-level atomic systems in which there are coherent routes for absorption that can destructively interfere, thus leading to the cancellation of absorption. Moreover, it was shown that when the excitation of an ensemble of atoms is swept in the direction of lasing (gain-swept superradiance, GSS), so that atoms are prepared in the excited state just as the radiation from previously excited atoms reaches them and yields superradiant emission [25, 26], which has common features with Dicke superradiance [27] for three-level system and can yield soft X-ray lasing in, for example, He-like ions of C^{4+} at 4.0 nm, schematically presented in Fig. 7.

It can be seen from Fig. 7 that strongly driving (with high Rabi frequency Ω) transition $|a\rangle \rightarrow |c\rangle$ in CV ions make feasible to coherently stimulate transition to

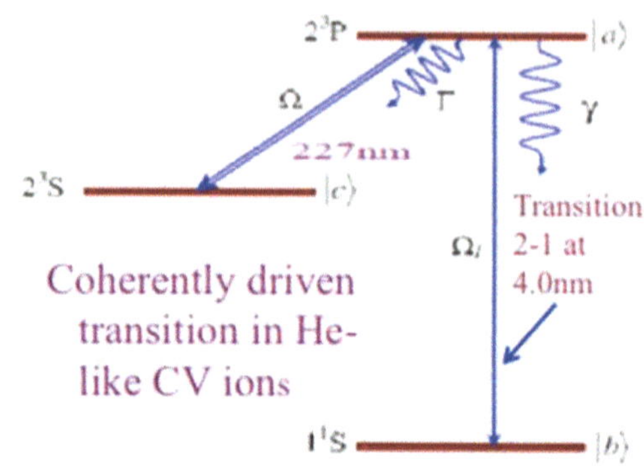

Fig. 7 Illustration of quantum coherence application to gain increase in recombination scheme for transition to ground state of He-like CV ions at wavelength of 4.0 nm

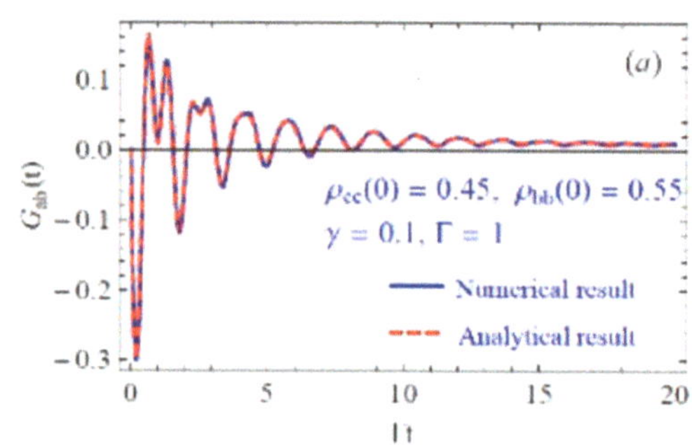

Fig. 8 Plots from [25] of dimensionless gain/absorption in the lasing transition |a⟩ →|b⟩ vs dimensionless time Γt. *Blue* and *red* curves show, respectively, numerical and analytical results, where total decay and radiative decay frequencies of *upper* state |a⟩ are $\Gamma = 1$ and $\gamma = 0.1$

ground state |a⟩ →|b⟩ with superradiance radiation and possible LWI at 4.0 nm, as it is illustrated in Fig. 8. The calculations presented in Fig. 8 were performed for total and radiative dimensionless decay frequencies $\Gamma = 1$ and $\gamma = 0.1$ (normalization factor is $\Gamma = 4.5 \times 10^7 s^{-1}$) of upper state |a⟩. The plots for analytical and numerical calculation results indicate excellent agreement between both of them and are very encouraging for significant gain increase for 4.0 nm radiation.

6 Ultraintense Femtosecond Plasma Laser for Pumping XRLs

This last section of the paper is related to the application of the ultraintense fs plasma laser, which is currently in the development process at Princeton, as the pump for compact XRLs operating in the "water window" and also at shorter wavelengths. The development of the ultraintense fs plasma laser, in which the large and expensive optics for pulse amplification and compression will be replaced by plasma

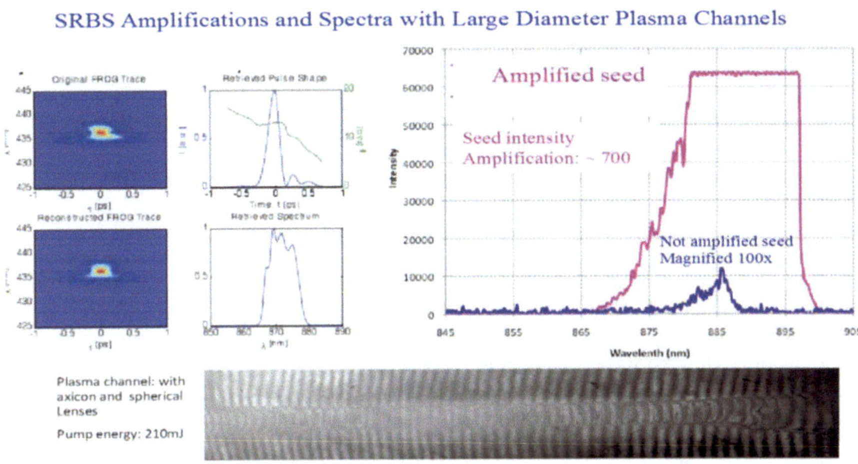

Fig. 9 Seed intensity amplification of ~700 and with a relatively low pump energy of 210 mJ in the plasma channel. Interferogram shows very good channel ~4 mm long, although effective length of ~2.5 mm was used here. The figure on the *right* shows the input and output seed spectra (input seed is magnified 100x to be visible, whereas the amplified seed saturated detector is due to insufficient signal attenuation In the *left* low corner the interferogram shows a plasma channel of good quality, and at the *left* the time profile of output amplified seed from single pulse autocolerator FROG measurement is shown

column, is based on stimulated Raman backscattering (SRBS) for creating a plasma amplifier and compressor. This experimental research was conducted for a number of years [28, 29] and was preceded by extensive theoretical research [30–32].

The very encouraging earlier results of seed intensity amplification of 10,000 and its compression from 550 fs to ~90 fs in a single pass, and increases this intensity by 2x and further decreases the pulse duration down to ~50 fs in the two passes experiment [33], stimulate present research towards practical fs plasma laser. With a much larger plasma column diameter of ~400–500 μm for waveguiding large diameter pump and seed pulses, which diameters of 200–250 μm are expected to ensure a compact plasma laser with intensities at the focus of ~10^{20} W/cm^2, the ideal pump for XRLs in the "water window" and down to a wavelength of ~1–2 nm.

Figure 9 shows recently obtained, still preliminary, seed intensity amplification in the order of ~700 using large diameter seed and pump pulses, propagating in a corresponding large diameter plasma channel, as discussed above. The pump energy of 210 mJ, hence its intensity, was still much below the required value to reach amplified output seed intensity in focus in the order of ~10^{20} W/cm^2, which is one of the main goals of the present research.

Acknowledgements We are thankful to Q.Chen and N.Tkach for their help with the experiments. The research was supported by NSF/Phys Grant 1068554, NSF/DOE Grant 1064465 and DOE/NNSA Grant DE-FOA-0000611

References

1. Jaeglé, P.: Coherent Sources of XUV Radiation. Springer, Berlin (2006)
2. Rocca, J., Kapteyn, H., Atwood, D., Murnane, M., Menoni, C., Anderson, E.: Tabletop Lasers in the Extreme Ultraviolet. Opt. Photon. News **17**, 30 (2006)
3. Daido, H.: Review of soft x-ray laser researches and developments. Rep. Prog. Phys. **65**, 1513 (2002)
4. Pert, G.J.: Recombination and population inversion in plasmas generated by tunneling ionization. Phys. Rev. E. **73**, 066401 (2006)
5. Suckewer, S., Jaeglé, P.: X-Ray laser: past, present, and future. Laser Phys. Lett. **6**, 411 (2009)
6. Rocca, J.: Table-top soft x-ray lasers. Rev. Sci. Instrum. **70**, 3799 (1999)
7. Attwood, D.: Soft X-Rays and Extreme Ultraviolet Radiation: Principles and Applications. Cambridge University Press, Cambridge (1999)
8. Elton, R.: X-Ray Lasers. Academic, Boston (1990)
9. Suckewer, S., Skinner, C.H.: Soft x-ray lasers and their applications. Science **247**, 1553 (1990)
10. Sebban, S., Gautier, J., Tissandier, F., et al.: Progress on Collisionaly-Pumped OFI Soft X-Ray Laser at LOA, This Conference, (May 2014)
11. Menoni, C.: Nanoscale Imaging with Table-Top Soft X-Ray Laser, This Conference, (May 2014)
12. Shlyaptsev, V., Avaria, G., Grisham, M., et al.: Capillary Discharge X-Ray Lasers; The Quest for the Sub—10 nm Lasing, This Conference, (May 2014)
13. Corkum, P., Burnett, N.: Short-Wavelength Coherent Radiation: Generation and Applications. Falcone, R.W. and Kirz, J. eds. of OSA Proceedings Series. vol.**2**, p.225, Washington. D.C (1988)
14. Burnett, N., Corkum, P.: Cold-plasma production for recombination extreme-ultraviolet lasers by optical-field-induced ionization. J. Opt. Soc. Am. B. **6**, 1195 (1989)
15. Korobkin, D., Nam, C.H., Suckewer, S., Goltsov, A.: Demonstration of Soft X-Ray Lasing to Ground State in Li III. Phys. Rev. Lett. **77**, 5206 (1996)
16. Goltsov, A., Morozov, A., Suckewer, S., Elton, R., Feldman, U., Krushelnick, K., Jones, T., Moore, C., Seely, J., Sprangle, P., Ting, A., Zigler, A.: Is efficiency of gain generation in Li III 13.5-nm laser with 0.25-μm subpicosecond pulses the same as with 1 μm?IEEE J. Sel. Top. Quantum Electron. **5**, 1453 (1999)
17. Avitzour, Y., Suckewer, S.: Numerical simulation of the effect of hydrogen on recombination gain in the transition to ground state of Li III . J. Opt. Soc. Am. B. **23**, 925 (2006)
18. Avitzour, Y., Suckewer, S.: Feasibility of achieving gain in transition to the ground state of C VI at 3.4 nm. J. Opt. Soc. Am. B. **24**, 819 (2007)
19. Durfee, C., Milchberg, H.: Light Pipe for High-Intensity Laser-Pulses. Phys. Rev. Lett. **71**, 2409 (1993)
20. Durfee, C., Lynch, J., Milchberg, H.: Development of a Plasma Wave-Guide for High-Intensity Laser-Pulses. Phys. Rev. E. **51**, 2368 (1995)
21. Sprangle, P., Ting, A., Esarey, E., Fisher, A.: Tunable, short pulse hard x-rays from a compact laser synchrotron source. J. Appl. Phys. **72**, 5032 (1995)
22. Harris, S.E.: Lasers without inversion: Interference of lifetime-broadened resonances. Phys. Rev. Lett. **62**, 1033 (1989)
23. Scully, M., Zhu, S.-Y., Gavrielides, A.: Degenerate quantum-beat laser: Lasing without inversion and inversion without lasing. Phys. Rev. Lett. **62**, 2813 (1989)
24. Svidzinsky, A., Yuan, L., Scully, M.: Transient lasing without inversion. New J. Phys. **15**, 053044 (2013); also Svidzinsky, A., Yuan, L., Scully, M.: Quantum Amplification by Superradiant Emission of Radiation. Phys. Rev. X. **3**, 041001 (2013)
25. Sete, E., Svidzinsky, A., Rostovtsev, Y., Eleuch, H., Jha, P., Suckewer, S., Scully, M.: Using Quantum Coherence to Generate Gain in the XUV and X-Ray: Gain-Swept Superradiance and Lasing Without Inversion. IEEE J. Sel. Top. Quantum Electron. **18**, 541 (2012)

26. Xia, H., Svidzinsky, A., Yuan, L., Lu, C., Suckewer, S., Scully, M.: Observing Superradiant Decay of Excited-State Helium Atoms Inside Helium Plasma. Phys. Rev. Lett. **109**, 093604 (2012)
27. Dicke, R.: Coherence in Spontaneous Radiation Processes. Phys. Rev. **93**, 99 (1954)
28. Ping, Y., Cheng, W., Suckewer, S., Clark, D., Fisch, N.: Amplification of ultrashort laser pulses by a resonant Raman scheme in a gas-jet plasma. Phys. Rev. Lett. **92**, 175007 (2004)
29. Cheng, W., Avitzour, Y., Ping, Y., Suckewer, S., Fisch, N., Hur, M., Wurtele, J.: Reaching the nonlinear regime of Raman amplification of ultrashort laser pulses. Phys. Rev. Lett. **94**, 045003, (2005)
30. Malkin, V., Shvets, G., Fisch, N.: Fast compression of laser beams to highly overcritical powers, beams to highly overcritical powers. Phys. Rev. Lett. **82**, 4448 (1999)
31. Malkin, V.M., Shvets, G., Fisch, N.: Ultra-powerful compact amplifiers for short laser pulses. J. Phys. Plasmas **7**, 2232 (2000)
32. Fisch, N., Malkin, V.: Generation of ultrahigh intensity laser pulses. Phys. Plasmas **10**, 2056 (2003)
33. Ren, J., Cheng, W., Li, S., Suckewer, S.: A new method for generating ultraintense and ultrashort laser pulses. Nature Phys. **3**, 732 (2007).

Overview of Development of Laser Driven Secondary Sources at PALS and ELI

M. Kozlova, J. Nejdl, M. Albrecht, S. Sebban, J. Gautier, K. Ta Phuoc, A. Klisnick, A. Le Marec and F. Tissandier

Abstract In this paper we report on development of the secondary X-ray sources at the PALS Centre and discuss the plan for the ELI Beamlines project. The spatial and temporal coherence of the most energetic quasi-steady state Ne-like Zn X-ray laser, which is operated at PALS Centre as standard user beamline, was examined proving that amplification of coherent EUV pulses with duration below 1 ps will be possible. Meanwhile, the first transient lasing at PALS Center was achieved using 10 Hz Ti:Sapphire laser chain with peak power of 20 TW as a driver. Finally, we discuss the recent design of laser driven secondary sources generating short coherent or incoherent EUV/X-ray pulses within the ELI Beamlines project.

1 Introduction

The past decade at PALS Center facility in Prague [1] has been dedicated to development of secondary sources with short-wavelength radiation. Currently two laser beamlines could be used as drivers for their generation. Moreover both laser beamlines could work in synchronized mode or separated. With iodine system we developed and operate, as standard user beamline, so far the most energetic soft X-ray laser in the world, Ne-like Zn X-ray laser emitting at 21.2 nm [2]. Recently, new X-ray sources were designed and experimentally investigated driven by the Ti:Sapphire laser system. In this paper the preliminary results of first transient Ni-like Mo X-ray laser at PALS Center are reported. The combination of the two driving

M. Kozlova (✉) · J. Nejdl · M. Albrecht · S. Sebban · J. Gautier · K. Ta Phuoc
IoP ASCR, Na Slovance 2, 182 00, Prague 8, Czech Republic
e-mail: kozlova@fzu.cz

M. Kozlova · J. Nejdl
IPP ASCR, Za Slovankou 3, 182 00, Prague 8, Czech Republic

S. Sebban · J. Gautier · K. Ta Phuoc · F. Tissandier
LOA, ENSTA, Chemin de la Hunière, 91761 Palaiseau cedex, France

A. Klisnick · A. Le Marec
ISMO, Université Paris-Sud, Orsay, France

M. Albrecht
FNSPE, CTU, Prague, Czech Republic

© Springer International Publishing Switzerland 2016
J. Rocca et al. (eds.), *X-Ray Lasers 2014*, Springer Proceedings in Physics 169,
DOI 10.1007/978-3-319-19521-6_4

systems gives an opportunity to investigate a three-stage amplifier chain for amplification of HHG in plasma amplifiers. Hence, the properties of temporal coherence of Ne-like Zn X-ray laser were studied for single and double pass regime. Also novel technique of investigating the spatial coherence in single shot of XRL radiation was tested.

2 PALS Beamlines

2.1 Iodine kJ Laser Beamline

The PALS (Prague Asterix Laser System) few terawatt iodine laser beamline is delivering 350 ps pulses at fundamental wavelength 1315 nm, with high-energy output in a single-beam configuration, up to 1 kJ, which makes it a powerful driver for various experiments. Moreover the main beam could be divided into two auxiliary beams with energy of 100 J combined, and it could be spatially and temporally overlapped with the main beam in interaction chambers. The splitting of the main beam enable the pump and probe experiments without any jitter. The repetition rate is 30 min with all amplifiers. Also the fundamental frequency of the Asterix laser can be up converted to its second and third harmonic. The output energy can be varied over two orders of magnitude with essentially unchanged beam geometry. Due to all the above features, the PALS is being intensely exploited for laser-plasma experiments at power density levels ranging from 10^{14} to few 10^{16} W/cm^2.

3 Ti:Sapphire 20 TW Laser Beamline

The Ti:Sapphire laser chain installed at the PALS laboratory is a commercial laser system with peak power of 20 TW [3]. The pulse duration is 40 fs and the total output energy is up to 900 mJ with maximum repetition rate of 10 Hz. Fundamental wavelength 812 nm could be converted into second harmonic. Laser chain is based on conventional CPA amplification scheme, consisting of oscillator, stretcher, regenerative amplifier, pulse picker, multi pass amplifier and compressor. The compressor is designed to use zero orders of Au reflective gratings to produce two auxiliary beams, 50 mJ each. Zero order, of the first grating, is subsequently compressed using second compressor. Also the zero order of second grating is extracted and propagated on its own beam path. All beams could be delayed with respect to each other in the range of 0–20 ns. For experiments all beams could be send into interaction area, which is equipped by three independent reconfigurable vacuum chambers.

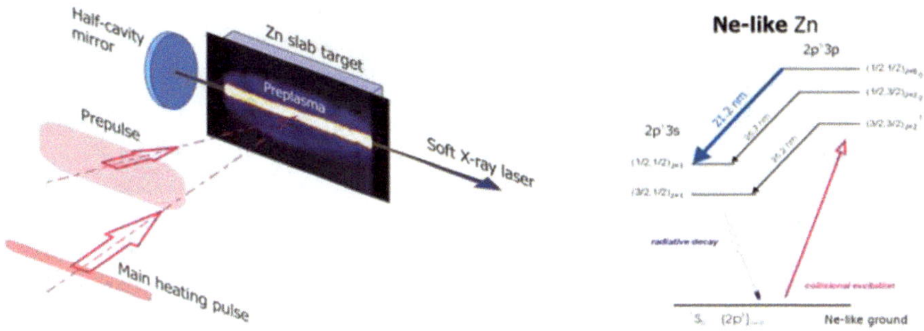

Fig. 1 Sequential pumping of the XRL amplifier, where the typical energy ratio of the prepulse ($< 10^{11}$ Wcm^{-2}) and the main pulse ($\sim 10^{13}$ Wcm^{-2}) is 2×10^{-3}. Half-cavity mirror is placed on one side of the Zn slab target to boots output energy

4 Secondary Sources and Its Specifications at PALS Facility

4.1 Ne-Like Zn X-ray Laser

Over past decade, the most energetic XRL is delivered at Prague Asterix Laser System (PALS) Centre [2]. The pump laser for XRL is the iodine laser chain ($\lambda = 1.3$ μm) with energy up to 1 kJ. To generate the XRL is necessary to split the pumping beam into two beams, prepulse and main pulse. The sequential pumping (see in Fig. 1, left) enables to provide output energies up to 10 mJ. The population inversion is produced between $2p^5$ 3p and $2p^5$ 3s fine-structure levels of Ne-like ions, through electron collisional excitation from the Ne-like ground state $2p^6$. The inversion is maintained by strong radiative dipole decay from the 3s levels back to the ground state while the upper 3p levels are metastable with respect to this process. Under appropriate plasma conditions, the strongest population inversion is generated (due to the highest monopole electron excitation rate from the ground state) for $(2p^5_{1/2}3p_{1/2})_{J=0} \rightarrow (2p^5_{1/2}3s_{1/2})_{J=1}$ transition in Ne-like Zn, corresponding to the wavelength of 21.22 nm (see Fig. 1 on the right). The intensity and beam profile emitted by this type of XRL strongly depends on the density profile of the amplifying plasma, as the largest population inversion on the $J = 0–1$transition occurs at high electron densities.

To reduce density gradients in the plasma ablated by a laser from a massive slab target, a technique, consisting of applying a prepulse typically a few nanoseconds prior to the main heating pulse, is employed [4–6]. The parameters of the prepulse relative to the main pulse are one of the0 key features determining the functionality of the XRL and characteristics of the emitted beam. Two configurations where a weak prepulse is applied 10 ns or 50 ns ahead of the main pulse were investigated. The prepulse is delivered and focused separately to produce a wide pre-plasma column into which the main pulse, focused much more tightly, is coupled (see in Fig. 1 left). The configuration where the prepulse is 10 ns ahead can be operated in single

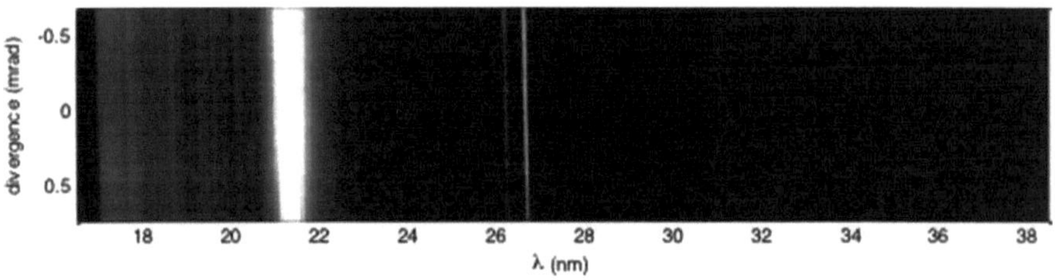

Fig. 2 Typical spectrum of the single pass Ne-like Zn X-ray laser. The prepulse is 10 ns ahead the main pulse

or double-pass regime. The typical value of energy (double-pass regime) is 4 mJ with peak power of 40 MW and with 3×10^{14} photons per pulse. The 10 ns configuration is the most stable and robust, and it is provided as the user beamline at PALS Centre. The typical spectrum of single pass regime with 10 ns sequence is shown in Fig. 2.

To increase the energy output of the XRL, the longer prepulse delays were investigated. The best results were observed for 50 ns prepulse delay operated in double-pass regime. This was studied with the streak camera and near-field measurements. While the streak camera records indicate comparable kinetics of the exited levels for both regimes, from the near-field data it is found, that the 50 ns prepulse delay can produce 2x larger (in the direction perpendicular to the target) amplification region than in 10 ns regime. Therefore the energy measured for 50 ns prepulse delay is 10 mJ with peak power of 100 MW and with 10^{15} photons per pulse.

5 Temporal Coherence and the Spectral Bandwidth

Using the Zn XRL as the user beamline requires a detailed characterization of its output parameters such as the energy, the pulse duration etc. In addition to parameters that are routinely monitored, the spectral profile and linewidth were explored. The relative bandwidth of the Zn XRL is extremely narrow ($\Delta\lambda/\lambda \sim 10^{-4}$), therefore direct measurement is beyond the capabilities of existing spectrometers in XUV range. Hence, a wavefront-splitting interferometer [7–9] was used to measure temporal coherence of the source from which the spectral linewidth is inferred. The experimental setup is shown in Fig. 3.

The XRL beam, operated in single-pass or double-pass regime, was send either to the footprint to monitor the beam intensity distribution and pointing, or to the interferometer. The direct light from the plasma was blocked by Al filter, placed after the multilayer mirror. Two sets of data were acquired for the single-pass and double-pass regime. The distance between the interferometer and the camera was about 1 m and the tilt of the camera was set to 35°. The spectral linewidth ($\Delta\lambda$) is given by

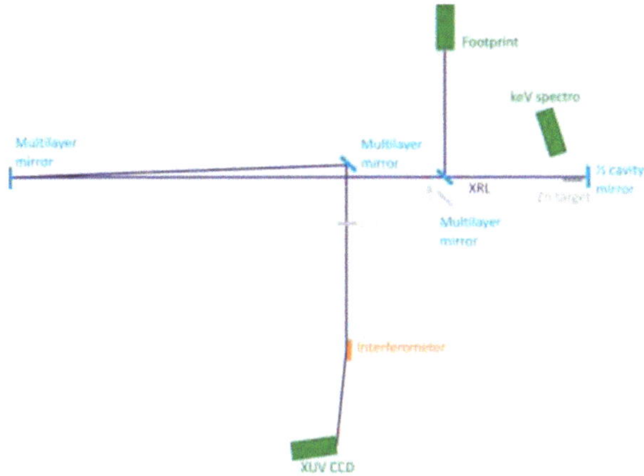

Fig. 3 The sketch of the experimental setup for temporal coherence measurement (not to scale)

$$\Delta\lambda = \lambda^2\, \Delta v/c, \tag{1}$$

where λ is XRL wavelength. The coherence time (τ_c) can be measured from fringe visibility change when varying the path difference between beams reflected by dihedron mirrors. This measurement resulted in 0.5 and 0.7 ps for single and double pass corresponding to spectral linewidths are 17 and 11 mÅ, respectively. The measured data of the coherence time leading to a Fourier transform limit of the pulse duration below 1 ps, which is much smaller than the duration of the duration of the XRL and potentially it indicates the possibility of the amplification of sub-picosecond pulses [9].

6 Demonstration of Transient Lasing of Ni-Like Mo

We have demonstrated lasing of Ni-like Mo in grazing incidence pumping arrangement using a shaped pulses of Ti:Sapphire laser chain with pulse energy 800 mJ. The transient population inversion pumped by collision excitation scheme in Molybdenum target is induced between energetic levels $(3d^9_{3/2}4d_{3/2})_{J=0} \rightarrow (3d^9_{5/2}4p_{3/2})_{J=1}$, which correspond to wavelength of 18.9 nm. A suitable amplifying medium has been created by a single shaped laser pulse containing a long prepulse with duration 4–5 ns that created a plasma column which was consequently heated by a short ~ ps pulse to produce transient population inversion. The experiment has been explored for pumping incidence angles $\alpha = 20$ and $\alpha = 25$. Experimental setup and typical on-axis spectrum with dominant lasing line is shown in Fig. 4. Changing the medium length we were able to evaluate the small signal gain coefficients

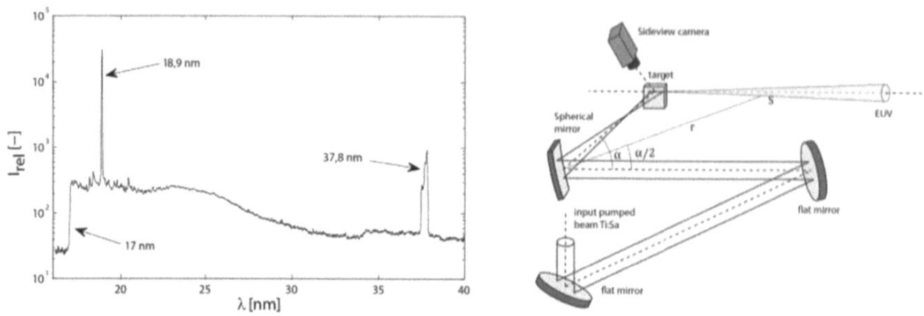

Fig. 4 Observed spectrum of the Ni-like Mo X-ray laser (18.9 nm) on the *left* and experimental setup showing GRIP geometry on the *right*

$G = 38.3\,\text{cm}^{-1}$ and $G = 29.3\,\text{cm}^{-1}$ for grazing angles $\alpha = 20$ and $\alpha = 25$, respectively. The corresponding gain-length products for those configurations are 13.4 and 12.6 indicating that the medium is close to saturation.

7 Single Shot Measurement of Spatial Coherence

We performed a single shot measurement of spatial coherence of the QSS Ne-like Zn XRL described above. The method is based on analysis of the diffraction pattern of an array of slits to sample the coherence function. The amplitude of the given spatial frequency in the recorded diffraction pattern is compared to the one obtained by calculating the pattern supposing a point source illuminating the array in the same geometry.

This method is particularly beneficial in the cases where the shot-to-shot fluctuations may affect the cumulative measurement when using the classical Young's double slit technique, where every point of the coherence function is measured within different laser shot. Plasma-based X-ray lasers are typical sources that can be characterized by our method, because of their narrow bandwidth (long longitudinal coherence) and their possible shot-to-shot fluctuations. We performed two measurements employing a mask with four $\sim 30\,\mu\text{m}$ wide vertical slits (see Fig. 5a) placed 2.5 and 1.7 m from the source. The values of four points of the degree of coherence were obtained for each measurement. Fitting the experimental data with Airy function we obtained the equivalent source size of $58\,\mu\text{m}$ which is in good agreement with previously measured near-filed images of the source [10]. Repeating this measurement we were able to evaluate the shot-to-shot fluctuations of the equivalent source size that stay below 10 % RMS.

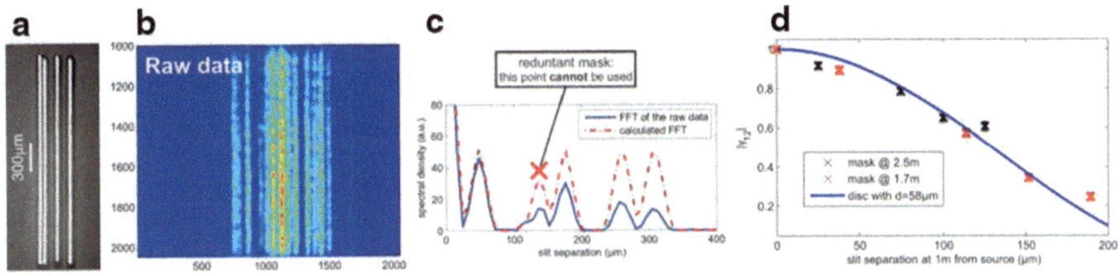

Fig. 5 Single shot measurement of the degree of spatial coherence. **a** multislit mask, **b** far field diffraction pattern, **c** comparison of spatial frequency amplitudes of the pattern with calculation of fully coherent source, **d** degree of spatial coherence as a function of slit separation for measurement with the mask 2.5 and 1.7 m from the source

Table 1 The expected performances of Betatron /Compton sources

Betatron			Compton		
Radiation features	Monochromatic regime	Broadband regime	Radiation features	For e-up to 500 MeV	For e-up to 1 GeV
Energy	Few tens keV to few tens MeV	Few tens keV to few tens MeV	Energy	Up to 50 keV	Up to 100 keV
Photons/shot	1.00E + 07	1.00E + 08	Photons/shot	1.00E + 08	1.00E + 08
Duration	< 10 fs	< 10 fs	Duration	< 10 fs	< 10 fs
Source size	< 2 microns	< 2 microns	Source size	< 5 microns	< 5 microns
Divergence	< 20 mrad FWHM	< 20 mrad FWHM	Divergence	< 20 mrad FWHM	< 10 mrad FWHM
Bandwidth	20 %	100 %	–	–	–

8 ELI Beamlines Project

Additionally to the well established PALS facility, and its X-ray programme, the new ELI Beamlines facility will give an opportunity for the future development and applications of ultrashort secondary sources in the Czech Republic [11]. In six experimental halls is planned to produce X-rays, coherent or incoherent, from the plasma based sources, by the high order harmonic generation (HHG) in the gas or from relativistic electron beams. More specifically, the experimental hall E1 will incorporate the k-alpha beamline and HHG beamline. As a driver will be used a kHz laser chain with 50 mJ (later upgradable to 100 mJ) output energy and 20 fs pulse duration. The HHG beamline is intended to deliver stable pulses from wavelength range 10–120 nm with 10^7–10^9 photons per shot. The k-alpha beam line will deliver pulses of hard X-rays at least 3 keV with 10^7 photons per pulse with time envelope shorter than 300 fs. In the experimental hall E2 there will be Betatron/Compton beamline. The Betatron beamline will use 10 Hz laser chain with 1 PW peak power and for Compton the second PW laser will be used. The expected performances of Betatron/Compton sources are summarized in Table 1.

The laser undulator X-ray (LUX) beamline will be placed in the E5 experimental hall. This user beamline will be focused to generate photons in "water window".

9 Summary and Future Prospects

In conclusion, the PALS facility has two laser chains, which are used, either in synchronized mode or separately, for generation of secondary sources. The iodine system with kJ output energy is used for QSS scheme. This scheme is generating the most energetic Ne-like Zn X-ray laser and is offered as standard beamline for users. To understand better the properties of the Ne-like Zn X-ray laser the investigation of its temporal coherence was tested for single and double pass regime. The results show that coherence time is 0.5 and 0.7 ps for single and double pass, respectively. The corresponding spectral linewidths are 17 and 11 mÅ. The Ti:Sapphire laser beamline with 20 TW peak power is commonly used for high repetition rate secondary sources. With this laser system the transient lasing at 18.9 nm of Ni-like Mo was successfully implemented. To study a spatial coherence of PALS X-ray lasers a single shot measurement method was employed to evaluate the equivalent source size and its shot-to-shot fluctuations. The future ELI Beamlines facility was presented, where three of five experimental halls will be dedicated to development and applications of secondary x-ray sources.

Acknowledgement This work benefits from LASERLAB-EUROPE—grant agreement no. 284464, EC's Seventh Framework Programme, ELI: EXTREME LIGHT INFRASTRUCTURE—project no. CZ.1.05/1.1.00/02.0061 and ECOP—project nos. CZ.1.07/2.3.00/30.0057 and CZ.1.07/2.3.00/20.0279.

References

1. Jungwirth, K., et al.: The Prague Asterix Laser System. Phys. Plasmas. **08**, 2495–2501 (2001)
2. Rus, B., et al.: Multimillijoule, highly coherent x-ray laser at 21 nm operating in deep saturation through double-pass amplification. Phys. Rev. A. **66**, 063806 (2002)
3. Hřebíček, J., et al.: 25TW Ti:sapphire laser chain at PALS. Proc. SPIE, 8080 (2011)
4. Nilsen, J., et al.: Prepulse technique for producing low-Z Ne-like x-ray lasers. Phys. Rev. A. **48**, (1993)
5. Cairns, G.F., et al.: Using low and high prepulses to enhance the $J = 0-1$ transition at 19.6 nm in the Ne-like germanium XUV laser. Proceedings of 4th International Colloquium on X-Ray Lasers. **332**, 289 (1994)
6. Rus, B., et al.: Experimental study of neonlike zinc $J = 0 - 1$ soft X-ray lasing at 21.2 nm. Proceedings of 4th International Colloquium on X-Ray Lasers. **332**, 152 (1994)
7. Klisnick, A., et al.: Experimental study of the temporal coherence and spectral profile of the 13.9 nm transient X-ray laser. J. Quant. Spectrosc. Radiat. Transf. **99**, 370–380 (2006)
8. Guilbaud, O., et al.: Longitudinal coherence and spectral profile of a nickel-like silver transient soft X-ray laser. Eur. Phys. J. D. **40**, 125–132 (2006)

9. Meng, L., et al.: Temporal Coherence and Spectral Linewidth of Neon-Like XUV Lasers Pumped in the Quasi-steady State Regime. Proceedings of 13th International Conference on X-Ray Lasers, 175–180 (2013)
10. Polan, J., et al.: Spatial and temporal profiles of the 21.2-nm saturated X-ray laser output. Proceedings of 10th International Conference on X-Ray Lasers, 139–145 (2006)
11. Extreme light infrastructure Whitebook. Science and technology with ultra-intense lasers. Issue 1, FZU (2010)

Spectral Properties of Collisional Xuv Lasers for the Amplification of Femtosecond Pulses

A. Klisnick, A. Le Marec, L. Meng, O. Larroche, O. Guilbaud, M. Kozlova, J. Nejdl and A. Calisti

Abstract We discuss the role of the spectral properties of XUV lasers in the amplification of femtosecond pulses in two different regimes. We present an overview of our recent spectral characterization of the four types of collisionally-pumped systems, also including other measurements from the literature. This is used to assess the potential of the different types of XUV lasers to amplify femtosecond pulses, either in the adiabatic or in the dynamic regime.

1 Introduction

Injection-seeding of plasma-based XUV lasers with a femtosecond high-order harmonic (HH) pulse is a promising technique for significantly improving the characteristics of these sources. Compared to the usual Amplification of Spontaneous Emission (ASE) mode, XUV lasers operated in the seeded mode have a significantly reduced beam divergence [1] with full spatial coherence [2], improved temporal coherence [3], and defined polarization [4]. The temporal duration of the output pulse is also reduced, but still remains limited to the picosecond timescale [5]. This is primarily due to the significant mismatch between the HH seed and XUV laser amplifier bandwidth. While the femtosecond HH pulse is associated with a broad bandwidth $\Delta\lambda/\lambda$ of typically 10^{-3}–10^{-2}, the gain bandwidth of XUV laser

A. Klisnick (✉) · A. Le Marec
ISMO, CNRS–Université Paris–Sud, 91405 Orsay, France
e-mail: annie.klisnick@u-psud.fr

L. Meng
Institute of Fluid Physics, China Acdemy of Engineering Physics, Mianyang, China

O. Larroche
CEA DIF, Bruyères le Châtel, Arpajon, France

O. Guilbaud
LPGP, CNRS–Université Paris–Sud, Orsay, France

M. Kozlova · J. Nejdl
IoP ASCR, ELI Beamlines projcct, Prague, Czech Repbublic

A. Calisti
PIIM, CNRS -Aix-Marseille Université, Marseille, France

© Springer International Publishing Switzerland 2016

J. Rocca et al. (eds.), *X-Ray Lasers 2014*, Springer Proceedings in Physics 169,
DOI 10.1007/978-3-319-19521-6_5

lines in Ne-like and Ni-like ions is typically 100 times narrower, as it will be discussed in this paper. As a result, when amplified in a linear, adiabatic regime, the seed HH pulse undergoes a strong spectral narrowing while it is amplified, with a corresponding temporal broadening. In order to progress towards the femtosecond timescale, it is thus necessary to enlarge the bandwidth of the XUV laser amplifier by a proper tailoring of the plasma parameters, which control the broadening of the lasing line [6]. Another approach towards the generation of femtosecond pulses with seeded XUV lasers is to exploit the non-linear dynamical response of the lasing ions to the amplified seed pulse, which can give rise to Rabi oscillations with ultrashort periods of 100 fs or below. The existence of Rabi oscillations in seeded XUV lasers was predicted from numerical simulations and was discussed recently by several authors [7–9]. However no evidence of such behaviour has been experimentally observed yet, for reasons that are not completely understood. There is thus a need to better understand the conditions that favour the development of Rabi oscillations in our systems. In a recent paper we have investigated the non-linear temporal behaviour of the amplification of a seed pulse, using a simplified model based on Maxwell-Bloch equations [10]. An expression for a gain threshold was derived, which discriminates between the adiabatic (linear) and dynamical (non-linear) amplification regimes. This gain threshold, above which Rabi oscillations can occur, is related to the dipole dephasing rate, ie to the homogenous spectral width of the lasing line.

In summary, the spectral bandwidth of the XUV laser amplifier is one of the essential parameters that control the temporal behaviour of the amplified HH seed, in both regimes of amplification. It is thus important to have an accurate knowledge of the bandwidth of the different types of existing XUV lasers. This relies on detailed numerical simulations accounting for the complex combination of processes, which influence the spectral characteristics of the lasing lines—in particular Stark and Doppler broadening, gain-narrowing and potential saturation effect. Experimental measurements of the linewidth of practical XUV lasers, which are currently developed in a number of laboratories worldwide, is also required to provide benchmarking. Such measurements are however experimentally challenging, because they can only be done with a specific instrument able to reach the very high spectral resolution required ($\lambda/\Delta\lambda > 10^5$).

Using a wavefront-division interferometer with variable path difference, specifically designed at Institut d'Optique (France), we have carried out a comprehensive characterization of the different types of existing collisional XUV lasers in several laboratories worldwide. In this paper we review the results obtained and we compare them to similar measurements performed elsewhere. We show that the different types of XUV lasers actually exhibit different values of bandwidth, with a clear influence of the electron density. We discuss this trend in terms of the contributions of the different plasma broadenings, as predicted from calculations, in relation with the corresponding local conditions in the gain zone. Finally, based on this experimental characterization we discuss the prospects of the different types of collisional XUV lasers towards the amplification of femtosecond pulses. In particular we show

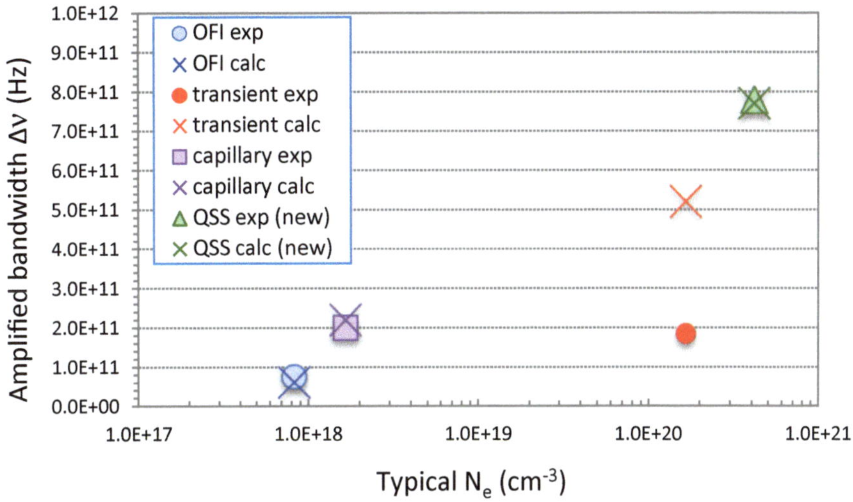

Fig. 1 Compilation of amplified linewidths (FWHM) obtained from the measurement of temporal coherence for four different types of XUV lasers, and corresponding calculated values. The horizontal scale shows the typical electron density at which each of these systems operates (see text)

that enlarging the bandwidth of the lasing line is not favorable to the generation of Rabi oscillations in the non linear amplification regime.

2 Summary of Measured Bandwidths

In Fig. 1 we have compiled the spectral linewidths that were inferred from the characterization of temporal coherence of the four currently existing types of collisional XUV lasers. These measurements were performed by implementing our wavefront-division interferometer at three different laboratories where these XUV lasers are developed by the respective local teams: Optical Field Ionized (OFI) Ni-like Kr laser (4d–4p J 0–1) at LOA, France [11]; transient Ni-like Mo [3] and capillary-discharge Ne-like Ar (3p–3s J 0–1) [12] at Colorado State University, USA and, more recently, quasi-steady state Ne-like Zn at PALS, Czech Republic. The latter experiment is described in a companion paper [13]. The amplified linewidth $\Delta\nu$ (FWHM) is plotted as a function of the typical electron density N_e at which pumping in each of these systems is achieved. It should be noted that for all measurements the investigated XUV laser was operated at saturation. The measured bandwidth is thus narrower than the intrinsic (optically thin) bandwidth of the lasing line, by a $(GL)^{1/2}$ factor of typically four at saturation.

One can see that the measured linewidths values $\Delta\nu$ are distributed over one order of magnitude, between 7.5 10^{10} Hz for the OFI-pumped, Ni-like Kr line [11]

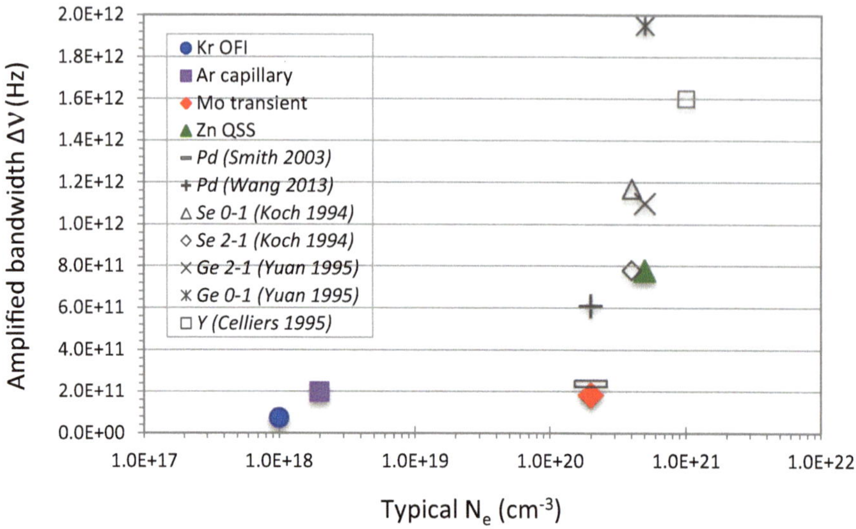

Fig. 2 Comparison of our experimental data with measurements reported by other authors: Ni-like Pd [14, 15]; Ne-like Se [16], Ne-like Ge [17]; Ne-like Y [18]. As in Fig. 1 the horizontal scale shows the typical (not exact) Ne value for each XUV laser. Note that the largest values displayed in the graph correspond to XUV lasers that were operated below saturation (see text)

generated in gas, and 7.8 10^{11} Hz for the QSS-pumped, Ne-like Zn line [13] generated from a solid target. The general trend is an increase of $\Delta\nu$ with the density, as expected from collisional broadening. Other parameters, such as the ion temperature, the charge of the emitting ion, or the wavelength of the line also play a role in the observed difference, but to a lesser extent. Also shown are the values predicted from numerical simulations performed independently in each case. These simulations generally involved line transfer calculations of the frequency-dependent amplified intensity, using calculated homogeneous and inhomogeneous components, as described in more details in the corresponding papers quoted above. The measured bandwidths are generally well reproduced by numerical simulations, except for the transient-pumping Ni-like Mo case, for which the predicted linewidth is larger by a factor $\sim$2.5 (a tentative explanation for this discrepancy is discussed in [3]). The QSS Zn laser thus exhibits the largest spectral width, with a corresponding Fourier-transform limit duration $\tau_{FL} = 0.44/\Delta\nu \sim 0.56$ ps. A Fourier-transform duration of 100 fs would require an even broader amplified linewidth of 4.4 10^{12} Hz (i.e. $\Delta\nu \sim 1.7$ 10^{13} Hz for the corresponding intrinsic linewidth).

Our experimental data are again plotted in Fig. 2 where they are compared to similar measurements performed by other authors, as available from the literature. There are no other measurements available for comparison, to our best knowledge, for OFI or capillary-discharge XUV lasers. For the transient pumping regime, two linewidth measurements of the Ni-like Pd lasing line (same transition: 4d–4p J 0–1) were performed independently. The first measurement was carried out by Smith and collaborators [14], using a Michelson-type, amplitude-division interferometer. The second measurement was reported recently by Wang and collaborators [15], using a Mach-Zehnder, wavefront-division interferometer, specifically designed to allow a single-shot measurement of the temporal coherence. The irradiation geometry for transient pumping was different in the two experiments, thus leading to different electron densities, which could explain the difference in the measured values.

For the QSS pumping, all the available measurements were performed in the mid 90's, at a time when transient and OFI pumping had not been demonstrated yet. The first characterization of the spectral broadening of an XUV laser line was performed in Ne-like Se by Koch et al., at Livermore laboratory [16]. This experiment was based on a direct measurement of the linewidth, using a high-resolution reflection grating spectrometer. Two lasing lines of the 3p–3s transition were investigated: (1/2 1/2) $J = 0 \rightarrow$ (1/2 1/2) $J = 1$ emitted at $\lambda = 18.2$ nm, and (3/2 3/2) $J = 2 \rightarrow$ (3/2 1/2) $J = 1$ emitted at 20.6 nm. One can see that the linewidth measured for the J $2 - 1$ component is close to our measurement in Ne-like Zn. The linewidth for the J $0 - 1$ component is slightly larger. However it should be noted that only the J $2 - 1$ lasing line was operated at saturation. The gain-length product achieved for the J $0 - 1$ line was limited to 10 [16], thus leading to a linewidth slightly larger than it would have been at saturation (i.e. GL ~ 15). A similar spectrometer-based measurement was performed soon after in Ne-like Ge by Yuan and collaborators at ILE, Japan [17]. Here the achieved gain-length products were even smaller: GL ~ 9 for the J $2 - 1$ line at $\lambda = 23.6$ nm and GL ~ 5 for the J $0 - 1$ line at $\lambda = 19.6$ nm. As a result the measured linewidths are broader than the saturated linewidths measured by Koch and by us. Finally the first characterization of the spectral linewidth of an XUV laser through the interferometric measurement of its temporal coherence was performed in Ne-like Y by Celliers et al. at LLNL in 1995 [18]. The measured linewidth ($\Delta\nu \sim 1.6\ 10^{12}$ Hz) is approximately twice as large as our measurement in Ne-like Zn, with a correspondingly shorter Fourier-transform duration of 0.27 ps. This seems consistent with the fact that saturated emission of the 15.5 nm lasing line was achieved at larger N_e ($\sim 10^{21}$ cm^{-3}), thanks to a 2ω driving laser pulse. However it should be reminded that the 15.5 nm lasing line is a particular case in the Ne-like XUV laser sequence, since the J $0 - 1$ component is blended with the J $2 - 1$ one mentioned above, which could complicate the interpretation of its behaviour [19].

In summary our measurements, together with data compiled from the literature, should be considered as "snapshots" of currently achieved linewidths, over a large variety of pumping and plasma conditions. The electron density at which the XUV laser operates plays a major role in controlling the amplified linewidth of the lasing line, in general agreement with predictions from simulations. Among the four types of currently operational XUV lasers, which were characterized with our wavefront-division interferometer, the QSS-pumped zinc laser generated at PALS exhibits the broadest amplified linewidth, compatible with the amplification of pulses with duration ~ 0.5 ps. A broader linewidth was measured at LLNL in Ne-like Y, indicating that increasing N_e above 10^{21} cm^{-3} might be a promising route towards the 100 fs timescale. This requires however that the gain coefficient can be kept large enough, since it is inversely proportional to the linewidth.

Table 1 Gain threshold G^*_{ad} for actual lasers

XUV laser type	G^*_{ad} [cm^{-1}]
Optical field ionization	2.3
Transient	12.8
QSS-capillary	0.9
QSS-laser	18.5

3 Spectral Width and Dynamical Amplification

In a recent paper [10] we have discussed the existence of a small-signal gain threshold, which must be surpassed in order to enter a dynamical regime of amplification, which will lead to Rabi oscillations (i.e. flopping of the population inversion between negative and positive values). Below the threshold the seed pulse injected in the inverted medium is amplified in an adiabatic regime, where the population inversion remains positive, while gradually decreasing at saturation.

An analytical expression for that gain threshold G^*_{ad} was derived from a stability analysis of the set of Maxwell-Bloch equations with simplifying assumptions, as described and discussed in detail by A. Le Marec in these Proceedings [20]. This expression reads as

$$G^*_{ad} = \frac{\gamma}{2c}$$

where γ is the dipole-dephasing rate, mainly due to collisions, leading to homogeneous broadening of the lasing transition, and c is the speed of light.

The condition that the adiabatic gain should be larger than G^*_{ad} for Rabi oscillations to occur is thus equivalent to write that the amplification rate should be larger than the dipole-dephasing rate.

Table 1 summarizes the values of the gain threshold G^*_{ad} for the four types of collisional XUV lasers that were discussed above. In these systems the γ value used to calculate G^*_{ad} is close to the intrinsic homogeneous linewidth. We have used the values predicted from the numerical simulations that were performed to support the experimental results. As it is evident from the expression of G^*_{ad}, the two XUV lasers, which operate at low electron density with a narrow collisional linewidth (OFI and capillary discharge), have a relatively low gain threshold. The gain coefficients achieved in OFI-pumped XUV lasers are typically larger that 50 cm^{-1} [8], thus significantly above threshold. Capillary-discharge XUV lasers operate at lower gain (typically ~1 cm^{-1} [12]) and are thus at the threshold limit. The calculated threshold values for the QSS and transient pumped lasers are significantly higher, due to their broader collisional linewidth. Due to the markedly different gains at which those two XUV lasers operate, they are however expected to behave differently. Transient XUV lasers operate at high gains (typically ~50 cm^{-1}) that are clearly above the calculated threshold. For QSS-pumped XUV lasers however, the achieved gains are generally smaller (typically ~ 2–5 cm^{-1}), significantly below the calculated threshold. In summary our analysis suggests that XUV laser systems with both high gain and narrow bandwidth are more favourable for the observation of Rabi oscillations.

4 Conclusions

We have presented an overview of XUV laser linewidth measurements. It provides an estimation of the typical Fourier-transform limit duration τ_{FL}, which could be reached with the four different existing types of X-ray lasers. Among the systems operational today, the QSS-pumped XUV lasers currently exhibits the largest bandwidth, corresponding to $\tau_{FL} \sim 500$ fs. Larger bandwidths could be obtained by shifting the gain region to higher density, possibly above 10^{21} cm^{-3} with a 2ω pump laser pulse.

An alternative way to progress towards femtosecond XUV lasers relies on the presence of Rabi oscillations, which can appear if the system behaves in a dynamical regime of amplification. Our analysis indicates that enlarging the bandwidth of the amplifier, as required in the adiabatic regime, is not favourable for this to occur.

Acknowledgements This work benefits from LASERLAB-EUROPE- grant agreement no. 284464, EC's Seventh Framework Programme.

References

1. Berrill, M., et al.: Improved beam characteristics of solid-target soft x-ray laser amplifiers by injection seeding with high harmonic pulses. Opt. Lett. **35,** 2317–2319 (2010)
2. Wang, Y., et al.: High-brightness injection-seeded soft-x-ray-laser amplifier using a solid target. Phys. Rev. Lett. **97,** 1–4 (2006)
3. Meng, L., et al.: Temporal coherence and spectral linewidth of an injection-seeded transient collisional soft-x-ray laser. Opt. Expr. **19,** 12087–12092 (2011)
4. Zeitoun, Ph, et al.: A high-intensity highly coherent soft x-ray femtosecond laser seeded by a high harmonic beam. Nature. **431,** 426 (2004)
5. Wang, Y., et al.: Measurement of 1-ps soft-x-ray laser pulses from an injection-seeded plasma amplifier. Phys. Rev. A. **79,** 1–4 (2009)
6. Calisti, A., et al.: Line profiles of Ni-like collisional XUV laser amplifiers: particle correlation effects. HEDP. **9,** 516–522 (2013)
7. Kim, C., et al.: Amplification of a high-order harmonic pulse in an active medium of a plasma-based x-ray laser. Phys. Rev. A. **80,** 053811 (2009)
8. Tissandier, F., et al.: Three-dimensional Maxwell- Bloch calculation of the temporal profile of a seeded soft-x-ray laser pulse. Appl. Phys. Lett. **101,** 251112 (2012)
9. Larroche, O., et al.: Two-dimensional Maxwell-Bloch simulation of quasi-p-pulse amplification in a seeded XUV laser. Phys. Rev. A. **88,** 033815 (2013)
10. Larroche, O., et al.: Inversion density threshold for Rabi oscillations and modified small-signal gain in extreme-ultraviolet lasers. Opt. Lett. **38,** 2505–2508 (2013)

11. Tissandier, F., et al.: Observation of spectral gain narrowing in a high-order harmonic seeded soft-x-ray amplifier. Phys. Rev. A. **81,** 063833 (2010)
12. Urbanski, L., et al.: Spectral linewidth of a Ne-like Ar capillary discharge soft-x-ray laser and its dependence on amplification beyond gain saturation. Phys. Rev. A. **85,** 033837 (2012)
13. Le Marec, A., et al.: Spectral and coherence properties of the Ne-like Zn X-ray laser at PALS, these proceedings
14. Smith, R.F., Dunn, J., et al.: Longitudinal coherence measurements of a transient collisional x-ray laser. Opt. Lett. **28,** 2261–2263 (2003)
15. Wang, Y., et al.: Single-shot soft x-ray laser linewidth measurement using a grating interferometer. Opt. Lett. **38,** 5004–5007 (2013)
16. Koch, J., et al.: Experimental and theoretical investigation of neonlike selenium x-ray laser spectral linewidths and their variration with amplification. Phys. Rev. A. **50,** 1877 (1994)
17. Yuan, G., et al.: Measurement of linewidths of Ne-like germanium soft-x-ray laser in slab targets. J. Appl. Phys. **78,** 3610–3616 (1995)
18. Celliers, P., et al.: Fringe formation and coherence of a soft-x-ray laser beam illuminating a Mach-Zehnder interferometer. Opt. Lett. **20,** 1907–1909 (1995)
19. Holden, P.B., et al.: The overlapping lasing lines in the collisional yttrium laser. J. Phys. B. **28,** 1369–1380 (1995)
20. Le Marec, A., et al.: Temporal response of seeded XUV lasers under different amplification regimes — Inversion density threshold. These Proceedings

Transient Collisionally Excited X-ray Lasers Pumped with One Long and Two Short Pulses

D. Ursescu, G. Cojocaru, R. Ungureanu, R Banici, L. Ionel, S. Simion, R. Dabu, J. Tümmler, R. Jung, H. Stiel, Olivier Delmas, Moana Pittman, Olivier Guilbaud, Sophie Kazamias, Kevin Cassou, Julien Demailly, Olivier Neveu, Elsa Baynard, David Ros, Andrea Le Marec, Sameh Daboussi, Li Lu, A. Klisnick and P. Zeitoun

Abstract X-ray laser simulations based on Ehybrid code have shown that enhanced plasma x-ray laser emission can be achieved mastering the ionization dynamics and plasma temperature using one long and two short pulses (Ursescu and Ionel, J Opto-electron Adv Mat 12:48–51, 2010). In parallel, two simple methods to generate multiple short pulses for pumping x-ray lasers were reported in conjunction with x-ray laser developments. Five to ten fold enhancement in the emission of the silver x-ray laser was demonstrated using the newly developed pump methods, when compared with the traditional one, based on a long pulse followed by one short

D. Ursescu (✉)
ELI-NP Department, Horia Hulubei National Institute for Physics and Nuclear Engineering, (IFIN-HH), Reactorului str. 30, 077125, Magurele, Romania
e-mail: daniel.ursescu@eli-np.ro

D. Ursescu · G. Cojocaru · R. Ungureanu · R Banici · L. Ionel · S. Simion · R. Dabu
National Institute for Lasers, Plasma and Radiation Physics (INFLPR), Atomistilor 409, Magurele, 077125, Romania

G. Cojocaru · R. Ungureanu
Faculty of Physics, University of Bucharest, 405 Atomistilor Str., Magurele 077125, Romania

G. Cojocaru · O. Delmas · M. Pittman · O. Guilbaud · S. Kazamias · J. Demailly · O. Neveu · D. Ros
Laboratoire de Physique des Gaz et des Plasmas, Universite Paris-Sud (UMR-CNRS 8578), Bat. 210, 91405 Cedex, France

J. Tümmler · R. Jung · H. Stiel
Max-Born-Institut (MBI), Max-Born-Strasse 2 A, Berlin 12489, Germany

M. Pittman · O. Guilbaud · S. Kazamias · K. Cassou · E. Baynard · D. Ros
LASERIX, Centre Laser Universite Paris Sud, FR2764

A. Le Marec · A. Klisnick
Univ Paris 11, CNRS, ISMO, F-91405 Orsay, France

S. Daboussi · Li Lu · P. Zeitoun
Laboratoire d'Optique Appliquée, ENSTA ParisTech - CNRS UMR 7639 - École Polytechnique, Chemin de la Hunière, F-91761 Palaiseau, France

© Springer International Publishing Switzerland 2016

J. Rocca et al. (eds.), *X-Ray Lasers 2014*, Springer Proceedings in Physics 169,
DOI 10.1007/978-3-319-19521-6_6

pump pulse. An overview of these recent experiments will be presented. The possible implementation of these novel x-ray laser pumping methods at Extreme Light Infrastructure—Nuclear Physics facility will be discussed.

1 Introduction

Various types of XUV sources, coherent and incoherent, at laboratory scale or at facility scale, are developed for their broad range of applications as reported in these conference proceedings. Among these, laboratory based x-ray lasers, with up to 100 μW at 13.9 nm recently reported [2], are now suitable for a broad range of applications.

Plasma x-ray lasers mastering involves control over about a dozen of parameters. Early work was focused in producing population inversion in highly ionized plasmas [3]; to succeed in this endeavor using laser pulses with duration in the nanosecond range was an outstanding achievement, as high intensity is needed to reach the ionization stage in spite of plasma expansion in vacuum. With the progress in short pulse lasers in eighties, the transient collisionally excited scheme [4] bypassed the problem by reducing the pump pulse duration. Subsequently, the energy deposition control in the density region with small gradients was introduced with the grazing incidence pumping (GRIP) scheme [5, 6]. All the methods above provided a significant reduction of pump energy through the mastering of plasma parameters.

Three development directions can be presently identified: higher repetition rate XRL, higher beam quality XRL and development of shorter wavelength XRL. They correspond to specific applications needs in terms of XUV photon flux, coherence/focusability and (bio-)materials investigations. In order to contribute along these lines, we followed the main idea: XRL pumping energy can be reduced through additional control of the plasma parameters. The intent of the following sections is to detail the specific issues in order to realize this desideratum.

2 Pump Energy Reduction

Most of the GRIP XRL were intensely optimized in order to achieve saturation with minimal pumping energy. One long pulse and one short pulse were tuned until optimal XRL operation was achieved, for example in [6]. Also, simulations of XRL plasma dynamics indicated that two short pulses might also be an interesting approach [7]. Experiments performed along this line have shown that this approach works also quite efficiently [8]. In both approaches, the first pulse was forming a plasma with the best possible electron density, density gradient and average ionization. In the case of long pulse for plasma formation, the significant disadvantage is related to the adiabatic cooling of the plasma due to expansion in vacuum and also conductive heat transport through the target; so the energy of the first, long pulse is unwisely spent in expensive processes in terms of energy. In the case of short pump pulse for plasma formation, there is a contradiction between creating a long plasma scale length for producing an extended gain region with good emission and a fast ionization time needed in order to avoid the adiabatic cooling and energy loss through heat conduction. It is important to mention other similar approaches [9, 10].

In order to bypass the energy loss associated with the adiabatic cooling and heat conduction, an intermediate approach was proposed [1] and subsequently experimentally demonstrated [11, 12]. In these works, the XRL plasma was formed using a low energy long pulse followed by a moderate energy short pulse at GRIP angle. This approach preserves the advantage of long scale length of the plasma generated with the low energy long pulse combined with the advantage of fast ionization induced by the short pulse in a region away from the target, hence with lower thermal conductivity and in a short time—associated with reduced adiabatic cooling.

Before discussing specific experimental results based on this novel approach, the production of the pumping pulses will be described in the following section.

3 Multiple Pulses for Additional Control

From the plasma formation considerations above, the necessity of generation of multiple short pump pulses with adjustable delay and energy ratio was derived. The first of these multiple short pulses will serve as the plasma forming moderate energy pumping pulse as proposed in the section above; the second short and energetic pulse is the one responsible with the generation of strong collisional excitation and population inversion, hence amplified spontaneous emission of the plasma.

The pump lasers for GRIP XRL are based on the chirped pulse amplification (CPA) technique. The straight-forward way to generate multiple pulses is at the end of such a CPA laser system, using a beam splitter in combination with a delay line [2]. The poor-man's solution based on the insertion of a thin film beam-splitter and of a halfwaveplate in the beam was proposed in [12]. However, if the pulses after compression are short and energetic, the short pulses transportation shall be performed under vacuum in order to avoid nonlinear effects that deteriorate the pulses, such as self phase modulation and filamentation. Also, depending on the beam size, the associated optical components are large and expensive; hence, alternative solutions were proposed to be implemented in the CPA amplification chain after the optical stretcher [8] or in the optical stretcher itself [13, 14].

The method reported in [13, 14] involves the insertion of a clipping mirror in the stretcher, at the place where the pulse has spatial chirp but no angular chirp, together with an additional roof mirror. This approach allows one to tweak the energy ratio of the two pulses after compression to any suitable value, with better than 1 % accuracy. The temporal delay can also be controlled with sub ps resolution from zero to few hundreds of ps. The downside of the approach is related to the fact that the two short pulses have complementary spectra and hence are slightly longer than the single short pulse of the CPA system.

The method we reported in [12], bypasses the pulse duration issue. The thin film beam-splitter is polarization sensitive, with significant reflectivity variation for s and p polarization and performs as a Fabry–Perot resonator in conjunction with a flat mirror, generating multiple pulses through multiple reflections. The halfwave-plate is inserted in front of the thin film beam splitter and controls the amount of s- and p-polarized light seen by the thin film beam-splitter, hence the energy ratio of the multiple pulses generated. It was possible to generate in this way energy ratios between 5 and 30 % with delay from 10 to 100 ps.

4 Experimental Results

In the first series of experiments partly reported in [11], Ni-like Zr, Mo, Pd, Ag and Sn XRL operation was demonstrated. The CPA pump laser was running at 810 nm central wavelength. The long pump pulse used at normal incidence had an energy of 87 mJ corresponding to an intensity of $2.5{\cdot}10^{11}$ W/cm^2 and ionization state at the arrival of the short pulse of Z=12+ in the case of Ag XRL. A larger average ionization of about Z=19+ (Ni-like) is optimum at the arrival of the short pulse. The long pulse was followed by one or two short pulses with a total energy of 125 mJ of energy. Ten fold increase in the XRL signal was observed in the later case.

For similar pumping scheme and energies, at least five fold enhanced emission was measured also for Ni-like Zr and Ni-like Mo XRL when the long pulse was followed by two short pulses (1L2S). For these optimization studies, best Mo XRL was obtained with 5 ps as delay between the short pulses while for Zr XRL a delay between 2 and 4 ps was required. Sn XRL at 10.9 nm was obtained only after pumping at full available energy on target (265 mJ) only in the configuration using one long and two short pulses and the lasing line was barely visible. This indicates that plasma reached the requested ionization stage in these specific conditions.

In the case of complimentary experiments partly reported in [12], Ni-like Ag XRL and Ne-like Ti XRL were studied. The long pulse was generated with a frequency doubled Nd:YAG laser with multimode pulses of about 145 mJ, 6 ns long, corresponding to an ionization state at the arrival of the short pulse of Z=6+ in the case of Ag XRL. The short pulse with an energy of 1.4 J followed the long pulse after few ns. In this case the short pulse duration was tuned from 1 to 20 ps.

An overview of these experimental results related to Ni-like 1L2S XRL is presented in Fig. 1. There, the optimal time delay between the two short pulses is depicted as a function of the number of single ionization steps needed to reach Ni-like state. It provides a hint on the local ionization dynamics of the plasma heated by short pulses in the ps range. The data are consistent with our proposed interpretation, namely, that the long pulse together with the first short pulse ionize efficiently the XRL plasma closer to the Ni-like state, required for XRL operation. The delay between the two short pulses provides hints related to the time until the Ni-like ionization state is reached; it is short for Zr, because only one or two ionization steps are required, while it is long for Ag XRL pumped with 125 mJ short pulse with 8 %

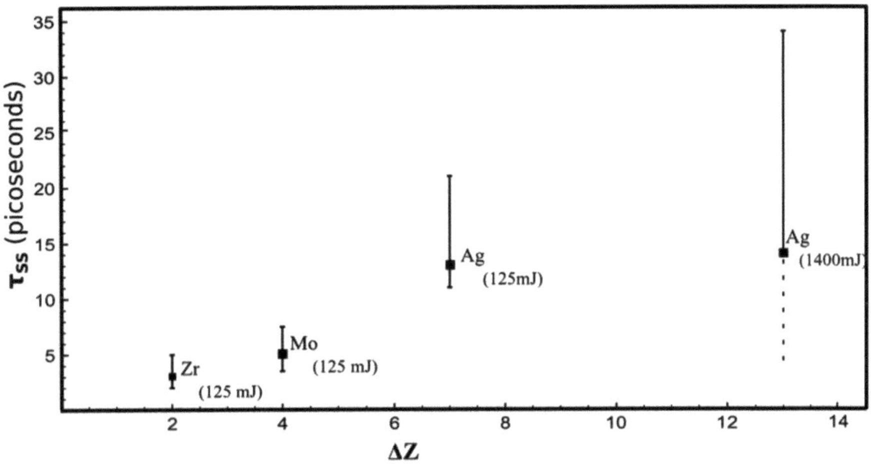

Fig. 1 Zr short pulse splitting 5 % and long to short pulses delay (τ_{LS}) 380 ps; Mo short pulse splitting 8 % and τ_{LS} = 266 ps; Ag(125 mJ) splitting 8 % τ_{LS} = 250 ps, Ag(1400 mJ) splitting 30 % τ_{LS} = 5 ns

split. For the Ag XRL pumped with 1.4 J short pulse with 30 % split, the preplasma ionization is very low but the energy of the first short pulse is significantly higher and this reduces the ionization time to a similar value. The bars represented in the Fig. 1 indicate the region of delay between the short pulses where the 1L2S XRL signal drops to half.

It has to be emphasized that the introduction of the moderate short pre-pulse with proper delay relative to the short main pulse acts as a mechanism to generate Ni-like ions earlier, closer to the arrival time of the strong short pulse. This process dramatically enhances the gain.

Preliminary results under evaluation show that the moderate short pulse can also bring improvement in the case of seeded XRL schemes.

5 Conclusions and Outlook

The introduction of a moderate short pulse is an effective way for pump energy reduction through indirect suppression of the cooling mechanisms and proper timing of the intense final pumping pulse. Thus, it provides a step towards the high average power XUV sources, as those reported in [2]. The technique proved its validity in different XRL pumping conditions, and for both Ni-like and Ne-like XRL. 100 Hz Mo XRL pumped with this novel scheme was also prospected; the results are promising but inconclusive for the time being.

The use of this technique remains to be experimentally proven in the case of shorter wavelength XRL.

A high average power XRL pumped at 100 Hz might be an interesting tool for preliminary studies of Compton scattering of XUV photons on ultra-relativistic electron bunches. A place to do such studies is now under construction at Extreme Light Infrastructure - Nuclear Physics pillar in Romania, where two lasers at 100Hz, 0.4J/pulse in 2ps pulse duration will be available.

6 Acknowledgements

The research leading to these results has received funding from the LASERLAB-EUROPE (grant agreement no. 284464, EC's Seventh Framework Programme) from UEFISCDI project PN2-Parteneriate-1/2012 and is supported by Extreme Light Infrastructure Nuclear Physics (ELI-NP) Phase I, a project co-financed by the Romanian Government and European Union through the European Regional Development Fund.

References

1. Ursescu, D., Ionel, L.: Gain and ionization dynamics in transient, collisionaly excited x-ray lasers. J. Optoelectron. Adv. Mat. **12**, 48–51 (2010)
2. Reagan, B.A., Berrill, M., Wernsing, K.A., Baumgarten, C., Woolston, M., Rocca, J.J.: High-average-power, 100-hz-repetition-rate, tabletop soft-x-ray lasers at sub-15-nm wavelengths. Phys. Rev. A. **89**, 053820 (2014)
3. Daido, H.: Review of soft x-ray laser researches and developments. Rep. Prog. Phys. **65**, 1513 (2002)
4. Nickles, P.V., Shlyaptsev, V.N., Kalachnikov, M., Schnürer, M., Will, I., Sandner, W.: Short pulse X-Ray laser at 32.6 nm based on transient gain in Ne-like titanium. Phys. Rev. Lett. **78**, 2748–2751 (1997)
5. Shlyaptsev, V.N., Dunn, J., Moon, S., Smith, R., Keenan, R., Nilsen, J., Fournier, K.B., Kuba, J., Osterheld, A.L., Rocca, J.J.G., Luther, B.M., Wang, Y., Marconi, M.C.: Numerical studies of transient and capillary x-ray lasers and their applications. In: Fill, E., Suckewer, S. (eds.) Soft X-Ray Lasers and Applications V, vol. 5197, pp. 221–228. SPIE, Bellingham (2003)
6. Keenan, R., Dunn, J., Patel, P.K., Price, D.F., Smith, R.F., Shlyaptsev, V.N.: High-repetition-rate grazing-incidence pumped X-ray laser operating at 18.9 nm. Phys. Rev. Lett. **94**, 103901 (2005)
7. Ursescu, D., Zimmer, D., Kuehl, T., Zielbauer, B., Pert, G.: Gain generation in the critical density region of a TCE XRL. In: Nickles, P., Janulewicz, K. (eds.) X-Ray Lasers 2006, Springer Proceedings in Physics, vol. 115, pp. 269–273 (2007)
8. Zimmer, D., Ros, D., Guilbaud, O., Habib, J., Kazamias, S., Zielbauer, B., Bagnoud, V., Ecker, B., Hochhaus, D.C., Aurand, B., Neumayer, P., Kuehl, T.: Short-wavelength soft-x-ray laser pumped in double-pulse single-beam non-normal incidence. Phys. Rev. A. **82**, 013803 (2010)
9. Luther, B.M., Wang, Y., Larotonda, M.A., Alessi, D., Berrill, M., Marconi, M.C., Rocca, J.J., Shlyaptsev, V.N.: Saturated high-repetition-rate 18.9 nm tabletop laser in nickel-like molybdenum. Opt. Lett. **30**, 165–167 (2005)
10. Kim, H.T., Choi, I.W., Hafz, N., Sung, J.H., Yu, T.J., Hong, K., Jeong, T.M., Noh, Y., Ko, D., Janulewicz, K.A., Tümmler, J., Nickles, P.V., Sandner, W., Lee, J.: Demonstration of a saturated Ni-like Ag X-ray laser pumped by a single profiled laser pulse from a 10-Hz ti:sapphire laser system. Phys. Rev. A. **77**, 023807 (2008)

11. Banici, R.A., Cojocaru, G.V., Ungureanu, R.G., Dabu, R., Ursescu, D., Stiel, H.: Pump energy reduction for a high gain ag x-ray laser using one long and two short pump pulses. Opt. Lett. **37**, 5130–5132 (2012)
12. Cojocaru, G.V., Ungureanu, R.G., Banici, R.A., Ursescu, D., Delmas, O., Pittman, M., Guilbaud, O., Kazamias, S., Cassou, K., Demailly, J., Neveu, O., Baynard, E., Ros, D.: Thin film beam splitter multiple short pulse generation for enhanced Ni-like Ag X-ray laser emission. Opt. Lett. **39**, 2246–2249 (2014)
13. Ursescu, D., Ionel, L., Banici, R., Dabu, R.: Multiple ultra-short pulses generation for collinear pump-probe experiments. J. Optoelectron. Adv. Mat. **12**, 100–104 (2010)
14. Banici, R., Ursescu, D.: Spectral combination of ultrashort laser pulses. EPL **94**, 44002 (2011)

Seeded Operation of a Ne-like Titanium Soft X-Ray Laser

O. Guilbaud, S. Kazamias, K. Cassou, O. Delmas, J. Demailly, O. Neveu, D. Ros, E. Baynard, M. Pittman, M. Shazad, A. Rossal, G. J. Tallents, A. Le Marec, A. Klisnick, L. Lu, Ph. Zeitoun, G. V. Cojocaru, R. G. Ungureanu, R. A. Banici and D. Ursescu

Abstract Seeded operation of a Neon-like Titanium plasma-based soft x-ray laser is described. The plasma amplifier is pumped using a variation of the classical grazing incidence pumping technique, combining a long low energy pulse followed by a main short pulse. Because the preformed plasma is underionized, a part of the main short pulse energy is used to ionized it to the lasing stage. Consequences of this feature on seeded laser properties are discussed.

1 Introduction

Multi-terawatt laser systems are ideal tools for jitter-free pump-probe experiments with interacting pulses of highly different wavelengths. In this context, plasma-based soft x-ray laser are interesting short wavelength probe for surface interferometry, plasma physics or warm dense matter studies. They usually require a low energy pulse train or pedestal [1] in order to reach a high yield or to increase the generating target lifetime [2]. However, high contrast or precisely tailored pulse shape can be desired for the other pump or probe beam [3]. We will present in this

O. Guilbaud (✉) · S. Kazamias · K. Cassou · O. Delmas · J. Demailly · O. Neveu · D. Ros
LPGP, CNRS, Universite Paris Sud, 91405 Orsay Cedex, France
e-mail: olivier.guilbaud@u-psud.fr

E. Baynard · M. Pittman
LUMAT, CNRS, Universite Paris-Sud, 91405 Orsay Cedex, France

M. Shazad · A. Rossal · G. J. Tallents
Department of Physics, University of York, York YO10 5DD, UK

A. Le Marec · A. Klisnick
Institut des Sciences Moleculaires d'Orsay, CNRS, Universite Paris-Sud, 91405 Orsay Cedex, France

L. Lu · Ph. Zeitoun
Laboratoire d'Optique Appliquee, ENSTA, Ecole Polytechnique, 91761 Palaiseau Cedex, France

G. V. Cojocaru · R. G. Ungureanu · R. A. Banici · D. Ursescu
National Institute for Lasers, Plasma and Radiation Physics (INFLPR), Atomistilor 409, 077125 Magurele, Romania

© Springer International Publishing Switzerland 2016

J. Rocca et al. (eds.), *X-Ray Lasers 2014*, Springer Proceedings in Physics 169,
DOI 10.1007/978-3-319-19521-6_7

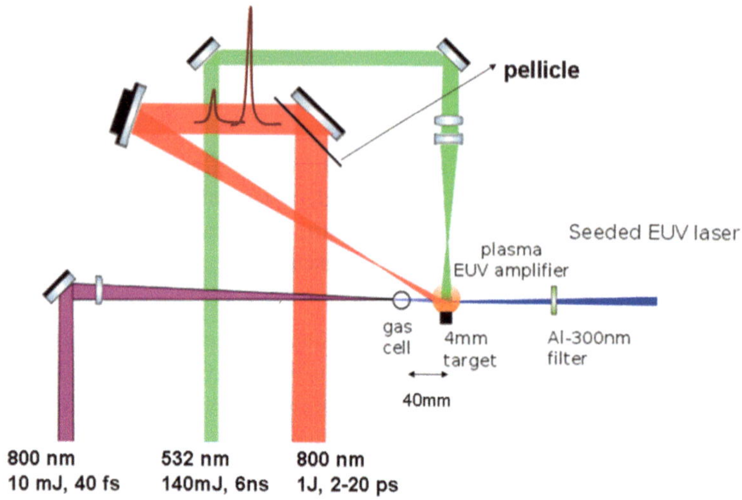

Fig. 1 Schematic of the harmonic seeded soft x-ray laser

paper a variant of the GRIP (Grazing Incident Pumping) transient soft x-ray laser adressing this issue. In this scheme, an electronically triggered frequency doubled commercial Q-switched Nd-YAG laser is replacing the ASE pedestal and prepulses used up to now to reach high soft x-ray laser through-output. This first laser pulse creates a low ionisation stage, expanded preformed plasma. After few nanosecond delay, a single Titanium Sapphire (Ti:Sa) compressed laser pulse is focussed on this pre-plasma in the GRIP configuration. This pulse is finishing the ionisation and is inducing the collisionnal gain generation. Successful lasing has been demonstrated from 13.9 nm (Ni-like Ag) to 32.6 nm (Ne-like Ti). Recently, the yield has been significantly upgraded by introducing an additional short pulse tens of picosecond prior to the main pulse [4]. This pulse structure is generated just before the last focussing optics. These pumping configurations are efficient while keeping the Ti:Sa laser contrast high. Besides, seeded operation of a Ne-like Titanium (32.6 nm) soft x-ray laser has been demonstrated in both configurations. First results on gain duration and beam propagation will be discussed. Further prospects on shorter soft x-ray laser pulses generation will finally be given.

2 Plasma Based Soft X-Ray Laser

2.1 Initial Setup

The soft x-ray laser generation scheme is presented in Fig. 1. A frequency doubled commercial Q-switched Nd-YAG laser delivering 6 ns, 140 mJ pulses is electronically triggered by the Ti:Sa laser chain masterclock. The beam is line focussed on a

4 mm long slab target by a combination of cylindrical lenses, forming a low ionised ($\langle Z\rangle = 7$ for Silver), preformed plasma. After an optimal peak-to peak delay of 4ns, a compressed, 800 nm, 1 J pulse is focussed on the preplasma in a GRIP geometry. Efficient lasing is observed for Ne-like Titanium ($\lambda = 32.6$ nm), Ni-like Molybdenum ($\lambda = 18.9$ nm) and Ni-like Silver $\lambda = 13.9$ nm). It has been observed that the optimal compressed pulse duration is increasing as the atomic number of the target element is higher. The optimal duration can reach a value of 15 ps for Ni-like Silver, which is longer than in classical GRIP pumping experiments. This fact can be related to the double role of this compressed pulse: finish the ionisation of the preplasma and heat free electrons in order to induce collisionnal pumping and laser gain [5]. As ionisation is a long process (tens of picoseconds) and high electron temperatures are decaying fast (few ps), a longer pulse helps the plasma to reach a high temperature when the ionisation level is reaching the desired one (Ne-like or Ni-like). This pumping configuration will later be referred as 1L1S (one long pulse, one short pulse).

2.2 *Double Short Pulse Pumping*

To increase further the soft x-ray laser efficiency, ionisation and heating steps should be uncoupled. This can be achieved by adding a low energy short pulse tens of picosecond prior to the main compressed pulse [6]. This additionnal pulse is bringing the preplasma to the correct ionisation level just before the pumping step. This configuration will be called 1L2S (one long, two short pulses). A simple, safe and high contrast preserving way of implementing this idea is depicted in Fig. 1. A pellicle is placed in front of the last plan mirror of the short pulse optical path. Optimization has been performed by varying the delay and energy balance between the two pulses. A five time increase in energy compared to 1L1S configuration has been observed for an energy balance of 30 % and a delay of 20 ps. The optimal pulse duration is 4 ps, shorter than in 1L1S, confirming that the plasma is ionised to the lasing ionisation level when the main short pulse is arriving.

In both cases, the short pulses are absorbed in a large expanded plasma. When the gain is building up, it is reasonable to think that an electron density bump is formed in the vicinity of the GRIP turning point region, where the laser energy is absorbed. Following this idea, it may be interesting to investigate if the gain is switched on in a region with a density larger than the GRIP density.

3 Seeded Operation of the X-Ray Laser Amplifier

Both soft x-ray laser schemes have been successfully seeded with high order harmonics [7]. The amplifier was a variable length target Ni-like Titanium soft x-ray laser. High order harmonics were generated in Argon, using a low energy (≤ 40 mJ)

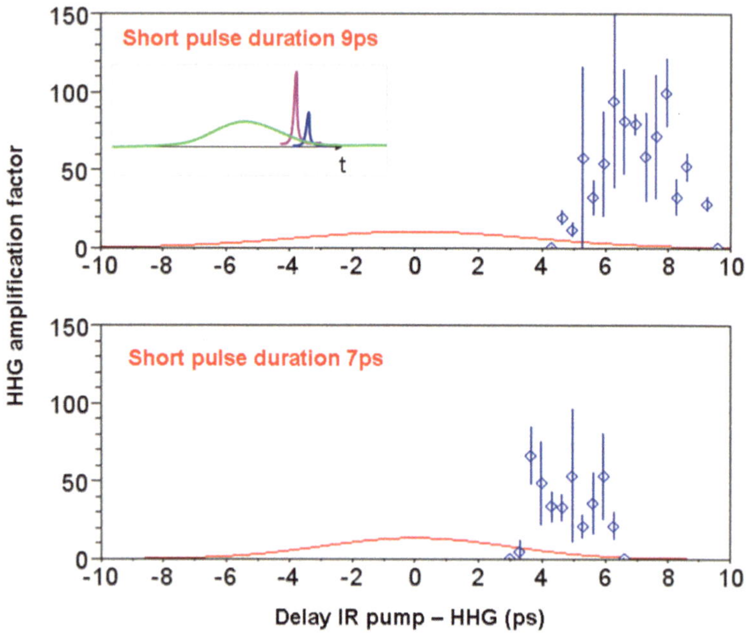

Fig. 2 Amplification factor as a function of delay between main pumping pulse and the harmonic seed

auxilliary beam compressed to 40 fs and focussed by a f = 1 m lens into a gas cell filled with 20 mbar of Argon. The gas cell was placed 40 mm from the target input. Near field images of the plasma exit at soft x-ray laser wavelength show a harmonic beam size of 100 μm × 50 μm in the vicinity of the target whereas the ASE emitting region is 20 μm × 40 μm. Harmonic spectral tunning has been achieved mainly by playing on the laser spectral phase with an acousto-optical modulator (Dazzler) placed in the laser chain front-end. Finally, picosecond range timing of the main short pulse compared to the harmonic generation pulse has been set using a non linear doubling crystal.

When the harmonic pulse is injected into the plasma tens of picosecond before the short pumping pulses, the beam profile and pointing are not affected. Only a fall in intensity is observed, probably due to the low ionisation plasma absorption. When the harmonic pulse is injected in the rising edge of the main pulse in 1L1S, or in the time interval between the two short pulses in 1L2S, the shape of the beam is affected, taking the shape of a crescent, and eventually being deflected away from the target. Finally, during a narrow delay window, the harmonic pulse can trigger the emission of an intense and collimated soft x-ray laser beam. The ratio between the seeded laser beam energy and the harmonic energy going through the gain region (estimated with near field images) has been measured as a function of the main short pulse - harmonic pulse delay in different configuration (Fig. 2). Delay zero refers to

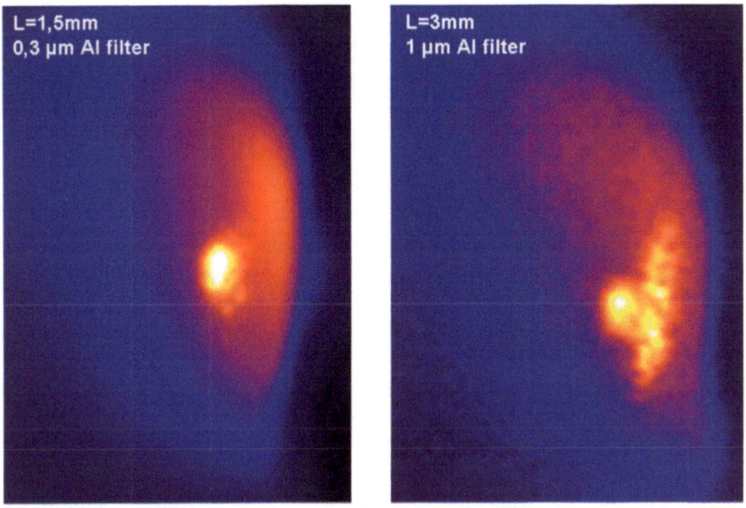

Fig. 3 Near field images of the amplifier exit plan at 32.6 nm. Bulk target is on the *right* and the plama (glow emission) is expanding towards the *left*. *Left*: for a target length of 1.5 mm the seeded and ASE beams are overlapped. *Right*: for a longer target (3 mm) and a thicker aluminum filter (1 µm instead of 0.3), the seeded beam is deflected away from the target and can be distinguished from ASE

the arrival of the main pulse intensity peak on target. The delay scan for the 1L1S configuration is represented in Fig. 2 for two main pulse durations (7 and 9 ps). The amplification window height and size depends on the pulse duration (at constant energy) and a shorter pulse is not optimum in this case. This can be again related to the necessity for the main pulse to further ionize the preplasma before generating any gain. This scenario is confirmed by the delay observed for the gain to take place. By contrast, in the 1L2S configuration, the gain is starting in the rising edge of the main short pulse. A higher amplification factor is observed, which is consistant with the observations described in the previous section.

The seeded laser is deflected away from the target compared to the initial harmonic beam direction. This deflection angle is increasing with the target length and is higher for 1L2S (5 mrad for a target length of 4 mm) than for 1L1S (2 mrad for a target length of 4 mm). This effect first observed in far field is also visible in near field. Figure 3 presents two near field images of the seeded laser (1L2S configuration) obtained for target length of 1.5 and 3 mm. In the first image, seeded beam is appearing at the same position as ASE emission in the plasma but ten times brighter. In the second image, the round bright spot corresponds to the seeded beam, shifted away from target and following a different path as compared to ASE (elongated spot on the right).

A more detailed analysis of the beam refraction may provide more information about the fast evolution of the electron density profile due to the rapid ionisation.

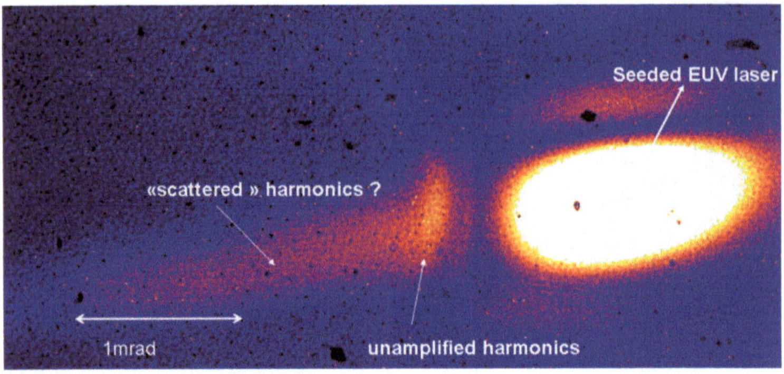

Fig. 4 Far field image of the seeded laser (1L1S configuration, 4 mm target length) after 1m propagation, an Aluminum filter and a 45° multilayer mirror

This aspect will be the object of future work. Some particular details might be related to the presence of an electron density bump in the plasma near the gain region. The general seeded laser far field pattern is composed of a bright collimated beam deflected by a few milliradians plus a weak spot at the position of the initial harmonic beam (see Fig. 4). This later spot might corresponds to a part of the harmonic beam passing far from the high density gradient region. For some particular conditions, an additional feature is visible, in the form of a few milliradians long stripe deflected towards the target. This might corresponds to a part of the harmonic beam passing between the target and the density bump formed near the gain region.

4 Conclusion

In conclusion, a new pumping technique, preserving the infrared laser contrast and easy to implement, has been demonstrated in Ne-like Ti, Ni-like Mo and Ag. Seeding operation at 32.6 nm is also reported. The role of rapid ionisation by short pulses is expected to play an important role on beam refraction, which is corroborated by experimental observation. The possible presence of a density bump in the plasma is of great interest as it might be a way to generate a simple plasma waveguide. Combined with frequency doubled laser pumping [8], a scheme where laser gain is generated at high density but where strong refraction is counteract by the waveguide would lead to larger bandwidth laser and shorter pulses in seeded mode.

Acknowledgements This work received funding from LASERLAB-EUROPE (grant agreement no. 284464, EC's Seventh Framework Programme). The support of the RTRA Triangle de la physique is also acknowledged.

References

1. Kim, H.T., et al.: Demonstration of a saturated Ni-like Ag x-ray laser pumped by a single profiled laser pulse from a 10-Hz Ti:sapphire laser system. Phys. Rev. A. **77**, 023807 (2008)
2. Habib, J., et al.: Low energy prepulse for 10 Hz operation of a soft-x-ray laser. Opt. Express **20**, 10128–10137 (2012)
3. Wilson, L.A., et al.: Energy transport in short-pulse-laser-heated targets measured using extreme ultraviolet laser backlighting. Phys. Rev. E. **86**, 026406 (2012)
4. Cojocaru, G.V., et al.: Thin film beam splitter multiple short pulse generation for enhanced Ni-like Ag x-ray laser emission. Opt. Lett. **39**, 2246–2249 (2014)
5. Berill, M., et al.: Pump pulse-width dependence of grazing-incidence pumped transient collisional soft-x-ray lasers. Phys. Rev. A. **75**, 063821 (2007)
6. Banici, A.R., et al.: Pump energy reduction for a high gain Ag X-ray laser using one long and two short pump pulses. Opt. Lett. **37**, 5130–5132 (2012)
7. Wang, Y., et al.: High-Brightness Injection-Seeded Soft-X-Ray-Laser Amplifier Using a Solid Target. Phys. Rev. Lett. **97**, 123901 (2006)
8. Kuba, J., et al.: Two-color transient pumping in Ni-like silver at 13.9 and 16.1 nm. Phys. Rev. A. **62**, 043808 (2000)

Output Beam Polarisation of X-ray Lasers with Transient Inversion

K. A. Janulewicz, C. M. Kim, B. Matouš, H. Stiel, M. Nishikino, N. Hasegawa and T. Kawachi

Abstract It is commonly accepted that X-ray lasers, as the devices based on amplified spontaneous emission (ASE), did not show any specific polarization in the output beam. The theoretical analysis within the uniform (single-mode) approximation suggested that the output radiation should show some defined polarization feature, but randomly changing from shot-to-shot. This hypothesis has been verified by experiment using traditional double-pulse scheme of transient inversion. Membrane beam-splitter was used as a polarization selector. It was found that the output radiation has a significant component of *p*-polarisation in each shot. To explain the effect and place it in the line with available, but scarce data, propagation and kinetic effects in the non-uniform plasma have been analysed.

1 Introduction

Polarisation of a generated radiation is a parameter of the paramount importance for application of any radiation source. X-ray lasers work nearly exclusively in the regime of amplified spontaneous emission (ASE) and it is commonly assumed that due to origination in noise, the polarization state of the output radiation is undefined. In spite of a some effort, the problem of the polarization state of the amplified radiation has not been clearly solved, yet. In both, the theory and the experiment there are contradictory opinions and results. The most prominent works delivering opposite conclusions are those by Kawachi et al. [1] and Rus et al. [2]. In the first of these works, the observed polarisation of a double-pulse-pumped X-ray laser has

K. A. Janulewicz (✉)· B. Matouš
Department of Physics and Photon Science, Gwangju Institute of Science and Technology, 123 Cheomdangwagi-ro, Buk-gu, Gwangju 500-712, Republic of Korea
e-mail: kaj@gist.ac.kr

C. M. Kim
Institute of Basic Research and Advanced Photonics Research Institute, Gwangju Institute of Science and Technology, 123 Cheomdangwagi-ro, Buk-gu, Gwangju 500-712, Republic of Korea

H. Stiel
Max Born Institute, 12489 Berlin, Germany

M. Nishikino · N. Hasegawa · T. Kawachi
Japan Atomic Energy Agency, 8-1-7 Umemidai, Kizugawa, Kyoto 619-0215, Japan

© Springer International Publishing Switzerland 2016
J. Rocca et al. (eds.), *X-Ray Lasers 2014*, Springer Proceedings in Physics 169,
DOI 10.1007/978-3-319-19521-6_8

been attributed to the anisotropic trapping of the resonance line radiation determining population of the magnetic sub-levels of the lower laser level 3 s. As a result, population of the sub-levels with $m = \pm 1$ was smaller (gain larger) than that for the sub-levels with $m = 0$ (smaller gain), and the s-polarised component has been under these specific conditions stronger pronounced. This effect disappeared with a higher pumping energy. On contrary, no polarisation was observed in the single-pulse, quasi-steady-state experiment reported in [2]. Theoretical approach to the problem was more popular, but also offered quite rich spectrum of explanations. The most convincing ones seemed to be those reported in [3–5]. The possibility of anisotropic electron impact excitation was considered as a source of the population imbalance in the magnetic sub-levels of the $(2p_{1/2}, 3s_{1/2})_1$ level. A substantial fraction of hot electrons (5–10 % of the bulk) should cause the 19.6 nm lasing line to be polarized. The effect in a laser plasma has also been confirmed experimentally, even if only for the light elements [4, 6]. In the transient collisional-excitation scheme, the heating pulse of a few picosecond duration and 10^{14}–10^{15} Wcm^{-2} intensity is strong enough to create an anisotropic electron velocity distribution; this would produce population imbalance in the relevant magnetic sublevels. The polarization destruction (alignment destruction) process could be important in such a scenario. Benredjem et al. [5] claimed that the population imbalance in the magnetic sub-levels is impossible owing to the fast (rate $\sim 10^{14}$ s^{-1}) misalignment process by electron-ion Coulomb elastic scattering. Stark broadening measurement of the levels under consideration showed that this rate is lower by 2–3 orders of magnitude [1]. Due to the fact that any kind of collision affects the linewidth, the measured values of the spectrum width support rather the latter.

2 Experiment

The X-ray laser used in thew experiment was pumped in the classic double-pulse arrangement of the transient inversion scheme with both pump laser pulses incident normally to target surface. Two targets were arranged to work as a master oscillator/power amplifier tandem, i.e. one worked as an injector and the second as an energy booster.

2.1 Set-up and Measurement

The output beam was sent to a membrane beam-splitter (BS) [7] placed about 4 m from the second (amplifier) target (see Fig. 1). The active area of the beam-splitter was 6×6 mm^2. Assuming typical for this type of laser divergence of the output XUV beam of 5–8 mrad the beam diameter at the BS plane was about 20–32 mm. Thus, only a minor part of the beam was used for diagnostic procedure introducing

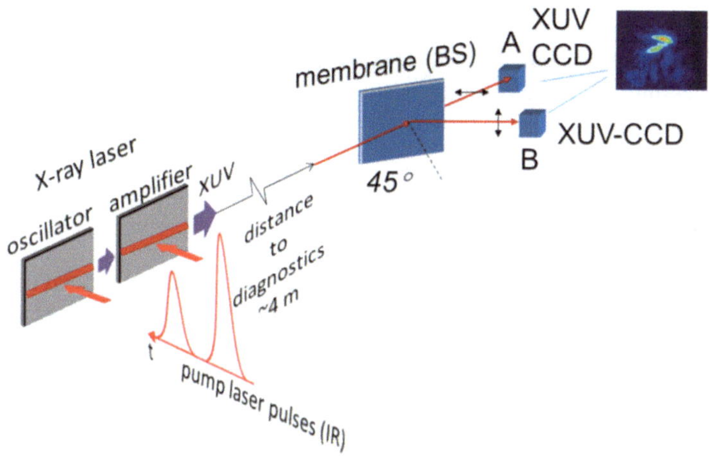

Fig. 1 Experimental set-up of beam polarisation measurement

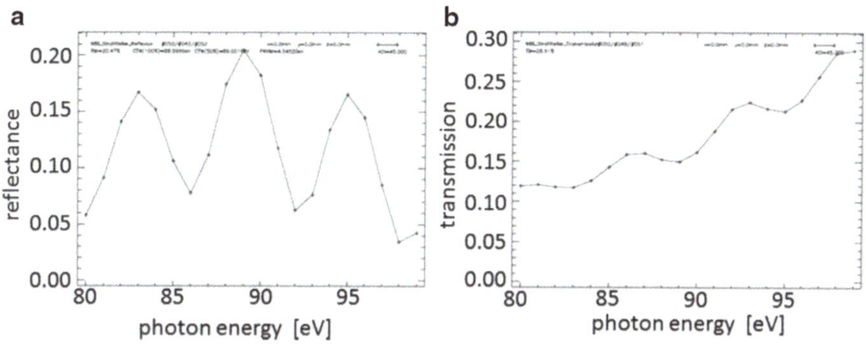

Fig. 2 Calibration curves of the membrane beam-splitter

an artificial mode selection. One of the crucial points of the measurement was verification of the beam splitter characteristics. This was checked first with an incoherent plasma source and gave results reasonably close to those assumed in the design [7]. The beam splitter parameters were again (directly after the experiment) tested on a synchrotron beamline of PTB Berlin. The synchrotron radiation was *s*-polarised. The result of this re-calibration of the beam-splitter gave: $T_s = 0.16$ and $R_s = 0.20$ (see Fig. 2). The multi-layer structure was deposited on a Si_3N_4 substrate and this introduced an additional, incidence-independent, transmission factor T_0 contributing to the transmission value quoted above. Hence, the effective transmission can be written as $T_s = T_0(1 - R_s)$. The substrate transmission does not depend on the polarisation state of the incident radiation and is equal to 0.20 [8]. As a result, one can write $T_p = T_0(1 - Rp)$. Here, R_p was estimated to be equal to 0.002 using the R_s value and the tabular values from [8]. It gave a T_p value equal to 0.20. It was found from the cross-correlated measurements with both CCD cameras (swapped cameras in the positions reflected and transmitted) a correction factor due to different sensitivities of the cameras equal to 1.3. This correction factor was taken into account in data processing. All these parameters are valid for an incidence angle of 45°.

3 Results and Analysis

The output of our measurement gives values of I^r and I^t, the integrated intensity of the reflected and transmitted signal, respectively. On the other hand one can express the polarization level of the incident beam P as

$$P = \frac{I_p - I_s}{I_p + I_s} = \frac{I_p/I_s - 1}{I_p/I_s + 1} \tag{1}$$

where $P = 1$ means full linear p-polarisation, P=0 the circular or natural (undefined) one and $P = -1$ the pure s-polarisation. One can express I^r and I^t as $I^r = I_s R_s + I_p R_p$ and $I^t = I_s T_s + I_p T_p$ to get

$$\frac{I_p}{I_s} = \frac{R_s - T_s(I^r/I^t)}{T_p(I^r/I^t) - R_p} \tag{2}$$

The data was collected for only the second target (denoted as the amplifier in Fig. 1). The signal was within a relatively limited area about 1000–2000 pixels concentrated around the maximum value of the signal. It should reduce influence of the constant background that was independently recorded in the first shot and subtracted from the analysed signal. The plots below (Figs. 3 and 4) show a ratio of I_p/I_s for a series of consecutive shots and the corresponding polarization state described by P=0.5–0.8 (pretty well working with the single-mode modeling of the coherent interaction). However, there are two problems to be cleared/explained. The coherent interaction model described in [9] assumes statistical distribution of the polarisation state on the shot-to-shot basis. Our measurements contradict this picture as the polarization state is elliptical with dominant p-component (field oscillates in the horizontal plane) in each of the consecutive shots. This is also in contradiction to the s-polarisation observed in [1]. Existence of highly energetic radiation about 1 keV for which BS is transparent and hence bringing a significant background in the transmitted signal is very unlikely due to geometrical relations determined by a 4 m distance between the source and the detector.

Generally, radiation incident obliquely on a plasma layer with a density gradient in the normal direction, shows distinctive difference in the interaction physics between p- and s- polarisation components. The most pronounced effect of such a behaviour is resonance absorption. Propagation can be described by two separate Helmholtz equations (each for the specific polarisation) [10]

$$\frac{\partial^2 E_z}{\partial x^2} + (\varepsilon - \sin^2 \vartheta) E_z = 0 \tag{3}$$

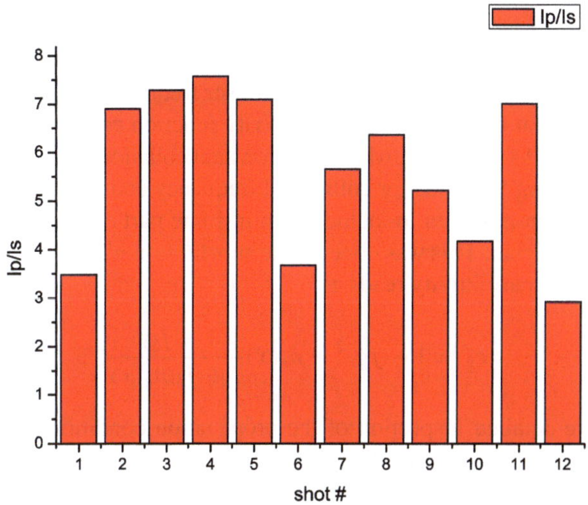

Fig. 3 Ratio of p-component and s-component intensities for 12 consecutive shots on the second target (amplifier)

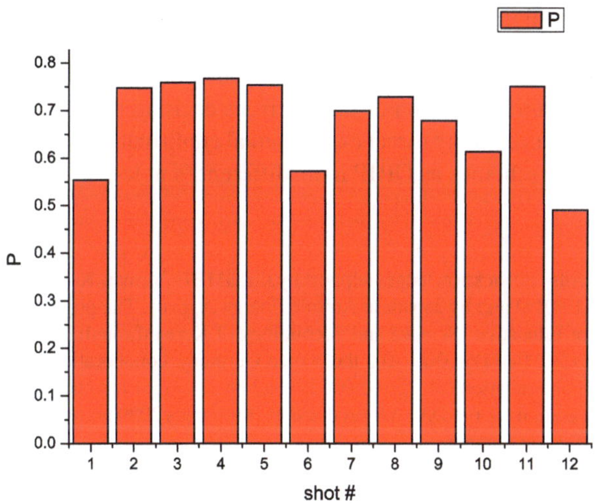

Fig. 4 Polarisation degree for 12 consecutive shots othe second target (amplifier)

$$\frac{\partial^2 B_z}{\partial x^2} - \frac{1}{\varepsilon}\frac{\partial \varepsilon}{\partial x}\frac{\partial B_z}{\partial x} + (\varepsilon - \sin^2\vartheta)B_z = 0 \tag{4}$$

We solved the equations in non-dimensional form for s- and p- polarised waves using the finite difference method in the way suggested by [10]. From given parameters as geometry and plasma characteristics all field components and the resulting energy densities ($W = |\vec{E}|^2 + |\vec{B}|^2$) have been computed for both polarisations.

Assuming relative permittivity in the typical form $\varepsilon(x) = 1 - n_e(x)/(1 + j\nu_{ie}(x))$, where n_e is plasma density normalised to the critical one, j is complex unit and ν_{ie} is the collisional frequency normalised to the laser frequency. Collisional frequency depends on n_e. That means that our permittivity function can be simplified to form $\varepsilon(x) = 1 - n_e(x)/(1 + j\mu n_e(x))$, where μ is constant collisional parameter which can be determined using formulae for collisional frequency and critical density. Permittivity has also been modified to include gain in some part of the medium traversed by the amplified signal. This was done by the substitution $n' = n - jg_r$ and the final expression for the permittivity reads as

$$\varepsilon_g = 1 - j{g_r}^2 - n_e(x)\frac{1 - jg_r}{1 + j\mu n_e(x)}, \tag{5}$$

However, more detailed inspection of the propagation conditions shows that even if the field discontinuity at the critical surface (n_e=n_{cr}) causes its dramatic increase only for p-polarised radiation, this sort of field penetration does not influence equal behaviour of both polarisation components before the turning point of the beam trajectory [11]. In the active medium of the X-ray lasers the efficiently amplified radiation propagates under very large incidence angle ($\theta \simeq \pi/2$) shifting the turning point far away from the critical surface. Hence influence of propagation is no more valid as the explanation of the recorded differences in the polarisation state. In this situation, the only reasonably argument in explanation the situation is the spatially anisotropic distribution function of the electrons exciting the active ions. However, this argument still need more direct experimental confirmation by experiments using different pumping schemes as GRIP in its different (single-, double- or triple pulse) versions.

Acknowledgements HS acknowledges support by the BMBF German-Korean Collaboration Program (No. KOR 10/016) and by Japanese Atomic Energy Agency. The project was also supported by the Ministry of Education, Science and Technology of Korea through Basic Science Research Program (No. R15-2008-006-03001-0), the Korea-Germany collaboration program of Korean National Research Foundation (No. 2010-00633) and by Gwangju Institute of Science and Technology through a grant from the DASAN fund and the Photonics 2020 project. CMK was supported by the Institute of Basic Science.

References

1. Kawachi, T., Murai, K., Yuan, G., Ninomiya, S., Kodama, R., Daido, H., Kato, Y., Fujimoto, T.: Observation of polarization of the soft x-ray laser line in Neonlike Germanium ions. Phys. Rev. Lett. **75**, 3826 (1995)
2. Rus, B., Lewis, C.L.S., Cairns, G.F., Dhez, P., Jaegle, P., Key, M.H., Neely, D., MacPhee, A.G., Ramsden, S.A., Smith, C.G., Sureau, A.: Demonstration of amplification of a polarized soft-x-ray laser beam in a neonlike germanium plasma. Phys. Rev. A. **51**, 2316 (1995)
3. Dubau, J., Inal, M.K., Benredjem, D., Cornille, A.: Theoretical predictions for the polarization of X- ray laser lines in the presence of a directed beam of hot electrons. J Phys IV France **11**, 2–277 (2001)

4. Kieffer, J-C., Matte, J.P., Pépin, H., Chaker, M., Baudouin, Y., Johnston, T.W., Chien, C.Y., Coe, S., Mourou, G., Dubau, J.: Electron distribution anisotropy in laser-produced plasmas from x-ray line polarization measurements. Phys. Rev. Lett. **68**, 480 (1992)
5. Benredjem, D., Sureau, A., Rus, B., Möller, C.: Polarization state of the output of soft-x-ray lasers through the paraxial Maxwell-Bloch approach. Phys. Rev. A. **56**, 5152 (1997)
6. Kieffer, J-C., Matte, J.P., Chaker, M., Baudouin, Y., Chien, C.Y., Coe, S., Mourou, G., Dubau, J., Inal, M.K.: X-ray-line polarization spectroscopy in laser-produced plasmas. Phys. Rev. E. **48**, 4648 (1993)
7. Delmotte, F., et al.: X-ray -ultraviolet beam splitters for the Michelson interferometer. Appl. Opt. **41**, 5905–5912 (2002)
8. Henke, B.L., Gullikson, E.M., Davis, J.C.: X-Ray Interactions: Photoabsorption, Scattering, Transmission, and Reflection at E=50 −30,000 eV, Z=1−92. Atomic Data Nucl. Data Table. **54**, 181–342 (1993)
9. Kim, C.M., Janulewicz, K.A., Lee, J.: Pulse buildup from noise and intrinsic polarization of plasma-based x-ray lasers. Phys. Rev. A. **84**, 013834 (2011)
10. Gibbon, P.: Short Pulse Laser Interactions with Matter. An Introduction. Imperial College Press, London (2005)
11. Ginzburg, V.L.: The Propagation of Electromagnetic Waves in Plasma. Pergamon, Oxford (1964)

Improved Sub-10 nm Ni-like Lasing by Varying the Slope of the Traveling-Wave Velocity

Fei Jia, Michael Siegrist, Felix Staub and Jürg E. Balmer

Abstract Large output variations have been observed in recent samarium (Sm, Z = 62) laser experiments under nominally identical pump condition. In this work, we focus on the spatio-temporal mismatch between the traveling-wave excitation (TWE) and the x-ray pulse group velocity. Through simulations incorporating the gain dynamics of the system we show that the group velocity is not constant but monotonously increases along the line focus due to gain saturation. We demonstrate a line-focusing configuration that results in TWE with the sweep velocity well matched to the spatially varying group velocity. A novel line-focusing configuration with adapted TWE velocity has been implemented into the existing 10-TW Nd:glass CPA laser system.

1 Introduction

Collisionally pumped coherent x-ray sources have been developed with remarkable progress since the first demonstration of neon-like selenium at the Lawrence Livermore National Laboratory in 1984 [1]. By using pre-pulses [2], short-duration pulses (~ps) [3] and the grazing-incident pumping scheme (GRIP) [4], the pumping efficiency has been significantly enhanced. The small signal gain of the transient collisional excitation (TCE) scheme has been predicted with a large value of gain coefficient (exceeding 100 cm^{-1}) but short lifetimes with duration in the picosecond regime [5, 6]. Since the GRIP introduces an excitation sweep velocity along the line focus at a speed higher than the speed of light, the amplification suffers from a severe velocity mismatch between amplified-spontaneous-emission (ASE) group velocity and excitation sweep velocity. Therefore, a traveling-wave excitation (TWE) technique was introduced to keep the two velocities synchronized [7]. The recent discussion on TWE suggested that it is beneficial to use TWE at velocities slower than c in high-gain plasma amplifier [8]. In this proceeding, we introduce a modified 'Ross' focusing scheme [9] which can produce an increasing TWE velocity.

F. Jia (✉) · M. Siegrist · F. Staub · J. E. Balmer
Institute of Applied Physics, University of Bern, Sidlerstrasse 5, 3012 Bern, Switzerland
e-mail: fei.jia@iap.unibe.ch

© Springer International Publishing Switzerland 2016

J. Rocca et al. (eds.), *X-Ray Lasers 2014*, Springer Proceedings in Physics 169,
DOI 10.1007/978-3-319-19521-6_9

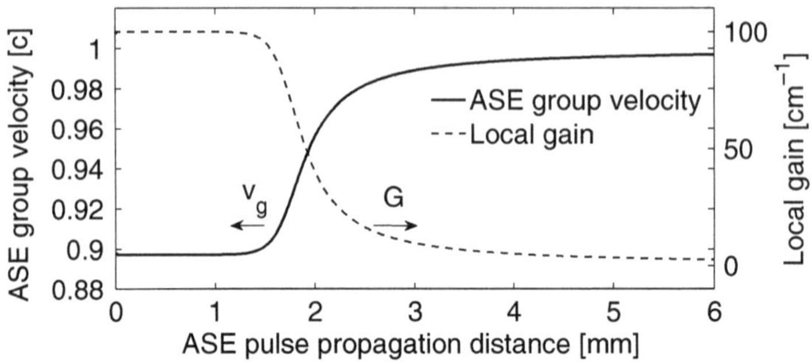

Fig. 1 ASE pulse propagation group velocity with respect to a small-signal gain value of 100 cm^{-1}

2 Discussion on the Spatio-Temporal Mismatch Between TWE and ASE

Detailed analytic solutions of the traveling-wave velocity for three commonly applied line-focusing configurations have been previously discussed in our report [10]. Here we will list some results to describe the phenomenon of the spatio-temporal mismatch between TWE and ASE signal group velocity. First, we assume that in collisionally excited soft-x-ray laser the broadening effect is dominated by the homogeneous component [11]. Thus the group velocity of the amplified signal v_g is a function of local gain [12]:

$$\frac{v_g}{c} = \frac{1}{1 + cG(0)/(2\pi\,\Delta v_h)} \tag{1}$$

Assuming the case of TWE velocity perfectly matched to local group velocity $v_g(x)$ of ASE, we plot the evolution of $v_g(x)$ with respect to the locally reduced gain in Fig. 1. However, this perfect scenario can not always be satisfied. We thus introduce the time-dependent local small-signal gain $g(z,t)$. We adopt the assumption of Strati [13] to describe the $g(z,t)$ with an exponential decay of 1/e lifetime τ as $g(z,t) = g(z,0)\exp[-t(z)_{mis}/\tau]$. Therefore, ASE amplification process including the mismatch contribution of the decay of the local gain can be described as [14]:

$$\begin{aligned}\frac{\partial}{\partial z} I(\nu,z) &= G(\nu,z)I(\nu,z) + E(\nu,z)\\ &= G(\nu,z)[I(\nu,z) + I_0 f(0)]\end{aligned}$$

$$G(\nu,z) = \frac{g(0,z)}{1+\overline{I}(z)/I_s}\varphi(\nu) \tag{2}$$

$$\text{if}\quad t(z)_{mis} \geq 0,$$

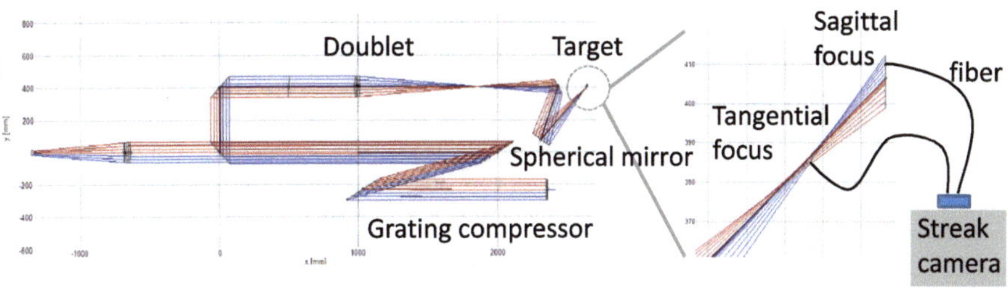

Fig. 2 *Left*: Raytracing analysis of the commonly applied 'Ross' focusing scheme. *Right*: The streak camera measurement of traveling wave

$$g(0, z) = g_0(0, z) \exp\left[-\frac{t(z)_{mis}}{\tau}\right] \tag{3}$$

$$t(z)_{mis} = \int_0^z \left[1/v_g(z) - 1/v_e(z)\right] dz$$

In the case of the spatially varying TWE velocity of the commonly used 'Ross' focusing scheme, the analytical solution as well as the raytracing result shows a monotonous decreasing TWE velocity $v_{TW(x)}$ along the propagation direction of the ASE. This unfavourable negative slope of the $v_{TW(x)}$ is confirmed by a direct measurement of the signal delay by the streak camera. We illustrate the raytracing model of the 'Ross' focusing scheme and the streak camera set-up in Fig. 2. This measured $v_{TW(x)}$ is plugged into Eq. (4) and a numerical integral of the partial differential equation (2) is conducted to estimate the signal amplification. The signal amplification and the group velocity of the ASE signal are plotted in Fig. 3. It is obvious that the $v_{TW(x)}$ and $v_g(x)$ are crossing due to the opposite sign of the slope, revealing the possibility of achieving more efficient amplification by reversing the slope of $v_{TW(x)}$.

3 Design of a Modified 'Ross' Focusing Scheme with Increasing Traveling-Wave Velocity

In order to reverse the sign of the slope of $v_{TW(x)}$, the sequence of the tangential and sagittal foci need to be swapped. Therefore, a cylindrical mirror is introduced to push the tangential focus behind the sagittal one. In Fig. 4 we show that by substituting the planar mirror with a cylindrical mirror (ROC = 5000 mm), a $v_{TW(x)}$ increasing along the line focus is achieved. Reduction of length of the line focus from 12 to 6 mm.We compare the line focus quality of the original and modified 'Ross' focusing by raytracing and plot the result in Fig. 5. Finally, this $v_{TW(x)}$ with positive slope is examined by the same numerical simulation, showing an enhanced amplification with a greater gain-length product (see Fig. 6). Compared with the

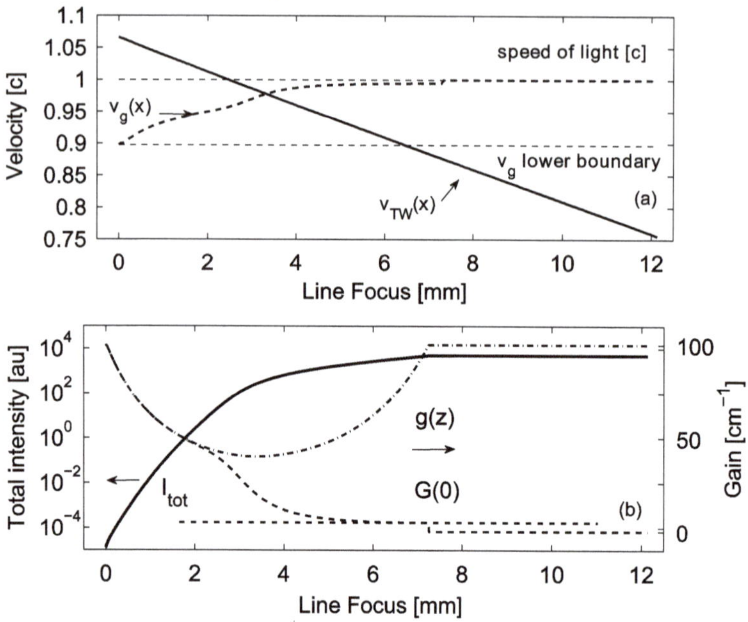

Fig. 3 Simulation of the signal amplification and $v_g(x)$ changing with measured $v_{TW(x)}$

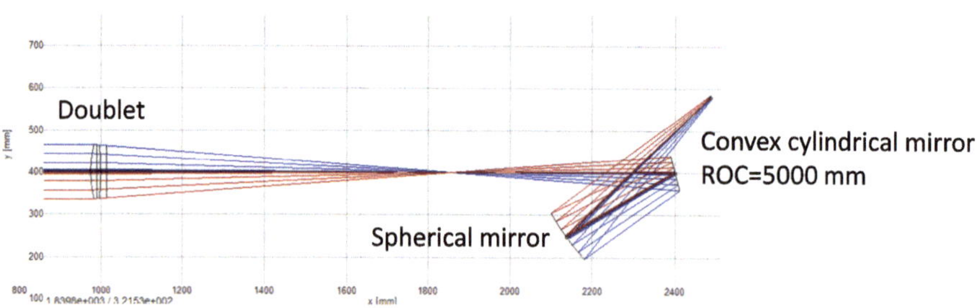

Fig. 4 Raytracing of the modified 'Ross' focusing scheme

result in Fig. 3, within 6 mm of propagation in the gain medium, the overall efficiency increased from 28.6 to 87.9 % (efficiencies are calculated as I_{tot}/I_p, where I_p denotes the maximum possible intensity obtained in the perfect matched velocities).

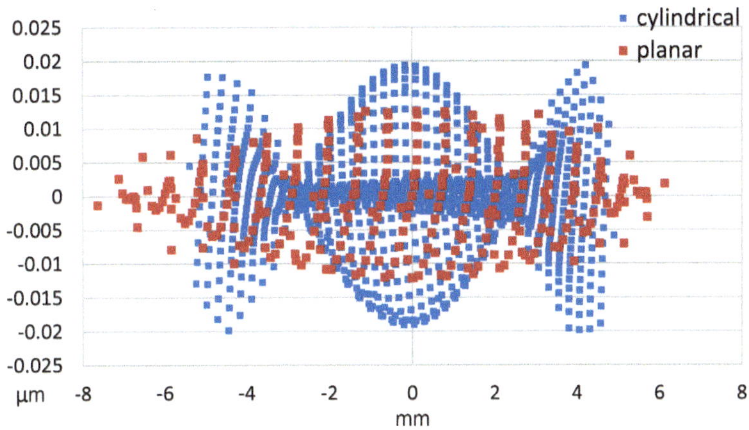

Fig. 5 Comparison of the line focus quality of the original and modified 'Ross' focusing

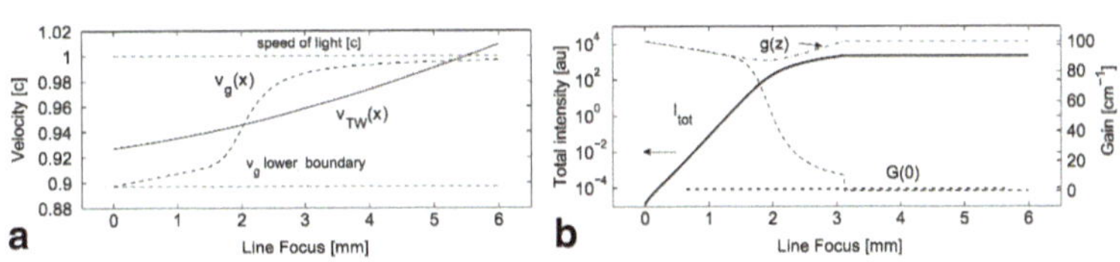

Fig. 6 Simulation of the signal amplification and $v_g(x)$ changing for the modified 'Ross' focusing configuration

4 Conclusion

We have shown that a mismatch between the ASE signal group velocity and the traveling-wave excitation velocity exists in the commonly used 'Ross' focusing scheme. Such mismatch is subjected to be the major reason contributing to the reduction of the gain-length product. We proposed a modified 'Ross' focusing scheme by adding one convex cylindrical mirror. An increasing traveling-wave velocity is obtained that better matches the spatially increasing group velocity of the amplified ASE signal.

References

1. Matthews, D.L., Hagelstein, P.L., Rosen, M.D., Eckart, M.J., Ceglio, N.M., Hazi, A.U., Medecki, H., Trebes, J.E., Whitten, B.L., MacGowan, A.J.: Demonstration of a Soft X-Ray Amplifier. Phys. Rev. Lett. **54**, 110–113 (1985)
2. Nilsen, J., MacGowan. B.J., Da Silva, L.B., Moreno, J.C.: Prepulse technique for producing low-Z Ne-like x-ray lasers. Phys. Rev. A. **48**, 4682–4685 (1993)
3. Nickes, P.V., Shlyaptsev, V.N., Kalachnikov, M., Schnrer, M., Will, I., Sandner, W.: Short Pulse X-Ray Laser at 32.6 nm Based on Transient Gain in Ne-like Titanium. Phys. Rev. Lett. **78**, 2748–2751 (1997)

4. Keenan, R., Dunn, J., Patel, P.K., Price, D.F., Smith, R.F., Shlyaptsev, V.N.: High-Repetition-Rate Grazing-Incidence Pumped X-Ray Laser Operating at 18.9 nm. Phys. Rev. Lett. **94** 103901 (2005)
5. Healy, S.B., Janulewicz, K.A., Plowes, J.A., Pert, G.J.: Transient high gains at 196 Å produced by picosecond pulse heating of a preformed germanium plasma. Opt. Commun. **132**, 442–448 (1996)
6. Nilsen, J.: Modeling picosecond-laser-driven neonlike titanium x-ray laser experiments. Phys. Rev. A. **55**, 3271–3274 (1997)
7. Kalachnikov, M.P., Nickles, P.V., Schnurer, M., Sandner, W., Shlyaptsev, V.N., Danson, C., Neely, D., Wolfrum, E., Zhang, J., Behjat, A., Demir, A., Tallent, G.J., Warwick, P.J., Lewis, C.L.S.: Saturated operation of a transient collisional x-ray laser. Phys. Rev. A. **57**, 4778–4783 (1998)
8. Tommasini, R., Fill, E.: Effective traveling-wave excitation below the speed of light. Opt. Lett. **26**, 689–691 (2001)
9. Ross, I.N., Hodgson, E.M.: Some optical designs for the generation of high quality line foci. J. Phys. E. **18**, 169–173 (1985)
10. Fei, J., Staub, F., Siegrist, M., Balmer, J.E.: Analytic study of traveling-wave velocity variation in line-focusing schemes for plasma xray lasers. Appl. Opt. **53**, 3247–3254 (2014)
11. Koch, J.A., MacGowan, B.J., Da Silva, L.B., Matthews, D.L., Underwood, J.H., Batson, P.J., Mrowka, S.: Observation of gain-narrowing and saturation behaviour in Se X-ray laser line profiles. Phys. Rev. Lett. **68**, 3291–3294 (1992)
12. Caperson, L.W., Yariv, A.: Pulse propagation in a high-gain medium. Phys. Rev. Lett. **26**, 293–295 (1997)
13. Strati, F., Tallents, G.J.: Analytical modeling of group-velocity effects in saturated soft-x-ray lasers pumped with a picosecond traveling-wave excitation. Phys. Rev. A. **64**, 013807 (2001)
14. Pert, G.J.: Output characteristics of amplified-stimulated-emission lasers. Opt. Soc. Am. B. **11**, 1425–1435 (1994)

Temporal Response of Seeded XUV Lasers Under Different Amplification Regimes—Inversion Density Threshold

A. LeMarec, O. Larroche, L. Meng and A. Klisnick

Abstract We present a simple model of amplification in XUV lasers which displays an inversion density threshold between a low-inversion, classical adiabatic regime and a high-inversion, dynamic regime involving Rabi oscillations. In the latter case, the small-signal gain is significantly smaller than the usually used adiabatic value, but the amplified pulse duration is no longer limited by the amplifier bandwidth. The dynamic regime can thus possibly lead to amplified femtosecond pulses.

1 Introduction

Injecting a resonant femtosecond high harmonic (HH) pulse into an amplifying plasma may cause a nonstationary behavior of the amplification involving Rabi oscillations, which are currently considered a promising way to generate femtosecond pulses with seeded X-ray lasers. In this work we investigate that point, starting from the Maxwell-Bloch formalism commonly used to model the propagation and amplification mechanisms in plasma-based X-ray laser media [1–5], presented in Sect. 2. We then focus in Sect. 3 on the temporal behavior of the system through a simple 0D reduction of the Maxwell–Bloch equations [6] in the limit of a long gain-recovery time. We show that the behavior of the system becomes qualitatively different when the inversion density increases above a threshold value, leading to both a modification of the small-signal gain and Rabi oscillations for high intensity. The inversion threshold value is discussed for various actual X-ray lasers. In Sect. 4 we present prospects for further work, including investigating the adiabatic/dynamic transition in the frame of a more realistic 1D model.

A. LeMarec (✉) · A. Klisnick
ISMO, CNRS, Université Paris-Sud, 91405 Orsay Cedex, France
e-mail: andrea.le-marec@u-psud.fr

O. Larroche
CEA DIF, Bruyères le Châtel, 91297 Arpajon Cedex, France

L. Meng
Institute of Fluid Physics, China Academy of Engineering Physics, Mianyang, China

© Springer International Publishing Switzerland 2016
J. Rocca et al. (eds.), *X-Ray Lasers 2014*, Springer Proceedings in Physics 169,
DOI 10.1007/978-3-319-19521-6_10

2 Theoretical Model—Maxwell–Bloch Equations

The complex amplitude of the z-polarized electric field E propagating in the (x, y) plane at small angles to the y-axis is described in the frame of the time-envelope approximation by the paraxial propagation equation:

$$\frac{2i\omega}{c}\left(\frac{1}{c}\frac{\partial E}{\partial t}+\frac{\partial E}{\partial y}\right)+\frac{\partial^2 E}{\partial x^2}=-\frac{\omega^2}{c^2}\left(\frac{-\omega_p(x)^2}{\omega(\omega+i\nu_{ei})}E+4\pi P\right). \tag{1}$$

The approximation is valid since the X-ray laser period $2\pi/\omega$ is a few tens of attoseconds while the seed pulse duration is a few tens of femtoseconds. The complex amplitude of the time-enveloped polarization density P, on the right-hand side of (1), is governed by:

$$\frac{\partial P}{\partial t}=-\gamma P-i\frac{d^2}{\hbar}(\rho_u-\rho_l)E. \tag{2}$$

where $\gamma = 1/T_2$ is the dipole dephasing rate, mainly due to collisions, leading to homogeneous broadening of the lasing transition, and d is the operator matrix element between the lasing levels u and l. The population densities of the laser transition ρ_u and ρ_l are governed by a single equation for the scaled inversion density D:

$$\frac{\partial D}{\partial t}=-\frac{D-D_0}{T_1}-\frac{d^2}{\hbar^2\omega}Re(iPE^*). \tag{3}$$

where $D=(d^2/\hbar)(\rho_u-\rho_l)$ and T_1 is the gain recovery time, taken as the reciprocal of the largest eigenvalue of the excitation/decay rate matrix (γ_{ij}).

Characteristic Times of the System

The Maxwell–Bloch system (1)–(3) displays different characteristic times: T_1 and T_2, used in classical laser theory (see e.g. [7]), τ the duration of radiation propagation through the plasma, and τ_c and T_{R0} defined hereafter:

- *The collisional damping time* $\tau_c = (\omega/\omega_p)^2\nu_{ei}^{-1}$ of the electromagnetic waves propagating through the plasma due to electron-ion collisions.
- *The asymptotic Rabi period* T_{R0} is the reciprocal of the Rabi frequency Ω_{R0} that would be reached if all the energy stored in the initial inversion density D_0 was transferred to the amplified wave, and

$$(\Omega_{R0})^2=4\pi\omega^2 D_0=\gamma c G_{ad}.$$

where G_{ad} is the usual adiabatic gain:

$$G_{ad}=\frac{4\pi\omega^2 D_0}{\gamma c}.$$

3 0D Study of the Temporal Response of the Inverted System

We focus the study on the temporal response of the system for a completely homogeneous case (0D) with no coupling with the external environment (under weak pumping conditions, i.e. large T_1). Equations (1)–(3) then read

$$\frac{dE}{dt} = 2\pi\omega iP. \tag{4}$$

$$\frac{diP}{dt} = -\gamma iP + \omega DE. \tag{5}$$

$$\frac{dD}{dt} = -\frac{d^2}{\hbar^2\omega} Re(iPE^*). \tag{6}$$

From (4) and (6) we can see that the system is conservative: the total energy is exchanged between the electrons of the transition and the amplified waves

$$\frac{d}{dt}\left[\left(\frac{\hbar\omega}{d}\right)^2 D + \frac{|E|^2}{4\pi}\right] = 0. \tag{7}$$

3.1 *Inversion Density Threshold*

We study the response of this system from an initial inverted density D_0 with no electric field and polarization ($E, iP = 0$) to a stationary state where all the energy stored in the initial inversion density has been transferred to the electric field reaching an amplitude E_0 ($D, iP = 0$). The conserved quantity in (7) is thus set equal to D_0.

We linearize the system around these initial ($D = D_0$, $E = 0$, $P = 0$) and final ($E = E_0$, $D = 0$, $P = 0$) states in order to respectively study the amplification rate and the temporal response behavior.

Small-Signal Gain

The linear stability analysis of the initial state under small perturbations $E, iP \propto e^{\Omega t}$ leads to the following equation

$$\Omega^2 + \gamma\,\Omega - \frac{(\Omega_{R0})^2}{2} = 0. \tag{8}$$

where the temporal rate of amplification Ω can be converted into the corresponding small signal gain $G = 2\Omega/c$, leading to

$$G^2 + \frac{2\gamma}{c}\,G - \frac{2\gamma}{c}\,G_{ad} = 0. \tag{9}$$

with the following solution

$$G = 2\,G_{ad}^{*}\left(\sqrt{1+\frac{G_{ad}}{G_{ad}^{*}}}-1\right). \tag{10}$$

where $G_{ad}^{*} = \gamma/(2c)$ is the adiabatic gain threshold. Let us study two asymptotic limits of the ratio G_{ad}/G_{ad}^{*} in (10).

When $G_{ad} \ll G_{ad}^{*}$, under weak population inversion, the amplification remains close to adiabatic and the adiabatic small-signal gain is recovered $G \approx G_{ad}$. In this regime, a broad-band HH pulse is adiabatically amplified only at the center of its spectrum (the part covered by the gain bandwidth). It results in the generation of a wake with a narrowed linewidth and the output duration is thus limited by the amplifier bandwidth.

When $G_{ad} \gg G_{ad}^{*}$, under strong population inversion, the real rate of amplification is lower than the adiabatic gain $G \approx 2\sqrt{G_{ad}G_{ad}^{*}} \equiv G_{dyn}$ and does not depend on γ. The amplification time scale is shorter than the dipole dephasing time T_2, hence the phrase "dynamic regime".

The ratio of the adiabatic gain to the threshold gain amounts to a comparison between the dipole dephasing rate and the asymptotic Rabi frequency:

$$\frac{G_{ad}}{G_{ad}^{*}} = 2\left(\frac{\Omega_{R0}}{\gamma}\right)^{2}. \tag{11}$$

Temporal Response Around the Stationary Solution

The linear stability analysis of the final state under small perturbations $D, iP \propto e^{\Omega t}$ leads to the following equation

$$\Omega^2 + \gamma\Omega + (\Omega_{R0})^2 = 0. \tag{12}$$

which is the classical equation for a damped harmonic oscillator with frequency Ω_{R0} and damping rate $\gamma/2$. One can see that the aperiodic case, when $\gamma > \Omega_{R0}$, corresponds to the adiabatic regime (11), while the damped oscillating case, with $\gamma < \Omega_{R0}$, corresponds to the dynamic regime.

3.2 Numerical Resolution in the Two Regimes

We present in Fig. 1 the results of simulations of system (1)–(3) using the code COLAX (see [8] and references therein) in 0D for two values of the ratio G_{ad}/G_{ad}^{*}: 0.1 (adiabatic regime) and 10 (dynamic regime). The initially amplifying system with an inversion density $\rho_u - \rho_l \approx 2 \times 10^{-16}$ cm^{-3} is seeded with an electric field with a small intensity $I_i \approx 10^5$ W/cm^2 (comparable to spontaneous emission). The

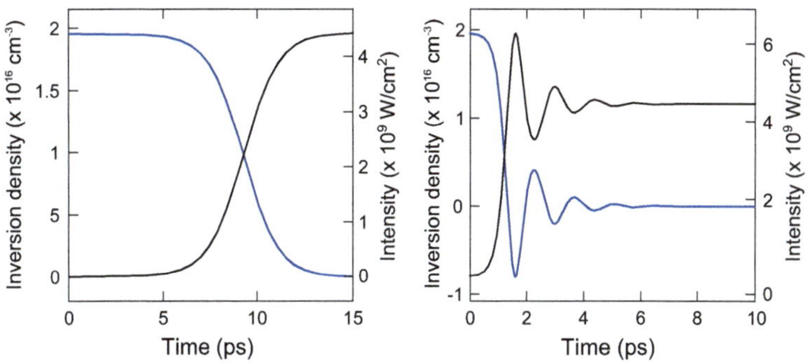

Fig. 1 Inversion density (*blue*) and output intensity (*black*) as functions of time for two values of the ratio $G_{ad}/G^*_{ad} = 0.1$ (adiabatic regime, *left*) and 10 (dynamic regime, *right*), starting from a small seed

energy stored in the inversion density is transferred to the wave which is amplified up to $|E| = |E_0|$. Saturation is reached in a slow and monotonous way in the first case, whereas in the second case Rabi oscillations appear, temporarily driving the inversion density to negative values before reaching the asymptotically steady state.

3.3 *Application to Actual X-Ray Laser Systems*

The gain ratio G_{ad}/G^*_{ad} has been computed from the relevant parameters (T_2, G, G_{ad}) collected from the literature for various X-ray laser systems (cited in square brackets). Results are displayed in Table 1. Systems are seen to range from the adiabatic to the deeply dynamic regimes. As a result, only the top two cases in the table are expected to lead to short-pulse amplification through Rabi oscillations.

Table 1 Relevant parameters (gains in cm^{-1} and T_2 in ps) for actual lasers

X-ray laser	$T_2 = \gamma^{-1}$	G	G_{ad}	G_{ad}/G^*_{ad}
Optical field ionization	2 [5]/7.2 [9]	60 [9]	168/451	20/196
Transient collis. excitation	1.3 [10, 11]	60 [12]	130	10
Quasi-st. state (capill. disch.)	18 [13]	1.2 [14]	1.6	1.8
Quasi-st. state (laser-based)	0.9 [11]	7 [15]	7.7	0.4

4 Conclusion and Prospects

From a simple 0D study, the widely spread parameters of current X-ray lasers are seen to lead to two different regimes of amplification, only one of which is expected to possibly amplify short pulses through Rabi oscillations. In this non-linear dynamic regime, the amplification rate can become much smaller than the usual adiabatic gain. Further investigations are currently in progress, taking radiation propagation into account in a 1D geometry. Our first results tend to confirm that the threshold discriminating between the two different regimes persist in this case. For simulations performed with $G_{ad}/G^*_{ad} = 10$ the population inversion density reaches negative values, a signature of the dynamic regime, whereas it remains positive for $G_{ad}/G^*_{ad} = 0.1$. The position of the threshold, which might differ from the one obtained in 0D, will be investigated. The next steps of our work will then aim at gradually adding the parameters put aside for the present study (pumping, 2D geometry effects, damping) to get closer to a realistic frame.

References

1. Larroche, O., Ros, D., Klisnick, A., Sureau, A., et al.: Phys. Rev. A. **62**, 043815 (2000)
2. Al'miev, I.R., Larroche, O., Benredjem, D., et al.: Phys. Rev. Lett. **99**, 123902 (2007)
3. Oliva, E., Zeitoun, P., Fajardo, M., et al.: Phys. Rev. A. **84**, 013811 (2011)
4. Kim, C.M., Janulewicz, K.A., Kim, H.T., Lee, J.: Phys. Rev. A. **80**, 053811 (2009)
5. Tissandier, F., Sebban, S., Gautier, J., et al.: Appl. Phys. Lett. **101**, 251112 (2012)
6. Larroche, O., Meng, L., Le Marec, A., Klisnick, A.: Opt. Lett. **38**, 2505–2508 (2013)
7. Siegman, A.E.: Lasers. University Science books, Sausalito (1986)
8. Larroche, O., Klisnick, A.: Phys. Rev. A. **88**, 033815 (2013)
9. Tissandier, F., Sebban, S., Ribière, M., et al.: Phys. Rev. A. **81**, 063833 (2010)
10. Calisti, A., Ferri, S., Mossé, C., Talin, B., Klisnick, A., et al.: HEDP **9**, 516 (2013)
11. Meng, L.M., Klisnick, A., Kozlová, M., et al.: Proceeding of 13th ICXRL (2012)
12. Luther, B.M., Wang, Y., Larotonda, M.A., Alessi, D., et al.: Opt. Lett. **30**, 165 (2005)
13. Urbanski, L., Marconi, M.C., Meng, L.M., et al.: Phys. Rev. A. **85**, 033837 (2012)
14. Rocca, J.J., Clark, D.P., Chilla, J.L.A., et al.: Phys. Rev. Lett. **77**, 1476 (1996)
15. Rus, B., Mocek, T., Präg, A.R., Kozlová, M., et al.: Phys. Rev. A. **66**, 063806 (2002)

Self-Photopumped X-Ray Lasers from Elements in the Ne-like and Ni-like Ionization State

Michael Siegrist, Fei Jia and Jürg Balmer

Abstract Self-photopumped x-ray lasers in laser-produced plasma have been previously proposed as an alternative scheme to electron collisional excitation. While, in general, self-photopumped lasers suffer from somewhat lower gain, their higher saturation intensity could in principle lead to higher peak irradiace. We have performed experiments on the 3d $^1P_1 \rightarrow$ 3p $^1P_1$ and 4f $^1P_1 \rightarrow$ 4d $^1P_1$ transitions for Ne-like and Ni-like ions, respectively. Lasing on the self-photopumped laser line has been observed for the first time for several elements including Ne-like V, Cr, Fe, and Co as well as Ni-like Ru, Pd, and Ag. We have investigated the lasing process by varying the prepulse delay, which shows a shift of the optimum main pulse to second prepulse delays towards lower values with higher atomic number Z. Experiments have been performed with the Bern Advanced Glass Laser (BeAGLE) laser using three pulses with total energy up to 15 J at a wavelength of 1054 nm.

1 Introduction

The Ne-like and Ni-like ionization states offer very similar energy level diagrams. Both ionization states have a monopole and a dipole transition leading to a "collisional" and a "self-photopumped" lasing line (see Fig. 1). The collisional laser line is preferentially pumped by electron collisions and is denoted as 3p $^1S_0 \rightarrow$ 3s $^1P_1$ and 4d $^1S_0 \rightarrow$ 4p $^1P_1$ for Ne-like and Ni-like ions, respectively. The self-photopumped line, which is the main focus of this work, is pumped by a hybrid pumping process [1]. Firstly, the population inversion is achieved by electron collisions. Secondly, the population inversion is preserved by radiation trapping, as the plasma is optically thick for this line. The self-photopumped laser lines are denoted as 3d $^1P_1 \rightarrow$ 3p $^1P_1$ and 4f $^1P_1 \rightarrow$ 4d $^1P_1$ for Ne-like and Ni-like ions, respectively. While the self-photopumped laser line suffers from somewhat lower gain compared to the monopole-pumped line, its higher saturation intensity gives rise to the potential of becoming the dominant lasing line [1]. Self-photopumped x-ray lasers have been shown before for Ne-like Ti as well as for Ni-like Zr, Nb and Mo [1,2]. Using the BeAGLE system we were able to demonstrate several new elements to lase on

M. Siegrist (✉) · F. Jia · J. Balmer
Institute of Applied Physics, University of Bern, Sidlerstrasse 5, Bern, Switzerland
e-mail: michael.siegrist@iap.unibe.ch

© Springer International Publishing Switzerland 2016

J. Rocca et al. (eds.), *X-Ray Lasers 2014*, Springer Proceedings in Physics 169,
DOI 10.1007/978-3-319-19521-6_11

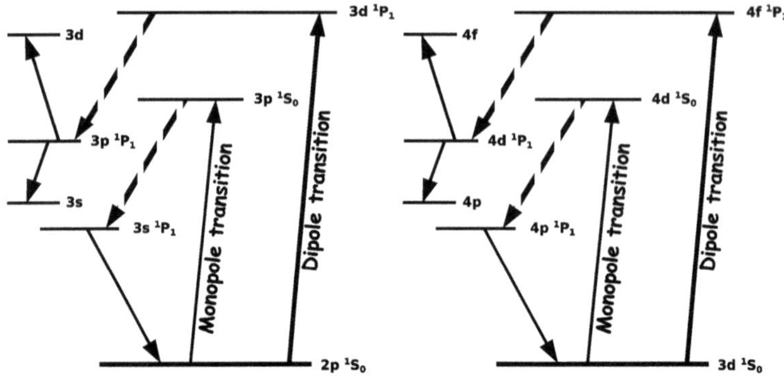

Fig. 1 Level diagram for the Ne-like (*left*) and Ni-like (*right*) ionization state respectively. The *dashed lines* mark the corresponding laser transitions, where the self-photopumped and the collisional transitions are pumped by the dipole and the monopole transition, respectively

this line, namely self-photopumped x-ray lasers for Ne-like V, Cr, Fe and Co as well as Ni-like Ru, Pd and Ag.

2 Experimental Setup—The BeAGLE System

The **Be**rn **A**dvanced **G**lass **L**aser **E**xperiment (**BeAGLE**) is a high-power laboratory-scale Nd:glass laser system that provides energies up to 25 J with 1.5 ps pulse duration at 1054 nm. In order to maximize the x-ray output, we use **gr**azing-**i**ncidence **p**umping (**GRIP**) at an angle of $\sim 50°$ and traveling-wave speeds ranging from 0.8 to 1 c. For the self-photopumping experiments we generated two pre-pulse by inserting beamsplitters with reflectivities of 0.5, 2.8, 4.5, 8 or 16 % (see Fig. 2) [3]. The BeAGLE system has two x-ray diagnostics to measure the plasma x-ray generation. Firstly, a pinhole camera pointing at a 45 angle to the target surface in order to measure the spontaneous emission from the plasma. Secondly, we measure the x-ray laser beam by using a CCD camera aligned with the beam axis. In the horizontal plane we measure the beam divergence and intensity distribution, whereas in the vertical plane, a flat field grating spectrally resolves the beam with a precision of ± 0.2 nm. Figure 3 shows a measurement of a Ne-like Ti shot. By using known laser lines to calibrate the spectrometer, the self-photopumped laser transition of the Ne-like Ti has been estimated to be at 30.1 ± 0.2 nm, which is in good agreement with literature data [2].

3 Experimental Results—Variation of the Prepulse Delays

Besides the energy, the prepulse configuration of the laser system is of high importance. In experiments we observed that the timing of the first prepulse (3–5 ns) has only a minor effect on the x-ray output. In contrast to this, the main pulse to second

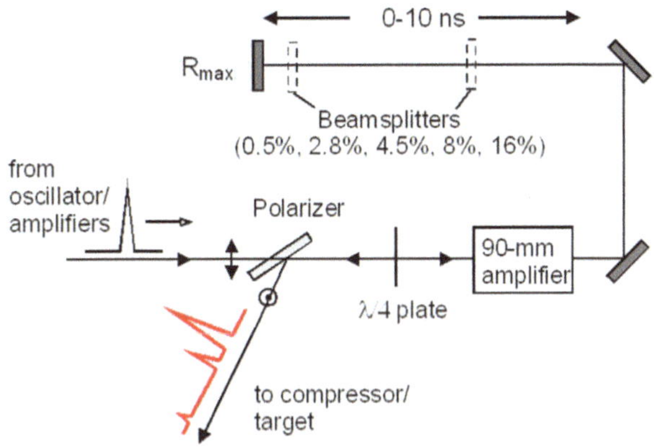

Fig. 2 Schematic of the prepulse setup

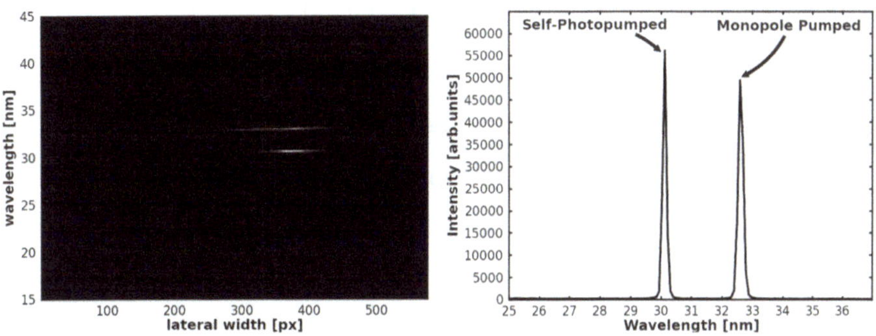

Fig. 3 Raw image (*left*) and the corresponding lineout spectrum (*right*) clearly showing the two distinct lines for the monopole and the self-photopumped lasing lines for Ne-like Ti

prepulse delay has been found to be the critical parameter on self-photopumped lasing. In Fig. 4, we show the dependence of the laser output on the second prepulse delay of Ne-like Ti and V and Ni-like Mo and Pd. Experiments have been performed with a 0.5 % first prepulse, which precedes the main pulse by 5 ns. For the second prepulse the 16 % beamsplitter was used. The total energy of the three pulses has been ~3 J.

We have investigated the time delay between the main pulse and the second prepulse for Ne-like Ti, V, Cr, Fe and Co as well as for Ni-like Mo, Ru, Pd and Ag. The parameters mentioned before have been used for all measurements, except for Ne-like Fe and Co, where we used a 2.8 % first prepulse, which preceded the main pulse by 4 ns with a total energy of 11 and 15 J, respectively. By plotting the delay for the highest output for each element against the atomic number Z, one observes a shift towards smaller delays with increasing atomic number Z for both the Ne-like and the Ni-like ionization scheme (see Fig. 5). The optimum delay seems to converge to a value around 100 ps for both the Ne-like and Ni-like lasing scheme. In addition, lasing has never been observed for a delay smaller than 100 ps, which could give rise to a explanation of the converging curve towards a value around 100 ps.

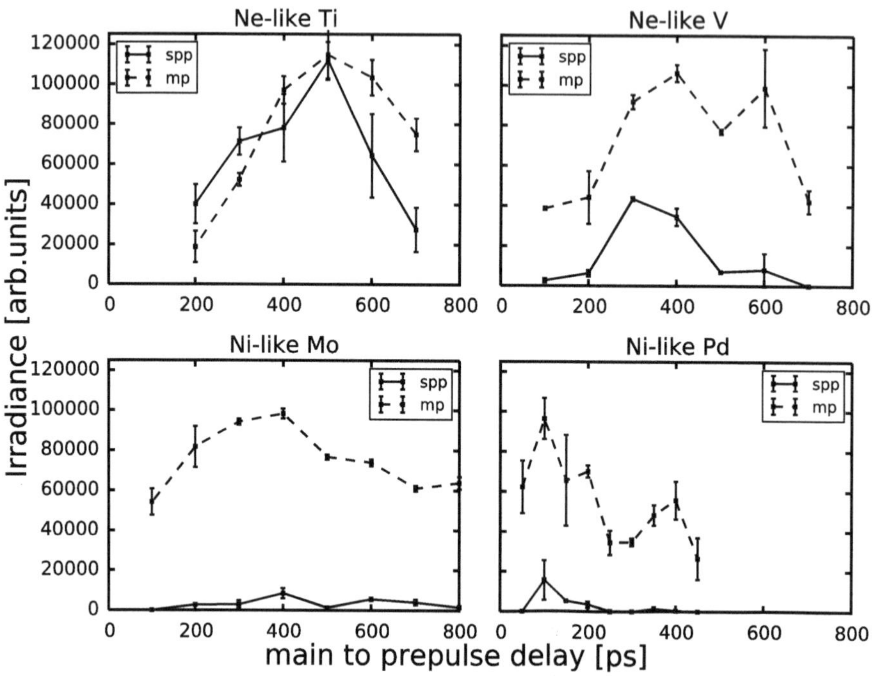

Fig. 4 Dependence of the self-photopumped (*spp*) and the monopole pumped (*mp*) on the main to second prepulse delay

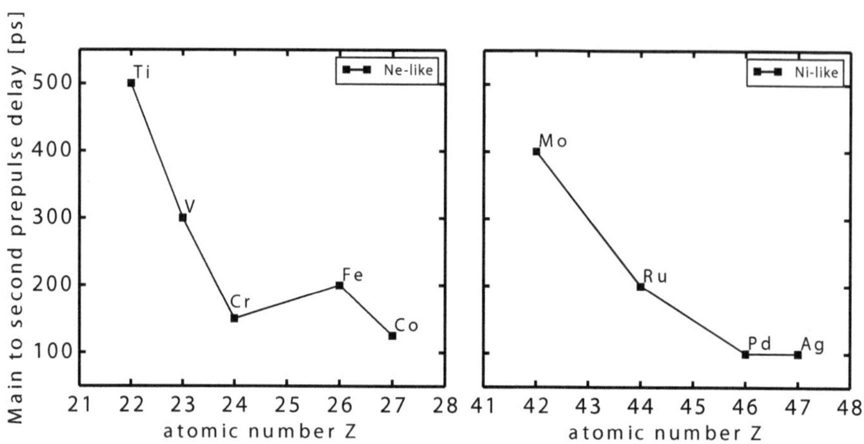

Fig. 5 Dependence of the optimum main to second prepulse delay on the atomic number Z

4 Conclusion and Outlook

The experiments on the self-photopumped transition demonstrate that the list of elements that lase on the self-photopumped transition can be extended much further than originally known. Namely, this includes lasing on the Ne-like 3p $^1S_0 \rightarrow$ 3s $^1P_1$ laser line for the first time for V, Cr, Fe and Co as well as on the Ni-like 4d $^1S_0 \rightarrow$ 4p $^1P_1$ line for Ru, Pd, and Ag. The desired characteristic of the self-photopumped lasing line to be the stronger has been only encountered so far in some shots in Ne-like Ti (see Fig. 3). Additionally, we found a strong dependence on the main to second prepulse delay. The optimum delay shifts towards smaller delays with increasing atomic number Z.

In order to get more insight into the self-photopumped lasing mechanism we plan to measure existing lasing elements more in detail and moreover try to expand the number of lasing elements even further. Besides that, time-resolved measurements comparing the collisional and the self-photopumped lasing lines are under way to provide more information on the time domain in which this lasing mechanism occurs.

Acknowledgements This work was supported in part by the Swiss National Science Foundation.

References

1. Nilsen, J., Dunn, J., Osterheld, A.L., Li, Y.: Lasing on the self-photopumped nickel-like $4f^1P_1$->$4d^1P_1$ x-ray transition. Phys. Rev. A. **4** R2677–2680 (1999)
2. Nilsen, J., Li, Y., Dunn, J.: Modeling picosecond-laser-driven neonlike titanium x-ray laser experiments. J. Opt. Soc. Am. **6** 1084–1092 (2000)
3. Balmer, J.E., Staub, F., Jia, F.: X-Ray Lasers 2012. Springer Proceedings in Physics, vol. 147, Chap. 5. Paris (2014)

Pump-Probe Experiment for Temporal Profile Measurement of Plasma X-Ray Laser

S. Namba, N. Hasegawa, M. Kishimoto, M. Nishikino and T. Kawachi

Abstract Temporal behavior of a soft x-ray laser pulse is measured by means of a pump-probe spectroscopy (cross correlation). In this scheme, first the innershell electron 4*d* of Xe atom is photoionized with 13.9 nm x-ray pulse (x-ray pump). Simultaneously, an ir probe laser pulse is irradiated to dress the electrons. By varying the time delay between both pulses, the sideband spectra associated with ir photon absorption/emission processes appear at a separation of the ir photon energy from the photo- and Auger main peaks. By analyzing the sideband intensity, we can evaluate the pulse width of the x-ray laser to be ~6.2 ps, which is in excellent agreement with the value measured by an x-ray streak camera

1 Introduction

Laser produced plasmas (LPPs) are expected to be a compact and bright light source in vacuum ultraviolet (vuv) and soft x-ray regime [1, 2]. Even hard x-ray up to 20 keV (gold Kα) caused by energetic electrons was obtained by an irradiation of ultra-intense laser pulses on the bulk target [3]. Various attractive research in physics [4, 5], chemistry [6] and biomedical science [7, 8] utilizing the alternative x-ray sources has been carried out. For an application of the LPP x-rays to radiation biology, the short pulse copper Kα (~ 8 keV) was exposed onto a biological cell and it was revealed that the LPP x-rays could exert similar influence on the nucleus of the cell with those obtained by conventional x-ray sources [9, 10].

By irradiating short pulse x-rays onto xenon (Xe) atom, we also intend to demonstrate an extreme state, where multiple vacancies in the innershell are created. The inner hole immediately decays into stable states in femtosecond time scale via the Auger process [11]. Therefore, the intense x-ray that has the photon energy sufficient to ionize the innershell electron and with very short pulse duration is essential to produce such multiply inner-ionized atoms. Although the information on the pulse width of x-rays has to be accurately examined, a conventional streak camera is not

S. Namba (✉)
Graduate School of Engineering, Hiroshima University, Higashihiroshima, Japan
e-mail: namba@hiroshima-u.ac.jp

N. Hasegawa · M. Kishimoto · M. Nishikino · T. Kawachi
Quantum Beam Science Center, Japan Atomic Energy Agency, Kyoto, Japan

© Springer International Publishing Switzerland 2016
J. Rocca et al. (eds.), *X-Ray Lasers 2014*, Springer Proceedings in Physics 169,
DOI 10.1007/978-3-319-19521-6_12

available to measure the x-ray pulse duration, because the time resolution is limited to around 1 ps.

In order to investigate the temporal characteristics of the x-rays in the subpicosecond domain, Woerkom et al. [12] and Schins et al. [13] adopted cross correlation techniques. In this scheme, single photon ionization by an x-ray photon (pump) and absorption or emission of optical photons of ir probe pulse occur simultaneously. The sideband structures with respect to the photo- and Auger peaks in the electron spectrum, separated by the ir photon energy, appear due to a two-color above threshold ionization (ATI) [14]. By analyzing the correlation of amplitude of the sidebands with the time delay between the ionizing pump and dressing probe pulses, we can evaluate the temporal shape of the x-ray pulse. By using similar technique, even xuv attosecond pulses can be characterized. Itatani et al. proposed the novel method to determine the pulse duration in the attosecond time domain, for which the energy and angular distribution of photoelectron in the presence of a strong femtosecond optical laser field were measured [15]. Here, the key is the phase between xuv pump for ionization and the carrier wave of the ir laser pulse for dressing. The single attosecond laser pulse has also been demonstrated for the first time by Hentschel et al. [16].

For pump-probe laser spectroscopy, an electron spectrometer capable of measuring energy spectrum by a single-shot base is indispensable. Therefore, we developed a magnetic bottle time-of-flight (TOF) electron spectrometer [17]. The magnetic field formed by a permanent magnet and solenoid coil was calculated by a finite element method. Accordingly, the dimension of the apparatus and magnetic field strength were chosen so as to meet the requirements. We examined the energy resolution of the spectrometer by measuring the electrons emitted from Xe atom irradiated with a laser-driven plasma soft x-ray laser pulse. Finally, the pulse width of the x-ray laser was measured by a x-ray pump and ir probe spectroscopy.

2 Magnetic Bottle Spectrometer

Figure 1a shows a schematic diagram of the magnetic bottle TOF electron spectrometer. The magnetic mirror field is generated by a taper-shaped permanent magnet and solenoid coil. The field strength of the permanent magnet (neodymium magnet, diameter: 20 mm, length: 30 mm and taper angle: 110°) is about 0.25 T at the taper tip. The weak and uniform field (~ 10 mT) is applied by the coil that is wound on a 160 mmφ stainless steel tube. The wire diameter is 2.4 mm and the coil length is 600 mm. The current applied in the coil is 2 A. The magnetic field outside the vacuum tube is shielded by a nanocrystalline soft magnetic material. A micro channel plate (MCP) having an effective area of 42 mmφ is located at 620 mm from the permanent magnet. As shown in Fig. 1a, the electrons are generated at 20 mm downstream from the permanent magnet, where the Xe atom is irradiated with the soft x-ray laser pulse. Subsequently, the electrons can arrive at the detector due to the guidance of the magnetic mirror field. In order to calculate the magnetic field, we

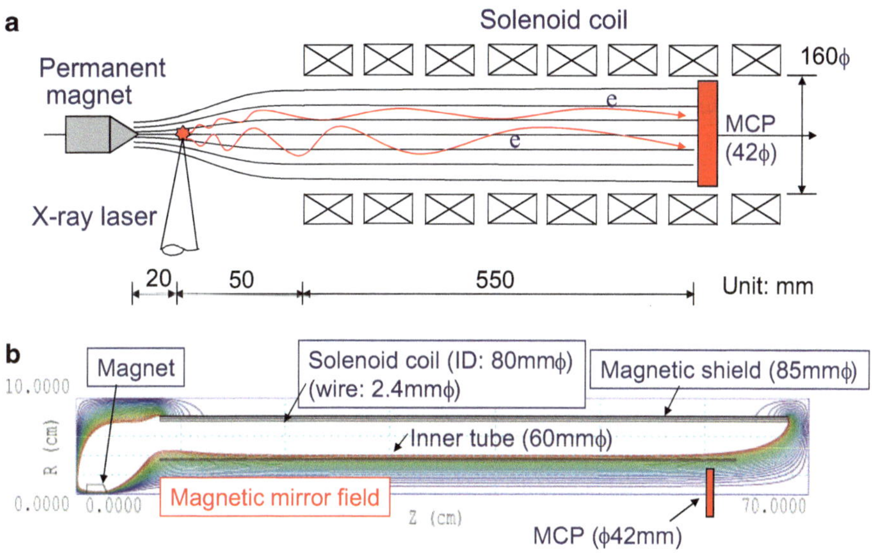

Fig. 1 Schematic diagram of the magnetic bottle TOF electron spectrometer (**a**), and the contour plot of the stream function of the magnetic field (rA_θ) calculate by using the finite element method (**b**).

perform a numerical calculation using a finite element method. Figure 1b shows the contour plot of the field strength, rA_θ,*stream function*) generated by the magnet and the solenoid coil, where r is the radius and A_θ is the vector potential.

The characterization of the spectrometer is examined by measuring the energy spectra of Xe atom subjected to the laser driven plasma soft x-ray laser pulse at the Japan Atomic Energy Agency (JAEA) [18–21]. The laser wavelength and photon energy are 13.9 nm (bandwidth: $\Delta\lambda/\lambda \leq 10^{-4}$) and 89.19 eV, respectively. The laser spot size and the highest intensity focused by a concave Mo/Si multilayer mirror ($R = 1000$ mm) are ~40 μmφ and 5 × $10^{8\mathrm{W}}/\mathrm{cm}^2$, respectively. As for the Xe gas target, the effusive atomic beam is created by expanding low pressure gas through an electromagnetic valve. The laser photon energy is high enough to photoionize Xe 4d core electrons (threshold: 69.54 and 67.55 eV for the spin-orbit states $j = 3/2$ and 5/2, respectively [22]). The cross section at 90 eV is as large as ~22 Mb due to a *giant resonance*, while for outer ones it is only 1.5 Mb [23–25].

Figure 2 shows the electron energy distribution function $f(E)/E^{3/2}$. The sensitivity of the spectrometer is not calibrated, thus low energy electrons being less sensitive. The photoelectron peaks attributed to $4d^{-1}$ and $5p^{-1}$ states and the Auger lines associated with the inner vacancy can readily be assigned. The spectral width of 4d $j = 5/2$ line ($E = 21.64$ eV) is around 1.1 eV, whereas for 8.26-eV Auger peak it is significantly improved to $\Delta E = 0.45$ eV due to longer flight time for slower electrons.

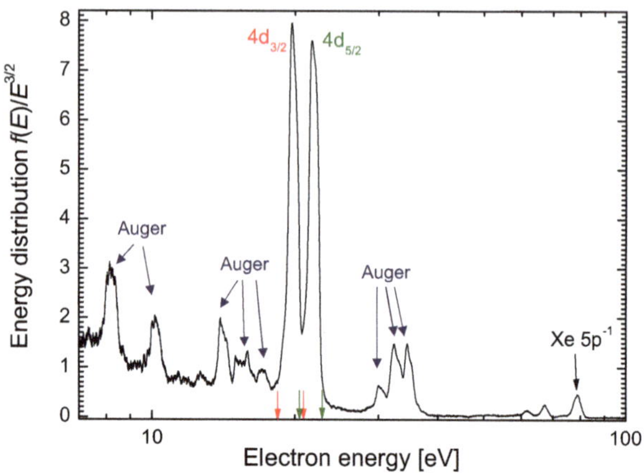

Fig. 2 Energy distribution function. Photo- and Auger peaks following the 4*d* ionization are observed. The sideband positions for the photo lines of $j = 3/2$ (19.65 eV) are 18.47 ($q = -1$) and 20.83 eV ($q = +1$), while for $j = 5/2$ (21.64 eV), they are 20.46 ($q = -1$) and 22.82 eV ($q = +1$), which are indicated by the *arrows*

3 Pump-probe Spectroscopy for Pulse Width Measurement

By using the electron spectrometer, we attempt to evaluate the pulse width of the x-ray laser. In this study, two Nd:glass laser pulses (1053 nm, $h\nu = 1.18$ eV) having a common oscillator and regenerative amplifier are used [21]. The one beam generates laser-driven x-ray laser pulse, whereas the other synchronizing with the x-ray laser (1053 nm, ~4 ps) is irradiated as an ir probe laser to dress the electrons. Consequently, the sidebands with the separation of 1.18 eV appear at both sides of the main peak, as described the above. The prominent sideband can be observed for the transition involving 4*d* photo electrons. The nearest dressed spectra for 4*d* $j = 5/2$ and 3/2 are located at 20.46 eV ($q = -1$), 22.82 eV ($q = +1$) and 18.47 eV ($q = -1$), 20.83 eV ($q = +1$), respectively, where the q is the number of the emitted ($q = -1$) and absorbed ($q = +1$) photons. The sideband positions corresponding to four lines are also shown in Fig. 2 by the arrows. However, the low energy resolution makes it difficult to decompose the sidebands of 20.46 and 20.83 eV. Thus, we adopt the $q = +2$ spectrum for $j = 5/2$.

Figure 3a shows the energy spectra for various time delays. For time delay of 0 ps, the prominent peak attributed to two ir photon ($2h\nu$) absorption can be observed. The temporal variation of the sideband intensity is plotted in Fig. 3b. Assuming that the instrumental width is given by the temporal profile of the probe pulse, the x-ray pulse width is determined to be 6.2 ps (FWHM), which is in good agreement with the value measured by an x-ray streak camera.

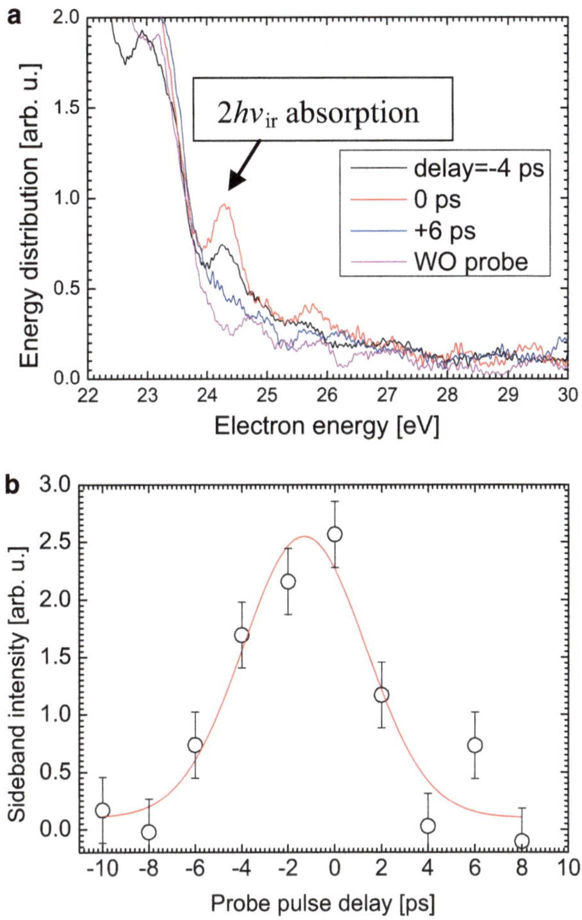

Fig. 3 Electron energy spectra (**a**) and sideband intensity (**b**) for various time delays. The red curve in **b** corresponds to the pulse shape of 6.2 ps

4 Summary

In order to perform the pump-probe experiment for x-ray pulse width measurement, we developed the magnetic bottle TOF electron spectrometer. The energy resolution was examined by irradiation of x-ray laser pulse onto Xe gas target. The energy resolution ΔE for 8.26-eV Auger electron following the photoionization of 4*d* core electron was ~ 0.45 eV.

The temporal profile of plasma x-ray laser was determined by the cross correlation, in which Xe atom was photoionized by 13.9 nm x-ray laser and the electrons generated were subsequently dressed by ir probe pulse, resulting in the appearance of sideband spectra. It was revealed that the pulse width of the x-ray laser was ~ 6.2 ps, which was reasonable compared with the value obtained by x-ray streak camera.

Acknowledgements This work was supported by a Grant-in-Aid for Scientific Research B (No. 22340174) from the Japanese Society for the Promotion of Science and the Murata Science Foundation. We would like to thank the JAEA x-ray laser research group for laser operations. This work was performed under the shared use program of JAEA Facilities.

References

1. Carroll, P., et al.: Table-top EUV continuum light source. IEEE J. Quantum Electron. **QE19**, 1807 (1983)
2. Bridges, J., Cromer, C., McIlrath, T.: Investigations of a Laser Produced Plasma VUV Light Source. Appl. Opt. **25**, 2208 (1986)
3. Zhang, Z., et al.: Efficient multi-keV x-ray generation from a high-Z target irradiated with a clean ultra-short laser pulse. Opt. Exp. **19**, 4560 (2011)
4. Rischel, C., et al.: Femtosecond time-resolved X-ray diffraction from laser-heated organic films. Nature. **390**, 490 (1997)
5. Siders, C.W., et al.: Detection of Nonthermal Melting by Ultrafast X-ray Diffraction. Science. **286**, 1340 (1999)
6. Lee, T., et al.: Ultrafast laboratory-based x-ray sources and their applications in chemical research. Proc. SPIE. **4978**, 77 (2003)
7. Frankel, R.: Nanosecond X-ray diffraction from biological samples with a laser-produced plasma source. Science. **204**, 622 (1979)
8. Laperle, C., et al.: Propagation based differential phase contrast imaging and tomography of murine tissue with a laser plasma x-ray source. Appl. Phys. Lett. **91**, 173901 (2007)
9. Nishikino, M., et al.: Application of laser produced plasma Kα x-ray probe in radiation biology. Rev. Sci. Instrum. **81**, 026107 (2011)
10. Sato, K., et al.: y-H2AX and phosphorylated ATM focus formation in cancer cells after laser plasma X irradiation. Radiat. Res. **174**, 436 (2010)
11. Jurvanssu, M., Kivimäki, A., Aksela, S.: Inherent lifetime widths of Ar 2p-1, Kr 3d-1, Xe 3d -1, and Xe 4d-1 states. Phy. Rev. A. **64**, 012502 (2001)
12. van Woerkom, L.D., et al.: High-Energy Density Physics with Subpicosecond Laser Pulses, Vol. 17 of 1989 SAS Technical Digest Series. Optical Society of America, Washington, DC (1989), postdeadline paper PDP3
13. Schins, J.M., et al.: Observation of Laser-Assisted Auger Decay in Argon. Phys. Rev. Lett. **73**, 2180 (1994)
14. Cionga, A., Florescu, V.: Target dressing effects in laser-assisted x-ray photoionization. Phys. Rev. A. **47**, 1830 (1993)
15. Itatani, J., et al.: Attosecond Streak Camera. Phys. Rev. Lett. **88**, 173903 (2002)
16. Hentschel, M., et al.: Attosecond metrology. Nature. **414**, 509 (2001)
17. Kruit, P., Read, F.: Magnetic field paralleliser for 2n electron-spectrometer and electron-image magnifier. J. Phys. E. **16**, 313 (1983)
18. Kawachi, T., et al.: Gain saturation of nickel-like silver and tin x-ray lasers by use of a tabletop pumping laser system. Phys. Rev. A. **66**, 033815 (2002)
19. Ochi, Y., et al.: Measurement of temporal durations of transient collisional excitation X-ray lasers. Appl. Phys. **B 78**, 961 (2004)
20. Nishikino, M., et al.: Demonstration of a soft-x-ray laser at 13.9 nm with full spatial coherence. Phys. Rev. A. **68**, 061802 (R) (2003)
21. Ochi, Y., et al.: Development of a chirped pulse amplification laser with zigzag slab Nd:glass amplifiers dedicated to x-ray laser research. Appl. Opt. **46**, 1500–1506 (2007)
22. Lablanquie, P., et al.: Dynamics and Post-Collision Interaction Effects in Two Electron Decay from the Xenon 4d Hole. Phys. Rev. Lett. **87**, 053001 (2001)

23. Holland, D.M., et al.: Multiple photoionisation in the rare gases from threshold to 280 eV. J. Phys. B: At. Mol. Opt. Phys. **12**, 2465 (1979)
24. Kivimäki, A., et al.: 'Intensities of the xenon N4,5OO Auger electron spectrum. J. Electron. Spectrosc. Relat. Phenom. **43**, 101–103, 43–47 (1999)
25. Luhmann, T., et al.: Final ion-charge resolving electron spectroscopy for the investigation of atomic photoionization processes: Xe in the region of the 4d ef resonance. Phys. Rev. A. **57**, 282 (1998)

Capillary Discharge X-Ray Lasers: The Quest for Sub-10 nm Lasers

V. N. Shlyaptsev, G. Avaria, M. Grisham, Jing Li, F. Tomasel, M. Busquet and J. J. Rocca

Abstract A new ultra-fast high power density micro-capillary discharge design has been demonstrated that is capable of achieving record temperatures and ionization degrees utilizing ~5 times smaller discharge currents than previously used. Temperatures in the range 0.3–0.7 keV were obtained with < 40 kA electric currents. Experiments conducted in Xe discharges show emission from Cr-like Xe XXXI ions, while discharges in Ne show He-like Al XII line emission. The major hydrodynamic, ionization, and spectral characteristics for these experiments were numerically reproduced. The studies also revealed new unusual spectral effects of such plasmas. This new development opens a path to search for efficient discharge driven soft x-ray lasers in sub-10 nm spectral ranges.

1 Introduction

After the demonstration of the capillary discharge soft x-ray laser in Ar IX in 1994 [1], further research concentrated in several directions, one of which took an evolutionary path by developing a more powerful pulse generator that increased the discharge current from the 40 kA of the initial experiments to 200 kA. In that design the current pulse duration was also shortened 2–3 times, reaching a 10–90 % rise-time of ~11 ns [2]. In experiments with 3–4 mm diameter capillaries these current pulses created ~4 times higher temperatures and densities than previously obtained with the 40 kA discharges. Line emission from higher Z ions, such as Be-like Ar XV and Ni-like Cd XXI, was routinely observed. In the latter case the 4d-4p lasing lines from Ni-like ions were observed in 13 nm wavelength region, but no substantial gain. Two obstacles for x-ray lasing in this discharge scheme are the difficulty to create homogeneous volume of metal vapors along the elongated capillary, and the required very high degree of ionization.

M. Busquet (✉)
APTEC Inc, Muskegon, USA,
e-mail: busquet@artepinc.com

V. N. Shlyaptsev · G. Avaria · M. Grisham · J. Li · F. Tomasel · J. J. Rocca
Colorado State University, Fort Collins, CO, USA

V. N. Shlyaptsev
e-mail: slava.shlyaptsev@colostate.edu

© Springer International Publishing Switzerland 2016
J. Rocca et al. (eds.), *X-Ray Lasers 2014*, Springer Proceedings in Physics 169,
DOI 10.1007/978-3-319-19521-6_13

Another approach consisted in investigations of microcapillaries. Although a number of microcapillary discharges has been previously studied [3–11], these discharges had either much longer excitation pulses leading to large wall ablation and cold plasmas, e.g. electron temperatures $T_e \sim 10$ eV for a 500 kA current pulse of 300 ns duration in a 20-μm-diameter capillary [5], or nanosecond duration but power density depositions orders of magnitude lower than those of interest here, resulting in $T_e \sim 30$ eV [10, 11]. Within our group we initially studied microcapillary discharges as plasmas for the development of soft x-ray laser interferometry and shadowgraphy [12]. These microcapillaries had diameters around 300 microns and were driven by small amplitude 1–2 kA current pulses ~ 150 ns in duration. The plasmas created by ablation of the capillary walls had high ionization despite both the small currents and proximity of the wall, and the fact that the kinetic energy conversion heating mechanism common for z-pinches compression played no significant role. Instead, Joule dissipation was the primary heating mechanism for these microcapillary plasmas. The results presented herein show for the first time that microcapillary discharges excited by current pulses of moderate amplitude can generate high-aspect-ratio plasma columns (> 300:1 length-to-diameter) with extreme degrees of ionization and temperatures of several 100 eV. In the next sections we describe specific differences micro-capillary and conventional capillary discharges of larger diameter, and describe experimental achievements resulting in generation of hot high-Z plasma.

2 Major Properties of Microcapillary Approach

a) Prepulse

Besides a significantly reduced diameter and ultra-fast currents, the high power density microcapillary approach described herein has an additional novel feature: a tailored prepulse. Over the history of x-ray lasers (based mostly in laser produced plasmas), the prepulse or multiple pulses were used for two purposes: to create smooth plasma gradients thereby mitigating refraction, and to create plasmas containing ions of the lasing species of interest in transient x-ray laser schemes.

In the current microcapillary approach the purpose of the prepulse somewhat resembles those used in transient schemes, but the goal here is to assist in achieving a high Z. Our numerical optimization studies showed that the best prepulse in microcapillary discharges has to have an amplitude of a few kA to assist the main pulse in ionizing the plasma to very high degrees without disturbing plasma homogeneity. Higher prepulse currents would substantially compress the plasma, and the column would become turbulent and inhomogeneous. Also, a higher initial pressure produced by a higher amplitude prepulse would also make difficult reaching higher compressions by the subsequent main pulse.

In our approach after the prepulse ends, the hot plasma expands and the temperature drops preserving ionization by the effect of "ionization freezing" until the

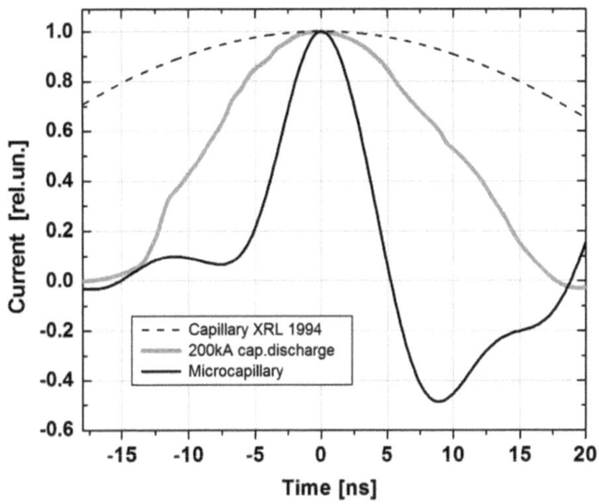

Fig. 1 Current pulse shapes of the 46.9 nm capillary discharge X-ray laser, a 200 kA capillary discharge, and the present microcapillary discharge

main pulse arrives (because recombination rates do not increase exponentially with decreasing temperature as opposed to the ionization rates). The right prepulse helps to obtain larger final temperatures and pressures and, along with a much higher fraction of hot electrons during this time, ionizes the plasma to higher Z. Modeling shows that during the prepulse the electron temperature reaches values sufficient to ionize heavy elements like Xe to Z $\sim$ 12–15.

b) Short current pulse

Figure 1 illustrates the trend in current pulse durations used to drive high power density capillary discharges in our laboratory. The figure shows how dramatically the pulse duration FWHM was shortened from $\sim$50 ns in 1994 [1], to $\sim$20 ns in 2002 [2], and to $\sim$7 ns in 2010. The leading front of the current pulse in the microcapillary discharges discussed in these proceedings has a 10–90 % risetime of 3–4 ns. The shorter pulse also decreases the thermal and radiation damage to the wall. We have to note that, despite the lower peak currents, the current density in this approach reaches values of the order of GigaAmpere/cm^2, much higher than in previous capillary discharges. This increase in current density favors Joule heating as heating mechanism during main pulse in microcapillary designs.

c) Design approaches: high power density microcapillary vs conventional capillary

The key difference between this high power density microcapillary design approach and previous setups is that in ultrafast capillary discharges, contrary to the case of conventional capillary discharges, Ohmic (Joule) heating becomes the dominant heating mechanism. It is known that magnetic hydrodynamic equations can be written in many different equivalent forms. The energy equation in cylindrical geometry may be for example written as [13]:

$$\frac{\partial}{\partial t}\left(\varepsilon+\frac{Mv^2}{2}+\frac{H^2}{8\pi N}\right)=-\frac{\partial}{\partial s}\left[\left(p+\frac{H^2}{8\pi}\right)rv\right]+\frac{\partial}{\partial s}\left(r\frac{EH}{4\pi}\right) \tag{1}$$

which expresses that the total energy of the system—internal ε [including thermal (T) and ionization (Π) energies, $\varepsilon = T + \Pi$], kinetic $Mv^2/2$, and magnetic $H^2/8\pi N$ per unit volume, are changing under the influence of the thermal and magnetic pressure work and the flow of electromagnetic energy $EH/4\pi$ (Pointing vector). This equation is written in 1D Lagrangian coordinates assuming only axial and tangential components of the electric E and the magnetic H field vectors exist. In this equation $s = Nrdr$ is the mass coordinate, N is the ion particle density, M is the ion mass, v is the speed, p is the thermal pressure, and r is the radius. The driving force of all changes in the system is of course the Lorentz force $f_L = -\partial/\partial s(H\hat{}2/8\pi)$. Note that the Ohmic heating is not explicitly present here, but it is implicitly accounted for via Maxwell's electromagnetic equations. Similarly the energy equation for just the internal and kinetic energy can be written as:

$$\frac{\partial}{\partial t}\left(\varepsilon+\frac{Mv^2}{2}\right)=-\frac{\partial}{\partial s}\left[prv\right]+f_L v+J \tag{2}$$

The absent magnetic energy, if needed, can be taken from Maxwell's equations. Here the influence of the Joule heating J and Lorentz force f_L is shown explicitly. However, the energy equation which is actually solved numerically by RADEX can be further simplified using the momentum equation and is written just for the electron thermal energy:

$$\frac{\partial \varepsilon}{\partial t}=-p\frac{\partial}{\partial s}[rv]+J \tag{3}$$

The transformation of electric energy into thermal and ionization energy defines how this method of plasma heating performs. We computed the space-time histories of all the forms of energy, to study how they transform into each other and what effects play the major roles. The energy equation, as part of an entire set of magneto-hydrodynamics equations including radiation, was solved over time and space by the code Radex using the conditions described in the caption of Fig. 2. This figure summarizes the spatio-temporal maps of the major computed quantities of these equations.

The first term on the right hand side of the Eq. (3) represents the pressure work per unit volume per second, which we will simply call pressure heating. It is expressed below in units of W/cm^3. In Fig. 2 we compare it with the second term, the Joule heating term, for two different types of capillary discharges: a micro-capillary discharge excited by an ultrafast current pulse, and larger diameter capillary discharge excited by a longer pulse current pulse of much higher amplitude. The relative contributions of these heating terms in the case of an ultrafast (4 ns rise time) microcapillary discharge are illustrated in Figs. 2a, b (left) in comparison to those corresponding to a large diameter capillary excited by a much larger current

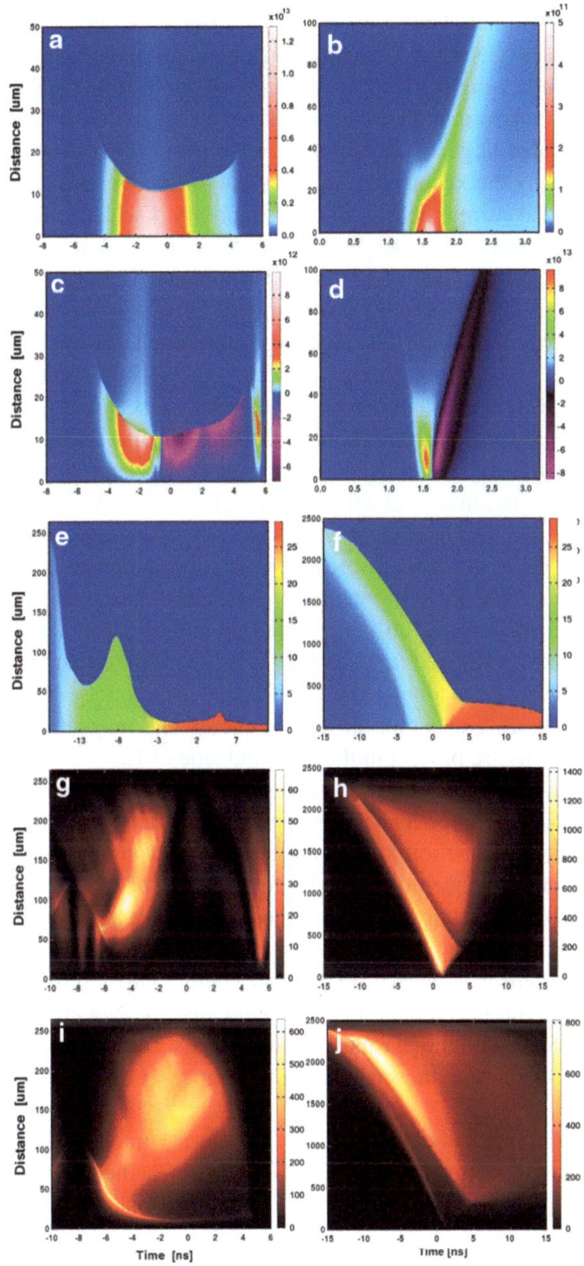

Fig. 2 Hydrodynamic simulation of microcapillary discharge versus conventional discharge driven by 400 kA current capillary Comparison of the magnitude of the Joule (*a, b*) and PV (*c, d*) heating terms in an ultrafast micro-capillary discharge (*left a, c*) compared to a conventional capillary discharge (*right b, d*). The units are in W/cm^3. The (*e, f*) plots correspond to ion charge. The (*g, h*) and (*i, j*) plots are kinetic energy and magnetic energy per electron in eV, respectively. The Xe microcapillary discharge parameters correspond to those used in the experiments (capillary diameter 520 μm, Xe pressure 0.5 Torr, peak current 39 kA, 7.5 ns FWHM). For the larger size capillary discharge the parameter values are: capillary diameter 5 mm, Xe pressure 0.5 Torr, current pulse of 400 kA peak amplitude and 21 ns FWHM

pulse [Fig. 2b, d (right)]. The sharp boundary clearly visible on all frames is the interface between the gas fill and the wall ablated material. In the microcapillary discharge, the Ohmic heating is found to be the dominant heating mechanism (Fig. 2a) followed by the transformation of plasma kinetic energy into heat (Fig. 2c). Not only is its peak value larger than that of pressure heating, but the integral over the entire space filled by the xenon (not shown) is larger too. As mentioned above, this happens because the current density of microcapillary discharges reach tremendous values (~ 0.3 GigaAmpere/cm^2), which are up to an order of magnitude larger than in the higher peak current and larger diameter capillary discharge.

The results are contrasted with those in Fig. 2b, d for a conventional 5-mm-diameter capillary with similar parameters (pulse width 21 ns FWHM, pressure 0.5 Torr) to those that we used in the experiments of ref. [2], but with twice the current, 400 kA. In the case of the larger diameter capillary discharge, the pressure heating term is two orders of magnitude larger than the Joule heating term. We used a higher peak current for the conventional capillary discharge here because larger diameter capillaries requires peak currents as high as 400 kA to achieve degrees of ionization similar to those reached in microcapillary discharges, which can be seen from Fig. 2e, c. These quantities were obtained solving the set of detailed atomic kinetic equations which run self-consistently with the hydrodynamics. We have to note that previous attempts of utilizing scaled-up conventional capillary designs to higher electric current allowed to reach only $Z \sim 20$ (Ni-like Cadmium) despite the high peak currents of up to almost 200 kA [2].

For completeness, we also plot the kinetic energy and magnetic energy r-t histories. These two terms are present in the left hand side of Eqs. (1) and (2). Typically, in many types of z-pinches the kinetic energy imparted to the radially accelerated plasma by the Lorentz force and the pressure work transforms at stagnation into the thermal energy. The r-t ion kinetic energy distribution on Fig. 2g, h illustrates how this process takes place in our two different types of capillary discharges (the kinetic energy of the ions is scaled per one electron to have a closer comparison with the plasma temperature). It can be seen that in the microcapillary discharge the ion kinetic energy per electron is very small everywhere compared to the peak thermal energy, while in the large capillary it is the dominant term. Not all the kinetic energy transforms into heat when the plasma reaches the axis, some is reflected, and also it takes some time for the ion thermal energy at stagnation to transform into electron thermal energy.

In contrast, the Ohmic heating directly transforms into the thermal energy of the electrons (some energy may go into hot electrons, the fraction of which we calculated can range from 0.2 to 1 % for different initial conditions, at prepulse time—up to 5 %).

The magnetic field energy density r-t plots, also per one electron, are shown on Fig. 2i, j. They are comparable in magnitude, but it is the gradient of the magnetic energy density that defines the Lorentz force, which is obviously doing larger work in accelerating plasma in the larger diameter capillary then in the smaller one, as it follows from Fig. 2g, h.

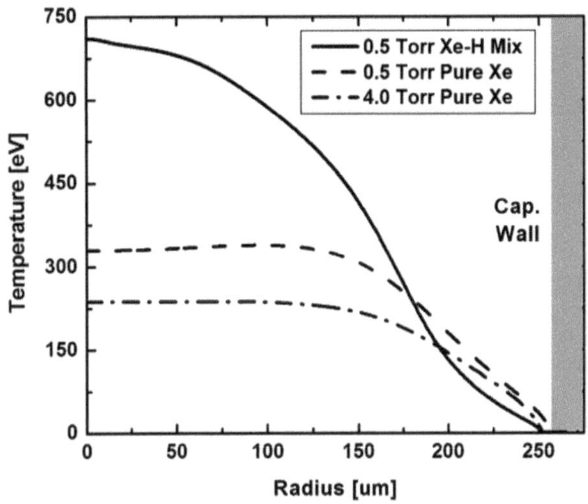

Fig. 3 Radex calculated temperature profile for three different gas fills of a xenon discharge on a 520-μm diameter alumina microcapillary. Driving current for these cases stayed in 33–39.6 kA range

d) Higher Te and Z at substantially lower currents

Several gases and mixtures were experimentally investigated, including hydrogen, helium, nitrogen, neon, silane and xenon. Spectra in the 1–3 keV range indicated that high temperatures are achieved despite small diameters and proximity of cold walls (see Fig. 3). The higher temperatures allowed to achieve record high degrees of ionization when high-Z materials were used as a fills.

For example Cr-like Xe^{30+} lines were observed in Xe-H mixtures (see Fig. 4 below). Transitions from ions up to Fe-like Xe^{28+} were seen in pure Xe discharges. The electron density was computed to reach $(2-5)\ x10^{19}\ cm^{-3}$ depending on the initial conditions. In all cases the peak current was in the 33–39.6 kA range. More details will be presented in [14].

The ability to reach higher Z is specifically beneficial for capillary discharge x-ray lasers because of properties of quasi steady-state (QSS) gain in Ni-like ions. Since the plasma lifetime in capillary discharges is relatively long (~ 1–2 ns) the transient interatomic processes which form transient gain succeed to relax ion level populations resulting in QSS conditions. But for lower elements ($Z < 45$) QSS gain becomes unpractically small, and for $Z < 40$ it disappears completely. Note that previous high current (200 kA) discharges in 3–4 mm capillaries were demonstrated in elements as heavy as Cd ($A = 48$)—if Ni-like ions could be produced in heavier elements like Xe, the gain in these large capillaries would be considerably higher [15]. With microcapillaries though we can rely on sizable QSS gain in such high Z elements as xenon, enough for lasing on 4–4 transitions of Xe XXVII ions below 10 nm. Using gas fills instead of metal vapors is also experimentally easier

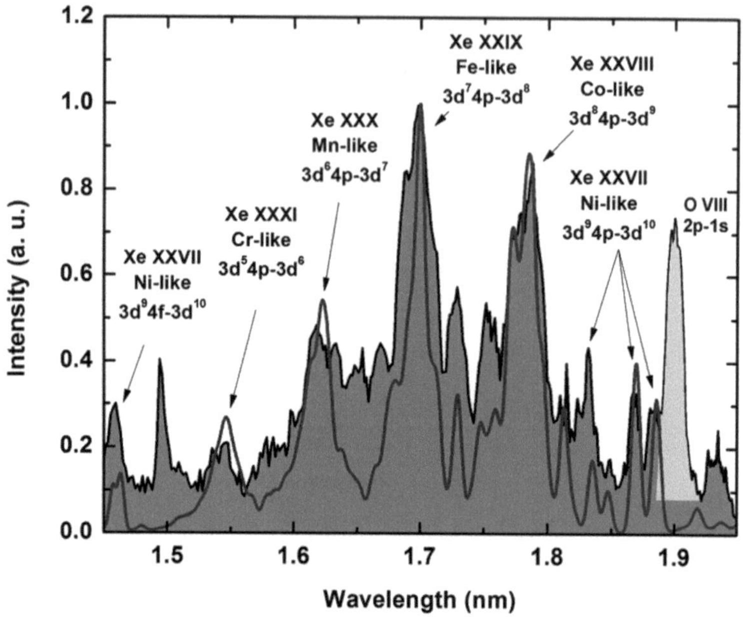

Fig. 4 Simulated spectra (*line*) versus experiment (*solid area*) for discharge plasmas in a 520-μm-diameter capillary for the 80:1 H2: Xe mixture at 0.5 Torr, peak discharge current 39.6 kA, 1.2 ns before peak of current pulse. Line emission from Cr-like Xe^{30+} is observed. The line at 1.9 nm is assigned to a blend of the O^{7+}2p-1 s and Xe^{25+} and Xe^{27+} lines. The simulations did not include O impurities, which in the experiment come from the wall ablated material

and allows to achieve better initial homogeneity along the volume. The electric current for lasing on Xe lies in 50–55 kA range, which is moderately larger than the 40 kA currently achieved.

3 Line Spectra Anomalies

We have recorded new spectroscopic phenomena, consisting in anomalies in the intensity of He-like intercombination line with respect to the resonance line for several elements in 1.5–2.5 keV spectral range. For aluminum spectra the intensity ratio of the intercombination $1s^{2\,1}S_0$-$1s2p\,^3P_1$ line respect to resonance $1s^{2\,1}S_0$-$1s2p\,^1P_1$ line reaches an order of magnitude. One previous experiment has reported similar line ratios in a laser produced plasma [16], but here the resemblance ends because it was observed in the tail of a steeply falling temperature, when the intensity of the resonance line decreases and is far from its peak value, while our observations were made not in the afterglow but in the spectra integrally over the whole pulse history, which includes the time of maximum emission when both current and temperature peak. Figure 5 shows an example of such spectra. At first glance, common sense

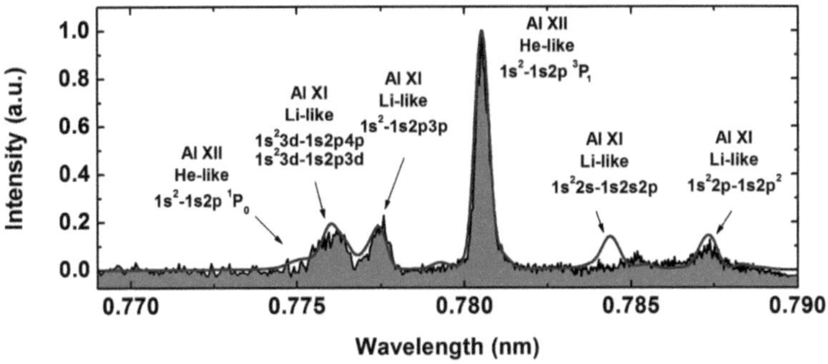

Fig. 5 Simulated (*line*) and measured (*filled background*) on-axis spectra of a neon discharge on a 520-μm-diameter alumina microcapillary showing the dominance of the $1s^{2\,1}S_0$-$1s2p\,^3P_1$ intercombination line of He-like Al over the $1s^{2\,1}S_0$-$1s2p\,^1P_1$ resonance line (barely visible at 0.775 nm). The peak current 36.0 kA and the neon pressure 0.5 Torr

seems to tell us that this effect can be explained just by the larger optical depth of resonance line. However, detailed analysis shows that this explanation is incorrect. Numerical simulations show that the cause of this effect is the very large aspect ratio geometry of the plasma, in which different optical depths in the transverse direction (optically thin for both lines) and longitudinal directions (optically thick in one for both lines) create a situation where the population of the $1s2p^3P_1$ level is favored over the $1s2p^1P_1$ upper level by up to two orders of magnitude, defining the line intensities in the axial direction and causing the intercombination line to dominate the spectrum.

The explanation for this effect then lies in the elongated geometry of our plasma, in which different optical depths in transverse and longitudinal directions create a situation where the atomic kinetics makes triplet level population significantly larger than the population of the singlet.

4 Summary

In summary, we experimentally realized and numerically modelled a new and improved method of plasma heating using a fast discharge. Record high temperatures approaching 1 keV and ion charges $Z \sim 30$ were observed and reproduced in the modeling. Unusually high intercombination to resonance line ratios in He-like ions obtained in the experiment were explained, and parameter values for future sub-10 nm x-ray lasers were computed. Calculations suggest that lasing on Ni-like ions is achievable with the same current pulse durations used in the experiment described here and an increase of 25–30 % in driving current.

Acknowledgements Work supported by NSF Award PHY-1004295. A.P acknowledges the support of DFG-funded project TR18. Authors also express their gratitude to A.V. Vinogradov, A. Faenov, O. Rozmey and M. Klapish for valuable discussions.

References

1. Rocca, J.J., Shlyaptsev, V.N., Tomasel, F.G., Cortazar, O.D., Hartshorn, D., Chilla, J.L.A.: Demonstration of a discharge pumped table-top soft-x-ray laser. Phys. Rev. Lett. **73,** 2192 (1994)
2. Gonzalez, J.J., Frati, M., Rocca, J.J., Shlyaptsev, V.N., Osterheld A.L.: High-power-density capillary discharge plasma columns for shorter wavelength discharge-pumped soft-x-ray lasers. Phys. Rev. E. **65,** 026404 (2002)
3. Bogen, P., Conrads, H., Gatti, G., Kohlhaas, W.: Continuum radiation source of high intensity. J. Opt. Soc. Am. **58,** 203 (1968)
4. McCorkle, R.A.: The high-power sliding-spark capillary discharge in vacuum: Variations and applications. Appl. Phys. **A 26,** 261 (1981)
5. Shepherd, R.L., Kania, D.R., Jones, L.A.: Measurement of the resistivity in a partially degenerate, strongly coupled plasma. Phys. Rev. Lett. **61,** 1278 (1988)
6. Ehrlich, Y., Cohen, C., Zigler, A., Krall, J., Sprangle, P., Esarey, E.: Guiding of high intensity laser pulses in straight and curved plasma channel experiments. Phys. Rev. Lett. **77,** 4186 (1996)
7. Kloshner, M.A., Silfvast, W.T.: Intense xenon capillary discharge extreme- ultraviolet source in the 10–16-nm- wavelength region. Optics Lett. **23,** 1609 (1998)
8. Wang, Y., Luther, M., Berrill, M., Marconi, M., Brizuela, F., Rocca, J.J., Shlyaptsev, V.N.: Capillary discharge-driven metal vapor plasma waveguides. Phys. Rev. E. **72,** 026413 (2005)
9. Choi, P., Favre, M.: Fast pulsed hollow cathode capillary discharge device. Rev. Sci. Inst. **69,** 3118 (1998)
10. Krisch, I., Choi, P., Larour, J., Favre, M., Rous, J., Leblanc, C.: Compact ultrafast capillary discharge for EUV projection lithography. Contrib. Plasma. Phys. **40,** 135 (2000)
11. Butler, A., Spence, D.J., Hooker, S.M.: Guiding of high-intensity laser pulses with a hydrogen-filled capillary discharge waveguide. Phys. Rev. Lett. **89,** 185003 (2002)
12. Marconi, M.C., Moreno, C.H., Rocca, J.J., Shlyaptsev, V.N., Osterheld, A.L.: Dynamics of a microcapillary discharge plasma using a soft x-ray laser backlighter. Phys. Rev. E. **62,** 7209 (2000)
13. Samarsky, A.A., Popov Yu.P.: The difference methods for gas dynamic problems. Nauka, Moscow (in Russian) (1992)
14. Avaria, G., Grisham, M., Li, J., Tomasel, F.G., Shlyaptsev, V.N., Busquet, M., Woolston, M., Rocca, J.J.: Extreme degree of ionization in homogenous microcapillary plasma columns heated by ultrafast current pulses, Phys. Rev. Lett. **114,** 095001 (2015,in press)
15. Tomasel, F.G., Shlyaptsev, V.N., Rocca, J.J.: Spectroscopically pure metal vapor source for highly charged ions spectroscopy and capillary discharge soft x-ray lasers. Rev. Sci. Inst. **79,** 013503 (2008)
16. Rozmej, F.B., Rozmej, O.N.: Transient formation of forbidden lines. J. Phys. B: At. Mol. Opt. Phys. **29,** L359 (1996)

Part II
New X-Ray Sources and Relativistic Phenomena

Progress Report on an Experiment to Clarify Collimated X-Ray Emission in Karabut's Experiment

Francis L. Tanzella, Peter L. Hagelstein, J. Bao and M. C. H. McKubre

Abstract Karabut has reported collimated x-ray emission in a high-current density glow discharge experiment, a result which has for some years remain unexplained. A model has been proposed to account for the emission based on the up-conversion of vibrational quanta to produce excitation of the 1565 eV excited state in ^{201}Hg. To test this idea, we are developing an experiment in which a Cu foil is vibrated. Evidence for charge emission is described; and preliminary x-ray measurements discussed.

1 Introduction

Karabut reported the observation of collimated x-ray emission from a high-current density glow discharge experiment in 2002 [1]. This experiment is not well known outside of the field of condensed matter nuclear science; however, in our view it is an important and fundamental experiment deserving of more attention.

Karabut and coworkers previously used a high-current glow discharge to load metals with hydrogen and deuterium, and reported observations of a variety of anomalies. After years of working with the system, it was noticed that about 10 watts of x-ray emission near 1.5–2.0 keV was emitted in a beam 9 mm in diameter; this corresponds to an efficiency of 20 % compared to the input discharge power [2]. The effect was reported for both hydrogen and deuterium gas, and for Pd, W and Ti cathodes. Subsequent experiments similar showed collimated x-ray emission near 1.5 keV with other gases, including He, Ar, Kr, and Xe; and with other cathode metals, including Al, Sc, Ni, Nb, Zr, Mo, Ta, Pb and Pt [3]. Consistent results were obtained with Al_2O_3 thermo-luminescent detectors, PMMA-based scintillator detectors, Kodak XBM x-ray film, and with a bent Mica crystal spectrometer and x-ray film.

The obvious question in connection with such astonishing experimental results is why such an effect should occur at all. Karabut conjectured that collimated x-rays

P. L. Hagelstein (✉)
Massachusetts Institute of Technology, Cambridge, MA, USA
e-mail: PHagelstein@aol.com

F. L. Tanzella · J. Bao · M. C. H. McKubre
SRI International, Menlo Park, CA, USA

© Springer International Publishing Switzerland 2016

J. Rocca et al. (eds.), *X-Ray Lasers 2014*, Springer Proceedings in Physics 169,
DOI 10.1007/978-3-319-19521-6_14

was due to an x-ray laser effect; however, there is no obvious mechanism present in the experiment that should produce a population inversion in the keV regime.

2 Theoretical Ideas

For years theoretical models had been pursued to account for other anomalies in condensed matter nuclear science, focused on the notion that it coherent energy exchange might occur between quantum systems with strongly mismatched characteristic energy quanta. The thought was the anomalies could be accounted for if it were possible to up-convert an extremely large number of vibrational quanta to produce nuclear excitation, and vice versa. Models were constructed which described such an effect. A "toy" mathematical model was developed which showed fast coherent energy exchange between an oscillator with a small characteristic energy, and two-level systems with a large transition energy [4–7]. Strong coupling between lattice vibrations and internal nuclear degrees of freedom is present at least initially in a relativistic condensed matter model. Usually this coupling can be removed by a generalized Foldy–Wouthuysen transformation; however, the same is true for the model considered in [4–7], except that when strong off-resonant loss is present a Foldy–Wouthuysen transformation becomes unhelpful [8].

It became clear that it would be important to isolate the effect experimentally if possible. Research focused on the design of an experiment to demonstrate the conversion of vibrational energy to nuclear excitation. To minimize the amount of up-conversion required, a search was carried out for the nuclear transition with the lowest excitation energy from the ground state in all stable nuclei. This led to the identification of the 1565 eV transition in ^{201}Hg as special (the next one is the 6240 eV transition in ^{181}Ta). The 1565 eV excited state decays primarily by internal conversion with a half life of 81 ns, with one radiative decay in $4 - 6 \times 10^4$ internal conversion decays. It was recognized that up-conversion with a uniform phase would result in directional x-ray emission near 1.5 keV with a broad line shape; the suggested the possibility that Karabut's experiment implemented the scheme.

Very bright short bursts of x-ray emission were observed on the order of a millisecond after the discharge was turned off [3], too long for electronic excitation to persist, but which is an appropriate time scale for vibrational excitation in the MHz regime. One could imagine that a small amount of Hg might be present as an impurity on the cathode surface. Attention should be drawn to the experiments of Kornilova et al in which directional x-ray emission is registered on x-ray film originating from a steel plate close to a high pressure water jet [9].

3 Vibrating Cu Foil Experiment

If the proposed interpretation of the Karabut experiment is right, then we should be able to implement an experiment in which vibrations are imposed on a foil, and hope to see directional x-ray emission near 1.5 keV from a small amount of Hg on the

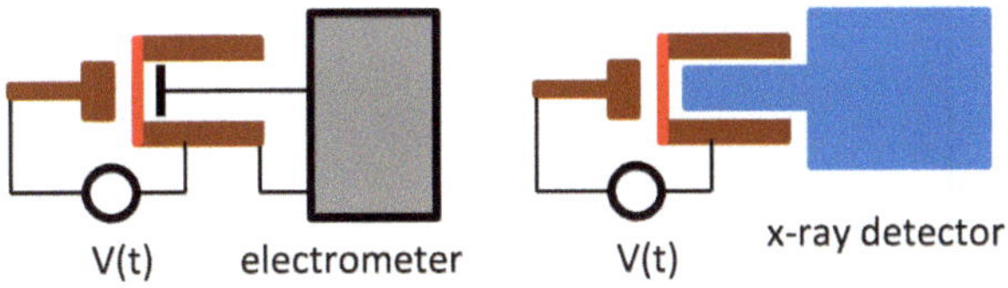

Fig. 1 *Left*: Vibrating Cu foil experiment with charge emission. *Right*: Vibrating Cu foil experiment with x-ray detection. The Cu foil is in *red*, across the air gap from the driver

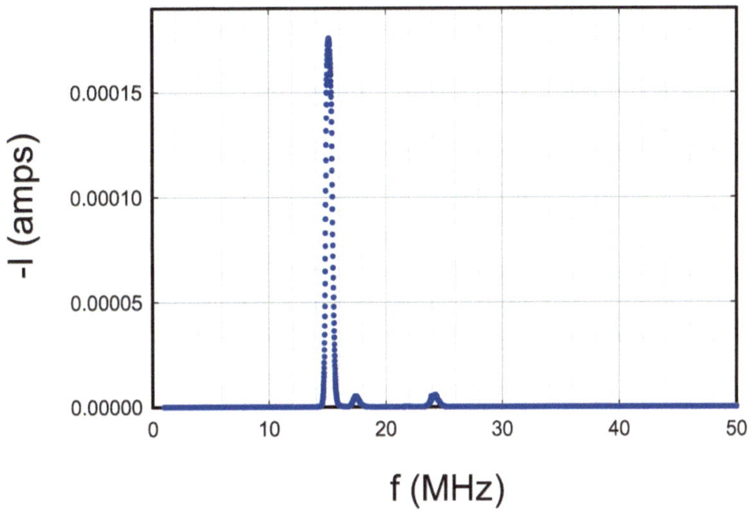

Fig. 2 Current as a function of drive frequency

surface. We selected 73 μ Cu for the foil, since Cu binds Hg well, and since these foils (which were available in our lab) were of a similar thickness to those used by Karabut. Vibrations were imposed through capacitive coupling with a driver, as illustrated in Fig. 1. In this arrangement vibrations are driven at twice the frequency of the drive voltage.

We first looked for charge emission as a possible precursor for x-ray emission, thinking that we might see electrons from the internal conversion of the 1565 eV ^{201}Hg state. A substantial (negative) charge emission signal was seen as shown in Fig. 2, with a strong resonance observed at the drive frequency of the second transverse mode (15.25 MHz), and a weaker resonance at the fundamental compressional mode (17.4 MHz). The charge emission at 15.25 MHz increases nonlinearly with drive voltage as shown in Fig. 3. Charge emission was found to occur with pure Cu, with similar results with surface Hg present; we conclude that this effect is not connected with excitation of the 1565 eV state in ^{201}Hg.

X-ray measurements were attempted with an AMPTEK X-123 x-ray detector. A cubic calibration curve was established using a ^{55}Fe source, and fluorescence of Al, Si and Fe with a ^{241}Am source (Fig. 4). Signals were seen in live runs which

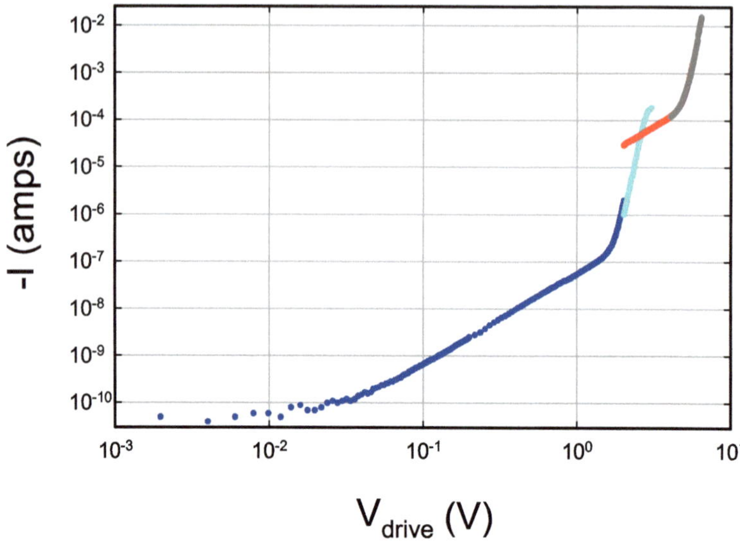

Fig. 3 Current as a function of drive voltage at 15.25 MHz drive frequency. The different colors indicate data taken on different registers of the Keithley 617 electrometer

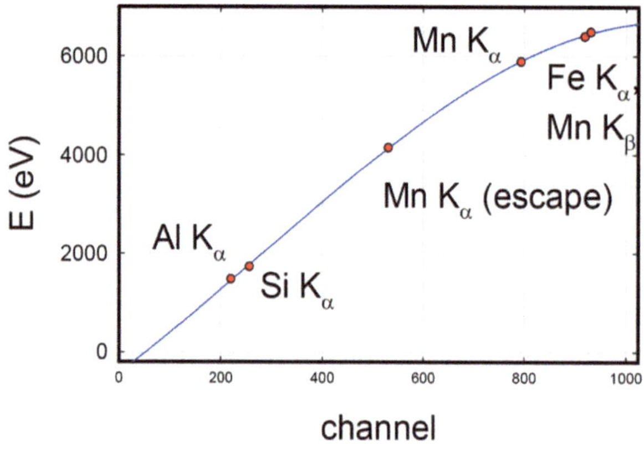

Fig. 4 Cubic calibration curve for the AMPTEK X-123 detector

appeared to give reproducible counts on the detector. The (early) spectrum in Fig. 5 is illustrative; a large number of counts are present below 1 keV, and there seems to be a Gaussian feature near 1.5 keV. The detector has a 0.5 mil Be window, so that there is very little sensitivity below 1 keV; hence the large number of counts below 1 keV is an artifact. This feature in the low channels has good reproducibility, and it can show up above 1 keV correlated with a vibrational resonance. Current thinking is that the detector is responding to charge emission. With a pure Cu foil and a well grounded detector, very few counts are seen above pile-up. Similar experiments with Hg on the surface and a well grounded detector give counts in the low channels

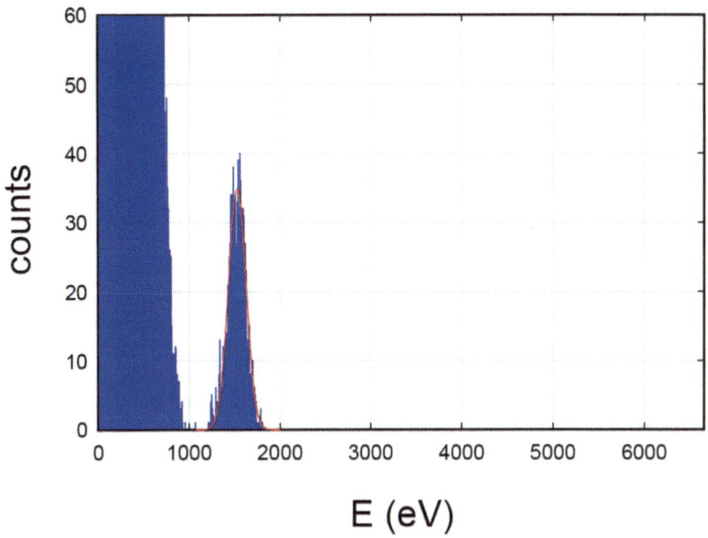

Fig. 5 400 s X-ray spectrum at 13.9 MHz with 90 V drive voltage

correlated with vibrational resonances. The feature near 1.5 keV in Fig. 5 is thought to be an extension of the low channel counts, and (unfortunately) probably not due to x-rays. In the experiments of Kornilova et al. [9] data was taken with an AMPTEK X-123 detector, and large numbers of counts are seen in the low channels. One wonders whether the detector might have responded to something other than x-rays in these experiments.

Subsequent experiments were done seeking to verify that the charge emission signal is due to charge, and not an artifact, by testing whether a capacitor close to the plate could be charged. The results from tests of this kind so far indicate that the large current signals are artifacts. A tentative conclusion from these recent results is that the small cathodes in the Karabut experiment is not causing the up-conversion, and attention has shifted to the more massive stainless steel cathode holder. Accordingly, the next round of experiments will focus on vibrating more substantial steel plates.

References

1. Karabut, A.B.: Research into powerful solid X-ray laser (wavelength is 0.8-1.2 nm) with excitation of high current glow discharge ions. Proceedings of 11th International Conference on Emerging Nuclear Energy Systems, 29 September to 4 October 2002, Albuquerque, New Mexico, USA, 374 (2002)
2. Karabut, A.B.: X-ray emission in the high-current glow discharge experiments. Proceedings of 9th International Conference on Cold Fusion. Tsinghua University, Beijing, China 155 (2002)

3. Karabut, A.B., Karabut, E.A., Hagelstein, P.L.: Spectral and temporal characteristics of X-ray emission from metal electrodes in a high-current glow discharge. J. Condensed Matter Nucl. Sci. **6**, 217 (2012)
4. Hagelstein, P.L., Chaudhary, I.U.: Energy exchange in the lossy spin-boson model. J. Condensed Matter Nucl. Sci. **5**, 52 (2011)
5. Hagelstein, P.L., Chaudhary, I.U.: Second-order formulation and scaling in the lossy spin-boson model. J. Condensed Matter Nucl. Sci. **5**, 87 (2011)
6. Hagelstein, P.L., Chaudhary, I.U.: Local approximation for the lossy spin-boson model. J. Condensed Matter Nucl. Sci. **5**, 102 (2011)
7. Hagelstein, P.L., Chaudhary, I.U.: Coherent energy exchange in the strong coupling limit of the lossy spin-boson model. J. Condensed Matter Nucl. Sci. **5**, 116 (2011)
8. Hagelstein, P.L., Chaudhary, I.U.: Phonon-nuclear coupling for anomalies in condensed matter nuclear science. J. Condensed Matter Nucl. Sci. **12**, 105 (2013)
9. Kornilova, A.A., Vysotskii, V.I., Sysoev, N.N., Litvin, N.K., Tomak, V.I., Barzov, A.A.: Generation of intense X-rays during ejection of a fast water jet from a metal channel to atmosphere. J. Surf. Invest. X-ray Synchrotron Neutron Techn. **4**, 1008 (2010)

Isolation of Coherent Synchrotron Emission During Relativistic Laser Plasma Interactions

B. Dromey, S. G. Rykovanov, C. L. S. Lewis and M. Zepf

Abstract Coherent Synchrotron Emission (CSE) from relativistic laser plasmas (Pukhov et al., Plas Phys Control Fusion 52:124039, 2010; Dromey et al., Nat Phys 8:804–808, 2012; Dromey et al., New J Phys 15:015025, 2013) has recently been identified as a unique platform for the generation of coherent extreme ultraviolet (XUV) and X-Ray radiation with clear potential for bright attosecond pulse production. Exploiting this potential requires careful selection of interaction geometry, spectral wavelength range and target characteristics to allow the generation of high fidelity single attosecond pulses. In the laboratory the first step on this road is to study the individual mechanisms driving the emission of coherent extreme ultraviolet and X-Ray radiation during laser solid interactions in isolation. Here we show how interactions can be tailored to permit the unambiguous observation of coherent synchrotron emission (CSE) and the implications of this geometry for the resulting harmonic spectrum over the duration of the interaction.

1 Introduction

Synchrotron radiation from insertion devices in large electron storage rings (> 10 m diameter) has been at the forefront of extreme ultraviolet/X-ray science for many decades. In principle, however, the instantaneous brightness of these machines could be increased substantially if dense nanometer (10^{-9} m) scale bunches of electrons could be produced to permit Coherent Synchrotron Emission (CSE) [4]. In this scheme the emission over the entire bunch length adds coherently (N^2 growth, where N is the number of emitters) leading to a dramatic enhancement of the emitted XUV signal. The problem is that current synchrotron/storage ring technology prohibits the maintenance of sufficiently small electron bunches over significant distance to permit coherent addition. Theoretical work, however, has shown that such nanobunches can in principle be formed and accelerated on timescales of less than an optical laser cycle ($\sim 10^{-15}$ s) during relativistic laser plasma interactions, opening up the potential for dramatically increased economy of scale for developing bright ultrafast

B. Dromey (✉) · C. L. S. Lewis · M. Zepf
Department of Physics, Queens University Belfast, Belfast, BT7 1NN, UK

S. G. Rykovanov · M. Zepf
Helmholtz-Institut Jena, Fröbelstieg 3, Jena, 07743, Germany

© Springer International Publishing Switzerland 2016

J. Rocca et al. (eds.), *X-Ray Lasers 2014*, Springer Proceedings in Physics 169,
DOI 10.1007/978-3-319-19521-6_15

extreme ultraviolet/X-ray sources. This work presents clear potential for a novel, bright source of attosecond XUV ($\sim 10^{-18}$ s) radiation [1].

Here we demonstrate, using a modified geometry to those studied theoretically before, how CSE generated during relativistically intense laser—ultrathin foil interactions can be isolated from competing harmonic generation mechanisms. Particle-in-cell code simulations demonstrate how periodic, sub-laser cycle acceleration of dense nanobunches of electrons formed during normal incidence interactions result in bursts of bright attosecond radiation in transmission. Experimentally, harmonic spectra ($I(n)$) are characterised by a slow decay ($n^{-1.62}$) before a rapid efficiency rollover to high harmonic orders, n [2, 3]. CSE has also been shown to provide harmonic order-dependent emission cone narrowing to higher photon energies ($\sim 10 mrad$ at $\sim$1keV), confirming excellent spatial coherence of the source [2]. This microscopic coherent synchrotron source ($< 5 \times 10^{-6}$ m) has the potential to significantly increase pulse brightness towards the goal of attosecond pump-probe experiments. Furthermore, the possibility of dramatically enhanced peak intensities beyond current state of the art systems [5] is another potential application of this efficient, ultrafast source of harmonic radiation.

For many cycle laser pulses, the experimental signature of this process is the emission of high-order harmonics. Together with other sources of high harmonic emission such as coherent wake emission (CWE) [6, 7] and relativistically oscillating mirror (ROM) [8], these mechanisms can be summarised as SHHG processes (Solid-target high harmonic generation). These mechanisms generally compete with each other but different requirements [1], for plasma scalelength, driving laser intensity and harmonic spectral range allow experimentalists to identify conditions where one mechanism is dominant over the others. Here we focus our discussion on identifying and demonstrating the method to observe laser-driven CSE extending to the X-ray spectral region—relativistically intense laser interactions from ultrathin foils from a 'normal-incidence-in-transmission' (NIIT) geometry and how this can affect the spectral emission characteristics.

2 Isolation of CSE with Respect Other SHHG Mechanisms

High harmonic generation (frequency domain) is simply a signature of periodic, sub-laser cycle emission of coherent extreme ultraviolet/X-ray pulses (time domain) and can equivalently be called Attosecond Interference. In the limit of single pulse emission, no interference occurs and a continuous spectrum is observed i.e. no harmonics.

Driven by relativistically intense laser pulses ($> 10^{18}$ Wcm^{-2}) at the surface of a solid density target, SHHG occurs due to electrons performing complex trajectories at the plasma vacuum boundary [9]. In this sense SHHG is an umbrella term covering multiple, distinct harmonic generation processes [6, 8]. To date, high harmonic spectra observed experimentally in specular reflection from solid targets are consistent with relativistically oscillating mirror (ROM) scaling, i.e. a $n^{\frac{8}{3}}$ power-law decay [8] being dominant. This scaling is applicable for harmonic orders n above the Coherent Wake Emission (CWE) cutoff order, n_{CWE}[7,8], which is equivalent to the harmonic order corresponding to the maximum plasma frequency $n_{\omega\max}$ of the target which can be calculated as

$$n_{\mathrm{CWE}} = n_{\omega\max} = \frac{\omega_{p\,\max}}{\omega_{\mathrm{L}}} = \sqrt{\frac{N_{\mathrm{e,max}}}{N_{\mathrm{c}}}}$$

where ω_{L} is the laser frequency, $N_{\mathrm{e,max}}$ is the peak electron density of the target and $N_{\mathrm{c}} = \frac{\varepsilon_0 m_e \omega_{\mathrm{L}}^2}{e^2}$ is called the critical density, which depends on the electron charge e and mass m_{e}

In principle, in the NIIT geometry CWE cannot be generated because electron bunches generated during the interaction cannot be injected into the plasma with a sufficient angle to phasematch the plasma density waves critical for linear mode conversion required for efficient harmonic emission into the specular direction. Moreover, Brunel absorption [6] (the underlying mechanism for CWE in oblique incidence) is not possible since there is no component of the electric field perpendicular to the plasma surface to accelerate electrons for suitable attosecond bunch formation. However, this is only true in the case of few driving laser cycle interactions. For longer (> 10 cycle) driving laser pulses, it is possible that significant oblique incidence components can be generated as the laser pulse deforms the foil during the interaction. Therefore we limit our discussion of harmonic orders $n > n_{\mathrm{CWE}}$ i.e. the spectral region where CWE does not emit.

In the NIIT geometry relativistic Doppler up-shifting (the physics underlying the ROM mechanism) is completely cancelled. This can be readily understood if the simple example of a 50:50 partial reflector travelling with relativistic velocities for $4\gamma^2$ up-shifting of a counterpropagating light pulse[14] is used. In this scenario the double Lorentz transformation that results in up-shifting for reflection, i.e. due to the change in the k-vector of the light, is cancelled in transmission.

As stated above, producing electron bunches of sufficiently small spatial dimension to extend the principle of CSE to the X-ray regime is currently beyond what is considered achievable with conventional large scale accelerators. Simulations investigating the dynamics of relativistic laser-plasma interactions have, however, shown that it is in principle possible for dense electron nanobunches (i.e. 10^{-9} m scale) to be formed on ultrafast timescales ($< 10^{-15}$ s) [5, 10]. Going one step further, recent work has shown the possibility of Coherent Synchrotron Emission (CSE) extending to the X-ray regime in specular reflection from laser-solid target interactions and has been discussed based on the acceleration of these nanobunches in the driving laser field [1].

Harmonic spectra with efficiency scaling characteristic of CSE ($I(n) \sim (n^{-1.62})$), strong wavelength dependent emission cone narrowing and importantly rapid scaling with changing plasma characteristics such as peak density and scalelength have been observed experimentally [2, 3]. This behaviour is unique to CSE generated from 'normal incidence in transmission' (NIIT) geometry (Fig. 1) and is distinct from other SHHG (Solid target High Harmonic Generation) mechanisms.

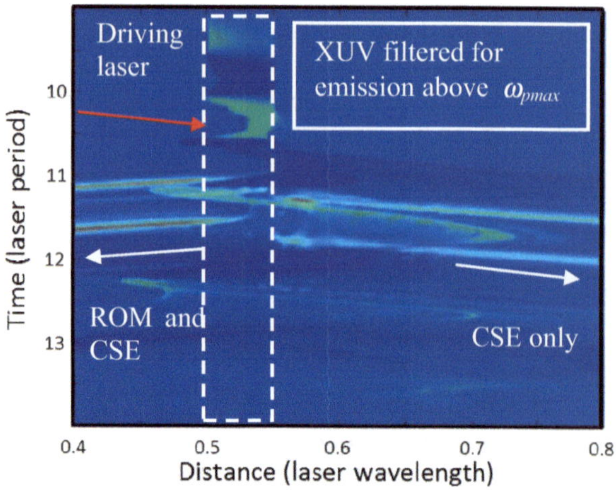

Fig. 1 1-D Particle in cell simulation (PICWIG, see [7] for more detail) to highlight the key NIIT interaction geometry. The dashed box shows the original target position

Figure 2 shows a 1-D PIC simulation outlining the mechanism for CSE in the NIIT geometry. Electrons (thick red lines, bottom panel, c) perform large synchrotron like trajectories in the intense laser field. The blue sections of these trajectories show the moments of XUV emission in the transmitted direction (cross ref top panel a) in Fig. 2. The middle panel (b) shows the components of the electron velocity in the laser field (V_x (transverse) and V_y (longitudinal)). It is clear from comparison of this figure with the top (a) and bottom (c) panels of Fig. 2 that only for instances of the electron trajectory where both V_x and V_y are approaching a maximum (i.e. circular motion) is there strong XUV emission in the transmitted direction.

A requirement for the observation of CSE in NIIT is the use of ultrathin foils. However the use of many cycle laser pulses leads to an evolution of the interaction conditions as the laser pulse interacts with the sub driving wavelength thick foil. For 500 fs pulses and < 500 nm scale foils this evolution has been shown to dramatically alter mechanisms for ion acceleration [11], due to the onset of relativistic transparency [12]. This evolution of the foil is also encoded in the emitted CSE spectrum (Fig. 3).

Initially the sharp well defined density profile of the target leads to the expected density dependent self-filtering of CSE in the transmitted direction (X, Fig. 3b) (1 and 2). However as the density evolves this sharp filter feature at X becomes less distinct (Fig. 3b) (3) and eventually rolls back to $n = 3$ in the emitted spectrum. The dashed lines in Fig. 3b) 1–6 are scaling guides to the eye to highlight the evolution of the spectral shape in a given panel.

The panels 1–6 represent the normalised spectral intensity in the corresponding time window and are representative of relative intensity. Examining this in greater detail is beyond the scope of this discussion. Here we demonstrate the evolution of the spectral shape with respect to interaction time for a many cycle laser pulse. Higher dimension (2-D, 3-D) PIC simulations and experiments will be required to

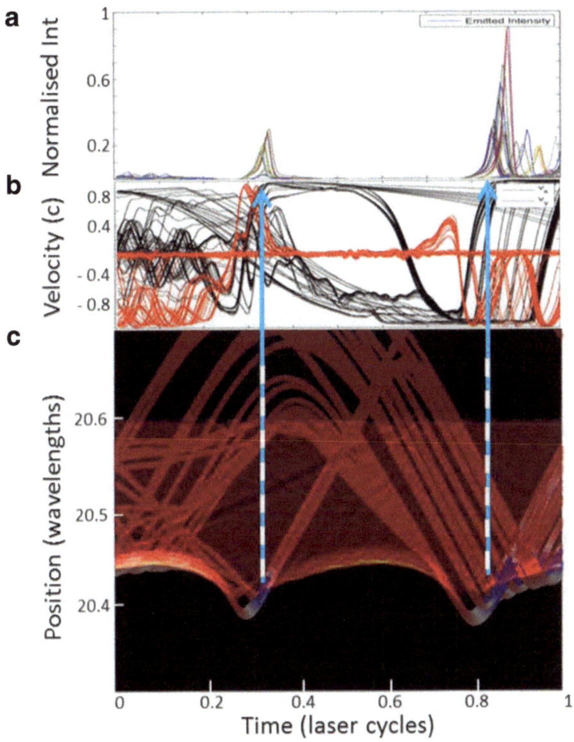

Fig. 2 Generation of CSE during NIIT for a 5 fs pulse interaction from 1-D PIC simulations for 500 fs pulse, wavelength 1 μm, intensity 10^{20}Wcm^{-2}

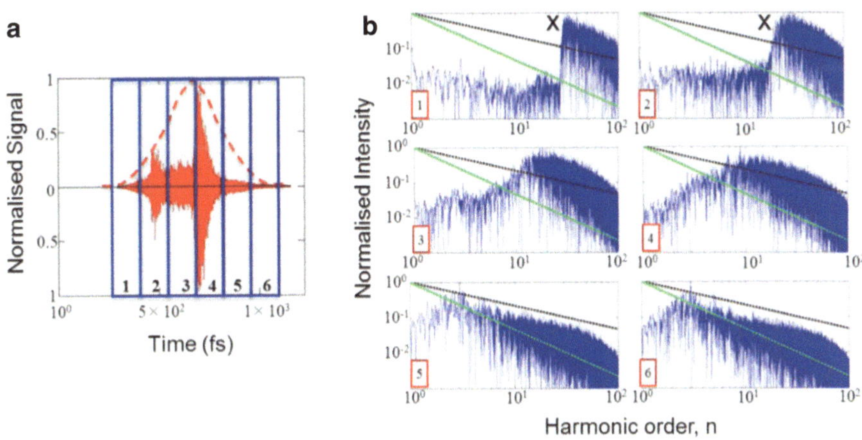

Fig. 3 Evolution of the CSE spectrum in the NIIT geometry from 1-D PIC simulations for 500 fs pulse, wavelength 1 μm, intensity 10^{20}Wcm^{-2}. The spectra shown in panels 1–6 in **b**) correspond tothe intevals 1–6 in **a**)

fully describe how the observed emission evolves with respect to target evolution. However it is worth noting that the results shown in Fig. 3 are in qualitative agreement with experiments examining the scaling of CSE in NIIT for different target thickness [2].

Future work will focus on the evolution of the angular distribution of the CSE emission and how this scales with pulse duration and target thickness to form complete picture of the evolution of CSE from ultrathin foils.

References

1. Pukhov, A., et al.: Relativistic laser plasmas for electron acceleration and short wavelength radiation generation. Plas. Phys. Control. Fusion. **52,** 124039, (2010)
2. Dromey, B., et al.: Coherent synchrotron emission from electron nanobunches formed in relativistic laser–plasma interactions. Nat. Phys. **8,** 804–808 (2012)
3. Dromey, B., et al.: Coherent synchrotron emission in transmission from ultrathin relativistic laser plasmas. New J Phys. **15,** 015025 (2013)
4. Schiff, L.I.: Production of particle energies beyond 200 MeV. Rev. Sci. Instrum. **17,** 6 (1946)
5. Naumova, N.M., et al.: Relativistic generation of isolated attosecond pulses in a λ^3 focal volume. Phys. Rev. Lett. **93,** 195003 (2004)
6. Quéré, F., Thaury, C., Monot, P., Dobosz, S., Martin, Ph, Geindre, J.-P., Audebert, P.: Coherent wake emission of high-order harmonics from overdense plasmas. Phys. Rev. Lett. **96,** 125004 (2006)
7. Dromey, B., et al.: Tunable enhancement of high harmonic emission from laser solid interactions. Phys. Rev. Lett. **102,** 225002 (2009)
8. Dromey, B., et al.: Harmonic generation in the relativistic limit. Nat. Phys. **2,** 456 (2006)
9. Wilks, S.C., et al.: Absorption of ultra-intense light pulses. Phys. Rev. Lett. **69,** 1383 (1992)
10. Kiefer, D., et al.: Relativistic electron mirrors from nanoscale foils for coherent frequency upshift to the extreme ultraviolet. Nat. Comms. **4,** 1763 (2013)
11. Jung, D., et al.: Efficient carbon ion beam generation from laser-driven volume acceleration. New J. Phys. **15,** 023007 (2013)
12. Palaniyappan, S., et al.: Dynamics of relativistic transparency and optical shuttering in expanding overdense plasmas. Nat. Phys. **8,** 763–769 (2012)

Polarization Gating in Relativistic Laser-Solid Interactions

M. Yeung, B. Dromey, S. Cousens, M. Coughlan, S. Rykovanov, C. Rödel, J. Bierbach, S. Kuschel, E. Eckner, M. Förster, G. Paulus and M. Zepf

Abstract High order harmonic generation from relativistic laser-solid interactions (focused intensity of $> 10^{18}$ Wcm^{-2}) has the potential to serve as a source of bright attosecond radiation. One key mechanism that can generate such radiation is the Relativistically Oscillating Mirror (ROM) where the overdense plasma surface oscillates at relativistic velocities leading to a Doppler upshift of the reflected laser radiation. A major obstacle to the application of such a harmonic source is that the radiation is emitted as a periodic pulse train with the frequency of the driving laser. One route to limiting this emission to a single pulse is to exploit the ellipticity dependence of these mechanisms by forming a pulse whose polarisation varies from circular to linear to circular—a technique known as polarization gating. At small angles of incidence it is expected that the efficiency of the ROM mechanism drops dramatically for circular polarization. Here we present a novel method of implementing this technique for high power laser pulses along with proof of principle experimental results.

1 Introduction

The generation of high-power attosecond light sources is a fast developing field as it offers a unique tool for investigating electron dynamics within atomic or molecular systems [1, 2]. The most common method of generating attosecond pulses is through laser interactions with gaseous media where high order harmonics of the

M. Yeung (✉) · B. Dromey · S. Cousens · M. Coughlan · M. Zepf
Department of Physics and Astronomy, Queen's University Belfast,
Belfast BT7 1NN, UK
e-mail: m.yeung@qub.ac.uk

M. Yeung · S. Rykovanov · C. Rödel · J. Bierbach · S. Kuschel · E. Eckner
G. Paulus · M. Zepf
Helmholtz-Institut Jena, Fröbelstieg 3, 07743 Jena, Germany
e-mail: m.yeung@gsi.de

C. Rödel · J. Bierbach · S. Kuschel · E. Eckner · M. Förster · G. Paulus
Friedrich-Schiller-Universität Jena, Max-Wien-Platz 1, 07743 Jena, Germany

© Springer International Publishing Switzerland 2016

J. Rocca et al. (eds.), *X-Ray Lasers 2014*, Springer Proceedings in Physics 169,
DOI 10.1007/978-3-319-19521-6_16

driving field are generated by the three-step model [3]. These sources have been limited in their applications, partly due to the low energies of the generated attosecond pulses. Generation of attosecond pulses by ultra-intense laser interactions with solid surfaces via the Relativistically Oscillating Mirror (ROM) mechanism [4, 5] can, in principle, allow much higher brightness pulses. Furthermore, for very high intensity driving laser pulses, spectra generated from this source can extend to multi-keV photon energies [5] with a sufficiently slow power decay law that bandwidths capable of supporting attosecond or even shorter duration pulses could be achieved [6].

As a harmonic radiation source, this mechanism produces a train of attosecond pulses which must be gated down to a single pulse for temporally resolved measurements. For harmonic generation from gaseous media this gating was achieved by exploiting the fact that the mechanism was suppressed when using elliptically polarised pulses [7]. By using a birefringent quartz plate, two time separated replicas of the laser pulse with orthogonal polarisations could be created. Passing these pulses through a second quartz quarter waveplate creates two circularly polarised pulses with opposite handedness that create a linear gate in the region where the pulses overlap [8]. Since the process is only efficient for linear polarisation this effectively gates the emission of the attosecond pulses with the gate time depending on the initial pulse duration and separation of the two pulse replicas.

It was shown in simulations that this method could be applied to ROM harmonics [9] which are heavily suppressed for circularly polarised pulses at normal incidence [4]. To date, the predictions of these simulations are yet to be realised, in part due to technical challenges in detecting reflected high harmonics in normal incidence interactions, and also due to problems with the gating method when using high pulse powers.

2 Gating Methods

The two birefringent plates method discussed above [8, 9] lacks easy tunability as the delay between the pulses is linked to the thickness of the first quartz plate. Furthermore, as the required laser systems often carry very high intensities even in the unfocused beam, it is preferable to avoid thick transmissive optics which may lead to a loss of focusability caused by accumulation of B-integral due to the non-linear refractive index of the material. Another technique is the interferometric method whereby the initial pulse is split into four replicas through the use of two interferometers [10]. The gate width can be finely tuned by adjustment of the delay of each of the pulses. While this method has the advantage of tunability, the system is difficult to implement for large beam diameter high power systems due to space constraints and the large diameter optics required for the interferometers.

Here we demonstrate a novel method of polarisation gating that builds on the ideas of the birefringent plate method and which is suitable for high power laser applications whilst only using optics that are relatively low cost even for large beam systems. The laser beam is split in two halves by placing a rotatable fused silica wafer over one beam half. Another static wafer is placed over the other beam half

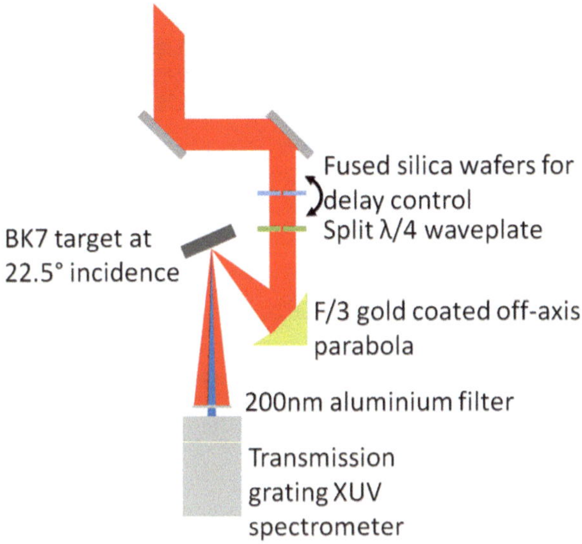

Fig. 1 Experimental setup. The laser beam passes through two 500 micron fused silica wafers, each covering one half of the beam. One of the wafers is rotatable to finely control the delay of one half beam. These two beam halves are passed through a split quarter waveplate with one beam half made left circularly polarised and the other made right circularly polarised. This beam is then focused using an F/3 parabola giving a peak intensity (in the case of zero delay between the two beam halves) of $\approx 2 \times 10^{19}$ Wcm^{-2}. The reflected harmonic radiation is filtered by a 200 nm Al foil and detected in a transmission grating spectrometer

so that the relative group delays give a total zero delay for no rotation of the first wafer. Rotation of this wafer then permits fine relative phase and group delay control over one beam half. The beam is then sent through a split quarter waveplate which has optical axes set so that one beam half becomes right circularly polarised and the other becomes left circularly polarised. When the beams focus, they will overlap spatially and will form a linear gate over the times when they overlap temporally. The width of the gate can then be tuned by controlling the delay of the beam halves relative to each other.

3 Experiment

In order to test this polarisation gating method, an experiment was conducted on the 30 fs JETI40 Ti:Sapphire system based at the Friedrich-Schiller-Universität Jena. Two 500 micron fused silica wafers were used for delay control and the split wave-plate was made of mica. The beam was focused using an F/3 off-axis parabola onto a fused silica target at an angle of incidence of 22.5°. The reflected harmonic radiation was detected in a extreme ultraviolet (XUV) spectrometer consisting of a 1-to-1 imaging toroidal mirror, a freestanding gold transmission grating and an Andor XUV CCD. The radiation is first filtered using a 200 nm thin aluminium filter. The setup is sketched in Fig. 1.

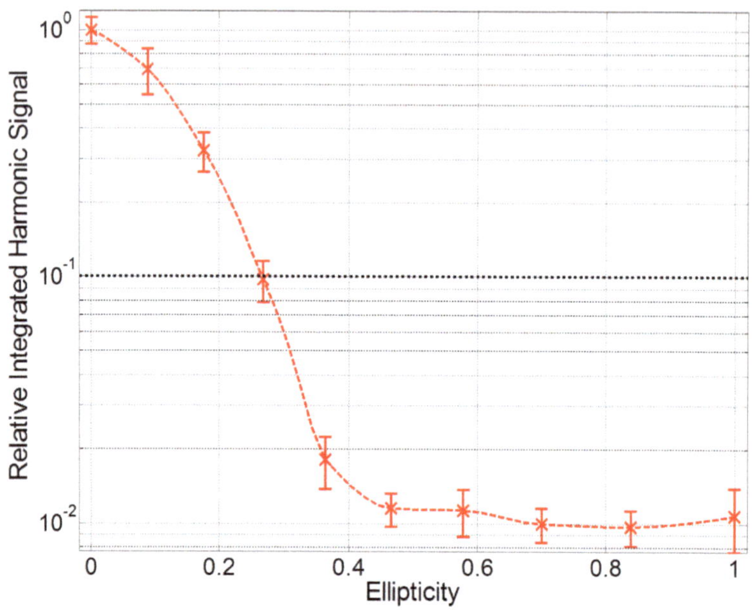

Fig. 2 Integrated signal for harmonics 13–28 averaged over 10 shots for varying driver ellipticity which was adjusted by rotation of a quarter waveplate. Errors are the standard error of the mean for the set of shots. Heavy suppression of the harmonic signal is seen for circular polarisation while the 10 % threshold ellipticity is 0.268

Although normal incidence would provide better suppression of the harmonic generation, it has also been shown in simulations that polarisation gating is also feasible at small angles of incidence [9]. To test this, a full beam quarter waveplate was used instead of the polarisation gating setup and was rotated in order to vary the ellipticity. In this configuration the on target intensity is $\approx 6 \times 10^{19}$ Wcm^{-2}. The results are shown in Fig. 2. We see a clear suppression of the harmonic signal for circular polarisation suggesting this angle of incidence is suitable for polarisation gating. Note that previous measurements of this dependence for larger angles of incidence and different laser conditions showed that the harmonic signal for a circularly polarised driver was still significant relative to the linear case [11, 12].

Results for the polarisation gating setup are shown in Fig. 3. Under these conditions the peak intensity is $\approx 2 \times 10^{19}$ Wcm^{-2} for a zero delay between the two half beams. The linear gate polarisation is chosen to be p-polarised, however due to the split waveplate 50 % of the energy will be in the s-polarisation which appears as two spots with a minimum at the single p-polarised spot. This is because, in the unfocussed beams, when the p-polarised components of the two beams are in phase then the s-polarised components are exactly out of phase and interfere in the focus.

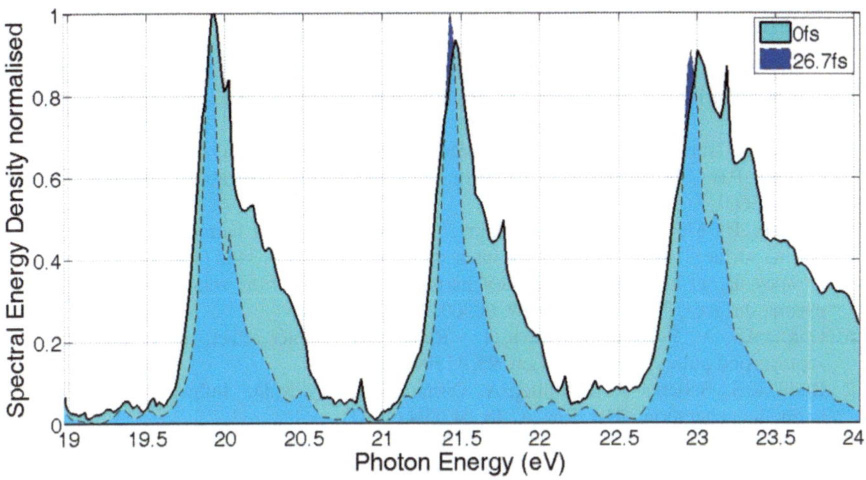

Fig. 3 Spectrum for the 13th–15th harmonics averaged over 10 shots for no delay and 26.7 fs delay. Clear spectral broadening of the harmonics is observed in the case where the second beam is delayed thus forming a shorter linear gate

The spectra in Fig. 3 have been normalised and the minima between harmonics set to zero to best compare the bandwidths. Going from the zero delay case to a 10 cycle delay we see clear harmonic broadening, consistent with gating of the pulse train. The harmonic bandwidth is consistent with a 3.5 cycle pulse train at the 10 % level which broadly agrees with the predicted gate width of $\approx$ 3 cycles following the analysis from Rykovanov et al. [9]. The asymmetry in the broadening is likely due to the lower intensity in the gated case ($\approx$50 % less) due to separation of the pulses. This would lead to different denting dynamics which are known to affect the redshift of the harmonics [13].

4 Conclusions

In conclusion, we have proposed a novel polarisation gating method utilising split beams with opposite handedness circular polarisation and tunable delay. The method has been demonstrated to gate a pulse train for a relativistic harmonic generation process—the ROM mechanism. To our knowledge, this is the first demonstration of polarisation gating for a relativistic surface high harmonic generation process. Further work will require the use of shorter pulse duration high power lasers in order to push the gate widths down to the single attosecond pulse level.

References

1. Sansone, G., Poletto, L., Nisoli, M.: High-energy attosecond light sources. Nat. Photon. **5**, 655 (2011)
2. Krausz, F., Ivanov, M.: Attosecond physics. Rev. Mod. Phys. **81**, 163 (2009)
3. Corkum, P.B.: Plasma perspective on strong-field multiphoton ionization. Phys. Rev. Lett. **71**, 1994 (1993)
4. Lichters, R., Meyer-ter-Vehn, J., Pukhov, A.: Short-pulse laser harmonics from oscillating plasma surfaces driven at relativistic intensity. Phys. Plasmas **3**, 3425 (1996)
5. Dromey, B., et al.: Bright multi-keV harmonic generation from relativistically oscillating plasma surfaces. Phys. Rev. Lett. **99**, 085001 (2007)
6. Gordienko, A., Pukhov, S., Baeva, T.: Relativistic doppler effect: universal spectra and zeptosecond pulses. Phys. Rev. Lett. **93**, 115002 (2004)
7. Budil, K.S., Salières, P., L'Huillier, A., Ditmire, T., Perry, M.D.: Influence of ellipticity on harmonic generation. Phys. Rev. A. **48**, 043804 (1993)
8. Tcherbakoff, O., Mevel, E., Descamps, D., Plumridge, J., Constant, E.: Time-gated high-order harmonic generation. Phys. Rev. A. **68**, 043804 (2003)
9. Rykovanov, S.G., Geissler, M., Meyer-ter-Vehn, J., Tsakiris, G.D.: Intense single attosecond pulses from surface harmonics using the polarization gating technique. New J. Phys. **10**, 025025 (2008)
10. Tzallas, P., et al.: Generation of intense continuum extreme-ultraviolet radiation by many-cycle laser fields. Nat. Phys. **3**, 846 (2007)
11. Heissler, P., et al.: Toward single attosecond pulses using harmonic emission from solid-density plasmas. Appl. Phys. B. **101**, 511 (2010)
12. Easter, J.H., Nees, J.A., Hou, B.X., Mordovanakis, A., Mourou, G., Thomas, A.G.R., Krushelnick, K.: Angular emission and polarization dependence of harmonics from laser-solid interactions. New J. Phys. **15**, 025035 (2013)
13. Vincenti, H., Monchocé, S., Kahaly, S., Bonnaud, G., Martin, Ph., Quéré, F.: Optical properties of relativistic plasma mirrors. Nat. Comm. **5**, 3403 (2014)

Progress and Prospects of a Compton X-ray Source Driven by a High-Power CO_2 Laser

I. V. Pogorelsky

Abstract X-ray sources based on inverse Compton scattering provide high peak-brightness combined with their well-controlled beam properties. We review recent progress in three research areas: Demonstrating the source's spatial coherence, so leading to single-shot, ultra-fast, phase-contrast tomography; high-average-brightness intra-cavity Compton source; and, exploring the relativistic regimes of electron oscillation within the laser field that produces multiple Compton harmonics. Next-generation Compton sources most likely will utilize all-optical schemes in which lasers serving as a virtual wiggler simultaneously will drive an electron beam from a plasma-wakefield accelerator. We address the possibility of reaching full coherency of all-optical Compton sources, analogous to free-electron lasers.

1 Introduction

Compton sources (CSs) of x-ray radiation can be viewed as a complement to conventional Synchrotron Light Sources (SLSs). Due to the micrometre-scale optical period of its virtual wiggler (the laser), a CS can be configured as a relatively compact and economical device that readily will fit to university and industrial settings. The natural synchronisation between the Compton x-ray (gamma) pulses and the driver laser pulses is ideal for pump-probe experiments. On-going efforts are directed towards bringing the CSs closer to practical applications.

This paper reviews recent progress achieved at the Brookhaven Accelerator Test Facility (ATF) in Compton studies that include: Achieving a high photon yield and fine x-ray beam quality that allows single-shot, pico- and femto-second phase-contrast tomography; studies of nonlinear phenomena in Compton scattering, and development of a high-average-power CS based on putting the Compton interaction point (IP) inside a laser cavity.

I. V. Pogorelsky (✉)
Accelerator Test Facility, Collider Accelerator Dept., Brookhaven National Laboratory, Upton, 11973, NY, US
e-mail: igor@bnl.gov

© Springer International Publishing Switzerland 2016

J. Rocca et al. (eds.), *X-Ray Lasers 2014*, Springer Proceedings in Physics 169,
DOI 10.1007/978-3-319-19521-6_17

The on-going ATF upgrade in the electron beam energy and laser peak power promises further breakthrough towards high-brightness gamma sources and development of a new-generation compact all-optical CSs where a laser-driven high-gradient laser accelerator will replace an RF accelerator.

2 Compton Research at BNL ATF

The maximum specific photon yield for matched counter-propagating laser- and electron-pulses is $\frac{N_x}{N_e} = \frac{N_L \sigma_T}{2\pi \sigma_L^2}$, where N_x, N_e, and N_L correspondingly are the number of Compton photons, electrons, and laser photons per pulse; $\sigma_T = 6.65 \times 10^{25} cm^2$ is the Thomson cross-section, and σ_L is the rms radius of the laser's focus. Using at BNL ATF a 10-μm CO_2 gas laser in place of the solid-state lasers employed elsewhere offers an obvious advantage in providing 10–12 times more photons per Joule. This approach allows us to reach a high yield of one photon per electron [1] with a moderate, 1-TW CO_2 laser peak power. With our low-emittance ($E_n \approx 1 \mu$m) pico- and femto-second electron bunches of a nano-Coulomb charge produced from the ATF's photo-cathode linac, billions of x-ray photons have been generated in a single shot at a *peak* brightness $B_p \approx 1.5 \times 10^{-3} \frac{N_x \gamma^2}{(2\pi)^2 \sigma_e^2 \tau_e} \approx 10^{20}$ph/s-mm^2-mrad^2- 0.1% BW, where, $\gamma = \frac{1}{\sqrt{1-\beta^2}} =$ E[MeV]/0.511 is the Lorentz factor, σ_e is rms radius of the electron beam, and τ_e is the electron bunch duration [2].

This high B_p allows obtaining single-shot x-ray images with a variety of recording devices thereby prompting exploring fine beam characteristics and applications that are obscured or impossible with multiple-shot techniques. One of such observations was the first demonstration of phase-contract imaging of biological objects using a single picosecond Compton pulse [3]. However, due to a relatively low duty cycle that is limited by the repetition rate, *f*, of high-power laser, the *average* brightness of CSs, $B_{avg} = B_p f \tau_e$ is many orders-of-magnitude below that of the SLSs. This limits the CS's potential for application. Efforts are underway to bridge this gap. One such research direction is multiplying the source's repetition rate and increasing its B_{avg} by placing the laser/e-beam's IP inside a laser cavity packed with fast-recurring laser pulses.

Several research groups in Japan [4], the United States [5] and France [6] have studied the value of a "coherent cavity" approach for a high-average-power CS. This concept is based on a well-known property of a Fabre-Perot interferometer: Thus, tuned in resonance with a continuous monochromatic beam, it builds up inside the power equal to, $P_c = \frac{F}{\pi} P_0$ where P_0 is the laser power outside the cavity $F = \frac{\pi\sqrt{R}}{1-R}$ is the cavity's finesse factor and R is the cavity's mirror reflectivity. With the demonstrated hundredfold multiplication of the acting laser power at the Compton IP, a 10^{12}photon/sec average x-ray flux can be achieved using 100-W mode-locked solid-state lasers. Further progress relies on the laser power increase and reduicing miror losses.

At BNL, we are investigating a different approach using an intra-cavity CO_2-laser amplifier to replenish the laser pulse's energy loss on the beam propagation. An active cavity experiment conducted by BNL and RadiaBeam Technologies LLC, demonstrated a train of picosecond pulses of the total 30 J energy, which is potentially available for multiple interaction with synchronized electron bunches over a time span of ~1μs [7]. The project's next goal is combining this laser-pulse train with a photocathode electron- linac to support a cumulative Compton photon yield of 4×10^7ph/0.1% BW per a single laser-shot. High-pressure CO_2 lasers operating at 1 kHz promise to raise the CS's flux to 4×10^{10}ph/s - 0.1% BW. This implies $B_{avg} = 3 \times 10^{10}$ at the photon peak energy of 6.4 keV is achievable with a 60-MeV linac. In considering the possibility of scaling the brightness with $\sim \gamma^3$ upon the ATF's linac upgrade to 500 MeV, we can project a hard x-ray source at 600 keV photon-energy with $B_{avg} \approx 2 \times 10^{13}$ that otherwise is not accessible in that spectral range.

Progress in ultra-intense lasers hastens explorations into strong-relativistic regimes, where multiple Compton harmonics are generated. The ATF provides the necessary combination of a high-brightness electron beams and relativistic-strong laser, $a_0 = 0.855 \times 10^{-9}\lambda_L[\mu m]\sqrt{I_L[Wcm^{-2}]} \geq 1$ where λ_L is the laser wavelength and I_L is the laser intensity. Following the 2nd -harmonic CS demonstration in 2006 at $a_0 = 0.4 - 0.5$ [8], the BNL-UCLA collaboration recently observed the 3rd harmonic and the mass-shift effect at $a_0 = 1 - 1.4$ [9]. For the yet unexplored $a_0 \gg 1$ regime, numerous harmonics will be generated increasing in intensity up to the critical harmonic number $n_c \cong \frac{3}{4}a_o^3$ [10] and thus yielding a Compton continuum. With the planned ATF upgrade to the 500-MeV electron linac and a_0= 10 from a CO_2 laser [11], the harmonic continuum will be extended to 6 MeV gamma rays reaching $n_c \simeq 1000$ (see Fig. 1).

3 Prospects for All-Optical Compton Sources

The pre-eminent way to miniaturizing the CS is to use the most space-efficient design for the electron accelerator. Here, lasers again offer a solution. Laser-driven plasma wake field electron accelerators (LWFAs) deliver acceleration gradients by three orders-of-magnitude above the conventional RF accelerators due to downscaling in the period of the accelerating structure to the plasma wavelength. The already demonstrated GeV-class LWFAs from centimetre-long plasmas deliver beam qualities competitive with RF accelerator parameters, such as the micron-scale σ_e and ε_n, ~ 100pC charge, sub-percent energy spread [12]; these beams can readily produce high-peak-brightness synchrotron radiation.

A variety of complementary synchrotron radiation mechanisms are associated with LWFA: The most straightforward approach is to send an electron beam from a compact several-centimetre plasma accelerator through a conventional undulator, thereby producing synchrotron radiation in UV and soft x-rays as has been proposed [13] and experimentally demonstrated [14].

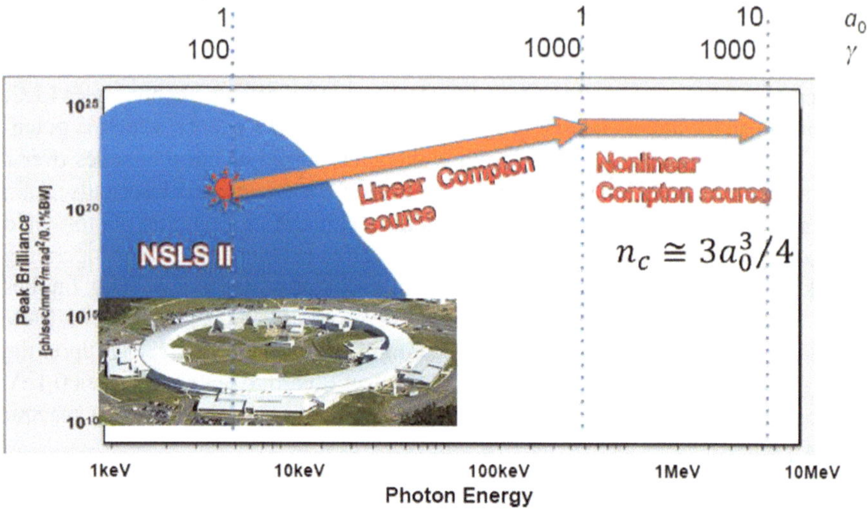

Fig. 1 Compton source is capable to achieve high peak brightness in the gamma region which is not attainable with conventional SLSs. A star marks the result achieved at ATF using 60 MeV linac [1], [2]; upward arrow illustrates the anticipated peak brightness scaling with the linac upgrade to 500 MeV; horizontal arrow outlines further progress towards the gamma continuum upon the proposed CO_2 laser upgrade to 100 TW peak power

At the same time, an electron bunch accelerated inside a plasma "bubble" radiates due to its betatron motion caused by transverse focusing plasma-fields [15]. Similar to other forms of synchrotron radiation, the fundamental wavelength due to the electron's betatron oscillation in a plasma column scales with γ into the x-ray- and gamma- region $\lambda = \lambda_b/2\gamma^2$, where $\lambda_b = \lambda_p\sqrt{2\gamma}$ is the betatron wavelength, and λ_p is the plasma wavelength. Inside the plasma wake field, electrons are being accelerated and radiating simultaneously, so resulting in a broad spread of the betatron spectrum. However, the very small micron size of the electron bunch in a plasma accelerator insures the spatial coherency of the synchrotron radiation, as was demonstrated by phase-contrast imaging [16]. Combined with the inherent femtosecond pulse- duration, the synchrotron radiation from the LWFA is a useful diagnostic tool and also suitable for applied exploitations [17].

Due to the short wavelength of a laser, CSs reach the highest photon-energy than do magnetic- or plasma-wigglers. Such approach has been experimentally investigated over the last few years, using laser beams synchronized for simultaneously driving LWFA electron bunches from a gas jet and bouncing Compton x-rays from these bunches [17, 18]. With further progress towards the kHz-class laser drivers for LWFA combined with CO_2 lasers for Compton scattering, we can realize compact all-optical Compton light sources tunable to the 10 MeV spectral range.

The electron beam-driven plasma wakefield accelerator (PWFA) offers us an opportunity of achieving the electron beam's ε_n not just comparable to contemporary RF linacs and LWFAs, but even highly superior to it. Unlike the LWFA

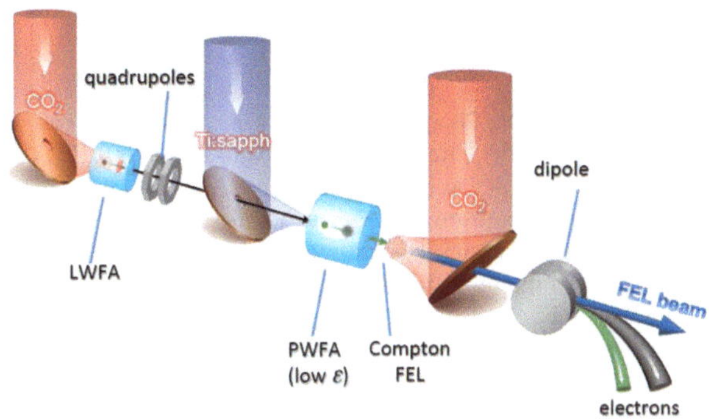

Fig. 2 Conceptual diagram of all-optical Compton FEL

where an accelerating cavity in plasma is produced by the *ponderomotive* electron expulsion, so resulting in the electrons being randomly heated to several MeV, in the PWFA, electrons experience negligible heating under the *Coulomb* force of the driver-bunch. Precisely injected into such "cold" plasma bubble with the micrometre-accurate laser-ionization technique named the Trojan Horse (a virtual photo-cathode) [19], the electrons can form an accelerated beam with a ultra-low emittance, ε_n. The impact of this predicted beam improvement on the CS's properties could be similar to the benefit gained from using low-emittance linacs instead of synchrotrons for SLSs, a concept that resulted in the inception of 4th -generation coherent light-sources—FELs. The Compton FEL's lasing threshold is attained more easily at lower electron's wiggling frequency with a CO_2 laser [20]. Assuming realizable PWFA parameters, $\sigma_E = 1\text{MeV}$, $\gamma = 60$, $I = 6\text{kA}$, $\varepsilon_n = 30\text{nm}$ and a 1-TW laser with a of 10 ps pulse and 10 J energy, focused to $\sigma_L = 50\mu\text{m}$, we anticipate FEL lasing at 7.6 Å, with a peak power of 10 MW, and $B_p = 3 \times 10^{26}$ (after [21]). Figure 2 shows the conceptual design of such an all-optical FEL.

4 Conclusions

We reviewed recent results in three areas of the CS explorations conducted at the ATF: High peak-brightness pico- and femtosecond regime, high-average-power intra-cavity CS, and the nonlinear regime. On-going ATF upgrade to 100 TW CO_2 laser power and 500 MeV linac energy will allow to extend these studies to the gamma-region and explore the highly nonlinear regime where a continuum of Compton harmonics will be produced. Substituting an RF linac with a laser-driven plasma accelerator opens the way to compact all-optical light sources and the opportunity to reach the regime of a coherent Compton FEL.

References

1. Yakimenko, V., Pogorelsky, I.V.: Polarized γ source based on Compton backscattering in a laser cavity. Phys. Rev. ST Accel. Beams. **9,** 091001 (2006)
2. Pogorelsky, I.V., et al.: Demonstration of 8×10^{18} photons/second peaked at 1.8 Å in a relativistic Thomson scattering experiment. Phys. Rev. ST Accel. Beams. **39,** 090702 (2000)
3. Endrizzi, M. at al.: Quantitative phase retrieval with picosecond x-ray pulses from the ATF inverse Compton scattering source. Opt. Express. **19,** 2748–2753 (2011)
4. Sakaue, K., Endo, A., Washio, M.: Design of high brightness laser-Compton source for extreme ultraviolet and soft x-ray wavelengths. J. Micro/Nanolith. **11,** 021124 (2012)
5. Graves, W.S., Broun, W., Kaertner, F.X., Moncton, D.E.: Intense super-radiant x-rays from a compact source using a nanocathode array and emittance exchange. Nucl. Instr. Meth. A. **608,** S103 (2009)
6. Akagi, T., et al.: "Production of gamma rays by pulsed laser beam Compton scattering off GeV-electrons using a non-planar four-mirror optical cavity" J. Instrum. **7**, P01021 (2012)
7. Pogorelsky, I.V., et al.: "High-brightness intra-cavity source of Compton radiation" J. Phys. B: At. Mol. Opt. Phys. **47**, 234014 (2014)
8. Babzien, M., et al.: Harmonic in Thomson Scattering from Relativistic Electrons. PRL. **96,** 054802 (2006)
9. Sakai, Y., et al.: Observation of redshifting and harmonic generation in inverse Compton scattering. Phys. Rev. ST Accel. Beams. **18**, 060702 (2015)
10. Esarey, E., et al.: Nonlinear Thomson scattering of intense laser pulses from beams and plasmas. Phys. Rev. E. **48,** 3003–3021 (1993)
11. Pogorelsky, I.V., Ben-Zvi, I.: Brookhaven National Laboratory's ATF—Research Highlights and Plans, Plasma Phys. Contr. Fus. **58**, 084017 (2014)
12. Esarey, E.: Physics of laser-driven plasma-based electron accelerators. Rev. Mod. Phys. **81,** 1229–1285 (2009)
13. Catravas, P., et al.: Femtosecond x-rays from Thomson scattering using laser wakefield accelerators. Meas. Sci. Technol. **12,** 1828–1834 (2001)
14. Schlenvoigt, H.P., et al.: A compact synchrotron radiation source driven by a laser-plasma wakefield accelerator. Nature Phys. **4,**130–133 (2008)
15. Kneip, S., et al.: Bright spatially coherent synchrotron x-rays from a table-top source. Nature Phys. **6,** 980–983 (2010)
16. Fourmaux, S., et al.: Single shot phase contrast imaging using laser-produced Betatron x-ray beams. Opt. Lett. **36,** 2426–2428 (2011)
17. Corde, S., et al.: Femtosecond x-rays from laser-plasma accelerators. Rev. Mod. Phys. **85,** 1–47 (2013)
18. Phuoc, K. Ta., et al: All-optical Compton gamma-ray source. Nat. Photon. **6,** 308–311 (2012)
19. Hidding, B., et al.: Ultracold electron bunch generation via plasma photocathode emission and acceleration in a Beam-driven plasma blowout. *PRL.* **108,** (2012)
20. Bacci, A., et al.: Compact x-ray free-electron laser based on an optical undulator. Nucl. Instrum. Meth. A. **587,** 388–397 (2008)
21. Bacci, A., et al.: Transverse effects in the production of x-rays with a free-electron laser based on an optical undulator. Phys. Rev. ST Accel. Beams, **9,** 060704, 2006

X-ray Generation from Ultra-High Energy Density Relativistic Plasmas by Ultrafast Laser Irradiation of Nanowire Arrays

Michael A. Purvis, Vyacheslav N. Shlyaptsev, Reed Hollinger, Clayton Bargsten, Alexander Pukhov, David Keiss, Amanda Towsend, Yong Wang, Shoujun Wang, Mark Berrill, Bradley Luther, Amy Prieto and Jorge. J. Rocca

Abstract We have demonstrated the volumetric heating of near-solid density plasmas to keV temperatures using ultra-high contrast femtosecond laser pulses of only 0.5 J energy to irradiate arrays of vertically aligned nanowires (Purvis et al. Nat Photonics 7:796–780, 2013). Our x-ray spectra and particle-in-cell (PIC) simulations show extremely highly ionized plasma volumes several micrometers in depth are generated by irradiation of Au and Ni nanowire arrays with femtosecond laser pulses of relativistic intensities. Arrays of vertically aligned Ni nanowires with an average density of 12 % solid were ionized to the He-like stage. The He-like line emission from the nanowire target exceeds the intensity of the Ni Kα line at this irradiation intensity. Similarly near-solid density Au nanowire arrays were ionized to the Co-like (Au^{52+}). This volumetric plasma heating approach creates a new laboratory plasma regime in which extreme plasma parameters can be accessed with table-top lasers. Scaling to higher laser intensities promises to create plasmas with temperatures and pressures similar to those in the center of the sun. The increased hydrodynamic-to-radiative lifetime ratio is responsible for a dramatic increase in the x-ray emission with respect to polished solid targets. As highly efficient X-ray emitters and sources of extreme plasma conditions, these plasmas could play a role in the development of new ultra-short pulse soft x-ray lasers.

J. J. Rocca (✉) · M. A. Purvis · V. N. Shlyaptsev · R. Hollinger · C. Bargsten · D. Keiss
A. Towsend · Y. Wang · S. Wang · M. Berrill · B. Luther · A. Prieto
Colorado State University, Fort Collins, CO, USA
e-mail: Jorge.rocca@colostate.edo

A. Pukhov
Institut für Theoretische Physik, Heinrich-Heine-Universität Düsseldorf,
Düsseldorf, Germany

M. Berrill · B. Luther
Oak Ridge National Laboratory, Oak Ridge, USA

© Springer International Publishing Switzerland 2016
J. Rocca et al. (eds.), *X-Ray Lasers 2014*, Springer Proceedings in Physics 169,
DOI 10.1007/978-3-319-19521-6_18

1 Introduction

The heating of dense matter to extreme temperatures motivates the development of powerful lasers. However, the barrier the critical electron density imposes to light penetration into ionized materials results in the deposition of most of the laser energy into a thin surface layer at typically only 0.1 % of solid density. The small plasma volume results in a short hydrodynamic lifetime that limits the efficiency with which x-rays are emitted. We demonstrate the volumetric heating of near-solid density plasmas to keV temperatures using ultra-high contrast femtosecond laser pulses of only 0.5 J energy to irradiate arrays of vertically aligned nanowires [1].

2 Demonstration of Volumetric Heating of Near Solid Density Plasmas to Multi-keV Temperatures

Our x-ray spectra and particle-in-cell (PIC) simulations show extremely highly ionized plasma volumes several micrometers in depth are generated by irradiation of Au and Ni nanowire arrays with femtosecond laser pulses of relativistic intensities (Fig. 1).

The sequence of PIC maps in Fig. 1a describe the phases of nanowire heating, explosion and volume plasma formation. The simulation shows that an intense femtosecond laser pulse can penetrate deeply into an array of vertically aligned Ni wires 55 nm diameter where is nearly totally absorbed (Fig. 1b). Optical field ionization is active at the boundary of the nanowires where vacuum heating of electron takes place. The hot electron population drives collisional ionization deep into the nanowire core, causing the plasma to rapidly reach an electron density of $2 \times 10^{24}\,\mathrm{cm}^{-3}$ (left frames in Fig. 1c). Despite the high average electron density, the laser pulse propagates deeply inside the nanostructured material as long as the gaps between the nanowires are free from plasma. The high energy deposited into the nanowire cores is transformed into a thermal energy density of $\sim$2GJcm^{-3}, with average electron energy of 4 keV over the penetration depth and $\sim$8 keV near the nanowire tips. The heated nanowire plasmas rapidly expand radially until the plasma fills the vacuum gaps. Beyond this point any further propagation of laser light deep into the target is effectively terminated by the formation of a critical density surface. The ultra-short duration of the laser pulse allows for the majority of the energy to be efficiently deposited before the gap closure.

The PIC simulations were complemented with hydrodynamic simulations that make use of a more detailed atomic mode (Fig. 2), that corroborate that the multi-KeV electron temperature in these near-solid-density plasmas results in extreme degree of ionization at the tip of the wire and propagates along the wires towards the supporting substrate (Fig. 1c). Our simulations show that 350 fs after the peak of laser irradiation a fiber length of 4–5 μm has "burned" (Fig. 1c), creating a plasma layer with an average density of 3×10^{23}cm^{-3} and an electron temperature of

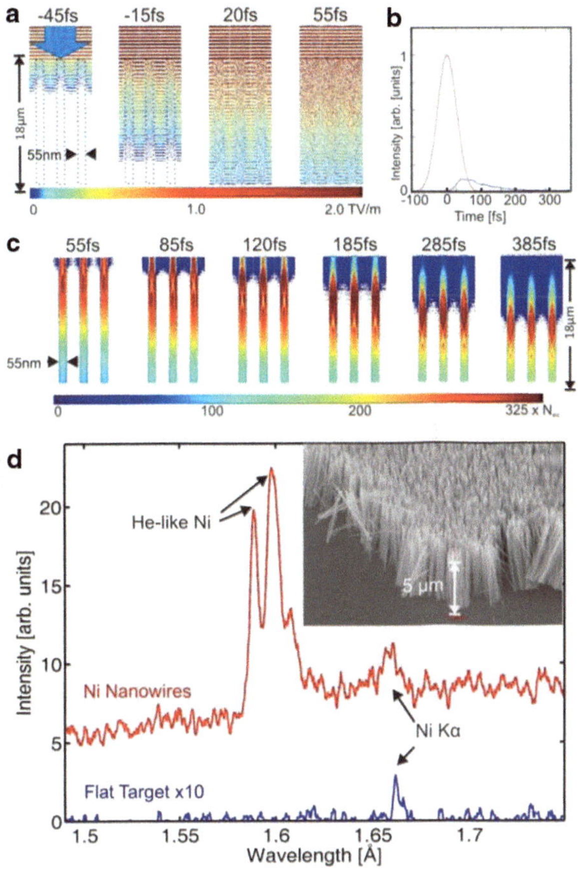

Fig. 1 Particle-in-cell simulation and spectra from plasma generated by fs pulse irradiation of a vertically aligned array of 55 nm diam. Ni nanowires. **a** PIC simulations of the electron density (N_e) evolution in an array of 18 μm long Ni wires with an average atomic density of 12 % solid density (inset) irradiated at 5×10^{18} W cm^{-2} by a $\lambda = 400$ nm, 60 fs FWHM laser pulse. N_e is in units of critical density ($N_{ec} = 6.8 \times 10^{21}$cm^{-3}). **b** Computed impinging (red) and reflected (blue) laser intensity. **c** Computed electron density (Ne) evolution. Ne is in units of in units of critical density ($N_{ec} = 7 \times 10^{21}$cm^{-3}). **d** Single-shot X-ray spectra comparing the emission from an irradiated array of 5 μm long, 55 nm diameter, Ni nanowires to that from a flat polished Ni target. The nanowire plasma spectra is dominated by He-like Ni resonance line, and by the intercombination line merged with $1s2p^2 - 1s^22p$ Li-like Ni lines

2–4 keV, in which an extremely high degree of ionization is rapidly reached by collisional electron impact ionization. (eg. He-like Ni, bottom right frame in Fig. 2) driven by the very high rates of collisional ionization.

We irradiated arrays of vertically aligned Ni and Au nanowires with ultra-high contrast ($> 1 \times 10^{11}$) pulses of ~60 fs FWHM duration from a frequency doubled ($\lambda = 400$ nm) high power Ti:Sa laser. The ultra-high contrast is critical to avoid destroying the array prior to the arrival of the high intensity ultra-short pulse. The laser beam was focused to reach an intensity of 5×10^{18} Wcm^{-2} onto arrays of 55

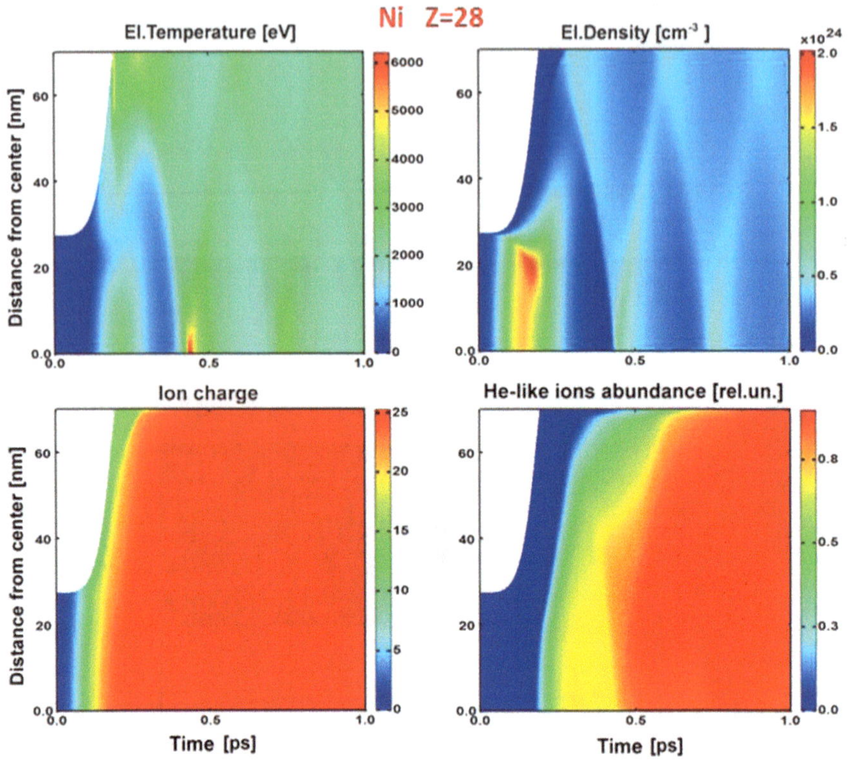

Fig. 2 Hydrodynamic simulation of nanowire plasma created by irradiation of an array of Ni nanowires 55 nm diameter with 50 fs pulse with an intensity of 5×10^{18} Wcm^{-2}. The plasma parameters are plotted as a function of distance from the nanowire target and of time respect to peak of the laser pulse. The four frames show the evolution of the electron temperature, electron density, mean ion charge and abundance of He-like Ni ions

and 80 nm diameter nanowires with an average density of approximately 12 % solid density with an off-axis parabolic mirror. The X-ray emission was simultaneously monitored with a von Hamos crystal spectrometer oriented at a 45° angle from the target surface and with an array of four filtered Si photodiodes.

Figure 1d shows a single-shot spectrum of the emission in the 1.5–1.75 Å region for a plasma created from vertically aligned Ni nanowires 55 nm diameter separated by 135 μm. Strong emission from the 2p-1s (= 1.588 Å) and intercombination lines of He-like Ni (Ni^{+26}) is observed. This spectrum differs dramatically from that corresponding to a polished flat target irradiated at the same conditions which only shows line emission from the Ni Kα line at 1.658 Å (with 10X magnified scale in Fig. 1d). The Kα emission is produced mainly by high energy electrons, while generation of the He-like ion transitions requires a hot thermal plasma to generate and excite the highly stripped ions. It is remarkable that the He-like line emission from the nanowire target exceeds the intensity of the Kα line at this irradiation intensity,

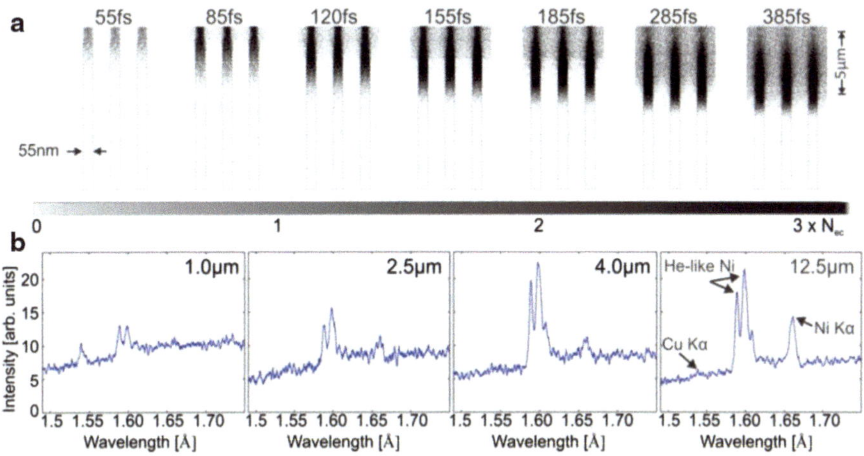

Fig. 3 Generation of He-like ions in the volumetrically heated Ni plasmas **a** Particle-in-cell simulation maps showing the evolution of the density of He-like nickel ions. The He-like ion density is in units of critical electron density (Nec = 6.8×10^{21} cm^{-3}). The wires are 55 nm diameter, and the array has an average atom density of 0.12 solid. The irradiation intensity is 5×10^{18} Wcm^{-2}, **b** X-ray emission of nanowire targets of four different lengths: 1, 2.5, 4, and 12.5 μm. Heα emission is observed to saturate for wire lengths > 4 μm, suggesting a plasma volume 3–4 μm deep is heated sufficiently for ionization to reach the He-like stage

since in previous work in which a Petawatt laser was used to irradiate Cu foils the emission from the Kα lines was only surpassed at irradiation intensities $> 2 \times 10^{20}$ Wcm^{-2} [2]. Similarly, a near solid-density Au plasma with an extraordinarily high degree of ionization was created by irradiating an array of 80 nm diameter Au wires 5 μm in length. The Au nanowire spectra ($\lambda = 4.5$–5.25 Å region) display strong Au M-shell emission with unresolved $4-3$ lines from ions ranging from Co-like (Au^{+52}) to Ga-like Au (Au^{+48}) [1].

The depth to which the Ni nanowire target is ionized to the He-like state was experimentally determined to be 3–4 μm by monitoring the spectrally resolved X-ray emission as a function of nanowire length. Figure 3b shows spectra from arrays of 55 nm diameter Ni nanowires of four different lengths between 1 and 12.5 μm. The intensity of the He-like line is observed to increase with nanowire length until it saturates for a length of about 4 μm, beyond which it remains virtually unchanged. In contrast, the K-α emission continues to increase beyond this length as the fast electrons interact with a larger volume of neutral atoms. PIC simulations reproduce the experimental results in showing a plasma depth of several μm is volumetrically heated to reach the He-like ion stage (Fig. 3a).

3 Efficient X-ray Generation

This large increase in heated plasma volume of these nanowire plasmas relative to plasmas from flat solid targets irradiated at the same conditions significantly lengthens the plasma hydrodynamic time. Simultaneously, the radiative lifetime is greatly decreased as a result of the nearly two orders of magnitude increase in plasma density, the higher temperature, and higher average Z of the nanowire plasma. The radiation rates for line radiation and the different continuum radiation channels were computed in comparison to the hydrodynamic time. As the electron density is increased from 1×10^{22} to $1 \times 10^{24}\text{cm}^{-3}$ the radiation cooling time, τ_{rad}, decreases up to 100 times. The combination of the increase in plasma size, ΔL, and decreased radiation time results in a greatly decreased cooling-time to hydrodynamic-time ratio $\tau_{\text{rad}}/\tau_{\text{hydro}}$, where:

$$\tau_{rad} \propto \frac{1}{N_e} < \frac{\Delta L}{C_s} \propto \tau_{hydro}$$

C_s is the speed of sound. This explains the observed dramatic increase in the X-ray yield respect to polished solid targets in agreement with spectra in Fig. 1 and published data [3]. The aligned Ni nanowire target produced an increase of more than 50 times in X-ray flux in this 7–8 keV spectral region. Similar enhancements were also observed in the 1 keV photon energy region in agreement with previous experiments. Au nanowire spectrum in the $\lambda = 4.5$–5.25 Å region displays strong Au M-shell emission with unresolved 4-3 lines from ions ranging from Co-like (Au^{+52}) to Ga-like Au (Au^{+48}). Both the spectra and the filtered photodiode array signals from the Au targets show a great increase in X-ray yield with respect to flat solid targets irradiated by the same pulse, up to 100X in the emission in the 2.3–2.75 keV photon region. Measurement of the X-ray yield conducting using the array of filtered silicon photodiodes yield a conversion efficiency in to $h\nu > 0.9$ KeV photons of $\sim 1 \times 10^{-4}$.

4 Summary and Discussion

In summary, electron densities nearly 100 times greater than the typical critical density and multi-keV temperatures are achieved using laser pulses of only 0.5 J energy. We obtained extraordinarily high degrees of ionization (eg. 52 times ionized Au), and gigabar pressures only exceeded in the central hot-spot of highly compressed thermonuclear fusion plasmas. The combination of the increase in plasma size, ΔL, and decreased radiation time results in a greatly decreased cooling-time to hydrodynamic-time ratio, that in turn results in a large increase in X-ray radiation yield. The increase in linewidth resulting from collisional broadening in these large volume high density plasmas offers a new type of gain medium with sufficient bandwidth to produce ultra-short pulse soft x-ray lasers. Also scaling to higher laser intensities promises to create plasmas with temperatures and pressures similar to those in the center of the sun.

Acknowledgements Work supported by DTRA grant HDTRA-1-10-1-0079 and the Fusion Energy program of the Office of Science of the US Department of Energy. A. Pukhov acknowledges the support of DFG-funded project TR18

References

1. Purvis, M.A., Shlyaptsev, V.N., Hollinger, R., Bargsten, C., Pukhov, A., Prieto, A., Wang, Y., Luther, B.M., Yin, L., Wang, S., Rocca, J.J.: Relativistic plasma nanophotonics for ultrahigh energy density physics. Nat. Photonics. **7,** 796–780 (2013)
2. Theobald, W., Akli, K., Clarke, R., Delettrez, J.A., Freeman, R.R., Glenzer, S., Green, J., Gregori, G., Heathcote, R., Izumi, N., King, J.A., Koch, J.A., Kuba, J., Lancaster, K., MacKinnon, A.J., Key, M., Mileham, C., Myatt, J., Neely, D., Norreys, P.A., Park, H.S., Pasely, J., Patel, P., Regan, S.P., Sawada, H., Shepherd, R., Snavely, R., Stephens, R.B., Stoeckl, C., Storm, M., Zhang, B., Sangster, T.C.: Hot surface ionic line emission and cold K-inner shell emission from petawatt-laser-irradiated Cu foil targets. Phys. Plasmas. **13,** 043102 (2006)
3. Kulcsár, G., AlMawlawi, D., Budnick, F.W., Herman, P.R., Moskovits, M., Zhao, L., Marjoribanks, R.S.: Intense Picosecond X-Ray Pulses from Laser Plasmas by Use of Nanostructured "Velvet" Targets. Phys. Rev. Lett. **84,** 5149–5152 (2000)

Ionization of Xenon to the Nickel-Like Stage and Beyond in Micro-Capillary Plasma Columns Heated by Ultrafast Current Pulses

G. Avaria, M. Grisham, J. Li, F. G. Tomasel, V. N. Shlyapstsev, M. Busquet, M. Woolston and J. J. Rocca

Abstract Homogeneous plasma columns with ionization levels typical of MA discharges were created by rapidly heating gas-filled 520 μm diameter channels with ns rise-time current pulses of unusually low amplitude, 40 kA. These conditions allow the generation of high aspect ratio (eg. > 300:1) plasma columns with very high degrees of ionization (e.g. Ni-like Xenon) of interest for soft x-ray lasers below $\lambda = 10$ nm. Spectra and simulations of plasmas generated in 520 μm diameter alumina capillaries driven by 35–40 kA current pulses with 4 ns rise time were obtained for discharges in Xenon and Neon discharges. The first shows the presence of lines corresponding to ionization stages up to Fe-like Xe. The latter show that Al impurities from the walls are ionized to the H-like and He-like stages. He-like Al spectra containing the resonance line significantly broaden by opacity, the intercombination line, and Li-like satellites are analyzed. For Xenon discharges, the spectral lines from the Ni-like transitions the $3d^9 4d\ (3/2, 3/2)_{J=0}$ to the $3d^9 4p(5/2, 3/2)_{J=1}$ and to $3d^9 4p(3/2, 1/2)_{J=1}$ are observed.

G. Avaria (✉) · M. Grisham · J. Li · V. N. Shlyapstsev · M. Woolston · J. J. Rocca
National Science Foundation ERC for Extreme Ultraviolet Science and Technology,
Fort Collins, CO 80523, USA
e-mail: gavaria@cchen.cl

G. Avaria · M. Grisham · J. Li · F. G. Tomasel · V. N. Shlyapstsev · M. Woolston · J. J. Rocca
Department of Electrical and Computing Engineering,
Colorado State University, Fort Collins, CO 80523, USA

F. G. Tomasel
Advanced Energy Industries, Fort Collis, CO 80525, USA

M. Busquet
ARTEP Inc, Ellicott City, MD 21042, USA

J. J. Rocca
Department of Physics, Colorado State University, Fort Collins, CO 80523, USA

G. Avaria
Comisi6n Chilena de Energía Nuclear, Casilla 188-D, Santiago, Chile

© Springer International Publishing Switzerland 2016
J. Rocca et al. (eds.), *X-Ray Lasers 2014*, Springer Proceedings in Physics 169,
DOI 10.1007/978-3-319-19521-6_19

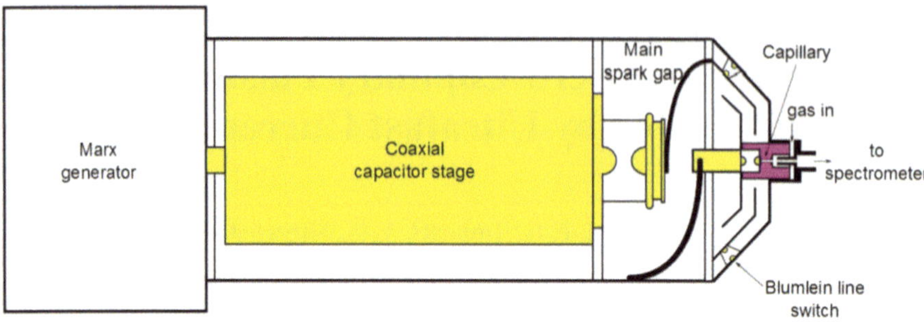

Fig. 1 Discharge schematic showing the different pulse compression stages

1 Introduction

The ability to produce hot and dense plasma columns with high degree of ionization in a compact setup is of interest to the plasma community in general and the x-ray laser community in particular. Dense homogeneous plasma columns are of interest to extend discharge pumped table-top soft x-ray lasers [1] to shorter wavelengths, to create plasma waveguides and generate neutrons, as well as to perform fundamental plasma studies. The observation of large laser amplification in capillary discharges [1] has allowed the development of high average power soft x-ray lasers at 46.9 nm in Ne-like Argon [2] that have enabled numerous applications. In the following work we present results from a micro-capillary plasma discharge heated by ultrafast current pulses that produces highly ionized Xenon plasmas with temperatures of hundreds of eV and densities close to $10^{20}\mathrm{cm}^{-3}$ [3]. These discharges use current pulses with ultrafast discharges and modest amplitude to generate plasma columns with conditions of interest for laser amplification at 9.9 nm in Ni-like Xe.

2 Experimental Setup

The experiments were carried out in $\sim 520\,\mu\mathrm{m}$ diameter alumina (99.8 %) capillaries of 3.3 cm in length. These capillaries were excited by a fast rise and moderate amplitude current pulse, generated from an eight stage Marx bank discharged through two pulse compression stages, as seen in Fig. 1. The first consists of an oil filled coaxial capacitor that is connected through a self-breakdown spark gap to the final pulse compression stage, a Blumlein line. The radial, oil filled Blumlein transmission line is then discharged into the capillary located at the axis of the Blumlein, decreasing the inductance of the system. This produces a 3.8–4.0 ns rise time (10–90 %) current pulse with amplitude of ~ 40 kA.

A typical current pulse measured with an air-core Rogowski coil can be seen in Fig. 2. The gas inside the capillary is pre-ionized by a µs-long pulse of ~ 20 A and the main current pulse is preceded by a 3–5 kA amplitude pre-pulse. The

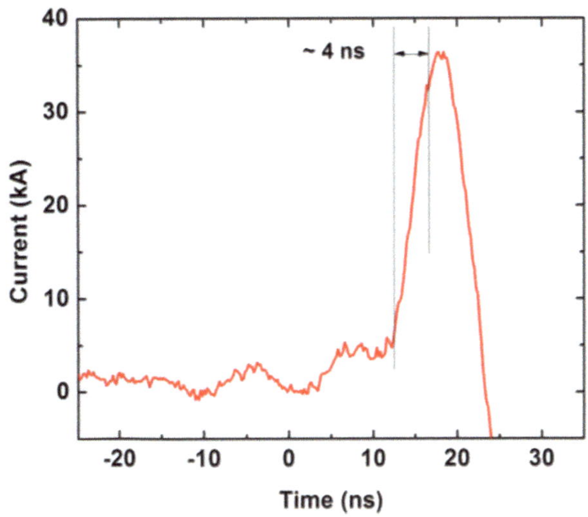

Fig. 2 Current pulse showing a fast rise time of 3.8 ns (10–90 %) and amplitude close to 35 kA. A 3–5 kA prepulse is observed

plasma properties were diagnosed by pinhole imaging and soft X-ray spectroscopy. Pinhole images were obtained by using a 38 μm diameter pinhole placed 14.6 cm from the capillary axis. Images of the plasma evolution at times earlier than the maximum current were obtained by using a stacked mylar/parylene foil filter (1/0.1 μm thick). At later times the emission of photons of energies higher than 0.6 keV was detected by using a stacked mylar/aluminium filter of 1/0.5 μm respectively. This setup allowed a spatial resolution of $\sim$42 μm. Time resolved images were acquired through a microchannel (MCP) plate/CCD detector system, with a time resolution $\sim$2 ns. Soft X-ray spectroscopy was done by means of a 2° grazing incidence flat field spectrometer equipped with a 600 l/mm variable spacing, gold coated diffraction grating. The spectral region observed ranged from 1.4 to 2.0 nm and was detected by the same MCP/CCD system used in the pinhole imaging setup.

Time integrated spectra were acquired by means of a convex KAP crystal spectrometer with a back illuminated CCD detector. This enabled the characterization of the spectral region around 0.78 nm. Hydrodynamic simulations were conducted using the RADEX code which simulated capillary plasmas using transient atomic physics with data from the HULLAC code.

3 Results and Discussion

Figure 3 shows a sequence of filtered pinhole images of a discharge in Xenon at a pressure of 0.5 Torr with a peak current of 35.4 ± 0.9 kA. The image at -5.9 ns was acquired using the mylar/parylene filter and shows a colder plasma

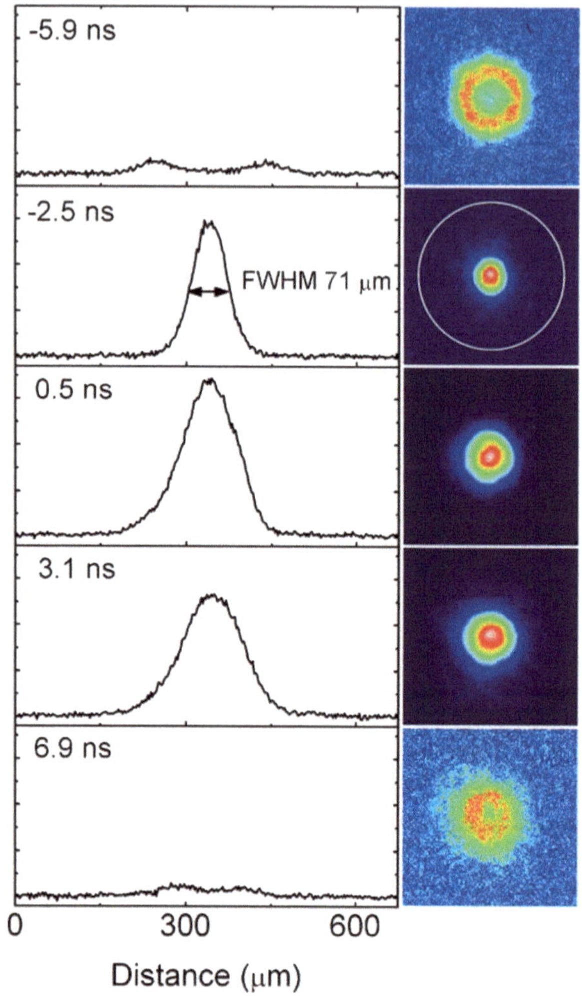

Fig. 3 Sequence of filtered pinhole images showing the compression of the plasma column at the centre of the capillary. The white circle indicates the approximate position of the capillary wall

with an annular shape moving towards the capillary axis. Later images were taken with the mylar/aluminium filter which shows a compressed column of around 70 μm in diameter, indicating the formation of a plasma column of > 300:1 length-to-diameter aspect ratio. The symmetry observed during the compression and expansion phase of the plasma column suggests that the plasma is free from any major instabilities as seen in larger diameter capillaries.

A time resolved sequence of the spectrum corresponding to a plasma column generated with 0.5 Torr of Xenon in a 520 μm diameter capillary with a peak current amplitude of 37.6 ± 0.6 kA, is detailed in Fig. 4 Simulations indicate that the 3–5

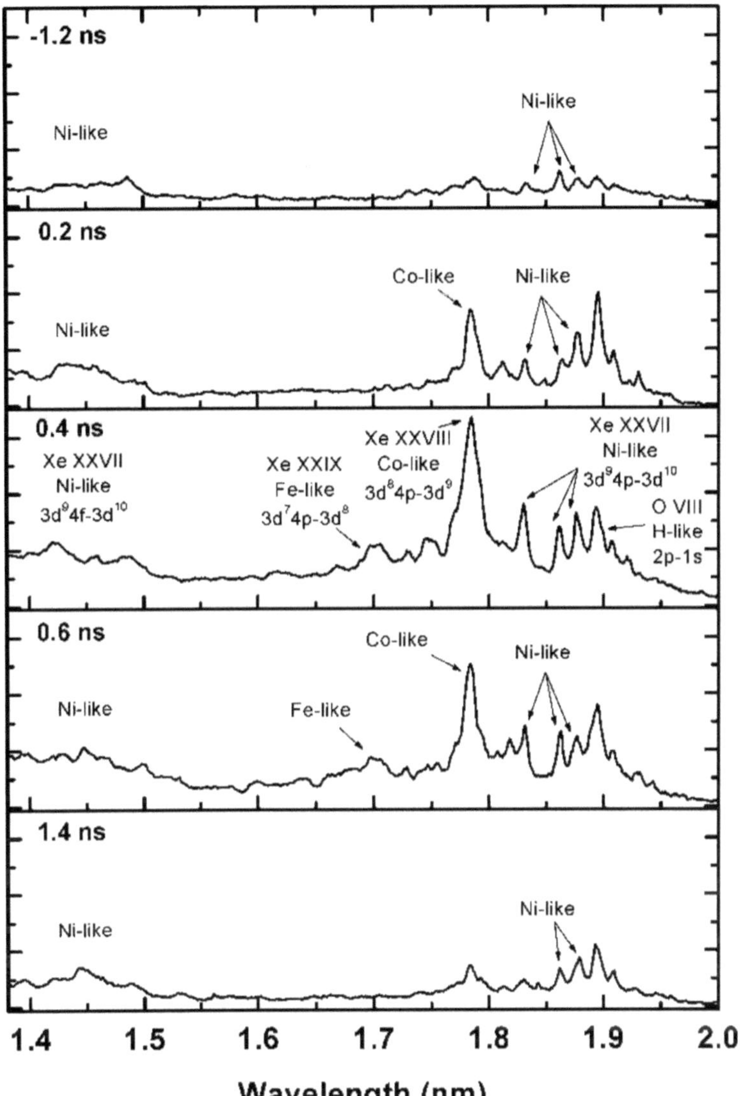

Fig. 4 Sequence of time-resolved single-shot spectra of Xe plasma in a 520 μm-diameter capillary channel excited by a 37.6 ± 0.6 kA current pulse, at a pressure of 0.5 Torr. Timing is referenced to the peak of the current pulse

kA pre-pulse produces a pre-heated plasma with Z ~ 12–15. From this stage, the ~ 40 kA pulse is able to ionize the plasma to levels up to Fe-like Xenon (Xe^{+28}). It can be seen that the plasma column rapidly reaches a stage in which Co-like Xenon (Xe^{+27}) lines dominate spectra followed by Ni-like Xenon (Xe^{+26}) lines. Fe-like Xenon (Xe^{+28}) is also detected at times close to the maximum current. Simulations

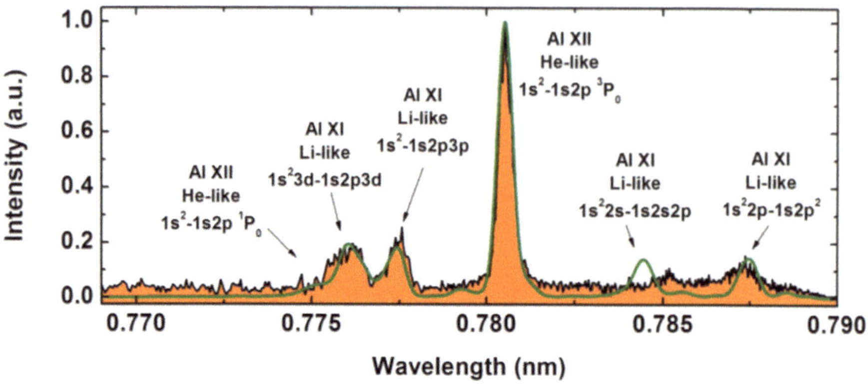

Fig. 5 Measured on axis and simulated spectrum of a neon discharge on a 520 μm-diameter alumina capillary. The intercombination line dominates over the resonance line. Current was 36.0 kA and the Neon pressure was 0.5 Torr

of these spectra predict an electron temperature of around 350 eV. When the filling gas pressure is increased to 4 Torr, similar ionization degrees are obtained where simulations indicate that the plasma reaches densities around $0.5 \times 10^{20}\text{cm}^{-3}$.

These results shows that the micro-capillary plasma column reaches the degree of ionization required to explore the development of a Ni-like Xe soft x-ray laser amplifier.

The extraordinary characteristics of the plasma generated in these capillaries also allow the observation of unexpected spectroscopic phenomena. As seen in Fig. 5, the emission originating from impurities extracted from the capillary wall show an Aluminum intercombination line dominating the spectra over the resonance line. Simulations of these results indicate that the plasma column large aspect ratio (> 300:1 length-to-diameter) is responsible for the unusual effect. The different optical depths,optically thin in the transverse direction and optically thick in the longitudinal direction for both transitions, create conditions in which the population of the $1s2p^3P_1$ level, upper level of the intercombination line, is favored by up to two orders of magnitude over the $1s2p^1P_1$ level, determining the ratio of the line intensities observed in the experiment.

4 Conclusion

The injection of ultrafast current pulses in micro-capillary channels is demonstrated to enable the efficient production of highly ionized and dense homogeneous plasma columns, which allow the observation of interesting new spectroscopic phenomena. These plasma columns are expected to lead to a strong monopole electron impact excitation of $4d^1S_0$ level of Ni-like Xenon, which could result in a large transient population inversion and amplification in the $4d^1S_0$-$4p^1P_1$ line at 9.9 nm.

Acknowledgements This work was supported by NSF Award PHY-1004295. G. Avaria was supported from CONICYT PAI Inserci6n 791100020, FONDECYT Iniciaci6n 11121587 and CONICYT PIA Anillo ACT1115.

References

1. Rocca, J.J., Shlyaptsev, V.N., Tomasel, F.G., Cortazar, O.D., Hartshorn, D., Chilla, J.L.A.: Demonstration of a discharge pumped table-top soft-X-ray laser. Phys. Rev. Lett. **73**, 2192 (1994)
2. Benware, B.R., Macchietto, C.D., Moreno, C.H., Rocca, J.J.: Demonstration of a high average Power Tabletop Soft X-ray Laser. Phys. Rev. Lett. **81**, 5804 (1998)
3. Avaria, G., Grisham, M., Li, J., Tomasel, F.G., Shlyaptsev, V.N., Busquet, M., Woolston, M., Rocca, J.J.: Extreme degree of ionization in homogenous microcapillary plasma columns heated by ultrafast current pulses. Accepted for publication Phys. Rev. Lett. **114**, 095001 (2015)

Progress of Laser-Driven Betatron Radiation in IOP

L. M. Chen, K. Huang, W. C. Yan, D. Z. Li, Y. Ma, W. M. Wang, Y. T. Li, Z. M. Sheng and J. Zhang

Abstract Betatron x-ray source generated in the process of laser wakefield acceleration has been studied in IOP using different species of gas. For helium plasmas, the generation of bright betatron radiation in electron acceleration via self-injection was studied in experiment and simulation. Two electron bunches with different qualities were injected sequentially into the wakefield driven by a super-intense laser pulse. Hosing instability causes the enhancement of transverse oscillation of the second electron bunch with GeV energy which results in the high yield of betatron radiation. For nitrogen plasmas, we observed the generation of stable quasi-monoenergetic electron beams having energy of 130 MeV and energy spread of $\sim 20\,\%$. And the betatron x-ray emission had a small divergence of 7.5 mrad and a critical energy of 4 keV. We believed that the laser wakefield acceleration process was stimulated in ionization injection.

1 Introduction

Hard x-ray sources from femtosecond laser produced plasmas have a number of interesting applications in the dynamic probing of matter and in medical/biological imaging. With the advent of table-top ultra-short and ultra-intense lasers, laser-plasma acceleration (LPA), which was proposed by Tajima and Dawson [1], has demonstrated its great potential as a compact accelerator and x-ray source. While accelerating longitudinally in the laser wakefield, the electron beams also oscillate transversally (wiggling motion) due to the transverse structure of the wakefield and emit well-collimated betatron x-rays. The betatron oscillation frequency is given

L. M. Chen (✉) · K. Huang · W. C. Yan · Y. Ma · W. M. Wang · Y. T. Li
Beijing National Laboratory of Condensed Matter Physics, Institute of Physics, CAS,Beijing, 100080, China
e-mail: lmchen@iphy.ac.cn

D. Z. Li
Institute of High Energy Physics, CAS,Beijing, 100049, China

Z. M. Sheng · J. Zhang
Laboratory for Laser Plasmas (Ministry of Education) and Department of Physics, Shanghai Jiao Tong University, Shanghai, 200240, China

© Springer International Publishing Switzerland 2016

J. Rocca et al. (eds.), *X-Ray Lasers 2014*, Springer Proceedings in Physics 169,
DOI 10.1007/978-3-319-19521-6_20

by $\omega_\beta = \omega_p(2\gamma)^{-1/2}$, where ω_p is the plasma frequency and γ is the Lorentz factor of the accelerated electron beams. For large-amplitude betatron oscillations (i.e., a few micrometers in size), the resulting broadband spectrum extends up to the critical energy at $\omega_c \approx 2\gamma^2\omega_\beta$, after which it drops exponentially. The radiation is emitted in the forward direction within a cone angle $\theta \approx K/\gamma$, where K is the strength parameter of the plasma wiggler given by $K = 2\pi(\gamma r_0)/\lambda_b = 1.33 \times 10^{-10}\gamma^{1/2}n_e{}^{1/2}[cm^{-3}]r_0[\mu m]$. The average photon number with mean energy $\hbar\omega_c$ emitted by an electron is given by $N_x = 5.6 \times 10^{-3}N_\beta K$, where N_x is the number of oscillation periods. Among several mechanisms to generate x-ray radiation from laser-plasma interactions, the betatron radiation is straightforward and able to deliver larger x-ray photon flux per shot (10^8 photons/shot) [2] and higher photon energy (up to gamma rays) [3]. However, yield of this source is always limited by contradictory between parameters during electron acceleration in the laser wakefield. We will present our recent progress in enhancing accelerating procedure to enrich the x-ray photon yield and optimizing the x-ray beam parameters.

2 Betatron X-Ray Generated from He Gas Via Self-Injection [4]

We demonstrate the generation of high-quality electron beams and collimated bright x-ray sources using a Helium gas jet and a hundred-terawatt (TW) laser. The experiments were performed at the Jupiter Laser Facility, Lawrence Livermore National Laboratory, using the Callisto laser system. The Callisto laser system is a Ti:sapphire laser with a central wavelength λ_0 of 806 nm. The laser beam, which delivers up to 200 TW in power in a 60 fs laser pulse, was focused with an $f/12$ off-axis parabolic mirror to a vacuum spot of 15 μm diameter with 35 % energy contained at the full-width-half-maximum (FWHM). During the experiments, the peak laser power ranged from 60 to 155 TW on the target. In the focused region, the laser intensity was $I = 1.1 \times 10^{19} - 3 \times 10^{19}\ Wcm^{-2}$, which corresponded to the normalised vector potential $a_0 = 2.4 \sim 3.8$. A supersonic nozzle with Mach number ~ 5, which was a rectangular slit nozzle 10 mm in length and 1.2 mm in width. The jets could generate well-defined uniform gas densities in the range of $1 \times 10^{18} \sim 3 \times 10^{19}\text{cm}^{-3}$ by changing the stagnation pressure. A 16-cm-long 0.98 T dipole magnet was set to deflect electrons. To monitor the laser plasma channel length, a charge coupled device (CCD) with a narrow-band-pass filter centred at 800 nm observed the top of the target (Fig. 1).

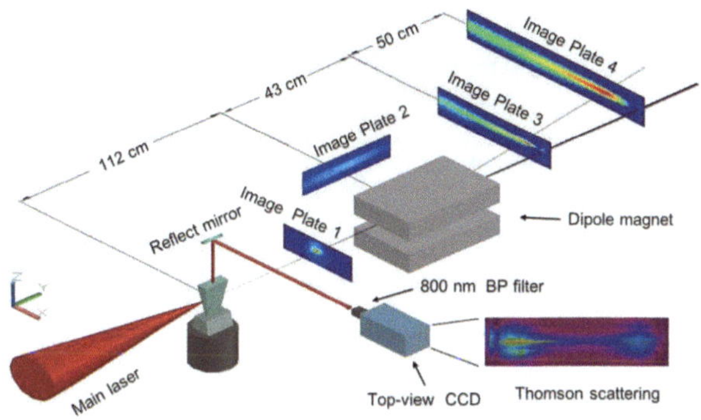

Fig. 1 Schematic of the experimental setup. All image plates were wrapped with 25 μm Al foil. A 1.5 mm-thick Beryllium vacuum window was set 100 cm away from the gas jet to seal the vacuum. The *Lower Right* Inset shows a Thomson scattering image of the plasma channel

2.1 Oscillation Injection Results in Enhanced Betatron X-Ray Emission

By increasing the laser intensity a_0 from 2.6 to 3.7, betatron emission enhancement was obviously observed in Fig. 2. The electron spectrum and betatron x-ray signal were recorded in Fig. 2a, b, c, d with a_0 of 2.65, 2.87, 3.35 and 3.67. The corresponding laser plasma channel was shown in Fig. 2e, f, g, h. We can clearly see that when the a_0 is small, only a bunch of monoenergetic electron was generated in Fig. 2a. As a_0 increased, the laser plasma channel length was elongated and a second bunch with continuous spectrum emerged. When a_0 was high enough, the second electron bunch surpassed the first bunch and reached to energy over 1 GeV. The corresponding betatron x-ray photon number in Fig. 2d reached to 4.5×10^8 for photon energy > 7 keV, which was about 20 times larger than that in Fig. 2a. The divergence angle of the second bunch increased with the betatron divergence and flux, which suggest the betatron oscillation amplitude enhancement of the accelerated electrons in the second bunch.

Two-dimensional particle-in-cell (2D PIC) simulations show that in the accelerating procedure, two bunches of electrons are injected and accelerated in sequentially. The first bunch is injected at early time and accelerated to form a stable quasi-monoenergetic electron beam, and the second bunch, which contains more electrons, is injected at a later time to form a continuous spectrum. As the interaction length gets longer, the second beam is accelerated to the GeV level, and the large transverse oscillation amplitude, which is triggered by the oscillation of the laser spot size, results in the enhancement of the betatron X-ray flux. The injection process was relative with the laser waist evolution as shown in Fig. 3. A distinguished feature appears after $t = 8$ ps when the oscillation period of the laser spot size reduces dramatically from 5 ps to 0.8 ps. Approximately at this time, the second bunch starts to be injected and accelerated. This type of injection is first observed here

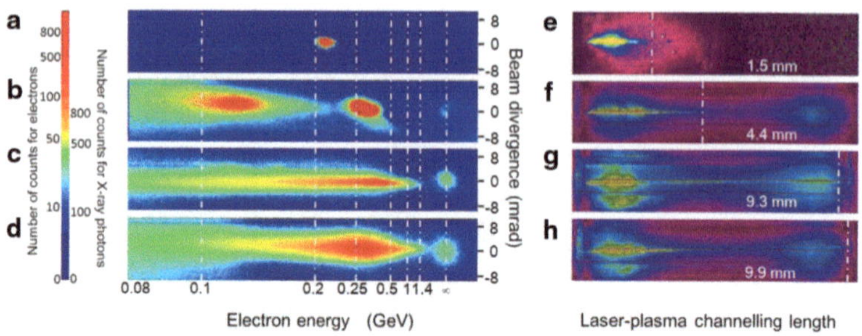

Fig. 2 Accelerated electron beams, betatron x-rays and plasma channel measurement. (**a–d**) show the spectrum characteristics of the electron beams and the betatron x-ray emission beam profile on Image Plate 3. (**e–h**) show the corresponding Thomson scattering The normalised vector potentials a0 used in this work were 2.65 for **a** and **e**, 2.87 for **b** and **f**, 3.35 for **c** and **g**, and 3.67 for **d** and **h**

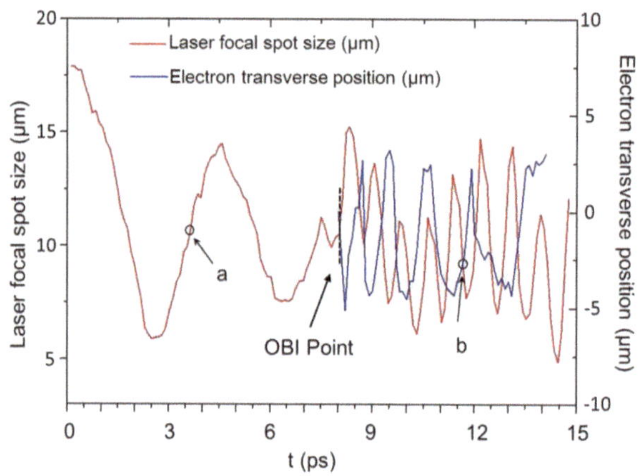

Fig. 3 Simulation on betatron oscillation mechanism. The laser spot oscillated in two completely different modes, and the oscillating bubble injection (for the second electron bunch) occurred exactly when the laser oscillating modes switched

and is therefore called the *oscillating bubble injection* (OBI) because the laser spot oscillation causes the transverse oscillation of the bubble. During the OBI, the bubble oscillation can effectively couple to the transverse motion of the second electron bunch, starting a betatron resonant and result in the enhanced betatron radiation.

2.2 *Electron Acceleration Diagnostics Via Betatron X-Ray Profile [5]*

In 2D-PIC simulations, we found that for different laser-plasma parameters and at different stages of the bubble evolution, there are four different distinct bubble structures. These four kinds of bubble structures result in four kinds of different electron

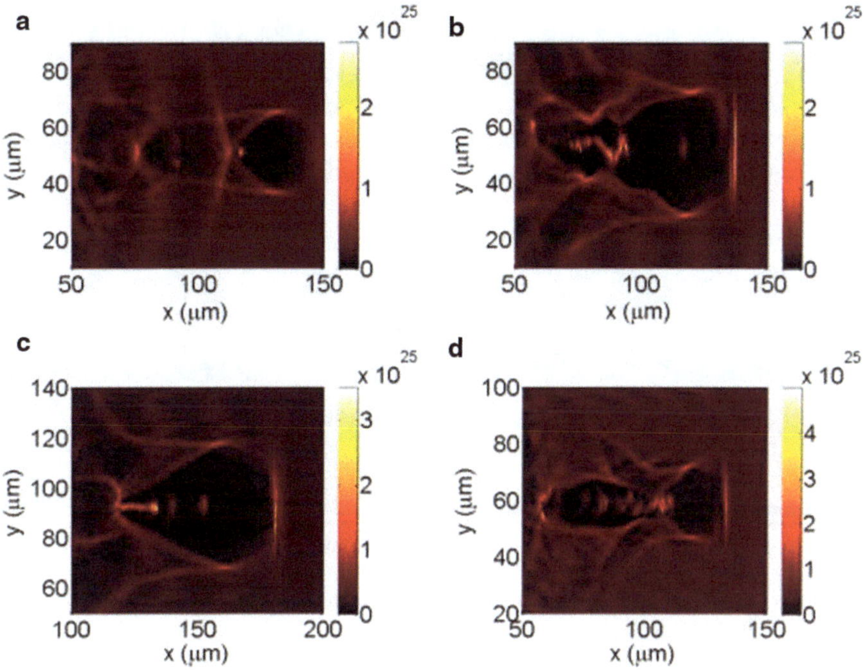

Fig. 4 Plasma density distributions in PIC simulations. With simulation parameters of **a** laser waist size $w_0 = 18\ \mu m$, plasma density $n_p = 1.5 \times 10^{18}\ cm^{-3}$, **b** $w_0 = 18\ \mu m$, $n_p = 2.0 \times 10^{18}\ cm^{-3}$, **c** $w_0 = 36\ \mu m$, $n_p = 2.0 \times 10^{18}\ cm^{-3}$, **d** $w_0 = 18\ \mu m$, $n_p = 5.0 \times 10^{18}\ cm^{-3}$

beam energy spectra and transverse oscillations and hence different angular distributions of the betatron radiation. These simulations are confirmed experimentally from the betatron radiation of multi-hundred MeV wakefield-accelerated electron beams. By analyzing the angular distributions of the betatron radiation and the electron beam energy spectra, the injection regime of the electron beams and their transverse oscillations behaviors were identified and the dynamic evolution of the bubble structure were reconstructed. The simulation and experiment results are listed in Fig. 4 and Fig. 5 respectively. Figure 4a, b, c, d respect to four different cases of laser wakefield acceleration. The corresponding experimental electron and betatron x-ray signal were listed in Fig. 5a, b, c, d. By comparison, we found that the betatron x-ray profile was related with the electron acceleration process. For Case 1: A monoenergetic electron beam and round betatron structure; Case 2: Two electron bunch, the first was monoenergetic and the second was continuous and performed large betatron oscillation. The betatron profile had strong round center and elliptical backgound; Case 3: Two electron bunch, the first was mono and the second was continuous but had small oscillation amplitude. The betatron was round; Case 4: continuous injection and had medium oscillation amplitude. The betatron profile

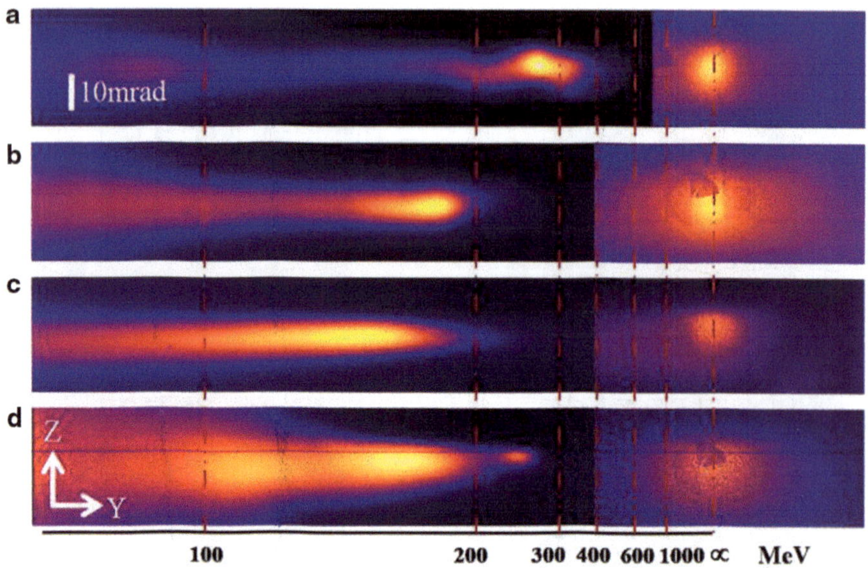

Fig. 5 Electron spectrum and betatron x-ray spectrum corresponding to the four cases shown in Fig. 4a, 4b, 4c, 4d respectively

was elliptical but had no strong center. This correspondent study provided a method to diagnose the electron acceleration process via betatron x-ray spatial structure.

3 Betatron X-Ray Generated from N_2 Gas Jet Via Ionization Injection [6]

Most betatron x-ray sources were generated from a plasma with a density above $10^{18}\mathrm{cm}^{-3}$ and the electron bunches were generated from the self-injection. Since self-injection regime relies on the nonlinear wave breaking of the plasma wave, usually the generated beams of electrons and betatron x-rays have large shot-to-shot fluctuations. Another drawback of the self-injection regime is that it usually requires a laser power much larger than the self-guiding power to get the electrons trapped in the bubble, thus reducing the energy transfer efficiency from laser to electrons and x-rays. Ionization injection using high Z gas is an alternative that has the advantage of lowering the injection threshold. Electron beams generated from ionization injection have been studies experimentally and theoretically in recent years. Quasi-monoenergetic electron beams have been observed from a single stage Nitrogen gas target. However, betatron x-rays emitted from ionization injection was merely studied and the simultaneous generation of high-quality quasimonoenergetic electron and x-ray beams was scarcely observed.

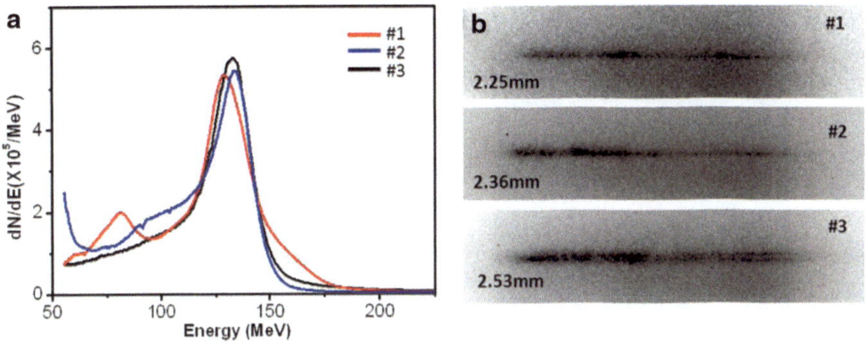

Fig. 6 Electron energy spectra (**a**), and corresponding plasma channel images (**b**)

We present a simultaneous generation of quasimonoenergetic electron and betatron x-ray beams in the ionization injection regime. By interacting ultra-short laser pulses ($a_0 = 1.8$) with low density pure nitrogen gas jet (electron plasma density $n_p = 5.4 \times 10^{17}$ cm^{-3}), no self-focusing was observed and the electrons were generated at the beginning of the process when the laser a_0 was large enough to cause ionization injection. The electron beams generated from ionization injection were highly stable and had quasimonoenergetic peaks around 130 MeV shown in Fig. 6a. The plasma channels corresponding to the electron spectrum in Fig. 6a were shown in Fig. 6b.

With stable channel formation and quasimonoenergetic electron beam generation, highly collimated x-ray beams were generated along the laser and electrons propagation direction. Typical x-ray transverse spatial profile is shown in Fig. 7a.The emitted x-ray beams were confined in a small divergence angle of 7.5 mrad (FWHM). With a γ of 260 (E = 130 MeV), the betatron oscillation strength parameter K = $\gamma\theta$ is 2, which is relatively small compared with previous studies. The spectrum of the x-ray source was measured by cut-off filters technique. Assuming the generated x-ray had a synchrotron-type spectrum $S(E) \sim N_\beta \frac{3e^2}{2\pi^3 \hbar c \varepsilon_0} \gamma^2 (E/E_c)^2 \cdot K_{2/3}^2(E/E_c)$, in which N_β is the number of oscillations, $K_{2/3}$ is a modified Bessel function of the second kind and E_c is the critical energy. When the photon energy $E > E_c$, the radiation decays exponentially. The filter pairs include: (1) 25 μm Al; (2) 25 μm Al + 14 μm Ti; (3) 25 μm Al + 25 μm W; (4) 25 μm Al + 25 μm Au. The x-ray transmission after each filter is T^i (E). The image plate response is $\eta(E)$. The calculated signal intensity after filter $I^i_{cal}(E_c) = \int S(E, E_c) T^i(E) \eta(E) dE$ should be proportional to the experimental data $I^i_{\exp}$. By performing least squares fitting between array I^i_{cal} and $I^i_{\exp}$,, we get a best fit for an x-ray critical energy of 4 keV. By integration of the PSL signal on IP, the total x-ray photon number in the FWHM of the spatial beam profile is 1.6×10^7.

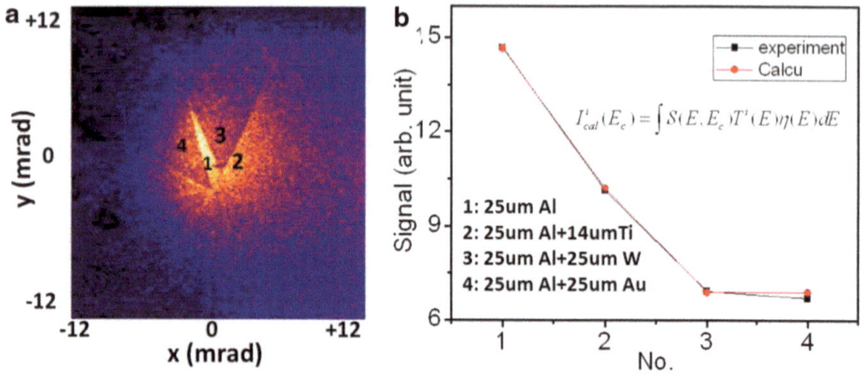

Fig. 7 X-ray measurement **a** Typical x-ray signal on IP, **b** x-ray signal fitting using filter technology, *red* circles are the calculated signal using a synchrotron type curve with $E_c = 4$ keV and *black* squares are the relative experimental signal strength

Since the x-ray critical energy, betatron strength parameter and electron energy have been measured, based on the definition of critical energy for the emitted betatron x-ray $E_c = 3\hbar K\gamma^2\varpi_b$, the oscillation frequency of the betatron oscillations is calculated to be $\varpi_b = 1.6 \times 10^{13} rad/s$. The oscillation period can be deduced as $\lambda_b = \frac{2\pi c}{\varpi_b} = 100\mu m$. The oscillation amplitude r_0 of the electrons performing the betatron motion is 0.1 μm deduced from $K = \gamma\theta = 2\pi\gamma r_0/\lambda_b$, which is smaller compared with previous experimental results and suggests a smaller x-ray source size. Based on the results presented here, it is anticipated that utilizing PW-class laser with a large focal spot (thus a longer Rayleigh length) propagating in low density nitrogen gas, it might be possible to simultaneously generate high-energy quasimonoenergetic electron beams and highly collimated x-ray source with good spatial coherence working in a small K- parameter regime.

4 Summary

This paper introduces recent progresses of laser plasma betatron x-ray sources at the Institute of Physics, Chinese Academy of Sciences. Different species of gases were used as target to study the betatron oscillation mechanism and betatron x-ray enhancement. For helium plasma, hosing instability causes the enhancement of transverse oscillation of the second electron bunch which results in the high flux betatron radiation. For nitrogen plasma, intense betatron x-ray emission was generated along with the generation of stable quasi-monoenergetic electron beams. The betatron x-ray emission had a small divergence and high critical energy in utilizing ionization injection.

References

1. Tajima, T., Dawson, J.: Phys. Rev. Lett. 43, 267–270 (1979)
2. Kneip, S., et al.: Nature Phys. 6, 980–983 (2010)
3. Cipiccia, S., et al.: Nature Phys. 7, 867–871 (2011)
4. Yan, W.C., Chen*, L.M., et al.: PNAS. 111, 5825 (2014)
5. Ma, Y., Chen*, L.M., et al.: Appl. Phys. Lett. 105, 161110 (2014)
6. Huang, K., et al.: Appl. Phys. Lett. 105, 204101 (2014)

A New Approach to Theory for the Rhodes Group Experiments with Channeled Pulse Excitation in Xe and Kr Clusters

Peter L. Hagelstein and Irfan U. Chaudhary

Abstract Rhodes and coworkers have reported a great many results over the last decade showing directional keV x-ray emission in laser pulse channeling experiments in Xe and in Kr clusters. Selective inner-shell ionization has been proposed as a mechanism to produce gain. We propose that directional emission might be due instead to a phased array emission effect, following up-conversion through a new quantum effect.

1 Introduction

Rhodes and coworkers have reported interesting results from experiments in which an intense KrF laser pulse is channeled in Xe clusters [1, 2] and in Kr clusters [3], where intense directional emission is observed (below 3 Å in Xe, and near 7.5 Å in Kr). This work has stimulated interest with the claim of x-ray gain and lasing at short wavelengths with only a modest experimental facility.

Our interest has been focused recently on the theoretical issue of pumping mechanism. Selective inner-shell collisional ionization has been discussed [4, 5], as well as inner-shell photoionization [6, 7]. Over the past two years we attempted to follow up on these proposals in an independent study, to see whether we could confirm either mechanism. We are able to derive a collective enhancement of the collisional ionization cross section, but the effect favors outer shell ionization. X-rays of an intensity required to produce sufficient inner-shell photoionization for gain should dominate the spectrum; but there is no evidence that such x-rays are present in the experiment.

Consequently it is not clear that a population inversion can develop. On the other hand, an inversion is not required for directional emission; it can occur instead through a phased array emission effect. Coherence among the phases could be established if the intense optical laser pulse were up-converted to produce excitation. From the experimental conditions one can estimate the rates for direct multi-photon

P. L. Hagelstein (✉)
Massachusetts Institute of Technology, Cambridge, MA, USA
e-mail: PHagelstein@aol.com

I. U. Chaudhary
University of Engineering and Technology, Lahore, Pakistan

© Springer International Publishing Switzerland 2016
J. Rocca et al. (eds.), *X-Ray Lasers 2014*, Springer Proceedings in Physics 169,
DOI 10.1007/978-3-319-19521-6_21

excitation; perhaps as expected we find (from brute force calculations) that no (conventional) up-conversion is predicted. A convincing case has been made that strong directional x-ray emission occurs in experiment; but how it works is so far unclear.

2 Fractionation and Inverse Fractionation

A number of experiments in condensed matter nuclear science exhibit anomalies, and have proven problematic for theorists; a relevant example is collimated x-ray emission near 1.5 keV in the Karabut high-current density glow discharge experiment [8]. There is no obvious mechanism available to produce a population inversion, yet directional emission is observed.

This experiment, and other anomalies as well, motivated us to explore quantum mechanical models for coherent energy exchange between systems with mismatched characteristic energies. For example, coherent energy exchange in the spin-boson model (two-level systems linearly coupled to an oscillator) is known for the multi-photon regime, but it is a weak effect that becomes ineffective when the large two-level system quantum is split into more than about 30 oscillator quanta. Of interest was why there should be such a limit. When the coupling is weak, perturbation theory can be used, and it became clear that the contributions from the different contributing paths showed strong destructive interference [9]. When we constructed a version of the spin-boson model augmented with a loss mechanism that removed the destructive interference, we found a much enhanced rate for coherent energy exchange even when the number of oscillator quanta was large [9–12].

3 Possibility of Up-Conversion by Inverse Fractionation

We consider the possibility that directional emission in the channeling experiments is a result of multi-photon up-conversion by inverse fractionation. For inverse fractionation, we require a highly-excited oscillator, which could be the channeled optical laser pulse; we require equivalent two-level systems with a large transition energy, which could be a dipole transition in the highly-stripped ions; and finally we need strong loss far off of resonance, which could be outer-shell photoionization.

If this proposal is correct, then the strong coupling between the x-ray transitions and the intense channeled pulse in the presence of strong off-resonant loss results in a strong multi-photon excitation effect, where excitation is produced at odd multiples of the drive laser frequency. For the Ne-like Kr and Co-like Xe transitions we have explored so far, the dimensionless coupling strength is so large that the model is pushed into a regime where the indirect coupling matrix element takes on it maximum value. This motivates us to carry out exploratory calculations of up-conversion by inverse fractionation.

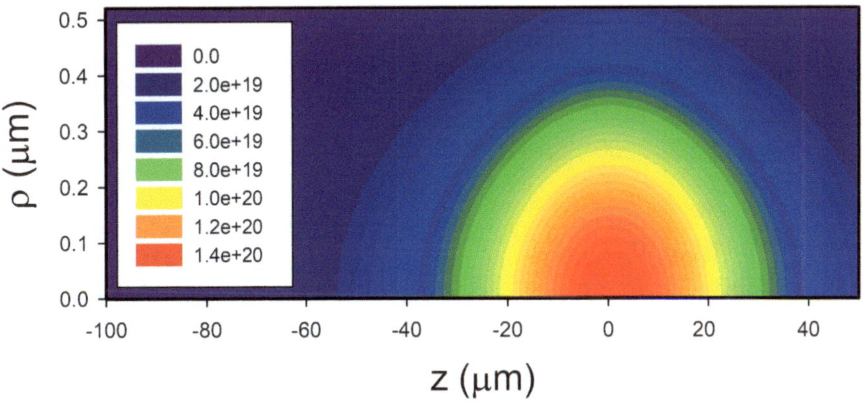

Fig. 1 Closeup of intensity of model pulse as a function of z and ρ

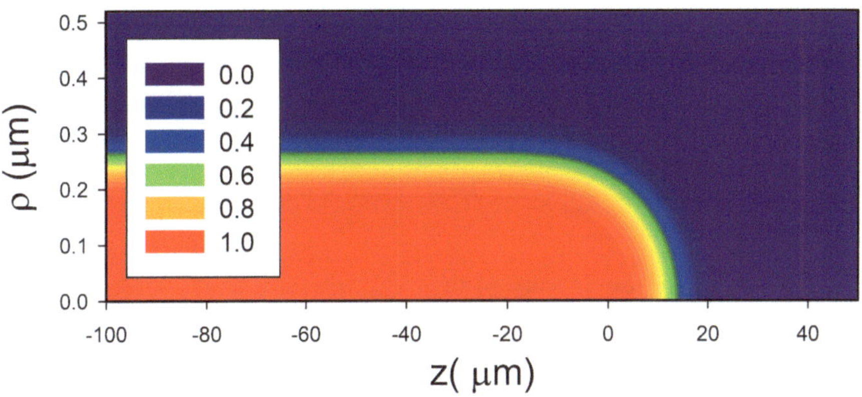

Fig. 2 Closeup of Ne-like fraction as a function of z and ρ

4 Inverse Fractionation in Ne-Like Kr

The simplicity of the directional x-ray spectrum in the case of the recent experiments with Kr clusters [3] allows for line identifications that we have confidence in, so we focused on the $3s-2p$ transitions first. We worked with a model intense hyperbolic secant pulse with a 230 fs FWHM, a peak intensity of 1.42×10^{20} W/cm^2, with radial confinement described by a Gaussian with $\langle\rho\rangle = 0.289\mu$. The total pulse energy is 164 mJ. A close up of the intensity distribution is shown in Fig. 1. We computed the transient ionization of the Kr using a simple Keldysh model for tunnel ionization, which gives a high fraction of Ne-like near and following the peak of the pulse (shown in Fig. 2). The excited states are strongly Stark shifted by the intense pulse. In Fig. 3 we show the transition energy for the $2p_{3/2}3s$ state, which in our model is

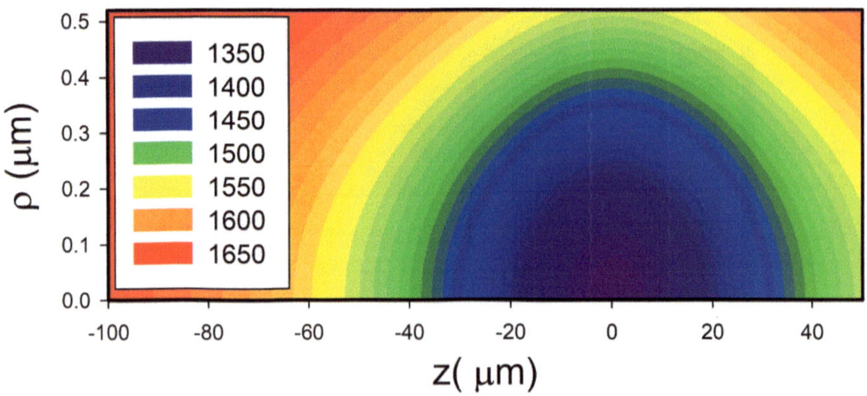

Fig. 3 Transition energy of the $\overline{2p_{3/2}}3s$ transition as a function of z and ρ

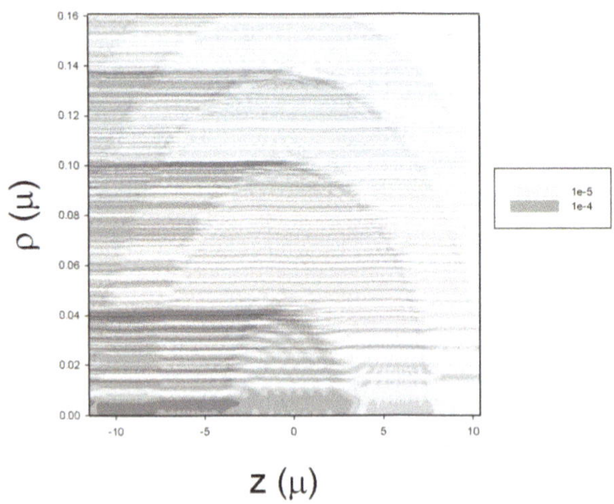

Fig. 4 Closeup of the occupation probability for the $\overline{2p_{3/2}}3s$ state as a function of z and ρ

calculated to be at 1651.6 eV with no field present. We see that this transition shifts down in this model by more than 300 eV at the peak of the pulse.

Upper state population accumulates in this model when the shifted transition energy matches an odd multiple of the laser energy (see Fig. 4). The directionality in this model is determined by the ion motion following excitation; if limited to on the order of an Angstrom, then most of the radiation will be forward directed; larger ion motion increases the isotropic fraction. The conjecture is that cluster formation leads to early ionization, which allows the ions to evolve into a Wigner crystal (which limits the ion motion following excitation). Sharp lines are predicted for the long-lived $3s - 2p$ transitions, which decay slowly enough to persist after the laser pulse passes. The short lifetime of the $3d - 3p$ and $3p - 2s$ transitions results in the lines strongly shifted and broadened, so as not to look like lines at all in the spectrum.

5 Inverse Fractionation in Co-like Xe

Exploratory calculations in Co-like Xe show that the strongest excitation occurs on the $3d - 4f$ transitions, which Stark shifts to higher energy. Since the lifetimes are short, emission is predicted to occur at ten or so odd harmonics of the laser transition above the zero-field energy values. For the strongest transition this leads to emission between 2.90 and 2.97 Å in the model. These sideband transitions could be confused with emission from more highly-stripped ionization stages. Wigner crystal formation leads to directionality, which could confused with a saturation effect.

References

1. Borisov, A.B., Boyer, K., Van Tassle, A., Song, X., Frigeni, F., Kado, M., Rhodes, C.K.: Observation of directed emission and spectral narrowing on Xe(L) hollow atom single-(2p) and (2s2p) double vacancy inner-shell transitions at 2.8-2.9 Angstroms. AIP Conf. Proc. **611**, 346 (2002)
2. Borisov, A.B., Davis, J., Song, X., Koshman, Y., Dai, Y., Boyer, K., Rhodes, C.K.: Saturated multikilovolt x-ray amplification with Xe clusters: single-pulse observation of Xe (L) spectral hole burning. J. Phys. B: At. Mol. Opt. Phys. **36**, L285 (2003)
3. Borisov, A.B., McCorkindale, J.C., Poopalasingham, S., Longworth, J.W., Rhodes C.K.: Demonstration of Kr(L) amplification at l = 7.5°A from Kr clusters in self-trapped plasma channels. J. Phys. B: At. Mol. Opt. Phys. **46**, 155601 (2013)
4. McPherson, A., Luk, T.S., Thompson, B.D., Boyer, K., Rhodes, C.K.: Multiphoton-induced X-Ray emission and amplification from clusters. Appl. Phys. B. **57**, 337 (1993)
5. Schroeder, W.A., Nelson, T.R., Borisov, A.B., Longworth, J.W., Boyer, K., Rhodes, C.K.: An efficient, selective collisional ejection mechanism for inner-shell population inversion in laser-driven plasmas. J. Phys. B: At. Mol. Opt. Phys. **34**, 297 (2001)
6. Petrova, Tz.B., Whitney, K.G., Davis, J.: J. Phys. B: A gain model for x-ray lasing at 2.8 Å in an intense laser irradiated gas of xenon clusters. At. Mol. Opt. Phys. **43**, 025601 (2010)
7. Davis, J., Whitney, K.G., Petrova, Tz.B., Petrov, G.M.: Intense ultrashort laser-Xe cluster interaction. High Energy Density Phys. **8**, 238 (2012)
8. Karabut, A.B., Karabut, E.A., Hagelstein, P.L.: Spectral and temporal characteristics of X-ray emission from metal electrodes in a high-current glow discharge. J Condens. Matter Nucl. Sci. **6**, 217 (2012)
9. Hagelstein, P.L., Chaudhary, I.U.: Energy exchange in the lossy spin-boson model. J. Condens. Matter Nucl. Sci. **5**, 52 (2011)
10. Hagelstein, P.L., Chaudhary, I.U.: Second-order formulation and scaling in the lossy spin-boson model. J. Condens. Matter Nucl. Sci. **5**, 87 (2011)
11. Hagelstein, P.L., Chaudhary, I.U.: Local approximation for the lossy spinboson model. J. Condens. Matter Nucl. Sci. **5**, 102 (2011)
12. Hagelstein, P.L., Chaudhary, I.U.: Coherent energy exchange in the strong coupling limit of the lossy spin-boson model. J. Condens. Matter Nucl. Sci. **5**, 116 (2011)

Part III
High Harmonics and Attosecond Science

Attosecond Dynamics of Parametric Amplification at 11 nm

J. Seres, E. Seres, B. Landgraf, B. Ecker, B. Aurand, A. Hoffmann, G. Winkler, S. Namba, T. Kuehl and C. Spielmann

Abstract X-ray parametric amplification is a high order parametric process with the capability to amplify ultrashort laser pulses with broad spectrum ranging from the extreme ultraviolet to the soft x-ray. In this contribution, the dynamics of this broadband amplification schema is studied with sub-20-as temporal resolution by seeding the amplifier with a train of 200-as-long pulses at 11 nm central wavelength.

1 Introduction

Intense short laser pulses focused into low density gases lead to the generation of coherent extreme ultraviolet pulses (XUV) via high-order harmonic (HH) generation with femtosecond or even attosecond pulse duration [1], [2]. The generated short wavelength pulses have unique parameters suited very well to probe physical phenomena with femtosecond temporal resolution [3], [4] in the spectral range up to a few keV. A major limitation for several experiments is the limited energy of the HH pulses. It can be boosted e.g. in plasma-base x-ray amplifiers [5] – [7]; however the widths of the suitable atomic transitions limit the pulse duration to the sub-picosecond range. On the other hand x-ray parametric amplifier (XPA) [8], [9] predicts amplification over a broad spectral range paving the way for realizing high energy sub-femtosecond XUV pulses.

J. Seres (✉) · E. Seres · B. Landgraf · A. Hoffmann · C. Spielmann
IOQ, Abbe Center of Photonics, Friedrich Schiller University, Jena, Germany

J. Seres · E. Seres · G. Winkler
Atominstitut (ATI), Vienna University of Technology, Stadionallee 2, 1020 Vienna, Austria
e-mail: j-seres@lycos.com

E. Seres · B. Landgraf · B. Ecker · B. Aurand · T. Kuehl · C. Spielmann
Helmholtz Institute Jena, Jena, Germany

B. Ecker · B. Aurand · T. Kuehl
GSI Helmholtz Centre for Heavy Ion Research, Darmstadt, Germany

S. Namba
Graduate School of Engineering, Hiroshima University, Hiroshima, Japan

B. Aurand
Department of Physics, Lund University, Lund, Sweden

© Springer International Publishing Switzerland 2016
J. Rocca et al. (eds.), *X-Ray Lasers 2014*, Springer Proceedings in Physics 169,
DOI 10.1007/978-3-319-19521-6_22

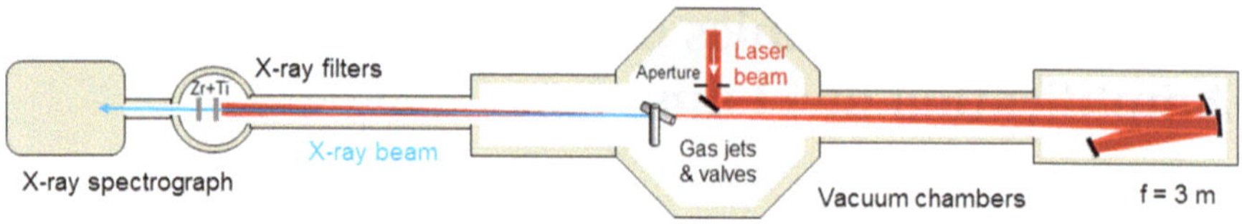

Fig. 1 Experimental setup. For details, see text

Here we demonstrate parametric amplification of attosecond pulse-trains [10] at around 11 nm in a two-stage (seed-amplifier) arrangement. We show that the amplification is very sensitive to the setting of the delay between the XUV pulses generated in the seed stage and the pumping laser field in the amplifier stage. Varying the delay, unprecedented sub-20-as temporal resolution has been achieved. Additionally we estimated the durations of the individual XUV pulses within the train to be in the order of 0.2 fs.

2 Experimental Setup

The reported measurements were performed with the 26-fs-long pulses of the JETI Ti:sapphire laser system (FSU, Jena, Germany) with a repetition rate of 10 Hz. The beam was loosely focused (confocal parameter was 160 mm) with a spherical mirror (f = 3 m, see Fig. 1). The pulse energy on the target was controlled with an adjustable iris to an intensity of 5×10^{15} W/cm^2 in the focus.

The targets consisted of two thin nickel tubes with diameters of 3 mm and were backed with Ne or He gas, respectively. In each tube, two holes with a diameter of 1 mm have been drilled for the laser beam. The targets were placed before the focus with adjustable position and distance. After blocking the fundamental laser light with thin Zr (200 nm) and Ti (200 nm) foils, we measured either the far field beam profiles with an x-ray CCD camera or the spectra with an x-ray spectrograph 2.6 m downstream from the gas jets.

3 The Underlying Process in the Amplifier Stage

In the two jets the laser interacts with the gas atoms in the following way: In the first jet (Fig. 2), the laser-field-accelerated and re-colliding electrons generate a train of attosecond XUV pulses [11] (Fig. 2a dark violet curve) perfectly synchronized to the electric field (Fig. 2a red curve) of the driving laser pulse [11]. The XUV radiation can be generated over many optical cycles (with slightly different maximum field strengths) and/or different long and short electron trajectories. As a consequence the measured spectrum (Fig. 2e, orange) represents the average and looks therefore more like a super-continuum than a line spectrum.

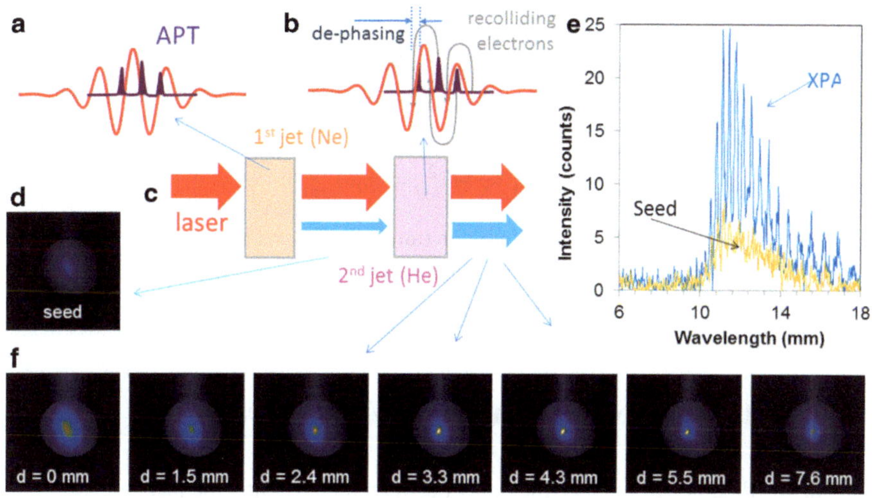

Fig. 2 (**a–c**) Schematic of the involved processes in the two-jet arrangement. The (**e**) spectra and (**d, f**) beam profiles with the XPA-specific hot spots at around 4.3 mm were measured after the (**d**) first and (**f**) second jet, respectively. For details, see text

In the second jet, due to the different propagation speed, the attosecond pulse train (APT) is no longer in phase with the re-colliding electrons (Fig. 2b, light gray curves). As it has been shown recently [12], amplification of XUV pulses by XPA [8] takes place at the time of the re-collision of electrons moved along the so-called long trajectories. In a next step, the temporal overlap has to be re-established by introducing a suitable delay via the Gouy- and atomic phase shift [11], [12] i.e. the two jets must be located before the laser focus at a well-defined position and in a well-defined distance (*d*). Then, amplification of the APT in the second jet becomes possible and a line spectrum is generated (Fig. 2e).

4 Beam Profile Narrowing and Optimal Jet Distance

Beside the increased energy of the XUV pulses, a clear indication of parametric amplification in the second jet is the appearance of a "hot spot" in the beam profile (see Fig. 2f). For the reported measurements, the second jet was kept at a fixed position, 15 mm before the laser focus and the position of the first (seed) jet (Ne, 6 mbar) was varied. First, the profile of the seed beam was measured without gas supply for the second jet (Fig. 2d). Switching on the gas in the second jet (He, 100 mbar), the signal was enhanced, even for merged jets, when the distance between the output of the first and the input of the second jet d = 0 mm (Fig. 2f). Increasing the distance between the gas jets, a small low divergent beam part appeared in the beam center and simultaneously the intensity of the remaining beam part continuously decreased down to the level of the pure seed beam. The smallest central beams were observed for separations between 3 and 5 mm. From the measurement we estimated the divergence of the seed and the amplified beams to $\Theta_{seed} = 2.0 \pm 0.3$ mrad and $\Theta_{XPA} = 0.40 \pm 0.03$ mrad, respectively. The divergence decreased due to the gain g_0 of the amplifier as [13]

$$\Theta_{XPA}^{-2} \approx \Theta_{seed}^{-2} + g_0 \Theta_{amp}^{-2} \tag{1}$$

From the measured ASE beam diameter (amplifier jet without seed) we obtained $\Theta_{amp} = 1.0 \pm 0.1$ mrad. Using these measured parameters and Eq. (1), the determined total gain was $g_0 = 6 \pm 2$.

5 Duration of the Pulses in the XUV Pulse Train

Beside the distance of the jets, the delay between the APT and the laser electric field can also be controlled by varying the gas pressure in the seed jet. This is because the dephasing (delay) between the XUV and NIR laser pulses is mainly governed by the plasma dispersion of the free electrons in the jet [10]:

$$\tau = \frac{e^2 n_e L}{2\varepsilon_0 m_e c \omega_1^2} \tag{2}$$

where n_e is the free electron density proportional to the gas pressure, L is the length of the gas jet and ω_1 is the angular frequency of fundamental laser beam. As described above, only if the seed APT is in phase with the driving laser field (i.e. with the re-colliding electrons) than the output signal and hence the gain is maximized. If the overlap is lost or the XUV electric field is out of phase, there is no enhancement or it is very small [13].

In another set of measurement series, we measured the on-axis fluence of the seed and the amplified XUV pulses (Fig. 3a) for a fixed jet distance of 3.3 mm. The XPA output fluence can be modelled as a saturated amplifier and scales with the well-known gain saturation formula [14]

$$j = \ln\left[1 + e^{g_0}\left(e^{j_0} - 1\right)\right] \tag{3}$$

Here, $j = J_{XPA}/J_{sat}$ and $j_0 = J_{seed}/J_{sat}$ are the normalized output and input fluence, J_{sat} is the saturation fluence. A more detailed inspection reveals an additional periodic modulation on the measured XPA output fluence. The observed signal and gain as a function of the delay can be described by a Gaussian envelope with a superimposed sinusoidal modulation [13]:

$$g_0 = g_a \exp\left[-4\ln 2 \frac{(\tau - \tau_0)^2}{\Delta\tau^2}\right]\left[1 + V\cos(\omega_1 \tau)\right] \tag{4}$$

where V is the magnitude of the periodic modulation, which is less than 1 due to the finite distribution of the re-colliding electrons and due to the fluctuations of the free

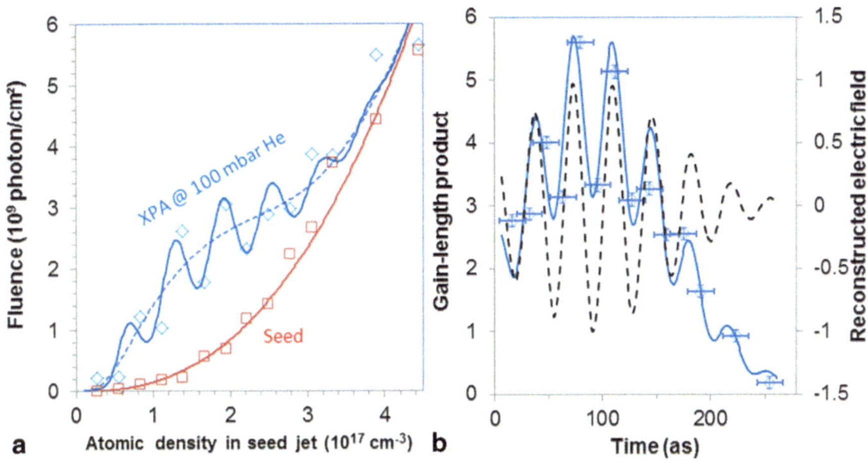

Fig. 3 **a** The measured fluence of the seed and amplified beams allows us to **b** reconstruct the time dependence of the gain in the amplifier jet (*blue*) and the electric field of the XUV pulses (*black* dashed)

electron density. Using Eq. (4), an average fluence curve (blue dashed in Fig. 3a) and a periodic fluence curve (blue in Fig. 3a and 3b) were fitted to the experimental data with fitting parameters of $J_{sat} = 5 \times 10^8$ photon/cm^2, $g_a = 4.5 \pm 0.1$, $V = 0.35 \pm 0.01$ and $\Delta\tau = 0.20 \pm 0.05$ fs. Note that the delay $\Delta\tau$ corresponds to the width of a single pulse in APT. The predicted curves not only fit very well to the measured data, but they are also in agreement with the period of the 11 nm XUV field as expected. Because of the large error bars in Fig. 3b, mainly caused by the shot-to-shot instability of the driving laser and pulsed gas valve, we are not able to fully reconstruct the exact electric field of the XUV pulses. It is only feasible to approximate the electric field within one pulse (black dashed curve in Fig. 3b) with a sinusoidal function.

6 Conclusion

Seed attosecond pulse trains were generated by high-order harmonic generation in a gas jet and were further amplified by x-ray parametric amplification (XPA) in another gas jet. Setting the delay between the seed pulses and the driving laser field correctly, we observed amplification accompanied by strong beam profile narrowing. By varying the delay, we were able to measure the gain dynamics of amplifier with sub-20-as temporal resolution.

These findings provide not only clarification of the operation of x-ray parametric amplification, but also a framework for operating XPA at substantially reduced pulse energy of the driving laser, paving the way to realize powerful laser-driven XUV pulses at higher photon energies [8], [15] and at kHz [8] or even MHz [16] repetition rate.

Acknowledgements This study was sponsored by DFG grants of TR18-P12, SE-1911/1–1; TMBWK grant of B715-08008; the European Fund for Regional Development (EFRE); the EC's 7th Framework Program (grant 284464, Laserlab Europe HIJ-FSU0019152) and the Machuo Fundation Japan. The authors acknowledge the support and the contributions of the JETI laser team.

References

1. Paul, P.M., et al.: Observation of a train of attosecond pulses from high harmonic generation. Science. **292,** 1689–1692 (2001)
2. Hentschel, M., et al.: Attosecond metrology. Nature. **414,** 509–513 (2001)
3. Seres, E., Seres, J., Spielmann, C.: Time resolved spectroscopy with femtosecond soft-x-ray pulses. Appl. Phys. A. **96,** 43–50 (2009)
4. Seres, E., Spielmann, C.: Time-resolved optical pump X-ray absorption probe spectroscopy in the range up to 1 keV with 20 fs resolution. J. Mod. Opt. **55,** 2643–2651 (2008)
5. Zeitoun, Ph., et al.: A high-intensity highly coherent soft X-ray femtosecond laser seeded by a high harmonic beam. Nature. **431,** 426–429 (2004)
6. Wang, Y., et al.: Phase-coherent, injection-seeded, table-top soft-X-ray lasers at 18.9 nm and 13.9 nm. Nature Phot. **2,** 94–98 (2008)
7. Hasegawa, N., et al.: Frequency filter of seed X-ray by use of X-ray laser medium: Toward the generation of the temporally coherent X-ray laser. Jap. J. Appl. Phys. **48,** 012503 (2009)
8. Seres, J., et al.: Laser driven amplification of soft-x-rays by parametric stimulated emission in neutral gases. Nature Phys. **6,** 455–461 (2010)
9. Aurand, B., et al.: Laser driven X-ray parametric amplification in neutral gases—a new brilliant light source in the XUV. Nucl. Inst. Meth. A. **653,** 130–133 (2011)
10. Seres, J., et al.: Parametric amplification of attosecond pulse trains at 11 nm. Sci. Rep. **4,** 4254 (2014)
11. Kohler, M.C., et al.: Frontiers of atomic high-harmonic generation. Adv. At. Mol. Opt. Phys. **61,** 159–207 (2012)
12. Seres, J., Seres, E., Spielmann, C.: Classical theory of strong-field parametric amplification of x-rays. Phys. Rev. A. **86,** 013822 (2012)
13. Serrat, C.: Coherent extreme ultraviolet light amplification by strong-field-enhanced forward scattering. Phys. Rev. Lett. **111,** 133902 (2013)
14. Stuart, B.C., Herman, S., Perry, M.D.: Chirped-pulse amplification in Ti:Sapphire beyond 1 μm. IEEE J. Quant. Electr. **31,** 528–538 (1995)
15. Seres, J., et al.: High-harmonic generation and parametric amplification in the soft X-rays from extended electron trajectories. Sci. Rep. **4,** 4234 (2014)
16. Seres, E., Seres, J., Spielmann, C.: Extreme ultraviolet light source based on intracavity high harmonic generation in a mode locked Ti:sapphire oscillator with 9.4 MHz repetition rate. Opt. Express. **20,** 6185–6190 (2012)

Coherent Soft-X-Ray Pulses at Multi MHz Repetition Rates Using Enhancement Cavities

E. Fill, I. Pupeza, S. Holzberger, H. Carstens, N. Lilienfein, A. Apolonski and F. Krausz

Abstract Enhancement cavities are passive optical resonators which allow increasing the power of multi-MHz pulse trains by orders of magnitude. Generating high harmonics from enhanced pulses of an Yb-laser ($\lambda = 1040$ nm) coherent XUV pulses with wavelengths down to 11.5 nm (108 eV) were obtained at a repetition rate of 78 MHz. Harmonics were coupled out of the cavity by a small hole in a mirror. With up to 5 kW of fundamental laser power in the cavity, average powers of 5.4 μW and 8 pW were measured at 38 and 11.5 nm respectively. New developments include scaling of fundamental laser powers in a cavity up to 670 kW and efforts aiming at the generation of attosecond pulses with multi-MHz repetition rates.

1 Introduction

The basic principle of an enhancement cavity involves coherent stacking of ultrashort pulses until the loss in the cavity is equal to the rate of input coupling [1–4]. A generic enhancement cavity is a ring resonator consisting of an input coupling mirror (typical reflectivity 99.5 %) and several other mirrors with much higher reflectivity. Two of these are curved with a radius of curvature and a separation chosen for providing a stable Gaussian beam circulating in the cavity.

The resonance conditions require that the roundtrip time of the cavity be matched to the repetition rate of the seed laser and that the phase of the input pulses be exactly synchronized to the phase of the circulating pulse. If the power reflectivity R of the input coupling mirror equals the round trip power transmission of the cavity the cavity is said to be "impedance matched". On resonance and with a lossless input coupling mirror (power transmission $T = 1 - R$) the enhancement factor is given by

$$e = 1/(1 - R). \tag{1}$$

Typical values for the repetition rate are around 100 MHz but cavities have been operated at 10 MHz, and up to 1 GHz. For the enhancement factor a value of a few

E. Fill (✉) · I. Pupeza · S. Holzberger · H. Carstens · N. Lilienfein · A. Apolonski · F. Krausz
Max-Planck-Institut für Quantenoptik, Hans-Kopfermann-Str. 1, 85748 Garching, Germany
e-mail: ernst.fill@mpq.mpg.de

Ludwig-Maximilians-Universität München, Am Coulombwall 1, 85748 Garching, Germany

© Springer International Publishing Switzerland 2016
J. Rocca et al. (eds.), *X-Ray Lasers 2014*, Springer Proceedings in Physics 169,
DOI 10.1007/978-3-319-19521-6_23

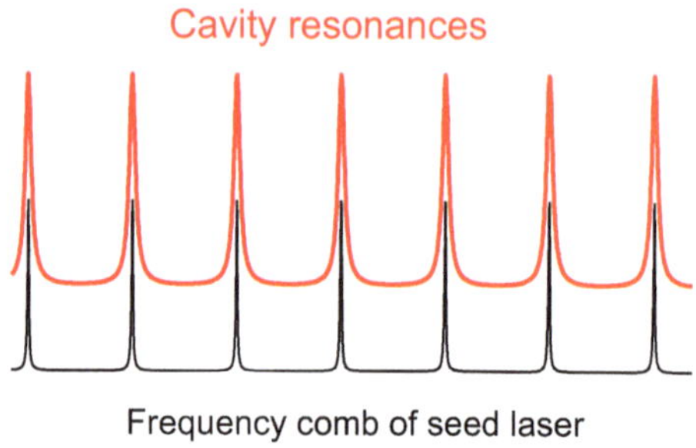

Fig. 1 Resonance conditions of an enhancement cavity in the frequency domain. The comb lines of the seed laser must be exactly aligned to the cavity resonances

hundred is routinely obtained but in special cases values of up to 10,000 have been demonstrated. However, for such a high enhancement the cavity becomes sensitive to small disturbances and its operation is demanding.

It is instructive to consider the situation in the frequency domain, as shown in Fig. 1. The resonance conditions require that the frequency comb lines of the seed laser exactly match the cavity resonances and that the carrier envelope offset frequency is equal to that of the cavity. The first condition is maintained by generating an error signal and locking the repetition rate of the seed laser by a piezoelectrically actuated mirror to the cavity. The carrier offset frequency is in most cases more stable and is manually adjusted.

2 Generating Harmonics in an Enhancement Cavity

Coherent harmonic pulses at multi-MHz repetition rates constitute a frequency comb in the XUV. This was first demonstrated by beating the third harmonic of a Ti:sapphire laser with the third harmonic generated by a nonlinear crystal [3]. In another experiment heterodyne beating between the third harmonic coupled out of an enhancement cavity and the fifth harmonic of a laser at 1064 nm was recorded [2]. Recently low-order harmonics have been used in direct frequency comb spectroscopy of argon and krypton resonance lines [5].

For harmonic generation a gas jet is placed at the focus between the curved mirrors. Intensities exceeding 10^{14}W/cm^2 are easily generated. Usually a so called "end-fire nozzle" consisting of a tube with a small opening at its end is used for producing the gas jet.

Harmonic radiation is emitted collinearly with the circulating fundamental pulses and therefore has to be coupled out of the cavity. Several methods have been used for this purpose, such as a Brewster window [3] or a nano-grating etched into a mirror placed after the focus [6]. These methods carry the risk of damage to these components and have the disadvantage of a low output coupling efficiency for the higher harmonics.

In our experiment output coupling was established by means of a small hole in the mirror after the focus [7]. This method had been suggested in combination with

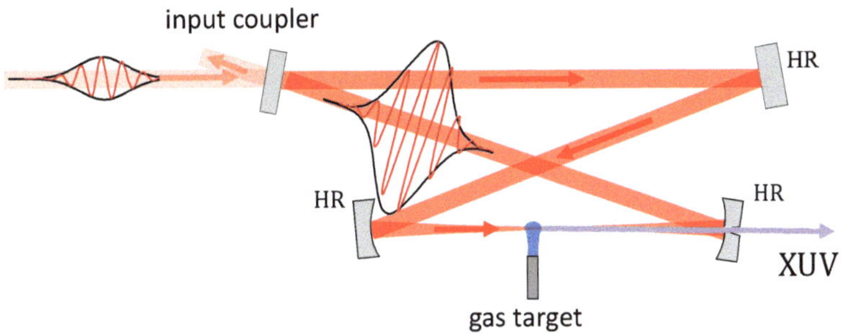

Fig. 2 Schematic layout of harmonic generation experiment. Input is a 175 or 57 fs pulse train with a 78 MHz repetition rate. A gas nozzle is placed close to the focus. Output coupling is realised through a small hole in the mirror following the focus. Focusing is asymmetric with curved mirror radii of 100 and 300 mm for argon and 38 and 150 mm for neon. XUV diagnostics is a grating followed by a calibrated diode or a grazing incidence spectrometer with a CCD camera on the Rowland circle

a donut-shaped mode circulating in the cavity, which avoids the hole [8]. However, we use the normal TEM_{00} mode and make the hole small enough so losses are below 0.5 % and the cavity can be operated with an enhancement factor of about 200.

Hole output coupling has the following advantages:

- Increased output coupling efficiency at shorter wavelengths
- No dispersion introduced
- Polarization independent
- Harmonics are collinear.

3 Experimental Results

A schematic layout of the experiment is shown in Fig. 2. The seed laser was an Yb-laser emitting a power of up to 57 W with a pulse duration of 175 fs. By focusing the pulse into a nonlinear fiber and shortening it by chirped mirrors pulses with duration down to 27 fs could be obtained [9].

First experiments were carried out with argon, using mirrors with radii of curvature of 100 and 300 mm and a focal spot size of 20 μm. Plasma generated in the focus severely affected the performance of the cavity. Dynamic ionization during pulse propagation resulted in a blue shift of the circulating pulse reducing the spectral overlap with the seed pulses. The situation was much improved when seeding the cavity with shorter pulses. The bandwidth of the cavity mirrors was not wide enough for enhancing 27 fs pulses. However, seeding with 57 fs pulses the intensity in the focus could be increased from 6×10^{13} W/cm^2 to 8×10^{13} W/cm^2. Higher order harmonics and higher power could be generated in this way (Fig. 3a). With a calibrated diode a power of 5.3 μW was measured at a wavelength of 38.5 nm

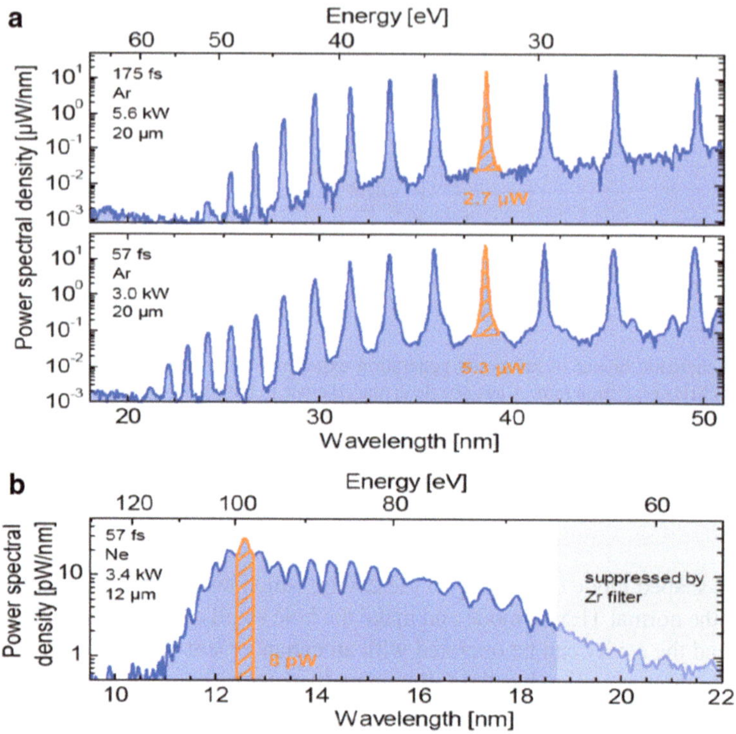

Fig. 3 Harmonic spectra obtained with argon (**a**) and neon (**b**). In argon, shorter pulses generate higher harmonics and more harmonic power. In neon a smaller focal spot size (12 μm instead of 20 μm) generates harmonics with wavelengths down to 11.5 nm. Figure reproduced from ref. [7] with permission

(27th harmonic). To generate higher harmonics intensities exceeding 10^{14} W/cm^2 are required. For this purpose the cavity mirrors were changed to radii of 38 and 150 mm, generating a 12 μm focal spot size. With 5 kW of average power in the cavity a harmonic spectrum as shown in Fig. 3b was obtained.

The data recorded by the CCD camera demonstrate spatial coherence of the generated harmonics. In Fig. 4 the horizontal axis is the spectrally dispersed coordinate and the vertical axis signifies the spatial dimension. For relatively low order harmonics the spatial pattern exhibits diffraction lobes and can well be simulated by a plane wave model. The higher order harmonics display a central core and a pedestal attributed to short and long quantum path harmonics [10].

4 New Developments

Recent projects conducted by the enhancement cavity group at MPQ/LMU include power scaling to MW-range circulating powers and the enhancement of few-cycle pulses in the cavity. Appropriately filtered harmonics are expected to generate isolated attosecond pulses.

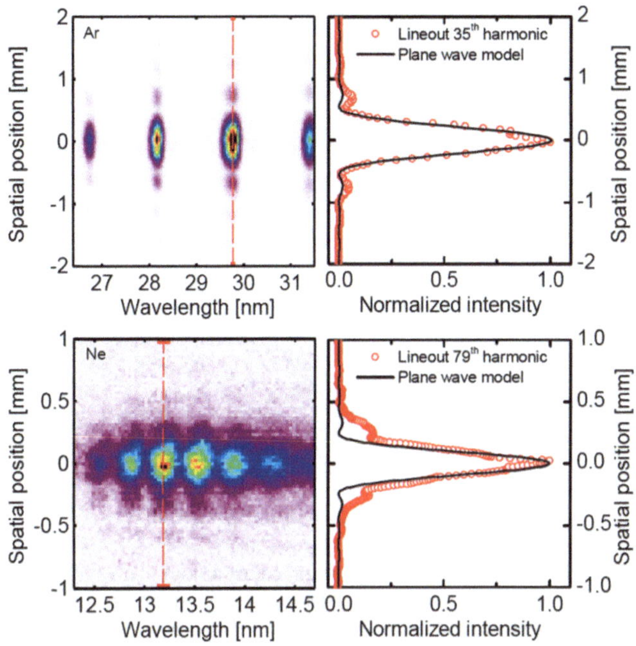

Fig. 4 Two-dimensional recording of harmonics demonstrating spatial coherence of *low*-order harmonics and short-long quantum path composition of *high*-order harmonics. Figure reproduced from ref. [7] with permission

4.1 Further Power Scaling

One of the goals of this research is to generate more powerful harmonics and to extend their spectral range to shorter wavelengths, possibly to the water window. For this purpose a new seed laser was installed, capable of delivering up to 420 W of power at a pulse duration of 250 fs and a repetition rate of 250 MHz. With such a high seed power two limitations are met in the enhancement cavity:

- Damage of the mirrors induced by the high intensity
- Thermal effects due to the high circulating power.

To circumvent these problems a new cavity was designed, which exhibits a considerably larger spot size on the mirrors. The cavity is operated close to the limits of stability, where the spot size of an optical resonator increases significantly. However, in turn an increased alignment instability is expected (Fig. 5).

In theory and experimentally it was found that at one of the stability edges, alignment of the cavity was very sensitive to small disturbances, whereas at the other (with the curved mirrors close to a confocal arrangement) alignment was straightforward. Operating the cavity close to that stability limit the illuminated area on the mirrors was increased by a factor of 15, allowing correspondingly higher circulating powers without the risk of mirror damage [11].

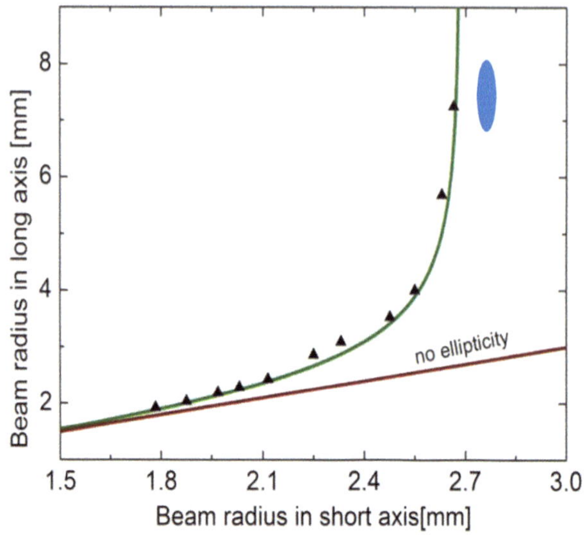

Fig. 5 Beam radius in long *axis* (*sagittal*) vs. beam radius in short *axis* (*tangential*). The diagram shows the increased ellipticity of the beam profile, resulting from oblique incidence on the curved mirrors. The experimental data (*triangles*) follow well the theoretical curve (*green*). The curve with no ellipticity would result for 0° angles of incidence

When running the cavity with high power thermal effects on the mirrors come into play. Absorption of the radiation in the first layers of the mirrors results in a temperature increase and a change in the radius of curvature. Using ultralow-expansion (ULE) glass for the mirrors and a coated sapphire substrate for the input coupler a circulating power of 670 W at a pulse duration of 10 ps could be realized [12]. For 250 fs pulses a circulating power of 400 kW was achieved, limited by mirror damage. Higher damage threshold mirrors allowing close to MW circulating power at a short pulse duration are being developed.

4.2 *Generation of Few-Cycle Pulses and Attosecond Harmonics*

One of the grand goals of this research is the generation of isolated attosecond pulses (IAPs) at multi-MHz repetition rates. In many investigations with low yield acquisition times could be dramatically reduced using such a source. Examples are angle resolved photoemission spectroscopy (ARPES) [13], and cold target recoil ion momentum spectroscopy (COLTRIMS) [14]. In these experiments the signal yield is limited by space charge effects, resulting in hours of acquisition time with kHz repetition rate lasers.

For generating IAPs the fundamental radiation must consist of few cycle pulses [15]. Their harmonic spectrum does not show isolated peaks but is continuous. To

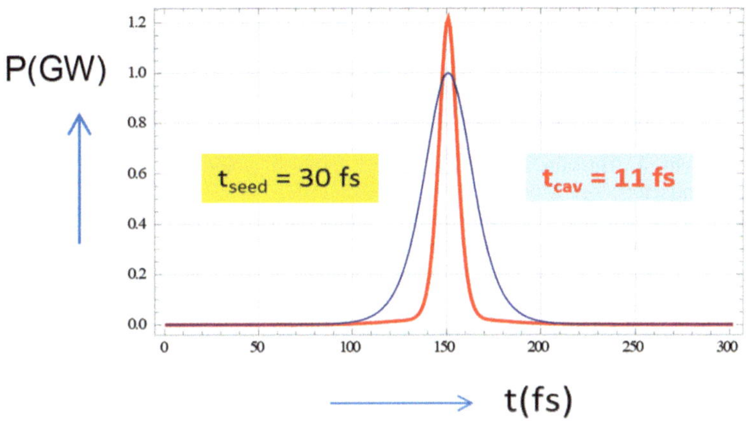

Fig. 6 Generating a soliton in an enhancement cavity. The 30 fs seed pulse is reduced to a 11 fs circulating pulse, shown in *red*. The parameters used in the simulation are: linear enhancement factor 200, seed peak power 5 MW, *second* order dispersion $\beta_2 = -1.2\text{fs}^2$, nonlinear medium: 100 μm sapphire plate. The seed pulse is scaled by the linear enhancement factor

achieve this goal a soliton may be launched in the enhancement cavity by introducing a Kerr medium (with a second order nonlinearity) and negative second order dispersion [16]. Numerically solving the "driven nonlinear Schrödinger equation" it appears that a soliton with significantly shorter pulse duration than that of the driver pulse can be generated in the cavity (Fig. 6). However, for obtaining an IAP, further pulse shortening must be achieved, possibly by applying a gating technique

References

1. Jones, R.J., Ye, J.: Femtosecond pulse amplification by coherent addition in a passive optical cavity. Opt. Lett. **27**, 1848–1850 (2002)
2. Gohle, C., et al.: A frequency comb in the extreme ultraviolet. Nature **436**, 234–237 (2005)
3. Jones, R.J., et al.: Phase coherent frequency combs in the vacuum ultraviolet via high-harmonic generation inside a femtosecond enhancement cavity. Phys. Rev. Lett. **94**, 193201 (2005)
4. Mills, A.K., et al.: XUV frequency combs via femtosecond enhancement cavities. J. Phys. B At. Mol. Opt. Phys. **45**, 142001–142023 (2012)
5. Cingöz, A., et al.: Direct frequency comb spectroscopy in the extreme ultraviolet. Nature **482**, 68–71 (2012)
6. Yost, D.C., Schibli, T.R., Ye, J.: Efficient output coupling of intracavity high-harmonic generation. Opt. Lett. **33**, 1099–1101 (2008)
7. Pupeza, I., Holzberger, S., et al.: Compact high-repetition-rate source of coherent 100 eV radiation. Nat. Photon **7**, 608–612 (2013)
8. Moll, K.D., Jones, R.J., Ye, J.: Output coupling methods for cavity-based high-harmonic generation. Opt. Express. **14**, 8189–8197 (2006)
9. Eidam, T., et al.: 57 W, 27 fs pulses from a fiber laser system using nonlinear compression. Appl. Phys. **92**, 9–12 (2008)

10. Zaïr, A., et al.: Quantum path interferences in high-order harmonic generation. Phys. Rev. Lett. **100**, 143902 (2008)
11. Carstens, H., et al.: Large-mode enhancement cavities. Opt. Express **21**, 11606–11617 (2013)
12. Carstens, H., et al.: Megawatt-scale average-power ultrashort pulses in an enhancement cavity. Opt. Lett. **39**, 2595–2598 (2014)
13. Damascelli, A., Hussain, Z., Shen, Z.: Angle–resolved photoemission studies of the cuprate superconductors. Rev. Mod. Phys. **75**, 473–541 (2003)
14. Ulrich, J., et al.: Recoil-ion and electron momentum spectroscopy: reaction-microscopes. Rep. Prog. Phys. **66**, 1463–1545 (2003)
15. Brabec, T., Krausz, F.: Intense few-cycle laser fields: frontiers of nonlinear optics. Rev. Mod. Phys. **72**, 545–591 (2000)
16. Kalashnikov, V.L.: Femtosecond pulse enhancement in an external resonator: impact of dispersive and nonlinear effects. Appl. Phys. **92**, 19–23 (2008)

Generation of Bright Circularly-Polarized High Harmonics for Magneto-Optical Investigations

P. Grychtol, O. Kfir, R. Knut, E. Turgut, D. Zusin, D. Popmintchev, T. Popmintchev, H. Nembach, J. Shaw, A. Fleischer, H. Kapteyn, M. Murnane and O. Cohen

Abstract We demonstrate the first generation of circularly-polarized high harmonics, which are bright enough for measurements of the x-ray magnetic circular dichroism (XMCD) at the *M* absorption edges of Fe, Co and Ni in the extreme ultraviolet range.

1 Introduction

Circularly polarized light in the extreme ultraviolet (EUV) and soft x-ray regions of the electromagnetic spectrum is extremely useful for exploring chirality-sensitive light-matter interactions. With respect to magneto-optics, x-ray magnetic circular dichroism (XMCD) makes it possible to extract element-specific information about the magnetic state of matter and its interaction with phononic and electronic degrees of freedom on femtosecond time scales and nanometer length scales. Specifically, XMCD can be used to distinguish between spin and orbital contributions to the atomic magnetic moment in ferromagnetic materials. To date, the generation of circularly polarized EUV and soft x-ray beams has been restricted to large-scale electron storage facilities, such as synchrotrons and x-ray free electron lasers. Such facilities have great advantages of high peak and average powers in the x-ray region. However, drawbacks include experimental complexity, limited access, temporal resolution, and a pump-probe timing jitter.

Tabletop short-wavelength sources based on high harmonic (HHG) up-conversion of femtosecond laser pulses represent a complementary and increasingly productive alternative to large scale sources, due to their unique ability to generate

P. Grychtol (✉) · R. Knut · E. Turgut · D. Zusin · D. Popmintchev · T. Popmintchev · H. Kapteyn · M. Murnane
JILA, University of Colorado, Boulder, CO 80309, USA
e-mail: p.grychtol@jila.colorado.edu

O. Kfir · A. Fleischer · O. Cohen
Solid State Institute and Physics Department, Technion, Haifa, 32000, Israel

R. Knut · H. Nembach · J. Shaw
Electromagnetics Division, NIST, Boulder, CO 80305, USA

A. Fleischer
Department of Physics and Optical Engineering, Ort Braude College, Karmiel 21982, Israel

© Springer International Publishing Switzerland 2016

J. Rocca et al. (eds.), *X-Ray Lasers 2014*, Springer Proceedings in Physics 169,
DOI 10.1007/978-3-319-19521-6_24

bright, broadband, ultrashort and coherent light from the UV to the keV region [1]. HHG sources have successfully exploited the resonantly-enhanced magnetic contrast at the M absorption edges of the 3d ferromagnets Fe, Co and Ni in the EUV region to study element-specific dynamics in magnetic materials on few femtosecond time scales [2, 3]. These studies have provided novel insights on ultrafast spin scattering and transport, as well as exchange interaction dynamics, in complex multi-species magnetic materials [4–6]. However, these investigations have so far been limited to linearly polarized HHG (sensitive only to the in-plane component of the sample magnetization), since generating circularly-polarized harmonics has been highly inefficient, reducing the photon flux as well as degree of circularity to a level that precludes scientific applications [7–9].

In this work, we present the first direct approach for generating bright circularly-polarized HHG under phase-matched conditions, based on a technique that was suggested almost two decades ago [10, 11] and was recently demonstrated by Fleischer *et al.* [12]. Surprisingly, although it was assumed for a long time that HHG from atoms was bright only when both the driving laser and HHG fields were linearly polarized, it can be shown that circularly-polarized HHG can be as bright [13]. In this work, we also demonstrate the first XMCD measurements using a tabletop light source at the M absorption edges of Fe, Co and Ni, thereby probing the magnetization component parallel to the light beam. This experiment thus represents the first application of circularly-polarized HHG and the first measurement proof of the helicity of circularly-polarized harmonics.

2 Theoretical Considerations

We generate bright circularly polarized harmonics by co-propagating bi-chromatic 790 nm and 395 nm driving laser beams with counter-rotating circular polarization in a gas-filled waveguide. When high harmonics are generated by a circularly polarized red driver with a frequency $\omega_0 = 2\pi/T$ rotating leftwards in combination with a blue driver rotating rightwards at a frequency $2\omega_0$, the electric fields of the two drivers $\vec{E}_{BC}(t)$ coalesce 3 times every cycle T. As schematically depicted in Fig. 1,

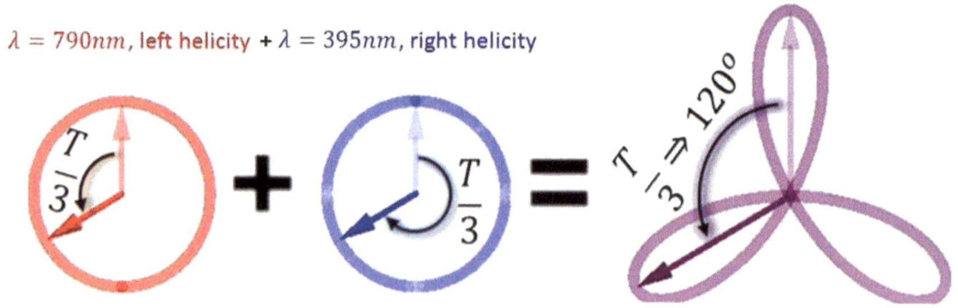

Fig. 1 Polarization scheme to generate circular polarized harmonics

the shape of the combined fields can be described by a cloverleaf and its dynamical symmetry suggests that the system is identical at t_0 and $t_0 + T/3$, *i.e.* by a spatial rotations of 120°. The emission of high harmonics emerging from such scheme must obey this symmetry and irrespective of the laser intensity ratio it has to satisfy the eigenvalue equation:

$$\vec{E}_q e^{-2\pi i q/3} = \hat{R}_{(120°)} \vec{E}_q,$$

where $\hat{R}_{(120°)}$ is the 120° rotation operator in the polarization plane and the phase term represent $T/3$ time delay. The solutions to this equation are a left circularly-polarized field with eigenvalue $e^{-2\pi i/3}$ and a right circularly-polarized field with eigenvalue $e^{+2\pi i/3}$. The left circularly-polarized qth-order harmonic, $\vec{E}_{q,L}$, satisfies $e^{-2\pi i/3 \cdot q} \vec{E}_{q,L} = e^{-2\pi i/3} \vec{E}_{q,L}$ which is fulfilled only when $q = 3m + 1$. Similarly, the right circularly-polarized qth-order harmonic, $\vec{E}_{q,R}$, satisfies $e^{-2\pi i/3 \cdot q} \vec{E}_{q,R} = e^{+2\pi i/3} \vec{E}_{q,R}$ which is fulfilled only when $q = 3m - 1$. Interestingly, $q = 3m$ harmonics do not satisfy the eigenvalue equation and are therefore forbidden and their emission is suppressed.

3 Experimental Details

As can be seen in Fig. 2, circular high harmonics are generated by focusing the fundamental and second harmonic of a Ti:sapphire laser amplifier with opposing helicities into a gas filled fiber. The polarization states of the two drivers can be independently adjusted in an interferometric Mach-Zehnder setup thereby controlling the helicity of the generated harmonics. Thus, the magnetic state of the sample

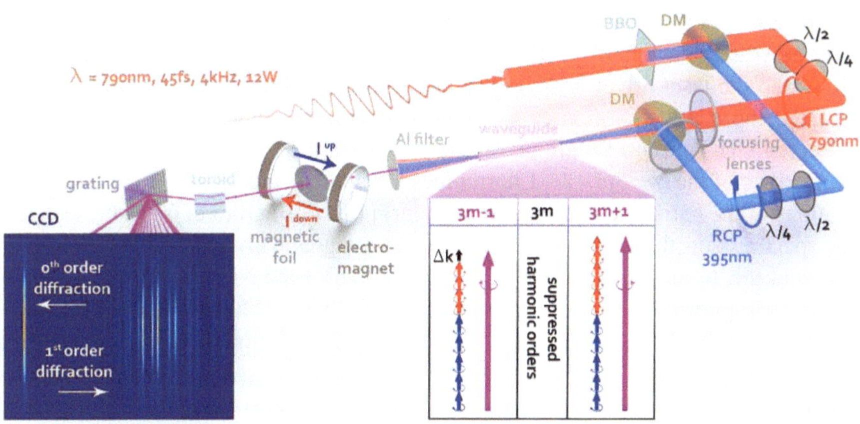

Fig. 2 Experimental setup to generate circular polarized high harmonics and measure the x-ray circular magnetic dichroism (XMCD) on the tabletop

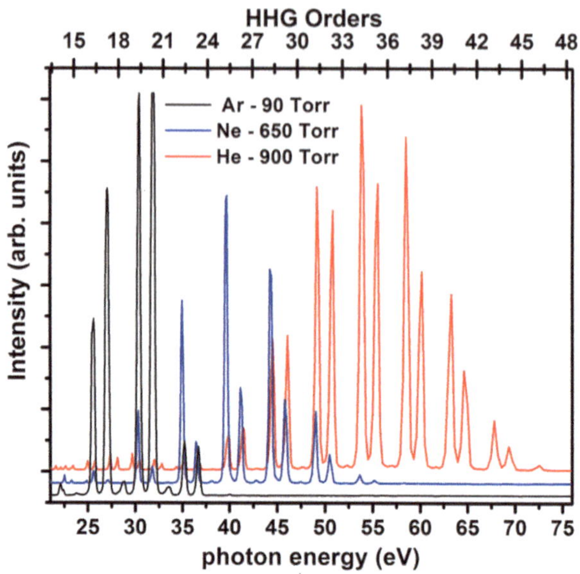

Fig. 3 Spectra of circularly-polarized high harmonics generated in Ar, Ne and He

foils can be probed element-selectively by the circularly-polarized HHG in a transmission geometry. We exploit the XMCD effect at the *M* absorption edges of Fe, Co and Ni in the extreme ultraviolet range.

4 Results

Figure 3 shows the observed HHG spectra when the hollow fiber was filled with 90 Torr of Argon (black line), 650 Torr of Neon (blue line) and 900 Torr of Helium (red line), and the laser driver beams were blocked by Al filters of ~800 nm thickness. Clearly, the $q = 3m$ harmonics are almost completely suppressed over the entire observed HHG spectra for all gas species. This feature is a clear indication that the ellipticities of the $q = 3m +/- 1$ harmonics are approximately $+/- 1$ (*i.e.* the polarizations of the harmonics are closely circular) [13]. Furthermore, it can be seen that the spectra spans the entire energy range of the magnetically sensitive *M* absorption edges of the 3d ferromagnets Fe (53 eV), Co (60 eV) and Ni (68 eV).

The top graph of Fig. 4 shows spectra of circularly polarized harmonics transmitted through 50 nm of magnetized Ni with magnetization direction either "Up" (I^{Up}, blue) or "Down" (I^{Down}, red). The black line shows the transmission spectrum of Ni. The bottom graph of Fig. 4 shows the normalized difference in signal between "Up" and "Down" magnetization. This so-called XMCD asymmetry defined by $(I^{Up} - I^{Down})/(I^{Up} + I^{Down})$, reaching values of 2 % at the 39th harmonic, clearly shows inverted helicities of adjacent harmonic orders, as convincingly pointed out by the alternating asymmetry.

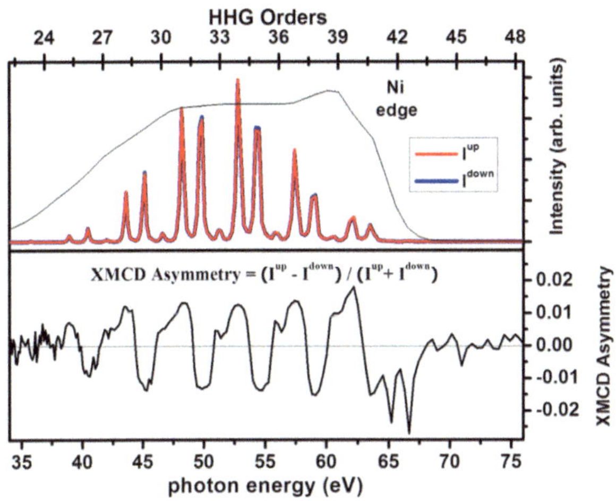

Fig. 4 XMCD of Ni with circular polarized high harmonics generated in He

5 Summary & Conclusion

We have demonstrated the first bright source of circularly-polarized high harmonics and used it for x-ray magnetic circular dichroism (XMCD) measurements of magnetic materials. Our HHG source emits a broad spectrum of coherent circularly-polarized harmonics under phase-matched conditions, where consecutive harmonics exhibit opposite helicity. This work removes a major constraint to date—that the polarization of bright high harmonics sources was limited to linear polarization. This work therefore will allow for investigations of ultrafast circular dichroism in magnetic samples, chiral molecules and nanostructures. In combination with coherent diffraction imaging techniques, this advance will enable ultrafast dynamics imaging of magnetic domains. It is important to note that this scheme is universal, and can be used to generate circularly-polarized high harmonics across broad spectral regions [14] Fan et al., and may be extended towards the water window and keV range for example through quasi phase matching of the HHG process, by mid-IR pumps or by using other nonlinear media (molecules, ions, etc.).

Acknowledgement This work was done at JILA and supported by the USA–Israel Binational Science Foundation (BSF). The Technion group is part of the Israeli Center of Research Excellence 'Circle of Light' supported by the I-CORE Program of the Planning and Budgeting Committee and The Israel Science Foundation. The JILA group gratefully acknowledges funding from the U.S. Department of Energy Office of Basic Energy Sciences, Award #DE-SC0002002, from the Deutsche Forschungsgemeinschaft #GR 4234/1–1, and from the Physics Frontiers Center Program. JILA also gratefully acknowledges support from an AFOSR DURIP award for the laser system used for this work.

References

1. Popmintchev, T., et al.: The attosecond nonlinear optics of bright coherent x-ray generation. Nat. Photonics **4**, 822 (2010)
2. La-O-Vorakiat, C., et al.: Ultrafast demagnetization dynamics at the M edges of magnetic elements observed using a tabletop high-harmonic soft x-ray source. Phys. Rev. Lett. **103**, 257402 (2009)
3. La-O-Vorakiat, C., et al.: Ultrafast demagnetization measurements using extreme ultraviolet light: Comparison of electronic and magnetic contributions. Phys. Rev. X **2**, 011005 (2012)
4. Mathias, S., et al.: Probing the timescale of the exchange interaction in a ferromagnetic alloy. Proc. Natl. Acad. Sci. **109**, 4792 (2012)
5. Rudolf, D., et al.: Ultrafast magnetization enhancement in metallic multilayers driven by superdiffusive spin current. Nat. Commun. **3**, 1037 (2012)
6. Turgut, E., et al.: Controlling the competition between optically induced ultrafast spin-flip scattering and spin transport in magnetic multilayers. Phys. Rev. Lett. **110**, 197201 (2013)
7. Liu, L., et al.: Optical rotation quasi-phase-matching for circularly polarized high harmonic generation. Opt. Lett. **37**, 2415 (2012)
8. Vodungbo, B., et al.: Polarization control of high order harmonics in the EUV photon energy range. Opt. Express **19**, 4346 (2011)
9. Husakou, A., et al.: Polarization gating and circularly-polarized high harmonic generation using plasmonic enhancement in metal nanostructures. Opt. Express **19**, 25346 (2011)
10. Long, S., et al.: Model calculations of polarization-dependent two-color high-harmonic generation. Phys. Rev. A **52**, 2262 (1995)
11. Eichmann, H., et al.: Polarization-dependent high-order two-color mixing. Phys. Rev. A **51**, R3414 (1995)
12. Fleischer, A., et al.: Spin angular momentum and tunable polarization in high-harmonic generation. Nat. Photonics (2014). doi: 10.1038/nphoton.2014.108
13. Fleischer, A., et al.: Spin angular momentum and tunable polarization in high-harmonic generation. Nat. Photonics (Supplementary Information) (2014). doi: 10.1038/nphoton.2014.108
14. Fan, T., Grychtol, P., Knut, R., Hernandez-Garcia, C., Hickstein, D., Gentry, C., Popmintchev, T.: Bright circularly polarized soft x-ray high harmonics for x-ray magnetic circular dichroism. In CLEO: 2015 Postdeadline Paper Digest. (2015). doi:10.1364/CLEO_AT.2015.JTh5C.1

Electron Motion Control in HHG Process with Multi-Color Laser Field

Zeng Zhinan, Wei Pengfei, Li Ruxin and Xu Zhizhan

Abstract The electron motion during the HHG process is sensitive to the waveform of the electric field. With multi-color laser field, a single high-order harmonic among the harmonic comb can be selectively enhanced by using a driving laser field with subcycle waveform control, which is synthesized by the fundamental 800 nm laser pulse and two controlling laser pulses at 400 and 267 nm with perpendicular polarizations. Such phenomena can be mainly attributed to the intra-atomic phase matching realized with the sub-cycle waveform controlled field.

1 Introduction

High-order harmonic generation (HHG) in gases has been intensively explored in the past decade [1, 2]. The high order harmonic source usually emits in the form of extreme ultraviolet (XUV) frequency comb in the plateau region with different harmonics at similar intensities, due to the nonperturbative nature of the HHG process and is described well by the three-step model. On one hand, the harmonic emission in the form of broadband supercontinuum is desired for the isolated attosecond pulse by using a few-cycle or a multicolor laser field. On the other hand, the selective generation of a single high-order harmonic emission from the harmonic comb as an intense monochromatic coherent source is also very important for many applications, which can be selected from the harmonic comb because of the phase matching effect including both macroscopic and intra-atomic phase matching [3]. The intra-atomic phase matching scheme was proposed [3, 4] to select a single harmonic emission from the harmonic comb. For intra-atomic phase matching, constructive

Z. Zhinan (✉) · W. Pengfei · L. Ruxin · Xu Zhizhan
State Key Laboratory of High Field Laser Physics, Shanghai Institute of Optics and Fine Mechanics, Chinese Academy of Sciences, Shanghai, 201800, China
e-mail: zhinan_zeng@mail.siom.ac.cn

W. Pengfei
College of Physics and Electronic Information Engineering, Wenzhou University, Wenzhou 325035, Zhejiang Province, China

L. Ruxin · Xu Zhizhan
School of Physical Science and Technology, Shanghai Tech University, Shanghai 200031, China

© Springer International Publishing Switzerland 2016
J. Rocca et al. (eds.), *X-Ray Lasers 2014*, Springer Proceedings in Physics 169,
DOI 10.1007/978-3-319-19521-6_25

interference between the emissions of different trajectories in each half-cycle is required [5]. The phases of the different trajectories can be optimized by controlling the delays and chirps of the multi-color laser field, and the phase differences of almost all trajectories for the selected harmonic can be optimized to be almost within 1 π, which signifies good intra-atomic phase matching.

2 Experimental Result and Discussion

In this paper, we report the controlling of the intra-atomic phase matching by using a driving laser field with sub-cycle waveform control. The laser field is synthesized with the fundamental, second and third harmonics of an 800 nm multicycle Ti:sapphire laser pulse. A specific harmonic can be selectively enhanced while the adjacent harmonics are dramatically suppressed, with both the harmonic intensity and the selectivity (the intensity contrast ratio between the target harmonic and the adjacent harmonics) significantly improved. We also selectively generate an intense single harmonic in argon from a long gas cell using a long-focus lens with a three-color scheme. When compared with single harmonic emission from a continuous gas jet, the intensity of the selected single harmonic emission from the long gas cell is more intense by as much as 1–2 orders of magnitude. Simultaneously, the contrast ratio (i.e., the spectral purity) is also increased by several times.

The schematic of the experimental setup is shown in Fig. 1a. A commercial Ti:sapphire femtosecond laser (Coherent, Inc.) is used to produce laser pulses at 800 nm center wavelength with 45 fs pulse duration at a repetition rate of 1 kHz. The output pulses are directed into the vacuum chamber for HHG. In the vacuum chamber, the laser beam is collimated and passes through sequentially a focusing lens, a SH-BBO crystal (β-barium borate crystal, $\theta = 29.2°$, $\varphi = 0°$, 0.3 mm thickness, type I phase matching, used for the second harmonic generation, the polarization of the generated 400 nm pulse is perpendicular to that of the fundamental 800 nm pulse), a $CaCO_3$ crystal ($\theta = 22.6°$, $\varphi = 0°$, 0.4 mm thickness, used for controlling the time delay between the 400 and 800 nm pulses), and a TH-BBO crystal (β-barium borate crystal, $\theta = 55.5°$, $\varphi = 30°$, 0.1 mm thickness, type II phase matching, for the third harmonic generation, the polarization of the generated 267 nm pulse is parallel to that of the fundamental 800 nm pulse), and is then focused onto the argon gas to generate the high order harmonics. The generated high-order harmonics are detected by a homemade flat-field grating spectrometer equipped with a soft-x-ray CCD (Princeton Instruments, SX 400). A 500 nm thick aluminum foil is used in the spectrometer to block the driving laser.

At first, we use a continuous gas jet and the pulse energy of 2 mJ. With a lens of 500 mm focal length, the total laser intensity on the argon gas is about 2.0×10^{14} W/cm^2. For comparison, we measure the harmonic yields by using the traditional two-color scheme with the orthogonally polarized 800/400 nm pulses (just remove the TH-BBO crystal, the measured energy ratio between 800 and 400 nm pulses is about 100:17). The measured harmonic yields are shown in Fig. 2a

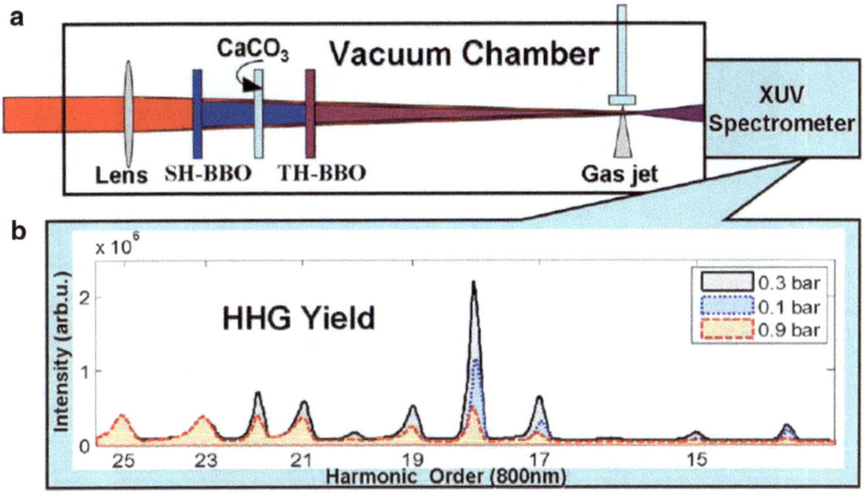

Fig. 1 **a** Schematic of the experimental setup. **b** Harmonic spectra obtained with different gas pressures [5]

as a function of the time delay between the two-color pulses. The measured harmonic yields by using the above three-color scheme are shown in Fig. 2b as a function of the time delay between the 800 and 400 nm fields. The modulation period of the generated harmonics in the two-color case is 1/4 cycle of 800 nm laser pulse. In the three-color case, the harmonic intensity is modulated with a period of about one cycle under the experimental condition since the time delay of the 267 nm pulse is correlated to the time delay between the 800 and 400 nm pulses.

For a more detailed comparison, we extract the maximum harmonic intensities at the optimum time delay from Fig. 2a and 2b, which are shown in Fig. 3a, together with the harmonic spectrum obtained by using only the fundamental 800 nm pulse. Compared to the harmonic generation in the fundamental 800 nm pulse, the addition of the 400 nm pulse of perpendicular polarization can enhance the conversion efficiency by about 2 orders of magnitude, and the further addition of a weak 267 nm pulse of parallel polarization further enhances the conversion efficiency by about 5 times. Moreover, compared to the traditional two-color scheme, the addition of the weak 267 nm pulse can greatly enhance the intensity contrast ratio between the 18th harmonic and the adjacent harmonics.

In order to identify the role of the relative phase delay between different colors in the three-color scheme, we extract the harmonic intensities from Fig. 2b at the time delays of 0 and 0.5 cycle, which are shown in Fig. 3b. By optimizing the relative phase, the 18th harmonic intensity is enhanced by 43.7 %, while the intensities of the adjacent harmonics are at least suppressed by 65.4 %. Then, the intensity contrast ratio between the 18th harmonic and the adjacent harmonics is improved from 3.8 to 15.9.

To increase the conversion efficiency further, we use the gas cell to do the work. The main differences are the focusing optics and the length of the lasing medium, while the other experimental conditions are all the same. We selected a longer focus

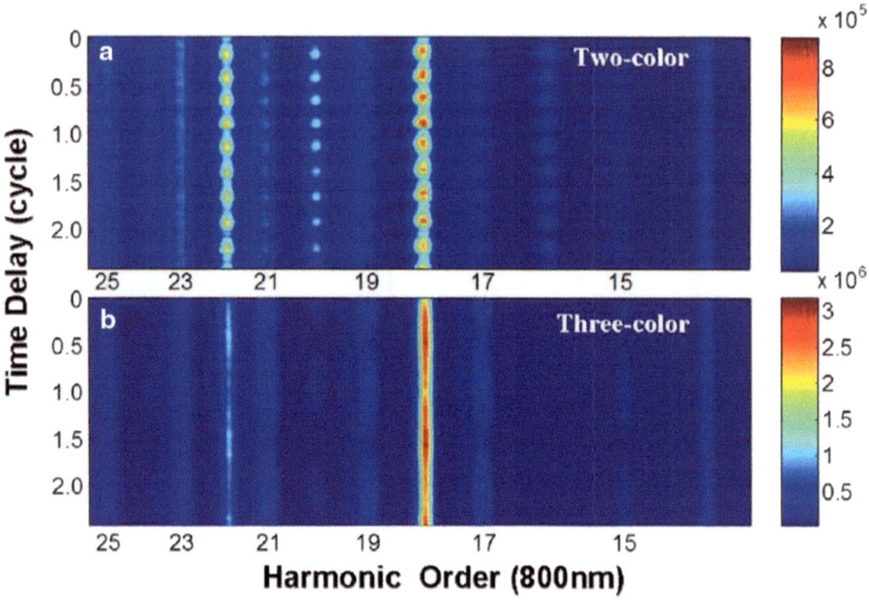

Fig. 2 Harmonic spectra driven by **a** the two color laser field, and **b** the three-color laser field as a function of the time delay, respectively [5]

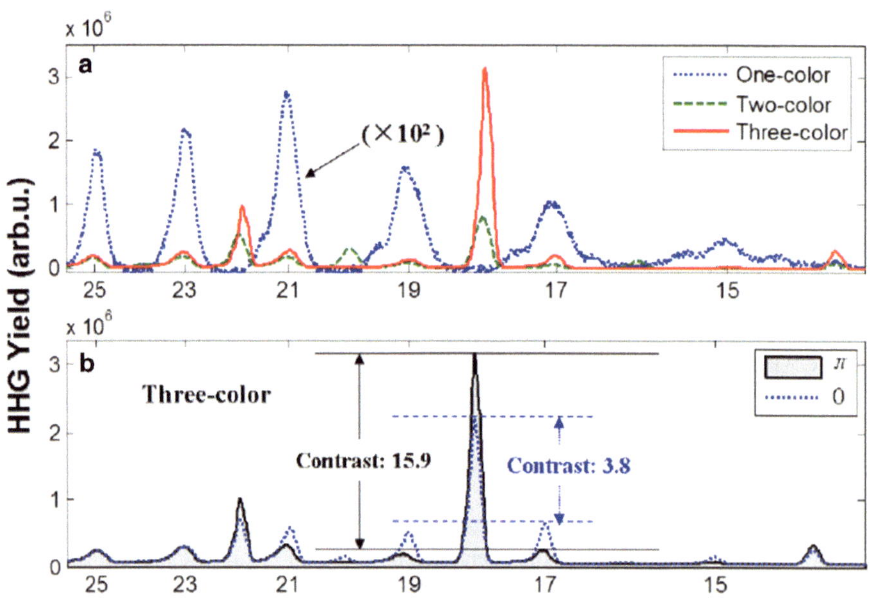

Fig. 3 **a** Comparison of the maximum harmonic yields driven by different laser fields. **b** Comparison of the harmonic yields driven by the three-color laser field with two different relative phases [5]

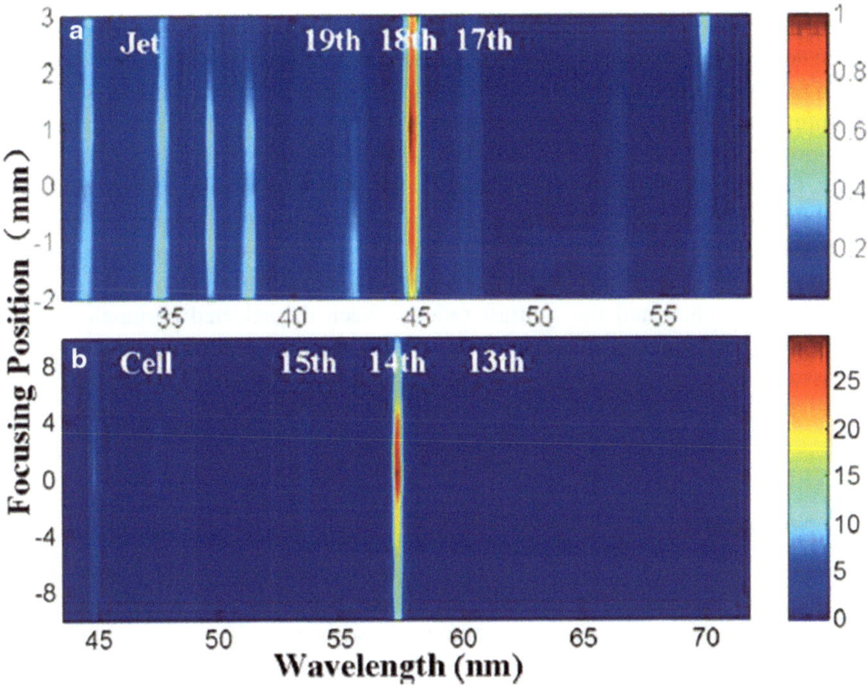

Fig. 4 Experimentally measured harmonic emissions as functions of focusing position in **a** the jet case, and **b** the cell case [6]

lens of 1100 mm and a much longer medium with a cell length of 25 mm. Also, a diaphragm of approximately 10 mm is used in front of the lens to further control the quantum trajectories by modifying the laser beam profile in the HHG process. Therefore, we can optimize the harmonic signal by modifying the laser beam profile, selecting the focusing position, adjusting the argon gas pressure, and controlling the time delays among the different colors.

Using the three-color scheme, the measured harmonic emissions from the long gas cell are shown as functions of the focusing position in Fig. 4, and the corresponding emissions from the gas jet are shown for comparison. We see that the 18th harmonic is selectively enhanced in the jet case, while the 14th harmonic is selectively enhanced in the cell case. This selectivity can be attributed to the fact that stronger laser intensity corresponds to a higher harmonic order for the selective enhancement. We can also see that the peak intensity of the selected harmonic from the cell is approximately 30 times more intense than that of the jet, which can be attributed to the use of the loosely focusing optics in the cell case and the fact that the much longer medium provides a much higher conversion efficiency.

Then, we further optimized the single harmonic emission by controlling the time delays between the 400 and 800 nm laser pulses, and the 267 nm laser pulse, which is from sum-frequency generation in a BBO (BaB_2O_4) crystal, has a time delay that is correlated to the time delay between the 800 and 400 nm laser pulses. When the time delay is optimized, the selected harmonic is enhanced, while most of the other harmonics are suppressed and even disappear at some particular time delays. Further optimization can be achieved by adjusting the gas pressure. In the cell case with gas pressure of 4×10^{-2} bars, the peak intensity is approximately 5 times of jet case, the contrast ratio between the selected harmonic and the adjacent harmonics is 41.4. In the cell case with a gas pressure of 1×10^{-2} bars (the maximum harmonic emission, as shown in the inset of Fig. 4b), the peak intensity is approximately 30 times of jet case, the contrast ratio between the selected harmonic and the adjacent harmonics is 48.5, and the contrast ratio between the selected harmonic and all remaining harmonics is 4.2. Therefore, when compared with the single harmonic emission in the jet case, both the harmonic intensity and the spectral purity have obviously been enhanced by use of the long gas cell and the long-focus lens.

3 Summary

In summary, we observe that a single high-order harmonic is selectively enhanced while the adjacent harmonics are dramatically suppressed by using the multicolor laser field with subcycle waveform control. This scheme is promising for the development of high-intensity monochromatic coherent XUV sources.

References

1. McPherson, A., et al.: Studies of multiphoton production of vacuum-ultraviolet radiation in the rare gases. J. Opt. Soc. Am. B **4**, 595 (1987)
2. Lewenstein, M., Balcou, P., Ivanov, M.Y., L'Huillier, A., Corkum, P.B.: Theory of hhg by low-frequency laser fields. Phys. Rev. A **49**, 2117 (1994)
3. Bartels, R., Backus, S., Zeek, E., Misoguti, L. Vdovin, G. Christov, I. P., Murnane, M.M., Kapteyn, H.C.: Shaped-pulse optimization of coherent emission of high-harmonic soft X-rays. Nature (London). **406**, 164 (2000)
4. Christov, I.P., Bartels, R., Kapteyn, H.C., Murnane, M.M.: Attosecond time-scale intra-atomic phase matching of high harmonic generation. Phys. Rev. Lett. **86**, 5458 (2001)
5. Wei, P., Miao, J., Zeng, Z., Li, C. Ge, X., Li, R., Xu, Z.: Selective enhancement of a single harmonic emission in a driving laser field with subcycle waveform control. Phys. Rev. Lett. **110**, 233903 (2013)
6. Wei, P., Zeng, Z. Jiang, J. et al.: Selective generation of an intense single harmonic from a long gas cell with loosely focusing optics based on a three-color laser field. Appl. Phys. Lett. **104**, 151101 (2014)

Part IV
FEL-Based X-Ray Sources

Stimulated X-Ray Raman Scattering with Free-Electron Laser Sources

N. Rohringer, V. Kimberg, C. Weninger, A. Sanchez-Gonzalez, A. Lutman, T. Maxwell, C. Bostedt, S. Carron Monterro, A. O. Lindahl, M. Ilchen, R. N. Coffee, J. D. Bozek, J. Krzywinski, T. Kierspel, T. Mullins, J. Küpper, B. Erk, D. Rolles, O. D. Mücke, R. A. London, M. Purvis, D. Ryan, J. J. Rocca, R. Feifel, R. Squibb, V. Zhaunerchyk, C. Såthe, M. Agåker, M. Mucke, J. Nordgren and J. E. Rubensson

Abstract Stimulated electronic x-ray Raman scattering is the building block for several proposed x-ray pump probe techniques, that would allow the study of electron dynamics at unprecedented timescales. We present high spectral resolution data on stimulated electronic x-ray Raman scattering in a gas sample of neon using a self-amplified spontaneous emission x-ray free-electron laser. Despite the limited spectral coherence and broad bandwidth of these sources, high-resolution spectra

N. Rohringer (✉) · V. Kimberg · C. Weninger
Max Planck Institute for the Physics of Complex Systems, Dresden, Germany
e-mail: nina@pks.mpg.de

N. Rohringer · V. Kimberg · C. Weninger · T. Kierspel · T. Mullins · J. Küpper
Center for Free-Electron Laser Science, Hamburg, Germany

A. Sanchez-Gonzalez
Imperial College, London, United Kingdom

A. Lutman · T. Maxwell · C. Bostedt · S. Carron Monterro · M. Ilchen · R. N. Coffee · J. D. Bozek
J. Krzywinski
Linac Coherent Light Source, SLAC National Accelerator Laboratory, Menlo Park, CA, USA

T. Kierspel · T. Mullins · J. Küpper
University of Hamburg, Hamburg, Germany

T. Kierspel · T. Mullins · J. Küpper · B. Erk · D. Rolles · O. D. Mücke
DESY, Hamburg, Germany

R. A. London
Lawrence Livermore National Laboratory, Livermore, CA, USA

M. Purvis · D. Ryan · J. J. Rocca
Colorado State University, Fort Collins, CO, USA

A. O. Lindahl · R. Feifel · R. Squibb · V. Zhaunerchyk
Department of Physics, University of Gothenburg, Gothenburg, Sweden

C. Såthe · M. Agåker · M. Mucke · J. Nordgren · J. E. Rubensson
Department of Physics and Astronomy, Uppsala University, Uppsala, Sweden

A. O. Lindahl · R. N. Coffee
Stanford PULSE Institute, SLAC National Accelerator Laboratory, Menlo Park, California

M. Ilchen
European XFEL GmbH, Hamburg, Germany

© Springer International Publishing Switzerland 2016

J. Rocca et al. (eds.), *X-Ray Lasers 2014*, Springer Proceedings in Physics 169,
DOI 10.1007/978-3-319-19521-6_26

can be obtained by statistical methods, opening the path to coherent stimulated x-ray Raman spectroscopy. An extension of these ideas to molecules and the results of a recent experiment in CO are discussed.

1 Introduction

X-ray free-electron lasers (XFELs) open the pathway to transfer nonlinear spectroscopic techniques to the x-ray domain. A promising all x-ray pump probe technique is based on coherent stimulated electronic x-ray Raman scattering. Similarly to optical impulsive stimulated Raman scattering, where coherent vibrational and rotational wave packets can be created and subsequently probed [1], stimulated x-ray Raman scattering with sources of a broad bandwidth covering the range of valence excitation energies would result in electronic wave packets, thus opening the pathway to coherently probe electronic and nuclear dynamics. The relatively small Raman-cross section at x-ray frequencies can be enhanced by near resonant excitation, which in addition gives the advantage of targeting specific elements in a molecular complex. The electronic wave packets that are created in such a stimulated Raman process are typically localized around the targeted atoms, and the evolution of the wave packet, resulting in phenomena like charge transport, charge migration or dissipation of the energy to vibrational modes, can be subsequently studied by a second, delayed x-ray pulse. A wealth of different x-ray pump-probe schemes has been suggested over the last decades [2–4], but the underlying building block of these sophisticated methods—stimulated electronic x-ray Raman scattering—was only recently demonstrated [5]. Here, we give a summary of our experimental results in atomic neon and, discuss a recent experiment in carbon monoxide (CO), where stimulated Raman scattering was attempted with two XFEL pulses of distinct wavelength.

2 Stimulated X-Ray Raman Scattering in Neon

The first demonstration of resonant stimulated electronic x-ray Raman scattering was achieved in neon at the Linac Coherent Light Source (LCLS) XFEL in 2011 [5]. A second independent demonstration followed in 2013 with a three-times better spectral resolution, i.e. 0.3 eV. The XFEL parameter regime and the setup were similar in both experiments. Pulses of 40–60 fs duration were focused to spot sizes in the range of 2–3 μm into a 1 cm long gas cell filled with neon at 500 Torr pressure. A grating spectrometer was fielded 4 m downstream and the spectrum of the transmitted XFEL and emitted line radiation from the gas was collected shot-by-shot. The photon energy was tuned from 865 to 875 eV across the pre-K edge, core-excited resonances and the K-edge of neon (see level structure in Fig. 1). Resonant stimulated x-ray Raman scattering from the ground state, with the intermediate $1s^{-1}3p^1$ core-excited to the final $2p^{-1}3p^1$ states results in emission lines at 849.3

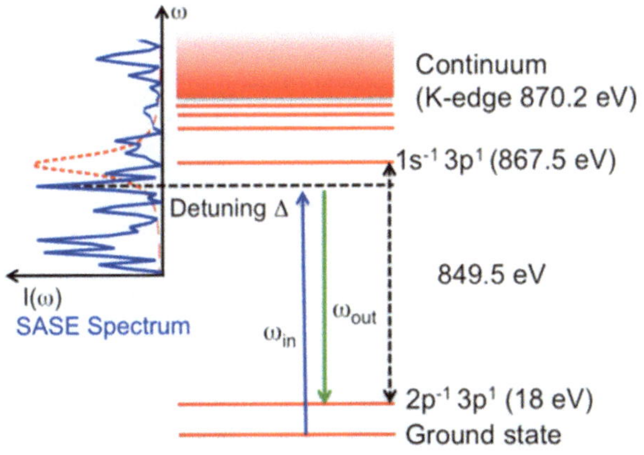

Fig. 1 Level scheme of the resonant scattering process with SASE pulses in neon

and 849.7 eV. These lines partly overlap with the stimulated K-α emission [6, 7] at 849.8 eV, resulting from inner-shell ionization. The core- excited and core-ionized states predominantly decay via Auger decay and have an extremely short lifetime of 2.4 fs. Similar to the photoionization K-α x-ray laser [7, 8], an avalanche of stimulated emission events sets in as the XFEL pulse propagates through the medium and prepares atoms in core-excited states, so that at saturation of the amplification process, stimulated Raman scattering dominates over the Auger decay.

Three measured single-shot spectra from our 2014 experiment are shown in Fig. 2a. The line emissions at around 850 eV show different energy positions and line profiles, highlighted in the inset. A summary of the recorded spectra is depicted in Fig. 2b. The spectra are ordered with respect to the incoming central photon energy, which was determined by the electron-bunch energy of the accelerator on a shot-by-shot basis. With the improved resolution compared to 2011, the $1s^{-1}3p^{1}$ resonant excitation can be discerned as a clear absorption dip of the XFEL pulse, 3 eV from the K-edge. Stimulated line emission is seen over the whole covered spectral range. Due to the relatively large XFEL bandwidth (5–7 eV) tuning the SASE pulses to the $1s^{-1}3p^{1}$ core-excited resonance in neon at 867.5 eV results in partial overlap with the continuum. Therefore resonant stimulated x-ray Raman scattering competes with stimulated emission following K-shell ionization for a typical SASE pulse and the processes can not be distinguished by simply analyzing the emission strength as a function of incoming photon energy. However, analyzing the emission-line profile as a function of the incoming photon energy (see Fig. 2c) reveals clear evidence for resonant stimulated electronic x-ray Raman scattering [5,6]. Whereas for energies well above the K-edge the emission profile and line position are stable, for photon energies overlapping with the resonance a shot-to-shot fluctuation of the emission energy and the line shape are observed. The emission energy of resonance scattering with a narrow-band source that is tuned through the resonance follows a linear dispersion. This is a consequence of the fact that energy conservation in the resonance scattering process is required only between initial and final states, whereas the intermediate core-resonant state does not have to be on the energy shell, as described by Kramers and Heisenberg [9]. The self-amplified spontaneous emission (SASE) XFEL pulses have spiky spectral intensity profiles.

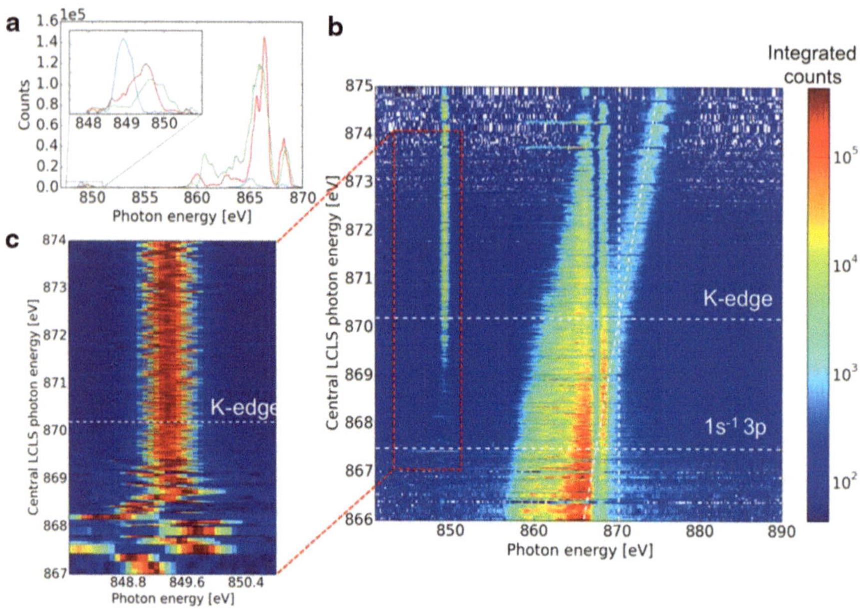

Fig. 2 **a** *Three* measured spectra in neon. The central photon energy of the SASE pulse was 868 eV and the pulse energy 1 mJ. The inset shows a magnification of the normalized *line* emission. **b** Series of measured single-shot spectra as a function of the incoming central SASE photon energy. The color code shows the number of integrated counts in the emission *line*. **c** Normalized emission spectrum as a function of the central SASE photon energy

The average width of each spectral spike (spectral coherence) scales as the inverse of the pulse duration and was estimated to 0.1 eV in our case, i.e. narrow compared to the Auger-width of the intermediate state. Individual spectral spikes of the SASE FEL therefore have random detuning from the resonance (see Fig. 1), which results in a stochastic line-shift from shot-to-shot.

A simulation of the emission spectrum based on a generalized Maxwell-Liouville-von-Neumann approach qualitatively reproduces this stochastic shift in the line emission [5, 6]. Studying the covariance of the recorded spectra would yield a highly resolved stimulated Raman spectrum, for which all the fine-structure levels and different emission lines can be uncovered [6]. The spectral resolution of resonant stimulated electronic x-ray Raman scattering with SASE sources is hence determined by the spectral coherence of the source, rather than the overall broad SASE bandwidth. A covariance analysis requires the recording of a large number of single shot spectra with a high dynamic range for measuring the SASE absorption profile, which was unfortunately beyond the scope of our experiment.

3 Resonant Stimulated Electronic X-Ray Raman Scattering in CO

In our recent experiment, an attempt was made to demonstrate stimulated Raman scattering in CO at 500 Torr pressure, by resonantly pumping the $\mathrm{O1s}^{-1}\pi^{*1}\Pi$ state at ≈ 534 eV (see Fig. 3b). In RIXS experiments, three partially overlapping spectator states are observed ($^1\Pi, ^1\Delta$ and $^1\sum^-$) [10], with oscillator strengths that are small compared to the $\mathrm{O1s}^{-1}\pi^{*1}\Pi$ transition. With a strong pump pulse, stimulated scattering back to the ground state is therefore the strongest channel. Seeding with a red-shifted second pulse ("dump" pulse) is therefore a possibility to enhance the electronically inelastic channel. To that end the XFEL was operated in a two-color mode, producing a strong pump pulse and a weak dump pulse shifted by 10 eV, to overlap with the transition to the states $1\pi^{-1}\pi^{*}l^1\sum^-$ at 525 eV. This is expected to be the strongest emission line for parallel polarizations of the two colors from the XFEL. Figure 3a shows the recorded spectra as a function of the central photon energy of the SASE-pump pulse. The broad absorption band to the $\mathrm{O1s}^{-1}\pi^{*1}\Pi$ state is clearly visible at a 534.2 eV (orange line). The pair of SASE pulses shows side bands to the low-energy side (the side bands are highlighted by white dashed lines). Strong stimulated electronic x-ray scattering would be visible as enhanced intensities in the black box at around 525 eV. Statistically significant evidence for enhanced intensities cannot be seen in the data. Numerical calculations for the actual experimental conditions (pump-pulse energies of 0.5–0.9 mJ) with a theory similar to that presented in reference [11] predict a stimulated Raman amplification of only $\approx 5\,\%$ (in contrast to 10^6 in neon) [12], unfortunately too small to be seen on top of the SASE shot-to-shot fluctuations of the spectral intensities. The experimental conditions were therefore in the best case 'at the onset' of stimulated Raman scattering and we have no evidence of the effect in molecules. Another important observation is clear absorption bands in the region from 526 to 533 eV (highlighted by red dashed lines). These absorption bands result from molecular ions in generally valence excited states that are produced during the course of the XFEL-pulse interaction: Competing processes to stimulated Raman scattering, i.e. core-ionization of C1s and core-excitation of O1s followed by Auger decay, valence ionization, etc. produce a substantial amount of excited molecular ions of different charge states, that subsequently absorb parts of the two SASE-XFEL pulses. This means that spectral regions of stimulated Raman emission in the seed pulse can overlap with absorption bands of the molecular ions that are produced in the sample. At small Raman gain-cross sections this can pose substantial challenges for the spectral analysis and for distinguishing between absorption dips and Raman emission peaks. In order to unveil the typically small stimulated Raman signal in molecular spectra, advanced statistical methods, beyond simple covariance methods, will have to be developed, to separate emission and absorption spectra.

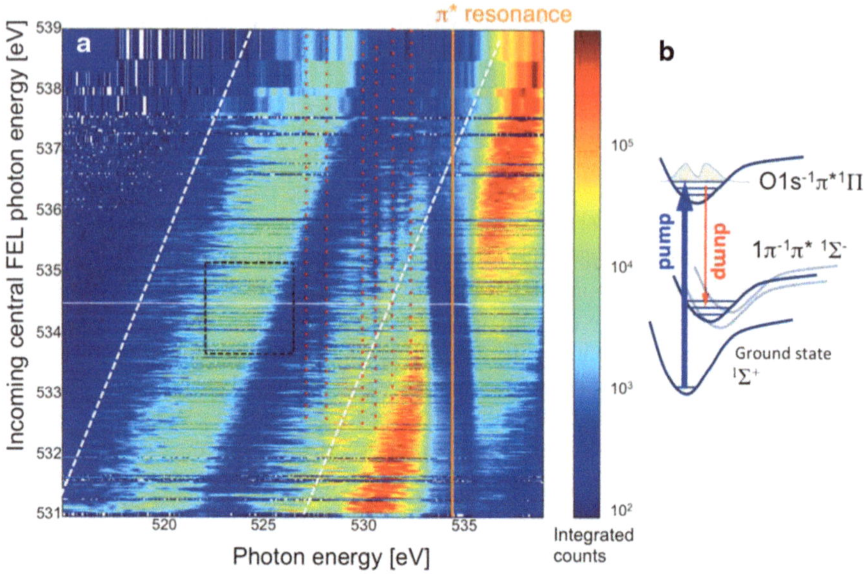

Fig. 3 **a** Measured single-shot spectra as a function of the central photon energy of the incoming pump pulse. The various *solid*, *dashed* and *dotted* lines are described in the text. **b** Level scheme of resonant stimulated electronic x-ray Raman scattering in CO

4 Conclusion

We have presented the results of a recent experiment on stimulated electronic x-ray Raman scattering in a dense gas of neon and CO. In neon, increased spectral resolution has enabled a more definitive identification of stimulated resonant Raman scattering by a stochastic shifts of the emission energies due to pumping with incoherent, structured SASE pulses. An attempt at demonstrating stimulated resonant Raman scattering by pumping the broad $\mathrm{O1s}^{-1}\pi^{*1}\Pi$ resonance and seeding the transition to the $1\pi^{-1}\pi^{*}l^{1}\sum^{-}$ states by a second SASE color was unsuccessful, since at the relatively low pulse-energies available in the experiment, the Raman gain was far below the SASE intensity fluctuations of the seed pulse. Moreover, several absorption bands of molecular ions that are produced by photoionization and Auger-decay of the targets have been identified in a broad energy range.

Acknowledgements Portions of this research were carried out at the Linac Coherent Light Source (LCLS) at the SLAC National Accelerator Laboratory. LCLS is an Office of Science User Facility operated for the U.S. Department of Energy Office of Science by Stanford University. Part of this work has been financially supported by the Swedish Research Council and the Knut and Alice Wallenberg Foundation, Sweden. Part of this work was performed under the auspices of the U.S. Department of Energy by Lawrence Livermore National Laboratory (Contract No. DE-AC52-07NA27344). Support for the work of M. P., D. R., and J. J. R. by the U.S. Department of Energy, Office of Science, Basic Energy Sciences AMOS Program is acknowledged.

References

1. Mukamel, S.: Principles of Nonlinear Optical Spectroscopy. Oxford University Press, New York (1999)
2. Tanaka, S., Mukamel, S.: Coherent x-ray raman spectroscopy: a nonlinear probe for electronic excitations. Phys. Rev. Lett. **89**, 043001 (2002)
3. Harbola, U., Mukamel, S.: Coherent stimulated x-ray raman spectroscopy: Attosecond extension of resonant inelastic x-ray Raman scattering. Phys. Rev. B **79**, 085108 (2009)
4. Healion, D., Wang, H., Mukamel, S.: Simulation and visualization of attosecond stimulated x-ray Raman spectroscopy signals in trans-N-methylacetamide at the nitrogen and oxygen K-edges. J. Chem. Phys. **134**, 1241 (2011)
5. Weninger, C., et al.: Stimulated electronic x-ray Raman scattering. Phys. Rev. Lett. **111**, 23 (2013)
6. Weninger, C., Rohringer, N.: Stimulated resonant x-ray Raman scattering with incoherent radiation, Phys. Rev. A **88**, 053421, 2013
7. Rohringer, N., et al.: Atomic inner-shell x-ray laser at 1.46 nm pumped by an x-ray free electron laser. Nature **481**, 488 (2012)
8. Weninger, C., Rohringer, N.: Transient-gain photoionization x-ray laser. Phys. Rev. A **90**, 063828 (2014)
9. Kramers, H.A., Heisenberg, W.: Über die Streuung von Strahlung durch Atome. Z. Phys. **31**, 681 (1925)
10. Skytt, P., Glans, P., Gunnelin, K., Guo, J., Nordgren, J.: Role of screening and angular distributions in resonant x-ray emission of CO. Phys. Rev. A **55**, 134 (1997)
11. Kimberg, V., Rohringer, N.: Amplified x-ray emission from core-ionized diatomic molecules. Phys. Rev. Lett. **110**, 043901 (2013)
12. Kimberg, V., Rohringer, N.: Manuscript in preparation

Comparing the Gain of the Ne K-α Inner-Shell X-Ray Laser Using the XFEL to Drive the Kinetics with Photo-Ionization Versus Photo-Excitation

Joseph Nilsen

Abstract Over the last four decades many photo-pumped X-ray laser schemes have been proposed. However, demonstrating these schemes in the laboratory has proved to be elusive because of the difficulty of finding a strong resonant pump line or X-ray source. With the advent of the X-ray free electron laser (XFEL) at the SLAC Linac Coherent Light Source (LCLS) we now have a tunable X-ray laser source that can be used to replace the pump line or X-ray source in previously proposed laser schemes and allow researchers to study the physics and feasibility of photo-pumped laser schemes. Many of these photo-pumped schemes are driven by photo-excitation from a resonant line source but others are driven by photo-ionization from a strong non-resonant X-ray source. Three years ago an inner-shell X-ray laser was demonstrated at 849 eV (1.46 nm) in singly ionized neon gas using the XFEL at 960 eV to photo-ionize the 1s electron in neutral neon followed by lasing on the 2p—1s transition in singly-ionized neon. In this paper we model the neon inner shell X-ray laser under similar conditions to those used at LCLS. We investigate how we can improve the efficiency of the neon laser and reduce the drive requirements by tuning the XFEL to the 1s-3p transition in neutral neon in order to create gain on the 2p-1s line in neutral neon. We explore the sensitivity to the drive intensity, pulse duration, and line-width of the XFEL to better understand how to optimize this inner shell laser by understanding the trade-offs between using photo-ionization versus photo-excitation to drive gain in these systems. We also discuss how photo-ionization of L-shell electrons can be used to create lasing on $n = 3 - 2$ transitions in materials such as Ar and Cu.

1 Introduction

Since the earliest days of the laser, scientists have proposed schemes to achieve lasing at shorter wavelengths. Five decades ago, working at Bell Laboratories where the laser was invented, Duguay and Rentzepis proposed using photo-ionization to create an X-ray laser on the inner shell K-α line in sodium vapour [1]. Forty-years

J. Nilsen (✉)
Lawrence Livermore National Laboratory, Livermore, CA, 94551, USA
e-mail: nilsen1@llnl.gov

© Springer International Publishing Switzerland 2016

J. Rocca et al. (eds.), *X-Ray Lasers 2014*, Springer Proceedings in Physics 169,
DOI 10.1007/978-3-319-19521-6_27

ago Ray Elton [2] discussed the challenges of making quasi steady state inner-shell K-α lasers in Si, Ca, and Cu. In 2011 the dream of demonstrating an inner-shell X-ray laser was realized at the SLAC Linac Coherent Light Source (LCLS) when the X-ray free electron laser (XFEL) at 960 eV was used to photo-ionize the K-shell of neutral neon gas and create lasing at 849 eV in singly ionized neon gas [3].

Another early approach for creating X-ray lasers was the idea of a resonantly photo-pumped laser where a strong emission line in one material could be used to photo-excite a transition in another material and create lasing. A classic example of this scheme is the Na-pumped Ne X-ray laser scheme proposed 40 years ago by Vinogradov and colleagues [4, 5]. This scheme used the strong Na He-α line at 1127 eV to resonantly photo-pump the Ne He-γ line and lase on the 4f—3d transition at 23.1 nm in He-like Ne. This scheme was studied extensively and numerous experiments were done to try to demonstrate lasing and measure gain [6]. While weak gain [6] was inferred in several experiments the difficulty with this type of scheme was creating a sufficiently strong pump line. With the availability of strong XFEL sources the pump line in the traditional photo-pumped schemes can be replaced with an XFEL that is tuned to the appropriate resonance. Since the resonant photo-pumped scheme selectively pumps a transition it offers the potential for higher gain and lower drive intensity than the photo-ionization pumping.

In this paper we look at the advantages and challenges of using the XFEL to resonantly photo-pump the 1s-3p line in neutral neon as a mechanism for creating gain on the K-α line in Ne and compare this with the photo-ionization pumping that has already been demonstrated. We show that with the use of a sufficiently short XFEL pulse (1-fs) the resonant photo-excitation could reduce the XFEL flux requirements by several orders of magnitude.

2 Modelling the Inner-Shell Ne Laser

Figure 1 shows the pumping mechanism used in the LCLS experiments that demonstrated lasing on the inner-shell neon laser. A strong XFEL beam tuned above the K-edge of neutral Ne I photo-ionizes the 1s electron. This creates an excited state of singly ionized Ne II that has a missing 1s electron. This excited state then lases to the ground state of Ne II by emitting an X-ray on the 2p—1s transition at 848.6 eV. The experiment starts with neon gas that is all in the Ne I ground state so the lower laser state is initially unoccupied. The natural lifetime of the laser transition is 135 fs. However the challenge with this scheme is that the Auger lifetime of the upper laser state is 2.3 fs so pumping this scheme requires a very short pulse duration in the fsec regime.

Figure 2 shows the resonant photo-pumping mechanism for driving the inner-shell neon laser. The XFEL is tuned to the 1s—3p transition in Ne I at 867.63 eV creating a large population in the $1s2s^22p^63p$ level. This level can then lase to the lower $1s^22s^22p^53p$ level state by emitting X-rays on the 2p—1s transition centered

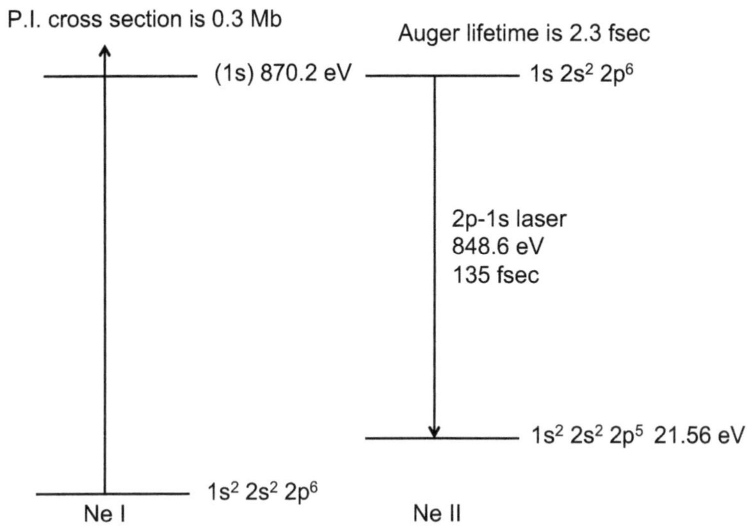

Fig. 1 Energy level diagram for the photo-ionization driven inner-shell neon X-ray laser

at 848.96 eV. Because of the splitting in the lower level state there are actually 5 X-ray lines emitted that are spread over a range of 848.67—849.25 eV. We calculate the total gain by summing the gain of the 5 lines. The upper laser level has a similar Auger lifetime of 2.3 fs that implies a line-width of 0.3 eV on the lasing transition. The photo-excitation scheme also requires a short pulse drive because of the very short Auger lifetime. The difference between this scheme and the photo-ionization scheme is that lasing in now in neutral Ne I. The lasing energies differ by about 0.4 eV. The potential advantage of this scheme is that the photo-excitation cross-section is about 18 Mbarns compared to 0.3 Mbarns for the photo-ionization scheme. The question we want to address in this paper is how we can take advantage of this much larger excitation rate to reduce the drive requirements on the XFEL source.

To model the photo-ionization and photo-excitation schemes we created a simple atomic model of the levels shown in Figs. 1 and 2. We then used the kinetics code Cretin [7] to model the kinetics and gain of the system under various conditions. For the baseline XFEL beam we assume the XFEL beam has 10^{12} photons in a 0.9 eV line-width focused to a 1-μm diameter. For the pulse duration we compare a Gaussian shape with 100-fs full-width half-maximum (FWHM) to a 1-fs FWHM. The XFEL was designed to have a bandwidth of 0.1 % as we are assuming even though the current bandwidth is larger by a factor of 5–10. The bandwidth has minimal impact on the photo-ionization mechanism but the strength of the photo-excitation mechanism is inversely proportional to the bandwidth. One challenge with the photo-excitation scheme is understanding the validity of the kinetics model and how to include the photo-excitation rate in the line-width calculation of the gain. Currently the stimulated rate is included in the kinetics but not in the line-width which means there are no Stark sidebands or broadening. The XFEL energy is set at

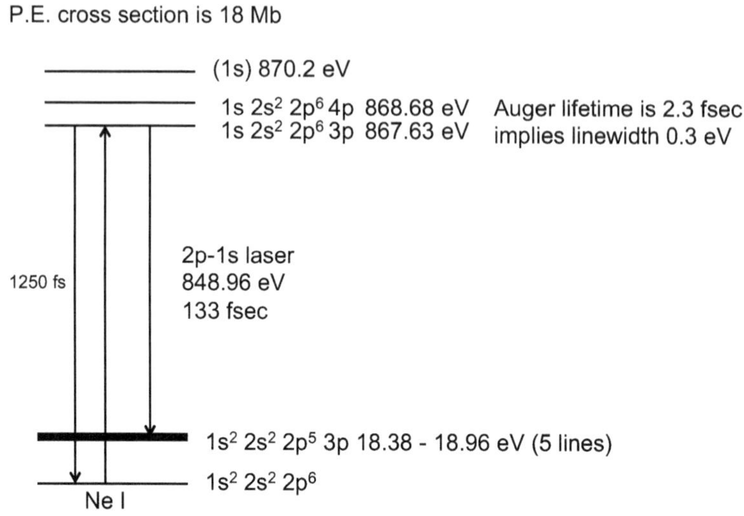

Fig. 2 Energy level diagram for the photo-excitation driven inner-shell neon X-ray laser

875 eV for modelling the photo-ionization scheme and 867.6 eV for modelling the photo-excitation scheme.

Starting with a 100-fs duration XFEL pulse, Fig. 3 shows the predicted gain versus time for both schemes. The time scale is set so that zero time is at the peak of the XFEL pulse. To understand the sensitivity to the XFEL flux a series of calculations were done using a multiplier between 1.0 (nominal) and 0.001 on the nominal XFEL flux described above. For the photo-ionization scheme we predict peak gain of 44/cm at $t = -81$ fs for the nominal case with multiplier of 1.0. In contrast the photo-excitation scheme has peak gain of 62/cm at -128 fs for the nominal case. The big difference between the behaviour of the two schemes is that the gain of the photo-excitation scheme falls much slower as the XFEL flux is reduced. With a multiplier of 0.001, which corresponds to 10^9 photons in the beam, the peak gain is still 12/cm at -12 fs as compared with 0.7/cm for the photo-ionization scheme. For both schemes the peak gain starts before the peak of the XFEL pulse and moves closer to $t = 0$ as the flux is reduced. To optimize the XFEL drive one wants the peak gain to occur at the peak of the XFEL drive pulse, otherwise some of the XFEL drive is not being used.

Now consider a 1-fs duration XFEL driving the Ne gas a shown in Fig. 4. For the nominal XFEL flux the peak gains are 910/cm at -0.5 fs for the photo-ionization and 703/cm at -1.1 fs for photo-excitation. As the flux intensity is reduced the gain for the photo-ionization drops quickly but one notices that the peak gain for the photo-excitation scheme drops from 703/cm to 639/cm as the XFEL drive flux is reduced by a factor of 100. Also the peak of the gain moves to a time of -0.2 fs, indicating near optimum drive conditions. As the XFEL flux drops further the gain drops more quickly and occurs after the peak of the XFEL flux indicating

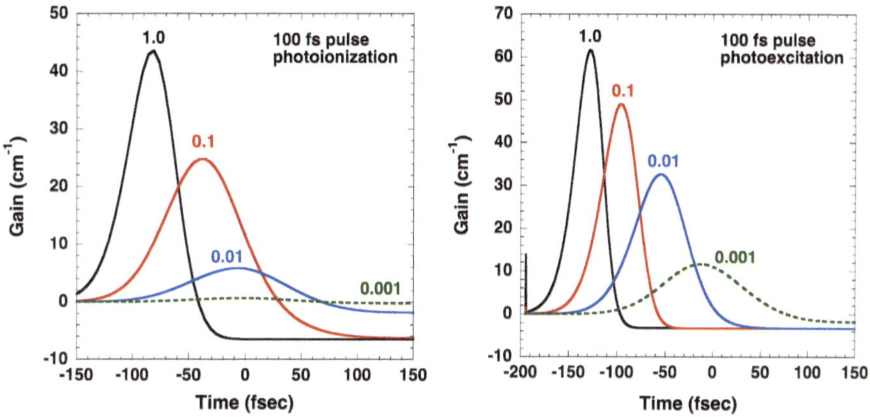

Fig. 3 Gain versus time for the neon X-ray laser driven by a 100-fs duration XFEL comparing the photo-ionization and photo-excitation mechanisms. The XFEL intensity is varied by using a multiplier between 1.0 (nominal) and 0.001

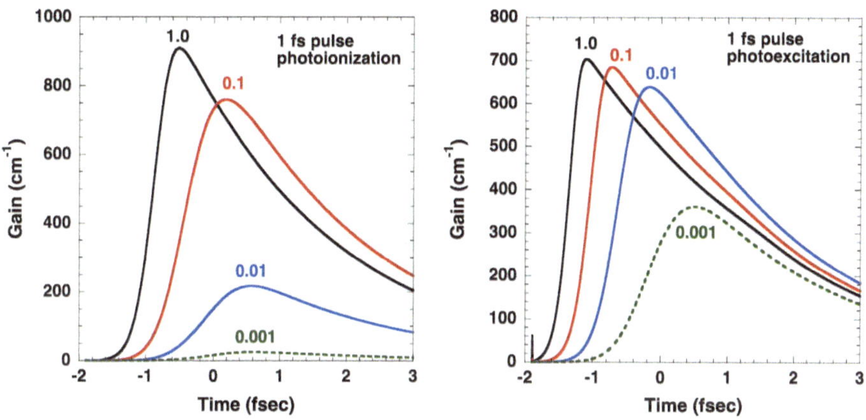

Fig. 4 Gain versus time for the neon X-ray laser driven by a 1-fs duration XFEL comparing the photo-ionization and photo-excitation mechanisms. The XFEL intensity is varied by using a multiplier between 1.0 (nominal) and 0.001

the flux is below ideal drive conditions. This figure shows that the photo-excitation mechanism offers the potential to reduce the XFEL drive by 1–2 orders of magnitude as compared with the photo-ionization mechanism. This advantage could enable smaller facilities to drive inner shell X-ray lasers or allow facilities such as LCLS to drive even higher energy X-ray lasers with the current XFEL fluxes.

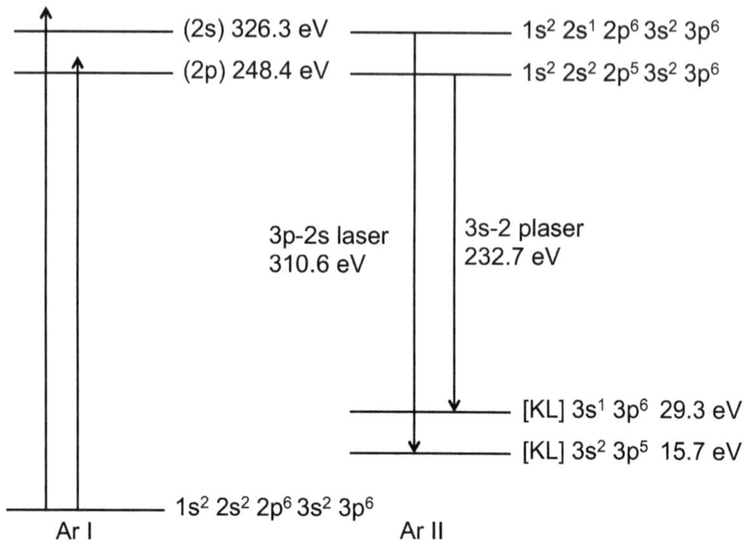

Fig. 5 Energy level diagram for the photo-ionization driven inner-shell argon X-ray laser

3 Modelling Ar and Cu L-Shell X-Ray Lasers

Given the success of the K-shell neon X-ray laser it should be possible to demonstrate inner-shell X-ray lasers in other principal shells such as the L and M shells. One promising candidate to consider is neutral argon gas. Figure 5 shows the energy level diagram for using an XFEL above the L-shell edge of neutral Ar I to create a L-shell hole in singly ionized Ar II. If an XFEL was tuned between the two L-edges at 248 and 326 eV one could create a 2p hole that would result in lasing on the 3s-2p transition at 232.7 eV. If the XFEL drive was tuned above the L-edge at 326.3 eV then one would have holes in both the 2s and 2p shells that would result in lasing on the 3p-2s transition at 310.6 eV as well as the 3s-2p transition. It would be very interesting to tune the XFEL from low to high energy and watch the 3s-2p lasing turn on followed by lasing on both lines.

Figure 6 shows the energy level diagram for using an XFEL above the L-shell edge of neutral Cu I to create a L-shell hole in singly ionized Cu II and create lasing on the strong 3d-2p line at 922 eV. As an alternative, photo-excitation of the 2p-4d transition in Cu I would also create lasing on the 3d-2p line in Cu I.

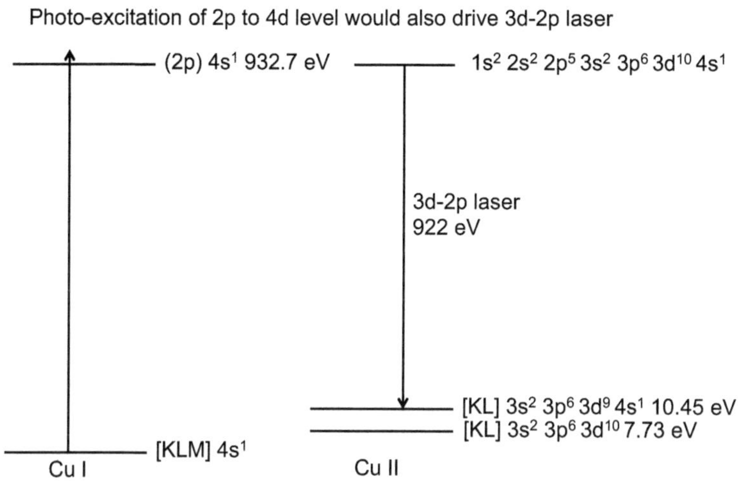

Fig. 6 Energy level diagram for the photo-ionization driven inner-shell copper X-ray laser

4 Conclusions

In this paper we model the neon inner shell X-ray laser under similar conditions to those used at LCLS. We show how we can improve the efficiency of the neon laser and reduce the drive requirements by tuning the XFEL to the 1s-3p transition in neutral neon in order to create gain on the 2p-1s line in neutral neon. We present the sensitivity to the drive intensity, pulse duration, and line-width of the XFEL to better understand how to optimize this inner shell laser by understanding the trade-offs between using photo-ionization versus photo-excitation to drive gain in these systems. We show that with the use of a sufficiently short XFEL pulse (1-fs) the resonant photo-excitation could reduce the XFEL flux requirements by several orders of magnitude. We also discuss how photo-ionization of L-shell electrons can be used to create lasing on $n = 3 - 2$ transitions in materials such as Ar at 232 and 310 eV and Cu at 922 eV.

Acknowledgements This work was performed under the auspices of the U.S. Department of Energy by Lawrence Livermore National Laboratory under Contract DE-AC52-07NA27344.

References

1. Duguay, M.A., Rentzepis, P.M.: Appl. Phys. Lett. **10**, 350–352 (1967)
2. Elton, R.C.: Appl. Opt. **14**, 2243–2249 (1975)
3. Rohringer, N., Ryan, D., London, R.A., Purvis, M., Albert, F., Dunn, J., Bozek, J.D., Bostedt, C., Graf, A., Hill, R., Hau-Riege, S.P., Rocca, J.J.: Nature **481**, 488–491 (2012)
4. Vinogradov, A.V., Sobelman, I.I., Yukov, E.A.: Sov. J. Quantum Electron **5**, 59–63 (1975)
5. Nilsen, J., Scofield, J.H., Chandler, E.A.: Appl. Opt. **31**, 4950–4956 (1992)
6. Nilsen, J., Chandler, E.: Phys. Rev. A **44**, 4591–4598 (1991)
7. Scott, H.A.: J. Quant. Spectrosc. Radiat. Transf. **71**, 689–701 (2001)

Part V
X-Ray/EUV Imaging

Reflection Mode Imaging with Extreme-Ultraviolet Light from a High Harmonic Source

Dennis F. Gardner, Bosheng Zhang, Matthew D. Seaberg, Elisabeth R. Shanblatt, Henry C. Kapteyn, Margaret M. Murnane and Daniel E. Adams

Abstract To date there have been few demonstrations of reflection-geometry coherent diffractive imaging (CDI). The work described here, using a high-harmonic source at 30 nm, is the first general reflection mode technique. We use a combination of ptychography and tilted plane correction to image an extended sample with no restriction on the incident angle or limitations on the numerical aperture. We find good agreement between our CDI images and images from scanning electron microscopy and atomic force microscopy.

1 Introduction

Since its first demonstration in 1999 [1], coherent diffractive imaging (CDI) has matured into a useful tool for nanoscale imaging. However, nearly all work with this technique has been in a transmission geometry (see for example the review by Chapman and Nugent [2]). In the last few years, the extension of CDI to reflection geometries has only begun to be explored [3–9].

The ability to image in a reflection geometry is critical when studying surfaces, for example in catalysis for energy harvesting systems or in next generation integrated circuits. Thus far, CDI reflection mode imaging with extreme-ultraviolet (EUV) or x-rays has had various limitations. These limitations include low numerical aperture (NA) [4, 7], normal incidence geometry requiring highly-reflective multilayer samples [8], and requiring isolated samples [7, 9].

To create a general CDI microscope, capable of imaging non-isolated samples with high NA at any angle of incidence, we use a combination of ptychography [10, 11] and tilted plane correction [5]. With 30 nm high harmonic generation (HHG) illumination we reconstruct images of Ti patterned on Si. The images display high fidelity and compare very well with images from a scanning electron microscope (SEM) and an atomic force microscope (AFM).

D. F. Gardner (✉) · B. Zhang · M. D. Seaberg · E. R. Shanblatt · H. C. Kapteyn
M. M. Murnane · D. E. Adams
JILA, University of Colorado at Boulder, Boulder, CO 80309, USA
e-mail: dennis.gardner@colorado.edu

© Springer International Publishing Switzerland 2016
J. Rocca et al. (eds.), *X-Ray Lasers 2014*, Springer Proceedings in Physics 169,
DOI 10.1007/978-3-319-19521-6_28

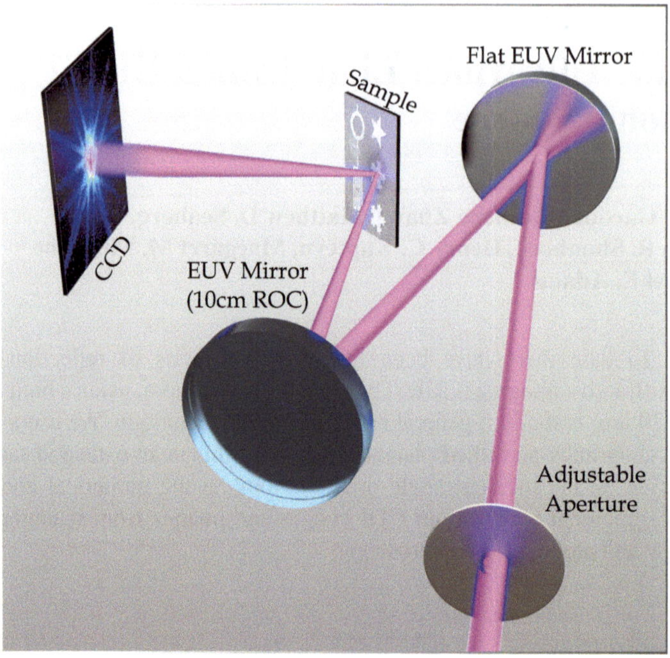

Fig. 1 A schematic of the experimental setup. The HHG beam is passed through an adjustable aperture to put a hard edge on the beam. Using two multilayer mirrors a single harmonic is focused onto the sample

2 Experimental Setup

We use a KM Labs Dragon Ti:Sapphire laser at 5 kHz, 1.5 mJ per pulse, and 22 fs pulse duration. The Ti:Sapphire driving laser is coupled into a 200 μm diameter fiber filled with 60 Torr of Ar gas. We generate harmonics of the fundamental driving wavelength (780 nm) through phase-matched, high harmonic up-conversion [12, 13].

After the generation of the harmonics, the co-propagating driving laser is attenuated by two 200 nm-thick Al filters. The 27th harmonic is selected using a pair of Mg/SiC multilayer mirrors, one flat and one with 10 cm radius of curvature (ROC). The curved mirror focuses the HHG beam onto the sample at a 45° angle of incidence (see Fig. 1).

Due to the off-axis incidence angle on the curved mirror ($\approx 2°$), slight astigmatism is introduced into the focus. As a result of the astigmatism, there is a horizontal focus and a vertical focus with the circle of least confusion (COLC) defined as halfway between. The sample is place 300 μm upstream of the COLC. At this position the beam size is about 10 μm in diameter.

The sample was fabricated using e-beam lithography and consisted of Ti shapes, approximately 10 μm in diameter and 30 nm tall, on Si (see supplementary material

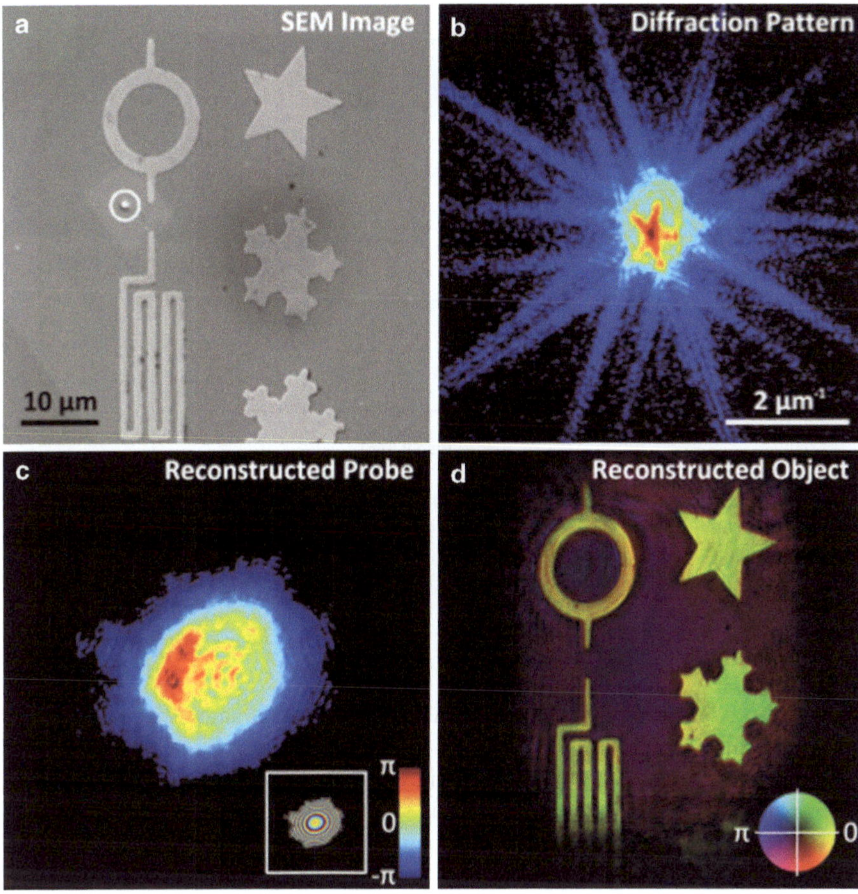

Fig. 2 From the diffraction data we reconstruct an image of the sample. **a** A SEM image of the sample. Note the circled piece of dust is contamination after performing the CDI experiment. **b** A representative diffraction pattern after applying tilted plane correction. **c** The recovered illumination probe (thresholded at 5 %) and **d** the reconstructed object. The inset in (c) is the phase of the beam. The same scale bar show in (**a**) is shared by (**c**) and (**d**)

of Ref. [14] for details). A SEM image of the sample is shown in Fig. 2a. Diffraction from the sample is recorded on a 2048 × 2048 Andor iKon camera pixel array with square 13.5 μm pixels placed 67 mm from the sample. With this geometry the NA is 0.1, enabling a half-pitch resolution of 150 nm.

The beam is scanned over the sample in 3 × 3 grids with 2.5 μm separation between scan positions. To avoid square raster artifacts in the image reconstruction, each scan position had a random offset of up to 1 μm [10]. Diffraction patterns at a total of 90 scan positions were acquired, with 30 second exposure time per position and 2 x 2 binning.

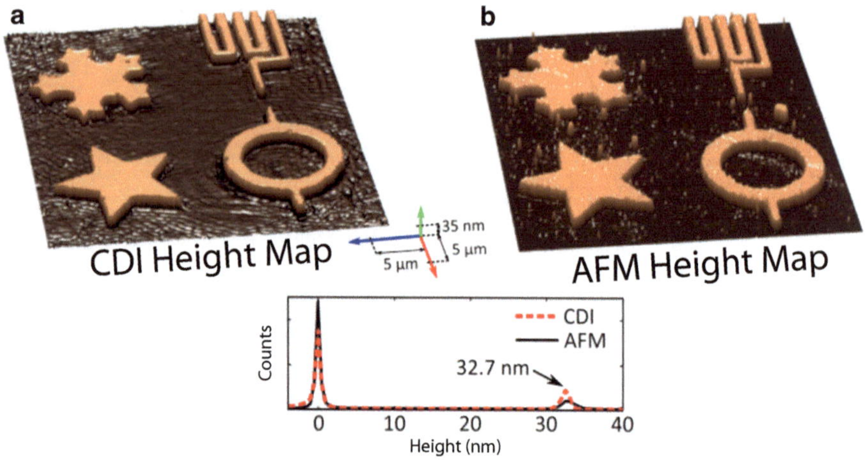

Fig. 3 Comparison between CDI and AFM height information. **a** A height map generated from the phase information of the object and **b** an AFM image of the sample. **c** A histogram of the height from the two techniques

3 Results and Discussion

Due to the combination of the high NA and the off-axis angle of incidence, the recorded diffraction patterns need to be interpolated onto a linear frequency grid in order to use fast Fourier transforms in the data analysis [5]. A representative tilted plane corrected diffraction pattern is shown in Fig. 2b. Due to the high beam curvature near the COLC, a Ronchigram can be seen at the center of the diffraction pattern [15].

To reconstruct an image from the diffraction patterns, we used the extended ptychographic iterative engine (ePIE) algorithm as described in Ref. [11]. The ePIE was initialized with an object guess of unity, and the probe guess was generated by using a model of the optical system. At first 20 iterations were performed with feedback on the object and the probe guess. Then we reset the object guess to unity and used the updated probe from the previous iterations as the starting guess. After 100 more iterations of feedback on the object and probe, the algorithm converged to a stable solution. The final solved probe is shown in Fig. 2c.

To improve the reconstruction, we used the position correction algorithm, as described by Zhang et al. in Ref. [16], to fix any translation stage errors. The ePIE algorithm was run for another 100 iterations with position correction and was only allowed to update the object (not the probe). The result of the reconstruction is shown in Fig. 2d and it agrees well with the SEM image shown in Fig. 2a.

With knowledge of the sample materials and the approximate height of the features, we can use the phase information to create a height map [14]. A height map of the sample from the CDI phase information is shown in Fig. 3a and an AFM comparison is shown in Fig. 3b. In Fig. 3c we compare the height of the features using a histogram and, as can be seen, our CDI image quantitatively agrees well with the AFM measurement.

4 Conclusion

We have demonstrated a general CDI technique for use in reflection geometry, capable of imaging extended samples at high NA with any angle of incidence. This general approach can provide a comprehensive and definitive characterization of how light, at any wavelength, scatters from a surface with an imminent feasibility of elemental imaging with few-nm resolution.

References

1. Miao, J., Charalambous, P., Kirz, J., Sayre, D.: Extending the methodology of X-ray crystallography to allow imaging of micrometre-sized noncrystalline specimens. Nature **400**, 342–344 (1999)
2. Chapman, H.N., Nugent, K.A.: Coherent lensless X-ray imaging. Nat. Photonics **4**, 833–839 (2010)
3. Marathe, S., Kim, S.S., Kim, S.N., Kim, C., Kang, H.C., Nickels, P.V., Noh, D.Y.: Coherent diffraction surface imaging in reflection geometry. Opt. Express **18**, 7253–7262 (2010)
4. Roy, S., Parks, D., Seu, K.A., Su, R., Turner, J.J., Chao, W., Anderson, E.H., Cabrini, S., Kevan, S.D.: Lensless X-ray imaging in reflection geometry. Nat. Photonics **5**, 243–245 (2011)
5. Gardner, D.F., Zhang, B., Seaberg, M.D., Martin, L.S., Adam, D.E., Salmassi, F., Gullikson, E., Kapteyn, H., Murnane, M.: High numerical aperture reflection mode coherent diffraction microscopy using off-axis apertured illumination. Opt. Express **20**, 19050–19059 (2012)
6. Claus, D., Robinson, D.J., Chetwynd, D.G., Shuo, Y., Pike, W.T., De J Toriz Garcia, J.J., Rodenburg, J.M.: Dual wavelength optical metrology using ptychography. J. Opt. **15**, 035702 (2013)
7. Zürch, M., Kern, C., Spielmann, C.: XUV coherent diffraction imaging in reflection geometry with low numerical aperture. Opt. Express **21**, 21131–21147 (2013)
8. Harada, T., Nakasuji, M., Nagata, Y., Watanabe, T., Kinoshita, H.: Phase imaging of extreme-ultraviolet mask using coherent extreme-ultraviolet scatterometry microscope. Jpn. J. Appl. Phys. **52** 06GB02 (2013)
9. Sun, T., Jiang, Z., Strzalka, J., Ocola, L., Wang, J.: Three-dimensional coherent X-ray surface scattering imaging near total external reflection. Nature Photonics **6**, 586–590 (2012)
10. Thibault, P., Dierolf, M., Bunk, O., Menzel, A., Pfeiffer, F.: Probe retrieval in ptychographic coherent diffractive imaging. Ultramicroscopy **109**, 338–343 (2009)
11. Maiden, A., Rodenburg, J.: An improved ptychographical phase retrieval algorithm for diffractive imaging. Ultramicroscopy **109**, 1256–1262 (2009)
12. Bartels, R.A., Paul, A., Green, H., Kapteyn, H.C., Murnane, M.M., Backus, S., Christov, I.P., Liu, Y., Attwood, D., Jacobsen, C.: Generation of spatially coherent light at extreme ultraviolet wavelengths. Science **297**, 376–378 (2002)
13. Popmintchev, T., et al.: Bright coherent ultrahigh harmonics in the keV x-ray regime from mid-infrared femtosecond lasers. Science **336**, 1287–1291 (2012)
14. Seaberg, M.D., Zhang, B., Gardner, D.F., Shanblatt, E.R., Murnane, M.M., Kapteyn, H.C., Adams, D.E.: Tabletop nanometer extreme ultraviolet imaging in an extended reflection mode using coherent Fresnel ptychography. Optica **1**, (in press) (2014)
15. Williams, G., Quiney, H., Dahl, B., Tran, C., Peele, A.G., Nugent, K., DeJonge M., Paterson, D.: Curved beam coherent diffractive imaging. Thin Solid Films **515**, 5553–5556 (2007)
16. Zhang, F., Peterson, I., Vila-Comamala, J., Diaz, A., Berenguer, F., Bean, R., Chen, B., Menzel, A., Robinson, I., Rodenburg, J.: Translation position determination in ptychographic coherent diffraction imaging. Opt. Express **21**, 13592–13606 (2013)

Volumetric Composition Imaging at the Nanoscale by Soft X-Ray Laser Ablation Mass Spectrometry

I. Kuznetsov, J. Filevich, M. Woolston, D. B. Carlton, W. Chao, E. H. Anderson, E. R. Bernstein, D. C. Crick, J. J. Rocca and C. S. Menoni

Abstract Single shot soft x-ray laser ablation in combination with mass spectrometry is shown to make it possible to detect intact molecular ions from organic and inorganic materials from atto-liter volumes with high sensitivity. The technique is demonstrated to record 3D composition images of heterogeneous samples.

1 Introduction

Laser ablation combined with mass spectrometry (MS) is an analytical method widely exploited to assess chemical composition of solid samples. In laser ablation MS a focused laser evaporates and ionizes the sample. Chemical specificity is obtained from the analyses of the ions in the plasma that are extracted and detected using a mass spectrometer. Two dimensional chemical images can be constructed from deconvolution of mass spectra obtained as the sample is laterally displaced with respect to the focused laser beam.

Laser ablation MS has been implemented using pulsed lasers with wavelengths ranging from the near infrared to the ultraviolet (UV) and pulse durations from nanoseconds down to femtoseconds [1–3]. The nature of the laser/material interactions influences the characteristics of the plasmas and in turn the ionization mechanisms. The size of the focused laser spot controls the extent of the probed

C. S. Menoni (✉) · I. Kuznetsov · J. Filevich · M. Woolston · D. B. Carlton · W. Chao · E. H. Anderson · E. R. Bernstein · J. J. Rocca
Engineering Research Center for Extreme Ultraviolet Science and Technology, Fort Collins, USA
e-mail: carmen@engr.colostate.edu

C. S. Menoni · I. Kuznetsov · J. Filevich · M. Woolston · J. J. Rocca
Electrical and Computer Engineering, Colorado State University, Fort Collins, USA

D. B. Carlton · W. Chao · E. H. Anderson
Center for X-Ray Optics, Lawrence Berkeley National Laboratory, Berkeley, CA, USA

C. S. Menoni · E. R. Bernstein
Department of Chemistry, Colorado State University, Fort Collins, USA

D. C. Crick
Department of Microbiology, Immunology, and Pathology, Colorado State University, Fort Collins, USA

© Springer International Publishing Switzerland 2016
J. Rocca et al. (eds.), *X-Ray Lasers 2014*, Springer Proceedings in Physics 169,
DOI 10.1007/978-3-319-19521-6_29

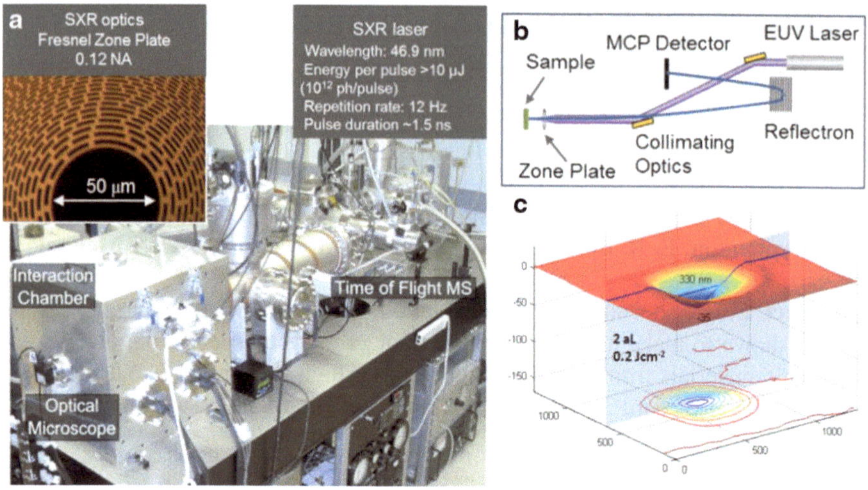

Fig. 1 Soft x-ray laser 3D mass spectrometry imaging system. **a** Photograph of the complete SXRL ablation mass spectrometry imaging system, **b** schematics showing the SXRL beam and ion trajectories, **c** 2aL crater ablated in PMMA with a single laser shot

sample region and thus the lateral spatial resolution, which is typically 10 μm when imaging.

We have recently demonstrated the use of SXRL ablation in MS for the first time, and showed the capability to image molecular composition in organic materials in three dimensions (3D) with nanoscale resolution and high sensitivity [4]. The 3D nanoscale resolution results from the ability to focus the SXRL beams to spots ∼100 nm in diameter in polymers and the high absorption of the SRXL which permits ablating craters ∼10 nm deep [4] in a single shot using a fluence of ∼100 mJcm^{-2} approaching the desorption fluence of polymethyl methacrylate (PMMA) [5]. In this paper we provide details of the method and give examples of its capability to determine the chemical makeup of organic and inorganic samples and show its capability to image composition in 3D with high sensitivity.

2 Experimental Details

The SXRL ablation mass spectrometry imaging method we have developed is illustrated in Fig. 1a. The experiments were conducted using a capillary discharge Ne-like Ar laser emitting at 46.9 nm wavelength [6]. The SXRL output is focused onto the solid sample using a 0.12 NA free standing zone plate lens. The ions in the laser-created plasma are extracted and analyzed with a time of flight (TOF) mass spectrometer operated in reflectron mode (Fig. 1b).

The 0.12 NA zone plate lens is a free standing structure fabricated by electron beam lithography with a central opening that facilitates ion extraction [7]. The ion

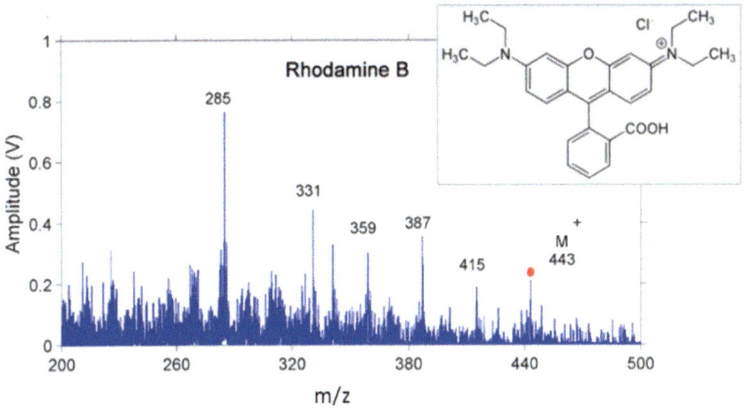

Fig. 2 Single shot positive ion mass spectrum of rhodamine B showing the intact molecular ion M^+ and large m/z fragments

optics can be biased to extract either positive or negative ions. A 2 attoliter (aL) crater ablated PMMA with a single laser shot using a fluence of 0.2 Jcm^{-2} is shown in Fig. 1b. The ablated volume in polymers is determined by the laser fluence which is controlled by attenuating the laser pulse energy.

3 Results

3.1 *Molecular Analysis of Heterogeneous Samples*

The challenge of laser ablation and ionization MS methods is in detecting intact molecular ions or large fragments to uniquely identify the chemical makeup of probed sample region. In the analysis of organic samples by laser ablation MS several strategies have been implemented. For example, when using UV laser pulses in the MS of organic samples, it is necessary to add a matrix to the analyte to allow for laser absorption and ionization [1]. Other approaches rely on OH ionization with near infrared (NIR) illumination or the use of femtosecond pulses for which the nonlinear laser-material interactions result in significantly different ion activation processes [3], [8].

The use of 46.9 nm SXRL illumination is ideal for the analysis of organic samples due to its naturally strong absorption (i.e. 20 nm absorption length) in carbonaceous materials, and the ability of the 26.4 eV photons to ionize any molecule [9]. Figure 2 shows the single shot mass spectrum of rhodamine B originating from an ablated analyte volume of 79 aL.

The rhodamine B mass spectrum contains a peak at m/z 443 that corresponds to the intact molecular ion and large fragments at lower m/z. In this analysis, as well as those previously reported, the mass spectra show singly ionized parent molecules.

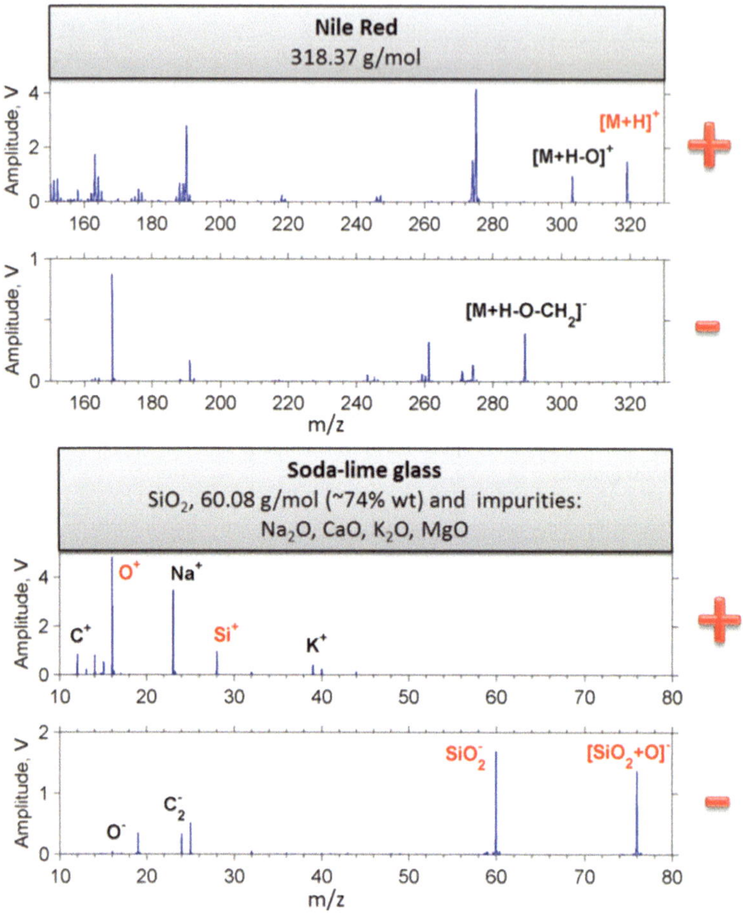

Fig. 3 Mass spectra of Nile red and soda-lime glass obtained with positive (+) and negative (−) ion extraction modes. The parent molecule ions $[M+H]^+$, M^-, and $[M+O]^-$ are identified

Furthermore, we showed the sensitivity of SXRL ablation MS in the detection of organic analytes within the range $m/z \leq 400$ is 0.01 amol, which is unprecedented for a laser ablation method [4].

Positive ion extraction is best suited for the detection of intact organic molecular ions, as shown in Fig. 2, and in results previously reported [4]. Instead, to get molecular information from inorganic samples, negative mode extraction might be necessary. A comparison of positive and negative ion extraction modes mass spectra of Nile red and glass is shown in Fig. 3. The parent ion $[M+H]^+$ is present in the positive ion spectra of Nile red but not when using negative mode extraction. The reverse occurs in inorganic materials, pointing to the difference in ionization mechanisms.

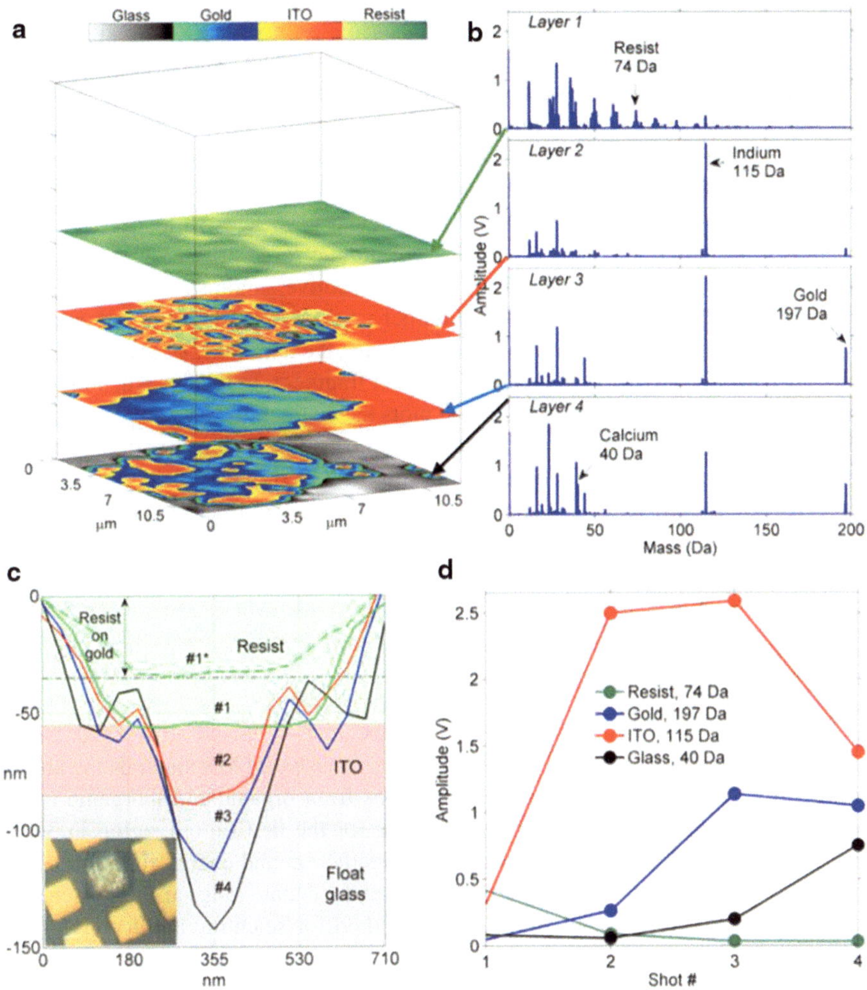

Fig. 4 Demonstration of 3D chemical imaging. **a** 3D ion image showing the distribution of resist, Au, ITO and SiO2 in the sample. The color scale indicates normalized concentration ranging from 0 (*light color*) to 1 (*dark color*), **b** Mass spectra for each of the surfaces in (a) obtained by adding up the mass spectrum of each voxel in the 15 × 15 array of probed sites, **d** Average amplitude for each of the characteristic ion peaks in (b) versus ablation event number. These data and the ablation depths of the craters created in consecutive ablation events shown in the AFM profiles of, **c** were used to construct the 3D image of (a). The inset on (c) is a micrograph of the sample

3.2 *Chemical Imaging of Heterogeneous Samples*

The high sensitivity of SXRL ablation MS in the detection of organic and inorganic analyte ions was exploited to demonstrate composition imaging with 20 nm depth and 75 nm lateral resolution in organic materials [4]. Here we show an example of

3D composition imaging of a heterogeneous sample consisting of a pattern of Au pillars covered by a layer of resist (Fig. 4). The 35 nm tall square Au pillars were fabricated on ITO-coated glass substrate and buried within a 55 nm thick resist layer. At a laser fluence of 0.7 Jcm^{-2} the resist was completely removed after the first shot. Instead, ITO and glass ablated at rates of ~ 20 nm/shot and Au at ~ 5 nm/shot. The ablation rate correlates with the absorption depth of the 46.9 nm laser light in the materials which is ~ 20 nm and 6 nm for ITO and glass and Au respectively. The profiles of the ablated craters obtained from atomic force microscope images are shown in Fig. 4d. Mass spectra from four consecutive single-shot ablation events were acquired on a given sample location before displacing the sample in steps of 750 nm. These spectra were analysed to determine the amplitude of the ion peaks at m/z 74, 115, 197 and 40, which identify a resist fragment, In in ITO, Au, and Ca (an impurity in glass) respectively. Figure 4a plots the content of each of the material in the sample versus shot number. The mass spectra used in the construction of the 3D image are shown in Fig. 4b. These are the sum-averaged over the 15 $\times$ 15 array of probed sites in each layer. The measured amplitude of the resist, Au, In and Ca ion peaks obtained from these spectra and plotted in Fig. 4d accurately represents the abundance of each of the materials in the sample.

4 Conclusions

SXRL ablation brings to the broad landscape of laser-based MS methods the ability to identify singly ionized intact molecular ions from organic and inorganic materials with extremely high sensitivity, enabling for the first time imaging in 3D at the nanoscale. Further studies of the SXRL interaction and characterization of the plasmas that are generated are necessary to extend the mass range and maximize the potential of this novel nanoscale 3D MSI method for scientific and technological applications.

Acknowledgements This work is supported by NIH/NIAID, Grant AI-065357. It made use of facilities of the NSF Engineering Research Center, supported through NSF grant EEC 0310717.

References

1. Hillenkamp, F., et al.: Matrix-assisted laser desorption/ionization mass spectrometry of biopolymers. Anal. Chem. **63**(24), A1193-A1202 (1991)
2. Karas, M., etal.: Ultraviolet-laser desorption/ionization mass spectrometry of femtomolar amounts of large proteins. Biomedical and Environ. Mass Spectrometry, **18**(9), 841–843 (1989)
3. Coello, Y., et al.: Atmospheric pressure femtosecond laser imaging mass spectrometry. Anal. Chem. **82**(7), 2753–2758 (2010)
4. Kuznetsov, I., et al.: Three-dimensional nanoscale molecular imaging by extreme ultraviolet laser ablation mass spectrometry. Nat. Commun. (2015). doi: 10.1038/ncomms8206

5. Juha, L., et al.: Short-wavelength ablation of molecular solids: pulse duration and wavelength effects. J. Microlithog. Microfab. Microsyst. **4**(3), 033007 (2005)
6. Heinbuch, S., et al.: Demonstration of a desk-top size high repetition rate soft x-ray laser. Opt. Express **13**(11), 4050–4055 (2005)
7. Anderson, E.H.: Specialized electron beam nanolithography for EUV and X-ray diffractive optics. IEEE J. Quantum Electron. **42**(1–2), 27–35 (2006)
8. Nemes, P., et al.: Three-dimensional imaging of metabolites in tissues under ambient conditions by laser ablation electrospray ionization mass spectrometry. Anal. Chem. **81**(16), 6668–6675 (2009)
9. Berrill, M., et al.: Warm photoionized plasmas created by soft-x-ray laser irradiation of solid targets. J. O.S.A. B-Opt. Phys. **25**(7), B32-B38 (2008)

Time-Resolved Holography with a Table Top Soft X-Ray Laser

M.C. Marconi, N. Monserud, E. Malm, P. Wachulak and W. Chao

Abstract We describe the implementation of a Fourier transform holographic setup capable to acquire flash images with 1 ns temporal resolution and 130 nm spatial resolution utilizing a soft X-ray (SXR) table-top laser. Sub-micron scale pillars oscillating at frequencies around 1 MHz were imaged by single shot laser pulses allowing for the composition of "movies" of the oscillating pillars by combining a sequence of single shot images.

1 Introduction

The development of intense highly coherent soft X-ray laser (SXRL) sources have opened the possibility to implement high resolution imaging techniques that so far were restricted to large synchrotron and free electron laser facilities [1] – [3]. Using a table top SXRL we have implemented a "flash" (single shot) time resolved Fourier holography setup to record images of nano-scale oscillating pillars. The final motivation of this project is to implement an imaging technique capable to detect intrinsic localized oscillation modes (ILM) in nano-pillars arrays, and to demonstrate a new single molecule detection technique [4]. ILMs consist of coherent oscillation of nano-pillars, set in stable and localized patterns. The key characteristic of ILMs is that if one pillar in an array has mechanical properties different from others, for example a variation in the mass, then there are regimes of oscillations when ILMs focus precisely on that defective pillars (*pinning*). Thus the lead idea to implement the single molecule detection technique is that the attachment of a molecule to one of the nanopillars oscillating in the array will introduce a "defect" that will concentrate the ILM. From the point of view of molecular detection, since one ILM

M.C. Marconi (✉) · N. Monserud · E. Malm
NSF Engineering Research Center for Extreme Ultraviolet Science and Technology and Electrical and Computer Engineering Department, Colorado State University, Fort Collins, USA
e-mail: marconi@engr.colostate.edu

P. Wachulak
Military University of Technology, Warsaw, Poland

W. Chao
Center for X-Ray Optics, Lawrence Berkeley National Laboratory, Berkeley, USA

© Springer International Publishing Switzerland 2016
J. Rocca et al. (eds.), *X-Ray Lasers 2014*, Springer Proceedings in Physics 169,
DOI 10.1007/978-3-319-19521-6_30

involves many oscillators, a change in ILM shape and location will predict attachment of a molecule with higher sensitivity than an individual resonator. However, the resonators for molecular detection of lighter molecules have to be excessively small (microns in length, nanometers in width), and conventional visualization techniques, based on visible or near-visible lasers, cannot resolve extremely fast and small oscillations of these nano-scale resonators. On the other hand, visualization techniques for nanometer scale like Atomic Force (AFM) and Scanning Electron Microscopes (SEM) cannot study the fast dynamics of the resonators in real time. An alternative imaging technique with high spatial and temporal resolutions is thus necessary to investigate the ILM dynamics and demonstrate the novel single molecule detection method. In this experiment it was demonstrated this capability using table-top 46.9 nm lasers developed at Colorado State University.

2 Experiment Set Up

To validate the ILM detection it was developed a holographic system for the visualization of the oscillations in the nano-pillar arrays. This required the capability to record images with high spatial and temporal resolutions. This experiment utilized a SXRL holography technique to detect the collective dynamics of the ILMs in real time.

To achieve high temporal and spatial resolution a Fourier Transform Holography (FTH) configuration was utilized. [5] The hologram was recorded in a charged-coupled device (CCD) camera and reconstructed numerically. With this experimental configuration it was possible to achieve time resolved holograms (or flash holograms) with a temporal resolution determined by the laser pulse width, in this case 1 nanosecond. Figure 1 is a schematic of the experimental set up for holographic recording. The EUV laser radiation was directed onto the Fresnel zone plate which was used as a beam splitter, generating the illumination and the reference beams. The nanopillar array was located in close proximity of the zone plate focal plane. The un-diffracted zero order beam illuminated the object (nanopillar array) producing a diffracted object beam. The object beam was superposed in the CCD detector plane with the spherical wave generated by the first order focal point of the zone plate producing an interference pattern. The hologram recorded in this way can be reconstructed using standard FFT numerical procedures.

The spatial resolution achievable is determined by the size of the point reference source, the precision with which the reference wavefront is known and the angular extent over which the hologram is recorded. The working distance is determined by the spatial sampling frequency and ultimately set by the detector pixel size and the acceptance angle of the recording. If the working distance z is defined as the distance between the reference source (focal point) and the detector plane, and Δx is the detector's pixel size, the numerical aperture of the recording is approximately

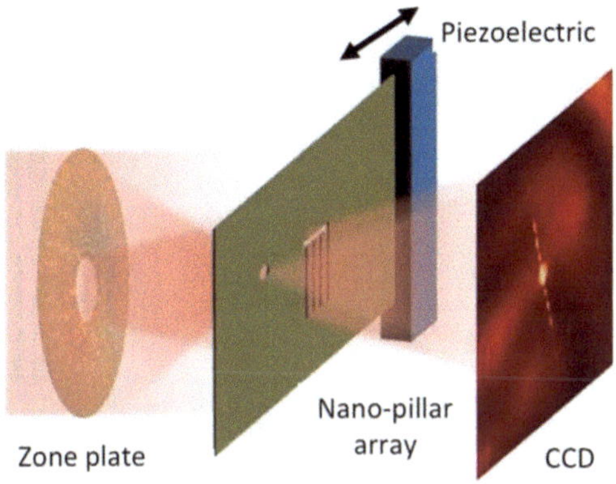

Fig. 1 Experimental setup for the Fourier transform "flash" holography

$NA = {}^{n\,\Delta x}/_{z}$, where n is the number of pixels in the detector. Approximating the object resolution by the transverse resolution of a zone plate, we get:

$$z = \frac{1.22\, n\, \Delta x\, \delta}{\lambda}$$

Here, δ denotes the thickness of the outermost zone in the Fresnel plate. For the parameters of the 46.9 nm laser and the available optics, this scheme allows to attain 100 nm spatial resolution with a working distance z of few centimeters. Regarding the temporal resolution, it was limited by the laser pulse width, which was measured to be approximately 1 ns.

3 Time-Resolved Imaging of Oscillating Pillars

The goal of the holographic imaging system is to determine the position and measure the amplitude of oscillation of nanopillars with sub-micron dimensions and that are oscillating with MHz range frequencies. The sample for this test was a set of three pillars fabricated with the focused ion beam tool in a 200 nm thick SiN membrane. The pillars were defined in a rectangular aperture. Each pillar has a length of 15 μm and a width of 200 nm. Figure 2 is a SEM image of the three pillars array used in the test.

The sample was attached to a ceramic piezoelectric that was excited with a sinusoidal wave 10 V peak to peak while the excitation frequency was scanned around the expected resonance frequency. For this experiment, the excitation frequency was 1.16 MHz. Several single shot holograms were recorded of the oscillating pillars shown in the sequence in Fig. 3. During the experiment the first (leftmost) pillar broke and the other two pillars collapsed together and got attached to each other in the tip. Nevertheless, the experiment is a proof of principle that the method is viable to detect minute oscillations in nano-pillars. The displacement of the pillars

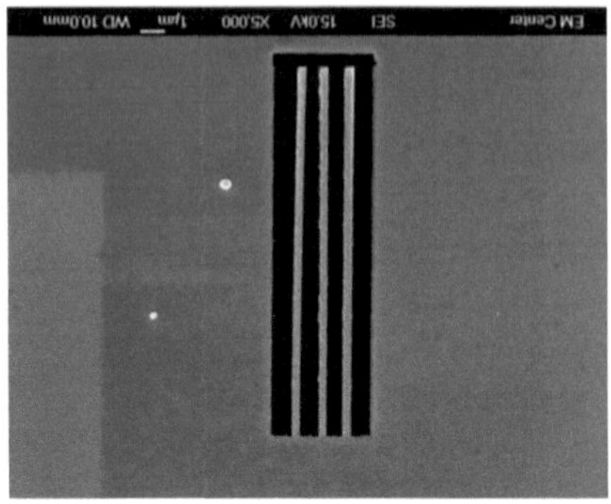

Fig. 2 SEM image of the three pillar array fabricated by focused ion beam milling. The pillars were defined in a 200 nm thick SiN membrane. The width of each pillar is 200 nm and the length 15 μm

is visible in the sequence displayed in Fig. 3. With this sequence of images it was possible to compose a "movie" of the oscillation by sequentially displaying the different images as frames in a film. The 60 nm amplitude movement is more evident in the movie that is available at http://www.engr.colostate.edu/ marconi/.

The results shown in Fig. 3 clearly demonstrate the capability of the time resolved holography to record the movement of nanometer size objects in real time. Also the versatility of the Fourier transform holography set up allows for a quick reconstruction enabling an almost "on-line" visualization of the nanopillar oscillations.

4 Out-of-Plane Imaging

Holographic imaging had the extra built-in advantage, allowing to simultaneously visualize the movements in the third dimension (the direction of the beam propagation, off the plane of the oscillation of the nanopillars). This imaging in the longitudinal direction has a lower spatial resolution, but nevertheless can be detected with a proper numerical treatment of the hologram. An example of this effect was evident when processing the hologram of a sample similar to the one described in Sect. 3 (Fig. 2). The first pillar of the sample stuck out of the sample plane, while the other two pillars remained in the aperture's plane. This displacement off the plane was visible in the hologram with the adequate processing, by numerically adding a spherical phase. This is equivalent to displacing the reference point source in the z direction (the beam propagation direction). The FFT reconstruction of the hologram automatically produces a sharp image of the object that is located in the same plane of the reference source. Thus, adding different phase shifts to the hologram, reconstructing the image and evaluating the sharpness of the image it is possible to determine the position of the object along the direction of the propagation of the beam. Figure 4 summarizes the result of this experiment. Figure 4a is the reconstruction after adding a phase shift corresponding to a reference source displacement of − 1 μm (i.e. a displacement of the point source of 1 μm behind the plane of the

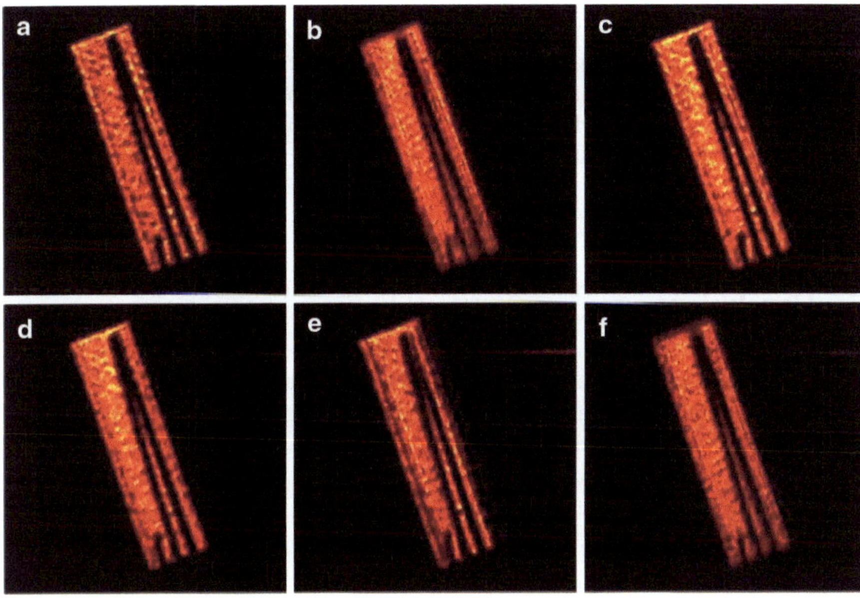

Fig. 3 Sequence of holograms of oscillating pillars. The images with the pillars in different locations can be combined in a "movie" of the oscillating pillars

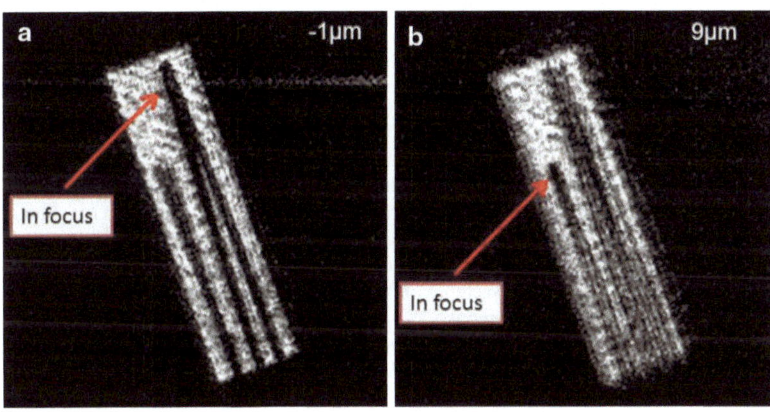

Fig. 4 Out of plane feature detection. **a** Reconstruction of the hologram after adding a phase corresponding to a displacement of the reference source of − 1 μm. The tip of the two collapsed pillars is in focus while the tip of the first pillar (from the *left*) if blurred. **b** Reconstruction of the hologram after adding a phase corresponding to a displacement of the reference source of 9 μm. The tip of the off plane pillar is in focus while the two in plane pillars are blurred

membrane). In this reconstruction it is clear that the tips of the two collapsed pillars (on the right of the aperture) are in focus while the tip of the off plane pillar is blurred. The base of the off-plane pillar, which is in the membrane's plane, shows a sharp image as well. Figure 4b is the reconstruction of the same hologram, but in this case a phase corresponding to a displacement of the point source of 9 μm above the membrane's plane was added. The image in Fig. 4b shows a sharp image of the tip of the off-plane pillar while the collapsed in-plane pillars are blurred. This analysis indicates that the tip of the off-plane pillar is sticking out 8 μm of the sample plane.

5 Conclusions

We have demonstrated a table-top soft-X-ray Fourier holography setup capable to acquire flash images of fast moving objects with high spatial resolution. These features are not available with SEM, TEM or AFM imaging. Soft X-ray flash holography can be a valuable complementary technique to the conventional nanoscale imaging approaches that may have a significant impact in nanosciences.

References

1. Eisebitt, S., Lüning, J., Schlotter, W.F., Lörgen, M., Hellwig, O., Eberhardt, W., Stöhr, J.: Lensless imaging of magnetic nanostructures by X-ray spectro-holography. Nature. **432**, 885–888 (2004)
2. Günther, C.M., Pfau, B., Mitzner, R., Siemer, B., Roling, S., Zacharias, H., Kutz, O., Rudolph, I., Schondelmaier, D., Treusch, R., Eisebitt, S.: Sequential femtosecond X-ray imaging. Nat. Photonics. **5**, 99–102 (2011)
3. Guehrs, E., Stadler, A.M., Flewett, S., Frömmel, S., Geilhufe, J., Pfau, B., Rander, T., Schaffert, S., Büldt, G., Eisebitt, S.: Soft x-ray tomoholography. New. J. Phys. **14**, 013022 (2012)
4. Brake, D., Xu, H., Hollowell, A., Balakhrisnan, G., Hains, C., Marconi, M., Putkaradze, V.: Intrinsic localized modes in two-dimensional vibrations of crystalline pillars and their application for sensing. J. Appl. Phys. **112**, 104326 (2012)
5. Malm, E.B., Monserud, N.C., Wachulak, P.W., Brown, C., Chao, W., Anderson, E., Xu, H., Hains, C.P., Balakrishnan, G., Menoni, C.S., Rocca, J.J., Marconi, M.C.: Time resolved holography scheme using a table top soft X-ray laser. Springer Proceedings in Physics Volume 147. S. Seban, J. Gautier, D. Ros, P. Zeitoun Editors, ISSN 0930–8989. (2013)

X-Ray Holography for Imaging Large Specimen with a Ag X-Ray Laser

Kyoung Hwan Lee, Hyeok Yun, Jae Hee Sung, Seong Ku Lee, Tae Moon Jeong, Hyung Taek Kim and Chang Hee Nam

Abstract We developed a new variant of Fourier transform holography (FTH) to overcome the separation condition of FTH by subtracting the autocorrelation signal, named as autocorrelation-subtracted FTH (AS-FHT). AS-FTH is advantageous in imaging large specimen using light sources of limited coherent photons because AS-FTH requires a much smaller coherent illumination area than conventional FTH. We experimentally demonstrated the AS-FTH using a Ni-like Ag x-ray laser at 13.9 nm. The hidden part of an image under its autocorrelation was successfully recovered by subtracting an independently measured autocorrelation signal.

1 Introduction

X-ray Fourier transform holography (FTH) is an attractive method to obtain high-resolution image without any sophisticated diffractive x-ray lenses and numerical ambiguities [1–4]. The hologram is generated by the interference between an object wave and a reference wave from pinholes. In order to reconstruct the holographic image, the reference pinhole should be located at a proper position fulfilling the separation condition, and the transmitted light through the pinhole should be sufficient strong to make the interference the object wave. Even though several types of reference structures were proposed to provide strong reference wave, any method to remove the separation condition has not been devised [5, 6]. Since the separation condition of FTH demands three times larger coherent illumination than the size of the object, it can be a crucial restriction to use an x-ray light source with limited coherent photons. Consequently, the separation condition should be eliminated, if possible, to image large specimen for a given coherent illumination condition.

H. T. Kim (✉) · K. H. Lee · H. Yun · J. H. Sung · S. K. Lee · T. M. Jeong · C. H. Nam
Center for Relativistic Laser Science, Institute for Basic Science (IBS), Gwangju 500-712, Korea
e-mail: htkim@gist.ac.kr

J. H. Sung · S. K. Lee · T. M. Jeong · H. T. Kim
Advanced Photonics Research Institute, Gwangju Institute of Science and Technology (GIST), Gwangju 500-712, Korea

C. H. Nam
Dept. of Physics and Photon Science, GIST, Gwangju 500-712, Korea

© Springer International Publishing Switzerland 2016
J. Rocca et al. (eds.), *X-Ray Lasers 2014*, Springer Proceedings in Physics 169,
DOI 10.1007/978-3-319-19521-6_31

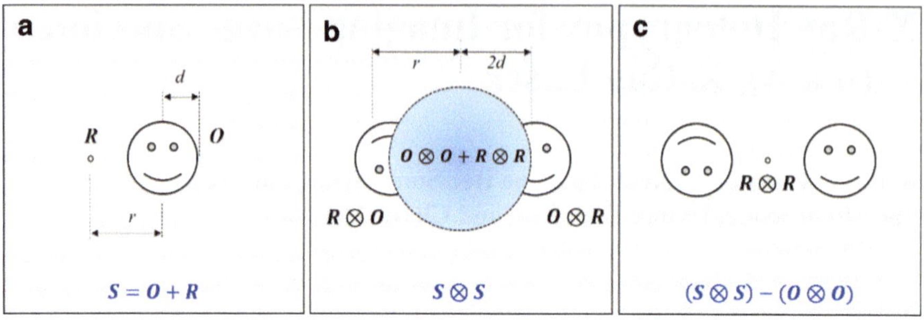

Fig. 1 **a** The sample plate (***S***) for FTH having an object (***O***) and a reference (***R***). **b** FTH image of sample plate violating separation condition. **c** Central autocorrelation removed image in (**b**) obtained by subtracting $O \otimes O$

In this proceeding, we report a variation of FTH, named as autocorrelation subtracted FTH (AS-FTH), for the elimination of the separation condition. AS-FTH overcomes the separation condition by removing the autocorrelation signal from a normal FTH image. Here we demonstrated AS-FTH using a Ni-like Ag x-ray laser [7]. We prepared a sample plate with a reference pinhole, violating the separation condition, that produced a hidden part of the object image in the case of a normal FTH method. The hidden part of object image was successfully recovered by subtracting the central autocorrelation signal. For the subtraction of the central autocorrelation, the diffraction pattern of an object without the reference pinhole was separately recorded. The unveiled part of the object image well matched the original sample shape, even though the AS-FTH result showed image deformation due to the imperfect subtraction of the autocorrelation signal.

2 Separation Condition of FTH

The separation condition is a required distance between an object and a reference pinhole for obtaining a proper FTH image. Figure 1a shows a nominal sample plate (S) for FTH containing an object (O) and a reference pinhole (R) with distance $\boldsymbol{r}$. The diffracted light from the sample plate generates the hologram, and the hologram can be reconstructed by applying inverse Fourier transform, as shown in Fig. 1b. The reconstructed FTH image is expressed as the autocorrelation of sample plate,

$$F^{-1}\{|F(S)|^2\} = 0 \otimes 0 + \mathrm{R} \otimes \mathrm{R} + 0 \otimes \mathrm{R} + \mathrm{R} \otimes 0, \tag{1}$$

where F, F^{-1}, $\otimes$ are Fourier transform, inverse Fourier transform, and cross-correlation operators, respectively. In the FTH image, O $\otimes$ O+R $\otimes$ R becomes the central autocorrelation signal, and O$\otimes$R and R$\otimes$O become the reconstructed object image placed with distance r from the center of the FTH image. Since the radius of the central autocorrelation is 2d when the radius of object is d, r > 3d should be satisfied to avoid an overlap among the central autocorrelation and reconstructed object images. This restriction of r is the separation condition of FTH, and it should be removed in order to reduce the size of required coherent illumination area.

3 Autocorrealtion-Subtracted FTH

We propose a variant of FTH named as autocorrelation-subtracted FTH (AS-FTH) to eliminate the separation condition. Since the separation condition is caused by the autocorrelation signal (O ⊗ O+R ⊗ R), the separation condition can be eliminated by removing the components, O⊗O + R⊗R. Since R⊗R is negligible with a pinhole reference, the central autocorrelation signal can be effectively erased by subtracting O⊗O. In order to remove O ⊗ O in Fig. 1b, AS-FTH separately records the diffraction signal of an object without a reference pinhole. Figure 1c shows the AS-FTH image obtained by subtracting independently produced O ⊗ O The hidden part of the object image in the Fig. 1b can be recovered through AS-FTH.

AS-FTH also can be applied to a differential holographic method using an extended reference. Recently, holography with extended reference by autocorrelation linear differential operation (HERALDO) has been proposed to obtain high-resolution x-ray images beyond the limitation of FTH by providing a strong reference wave [6]. HERALDO can obtain high resolution corresponding to the sharpness of a reference edge, but it is still governed by the separation condition. If the contribution of an extended reference, such as a line reference, is small, the effective elimination of the separation condition of HERALDO is possible by subtracting O⊗O.

4 Experimental Demonstration

We have demonstrated AS-FTH using the setup shown in Fig. 2. The Ni-like Ag x-ray laser at 13.9 nm was generated using a pumping laser with 8-ps pulse duration and 2-J energy. The energy of x-ray laser was 1.5 μJ, corresponding to 10^{11} photons [8]. The x-ray laser was focused into a sample plate using a concave Mo/Si x-ray mirror. A 200-nm Zr filter was placed in front of the sample plate to block a stray light. Then the diffracted signal from the sample plate was recorded using an x-ray CCD placed at 42-mm away from the sample plate. The x-ray CCD consists of 2048 × 2048 pixels with unit pixel size of 13.5 μm. Insets in Fig. 2 show the scanning electron microscope images of the sample plates with 'ħ' pattern for AS-FTH without and with two reference pinholes. The 'ħ' pattern was fabricated on 300 nm-thick Au-coated 50-nm thick Si_3N_4 membrane using a focused ion beam.

For the AS-FTH, two diffraction signals of the sample plate were obtained. The first diffraction image was used to produce the FTH image, and the other was used to subtract the central autocorrelation. First, the central autocorrelation signal in

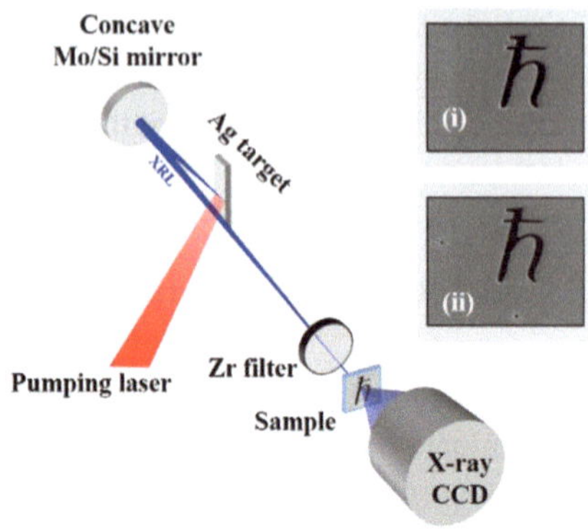

Fig. 2 Experimental layout of AS-FTH using a Ag x-ray laser. Insets (i) and (ii) show the scanning electron microscope images of sample plates of 'ħ' pattern without and with reference pinholes, respectively

Fig. 3b was prepared using the diffraction signal of the sample plate having 'ħ' pattern without any reference pinhole, as shown in Fig. 3a. Then, two reference pinholes were fabricated using a focused ion beam in the sample plate, and the diffraction signal from the sample plate was recorded as shown in Fig. 3c. Figure 3d shows the FTH image of Fig. 3c. The FTH image contains two reconstructed 'ħ' pattern images coming from the two reference pinholes; one is fully isolated while the other is partly blocked by the central autocorrelation signal due to the reference pinhole violating the separation condition.

The hidden part of 'ħ' pattern image was successfully recovered by subtracting the autocorrelation signal (Fig. 3b) from the FTH image (Fig. 3d), as shown in Fig. 4a. The 'ħ' pattern image recovered through AS-FTH matched well the isolated 'ħ' pattern image, even though the AS-FTH image contains negative background signals due to the imperfect subtraction of autocorrelation. Figure 4b shows the intensity profiles of hidden part of the image and isolated image. The black (triangles) and red (squares) lines in Fig. 4b were measured along the black and red arrows in Fig. 4a. Despite the black line has negative background in the left edge, it is very similar to the red line. Furthermore, the black line shows a relevant resolution of 140 nm compared to the red line of 120 nm. Since the imperfect subtraction mostly occurred due to the change of illumination conditions between the acquisitions of two diffraction signals, it can be reduced by introducing an in-line blocker which can open and close the reference wave in a short time interval for minimizing the change of illumination conditions.

5 Conclusion

We succeeded in obtaining the FTH image of a sample violating the separation condition using the AS-FTH method. The hidden part of the object image could be recovered to the original object structure by applying AS-FTH. Since the AS-FTH technique can eliminate the separation condition by removing the central autocorrelation signal, it allows obtaining FTH images with an illumination area comparable

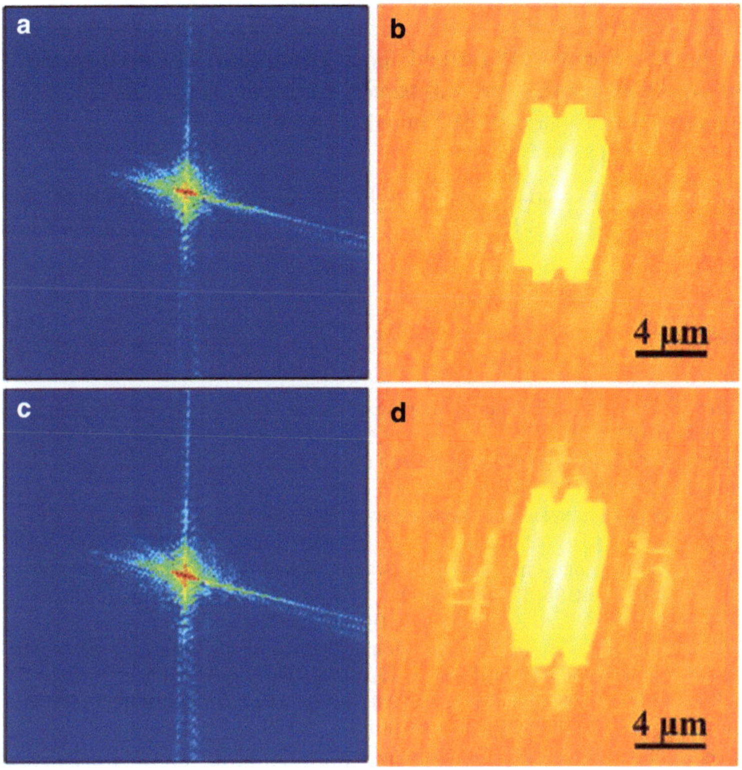

Fig. 3 **a** Diffraction signal of the 'ħ' pattern without reference pinholes. **b** Inverse Fourier transform of (**a**) (logarithmic scale). **c** Diffraction signal of the 'ħ' pattern with reference pinholes. **d** Inverse Fourier transform of (**c**) (logarithmic scale)

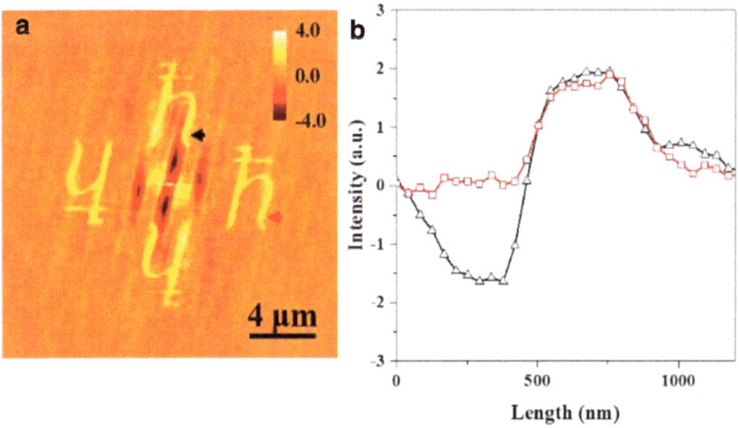

Fig. 4 **a** Central autocorrelation subtracted image of Fig. 3d. **b** Intensity profiles measured along the *black* (*black triangles*) and *red* (*red squares*) arrows in (**a**)

to the object size. It means that AS-FTH can image larger samples than those allowed in the normal FTH method for a given illumination condition. Consequently, the AS-FTH technique will be very useful in investigating large specimen using table-top x-ray sources with limited coherent photons.

Acknowledgments This work was supported by IBS (Institute for Basic Science) under IBS-R012-D1.

References

1. McNulty, I., et al.: High-Resolution imaging by fourier transform X-ray Holography. Science. **256**, 1009–1012 (1992)
2. Chao, W., et al.: Soft X-ray microscopy at a spatial resolution better than 15 nm. Nature. **435**, 1210–1213 (2005)
3. Lee, K.H., et al.: Ultrafast direct imaging using a single high harmonic burst. Opt. Lett. **38**, 1253–1255 (2013)
4. Kim, H.T., et al.: Single-shot nanometer-scale holographic imaging with laser-driven x-ray laser. Appl. Phys. Lett. **98**, 121105 (2011)
5. Marchesini, M., et al.: Nat. Photonics. **2**, 560–563 (2008)
6. Guizar-Sicairos, M., Fienup, J.R.: Holography with extended reference by autocorrelation linear differentialoperation. Opt. Express. **15**, 17592–17612 (2007)
7. Lee, K.H., et al.: Autocorrelation-subtracted Fourier transform holography method for large specimen imaging. Appl. Phys. Lett. **106**, 061103 (2015)
8. Kim, H.T., et al.: Demonstration of a saturated Ni-like Ag x-ray laser pumped by a single profiled laser pulse from a 10-Hz Ti:sapphire laser system. Phys. Rev. A. **77**, 023807 (2008)

Development of X-Ray Sources Using Intense Laser Pulses and Their Applications to X-Ray Microscopy

H. T. Kim, K. H. Lee, H. Yun, I. J. Kim, C. M. Kim, K. H. Pae, J. H. Sung, S. K. Lee, T. M. Jeong and C. H. Nam

Abstract We report the development of x-ray sources and their applications to x-ray microscopy. We applied a Ni-like Ag x-ray laser at 13.9 nm to Fourier transform holography (FTH) and obtained a reconstructed image with a resolution of 87 nm using a single-shot x-ray laser pulse. As the achievable spatial resolution of the single-shot FTH is limited by the strength of the reference wave, we enhanced the resolution of the x-ray imaging using the holography with extended reference by autocorrelation linear differential operation (HERALDO). Clear holograms could be recorded over the full size of an x-ray CCD, and the hologram reconstruction provided an image with a minimum resolution of 24 nm, close to the diffraction limited resolution of the imaging system.

1 Introduction

X-ray lasers using intense laser pulses [1] have unique properties of short pulse duration, high brightness and coherence, which can be suitable for time-resolved x-ray imaging. Since the laser-driven x-ray lasers have a compact size compared to an x-ray free electron lasers, the table-top x-ray lasers can be powerful in realizing a compact ultrafast x-ray microscope for various applications in nanoscience. To this end, the table-top x-ray lasers have been applied to imaging techniques using diffractive optics [2], coherence diffraction imaging (CDI) [3], and holography [4].

H. T. Kim (✉) · K. H. Lee · H. Yun · I. J. Kim · C. M. Kim · J. H. Sung · S. K. Lee · T. M. Jeong · C. H. Nam
Center for Relativistic Laser Science, Institute for Basic Science (IBS), Gwangju, Korea
e-mail: htkim@gist.ac.kr

H. T. Kim · I. J. Kim · C. M. Kim · K. H. Pae · J. H. Sung · S. K. Lee · T. M. Jeong
Advanced Photonics Research Institute, Gwangju Institute of Science and Technology (GIST), Gwangju, Korea

C. H. Nam
Dept. of Physics and Photon Science, GIST, Gwangju, Korea

© Springer International Publishing Switzerland 2016
J. Rocca et al. (eds.), *X-Ray Lasers 2014*, Springer Proceedings in Physics 169,
DOI 10.1007/978-3-319-19521-6_32

For the realization of table-top ultrafast x-ray microscopy, we have performed research on x-ray laser development [5, 6] and soft x-ray imaging [3, 4]. We successfully demonstrated a Ni-like silver x-ray laser at 13.9 nm by developing a new variation of the grazing incidence pumping (GRIP) scheme [7] using a single-profiled pumping pulse [5]. The x-ray laser was applied to single-shot Fourier transform holography (FTH) with the spatial resolution of 87 nm [4]. The spatial resolution of the single-shot FTH was however limited by the strength of the reference wave. In order to overcome the limitation of the FTH result, we performed holographic x-ray imaging with an extended reference by applying the autocorrelation linear differential operator (HERALDO) [8] and succeeded in achieving near diffraction-limited spatial resolution of 24 nm.

In this proceeding, our research on x-ray laser and their application to x-ray imaging is summarized. The proceeding is organized as follows. The development of x-ray laser is described in Sect. 2, and the Fourier transform holography using x-ray laser pulses is presented in Sect. 3. The enhanced spatial resolution of holographic imaging using extended reference is presented in Sect. 4. Section 5 delivers a summary.

2 Development of X-Ray Laser

The development of table-top x-ray laser is a critical step for x-ray microscopy. Among various methods to realize x-ray lasers [ref], the GRIP scheme is one of the most successful methods [ref]. The GRIP scheme usually uses two pumping beams; the first beam of a nanosecond pre-pulse is applied to create plasma and the other beam of a picosecond main pulse generates the population inversion. However, the GRIP scheme has a difficulty to overlap two laser beams, inducing instability in output power. In order to solve the problem, we developed a new variant of the GRIP scheme using a single profiled pumping pulse [5]. Using this scheme, we succeeded in achieving a saturated output from the Ni-like Ag x-ray laser at 13.9 nm pumped by a temporally profiled laser pulse with 1.5-J energy. For the temporal profiling of the pumping pulse, we controlled the injection time of the oscillator to amplifier stages in 10-Hz Ti:Sapphire laser to adjust prepulse and amplified-spontaneous-emission (ASE) background for proper pre-plasma conditions.

We measured the temporal profile of the pumping pulse and x-ray laser output as shown in Fig. 1. By controlling the pulse profile, we could generate the Ni-like Ag x-ray laser output with energy of about 1.5 μJ per pulse. The temporal profiles of the laser pulse were monitored during the experiment before the pulse compressor using a photodiode and an oscilloscope with temporal resolution of about 500 ps, shown as the dotted line in Fig. 1a. In order to measure the details of pulse profile, we installed a very fast photodiode and an oscilloscope having a temporal resolution below 100 ps after the pulse compressor. The black filled graph in Fig. 1a shows the complex temporal structure of the pumping pulse. The pumping pulse has four distinguishable peaks; the main pulse, a strong prepulse at—400 ps and two peaks

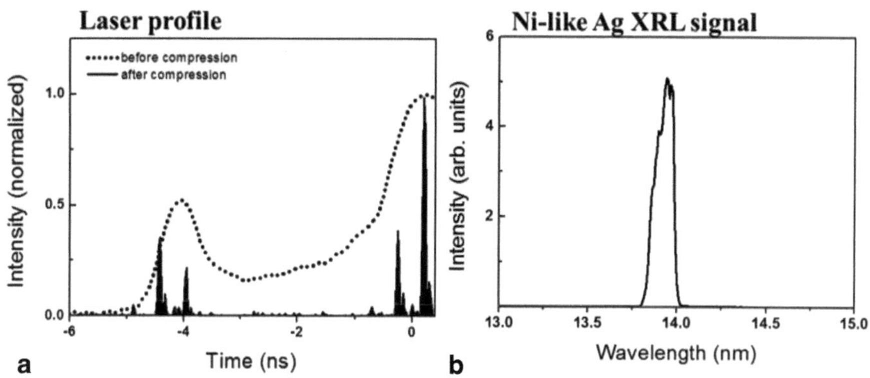

Fig. 1 **a** Temporal profiles of pumping laser pulse and, **b** output of the Ni-like Ag x-ray laser pumped with a single profiled laser pulse

at—4 ns. This complex temporal structure of the pumping pulse may provide the condition for efficient pumping—formation of multiple pumping pulses irradiating the target at an oblique angle [9]. Consequently, the single-profiled pumping scheme can be a simple and efficient method to generate an x-ray laser with stable output.

We have pursued also the development of new x-ray/gamma-ray sources using PW laser pulses. Using two PW Ti:Sapphire laser beam lines with outputs of 1.0 and 1.5 PW at 30 fs [10, 11], we examined OFI recombination x-ray lasers from high-density Ne and Nitrogen plasma [12]. We could not however obtain proper amplification of lasing lines by OFI recombination process due to strong reabsorption and short propagation distance in the dense medium [13]. In addition, we are planning to develop hard x-ray/gamma-ray sources from an inverse Compton scattering [14] between an electron beam, generated from laser waked field acceleration, and a PW laser pulse. These developments of x-ray sources based on PW lasers can provide new possibilities in x-ray imaging for living cells, non-destructive inspections and photo-nuclear reactions.

3 Fourier Transform Holography Using a Ni-like Ag X-Ray Laser

Lensless imaging methods using coherent x-ray sources have advantages of aberration-free imaging and diffraction-limited resolution. Among lensless imaging techniques, the CDI has simplest optical layout and sample preparation. The CDI however uses a time-consuming iterative phase retrieval algorithm that can cause problems of stagnation and ambiguity. On the other hand, FTH is beneficial in obtaining an instant reconstruction of images using only Fourier transform (FT). We applied the x-ray laser pulses to single-shot x-ray imaging obtained using FTH.

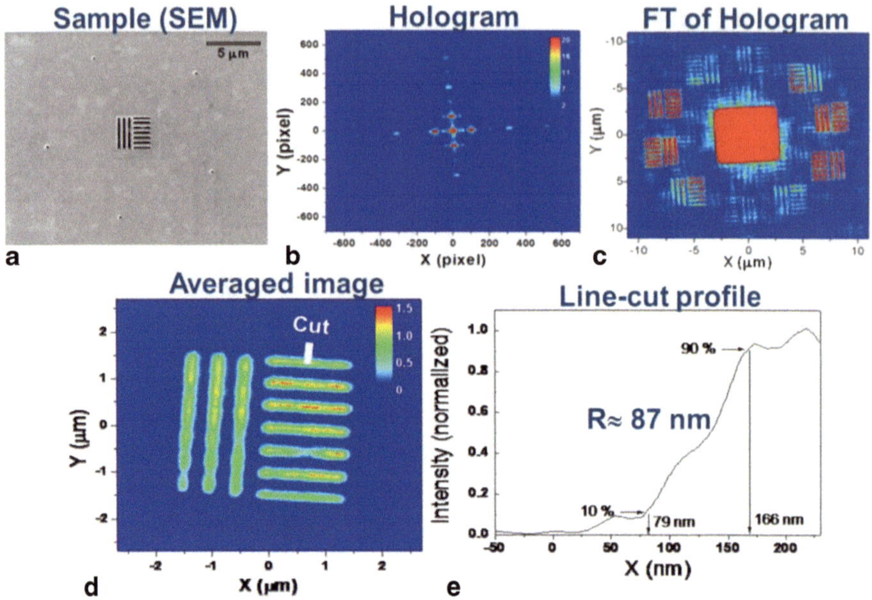

Fig. 2 Single-shot FTH imaging performed using an x-ray laser pulse. **a** SEM image of a sample, **b** recorded hologram, **c** FT of the hologram, **d** averaged image of ten sub-images, and **e** line-cut of the image

The FTH image was successfully reconstructed with a single exposure of the x-ray laser pulse. As a light source, we used the Ni-like Ag x-ray laser at 13.9 nm driven by a single profiled pumping pulse as described in Sect. 2. In the experiment, the x-ray laser beam was collected by a Mo/Si multilayer mirror and focused on a sample plate. The sample plate was composed of a µm-scale figure of a bar pattern and 5 reference holes surrounding the central figure as shown in Fig. 2a. The diffracted signal was observed with a soft x-ray CCD installed 48 mm after the sample plate. The details of the experiment were described in Ref [4]. The hologram, i.e. diffraction signal from the sample plate, was recorded clearly on the soft x-ray CCD, and the FT of the hologram shows the ten sub-images from 5 reference holes. Figure 2d shows the averaged image of reconstructed ten sub-images, and Fig. 2e is the line-cut profile of the image, showing the knife-edge resolution of about 87 nm. Consequently, single-shot FTH imaging was realized using the Ag x-ray laser at 13.9 nm.

In order to enhance the spatial resolution of the FTH image, we tried to reduce the size of reference hole below 50 nm. The electron-beam lithography was used for the fabrication of the reference hole below 50 nm. Figure 3a shows the SEM image of the sample plate and the inset is the enlarged image of the reference hole with size below 50 nm. In the experiment, we could observe the diffraction signal, but we could not obtain the reconstructed image, because the transmitted light from reference hole was so drastically reduced. This can be a fundamental limitation in enhancing the resolution of FTH. In order to overcome the problem, a new method using an extended reference can be a solution.

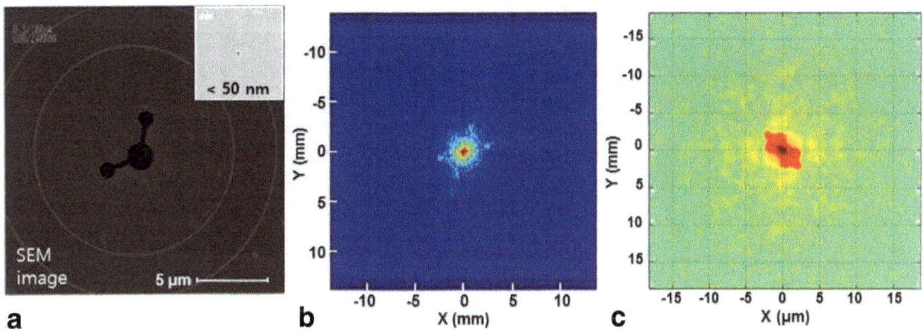

Fig. 3 FTH imaging with reference holes smaller than 50 nm. **a** SEM image of the sample. The inset shows the image of the reference hole, **b** hologram signal, and **c** FT of the hologram

4 Fourier Holography with Extended References

The enhancement of the spatial resolution of FTH can be limited by the weak reference wave from a tiny reference hole. Recently, a variation of FTH using an extended reference, called holography with extended reference by autocorrelation linear differential operation (HERALDO), was proposed to solve the problem of FTH [8]. Since the resolution of the FTH image is determined by the size of the reference hole, the reference wave is getting weaker as with a smaller reference used for obtaining better resolution. In the HERALDO method, the spatial resolution is determined by the sharpness of a reference edge, not by the size of the reference [8]. Thus, the spatial resolution of holographic imaging can be enhanced by employing an extended reference with a sharp edge.

We performed the single-shot coherent x-ray imaging using the HERALDO method with an illumination source of the Ni-like Ag x-ray laser. The sample plate for HERALDO was fabricated with a focused ion beam on a 300-nm thick Au film. The sample plate was composed of a sample feature of 'ħ'-pattern of 3 µm × 5 µm and two reference lines, as shown in Fig. 4a. Figure 4b shows the clear hologram from the sample plate with high visibility over the full x-ray CCD using a single x-ray laser pulse. The diffraction-limited resolution, corresponding to the numerical aperture of 0.33, is 21 nm. We reconstructed images of the sample feature applying a differential operator to the FT of the diffraction signal. The vertical differentiation, along the 45-nm reference line, provides a reconstructed ħ-pattern image. The HERALDO method can provide a higher flux of a reference wave than that of the conventional FTH, which enabled the reconstruction from the hologram recorded with the 45-nm reference line as shown in Fig. 4d. On the other hand, the reconstruction of hologram with the 15-nm reference line was not successful due to the noise component generated by the differential operator.

The spatial resolution of the single-shot HERALDO was estimated from the intensity profile of the reconstructed image obtained from the 45-nm reference line. Figures 4d and 4e show the intensity profiles measured along horizontal and vertical line on Fig. 4c, respectively. The resolution of the reconstructed ħ-image was estimated to be 42 nm for the horizontal direction (Fig. 4d) and 24 nm for the vertical

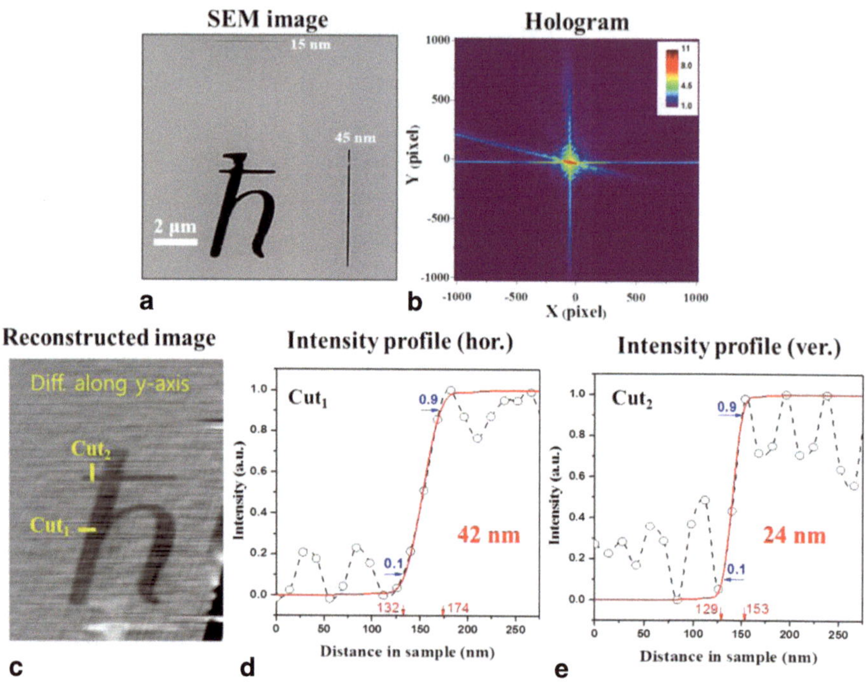

Fig. 4 HERALDO x-ray imaging using the Ni-like Ag x-ray laser **a** SEM image of the sample, **b** hologram, **c** Reconstructed image, and the line-cut profiles in **d** horizontal direction and **e** vertical direction

direction (Fig. 4e). In the HERALDO method, the resolution of images can depends on the direction of measurement. In the case of HERALDO with a line reference, the spatial resolution is determined by the sharpness of line end in the direction along the reference line. For the direction perpendicular to the line reference, the spatial resolution is determined by the line width. In this experiment, the width of the reference line was 45 nm, while the sharpness of the line end should be narrower than the line width. Thus, we could obtain the spatial resolution of 24 nm along the reference line, close to the diffraction-limited resolution of 21 nm. In addition, the resolution of the reconstructed image could be also improved by sharp phase edge in the sample, which can provide broad asymmetric diffraction signal. Consequently, we significantly enhanced the spatial resolution of single-shot x-ray imaging with the table-top Ag x-ray laser by adopting the HERALDO method.

5 Summary

We presented our research on the development of an x-ray laser source and their applications to x-ray microscopy. We successfully developed a Ni-like Ag x-ray laser using a variant of GRIP scheme by employing a single profiled pumping pulse, and applied the x-ray laser pulse to coherent x-ray imaging. The single-shot FTH image was successfully reconstructed with spatial resolution of 87 nm, while further enhancement of spatial resolution of FTH image using smaller reference pinhole was failed due to a weak reference wave. In order to solve the problem of the weak reference wave in FTH, we examine the x-ray imaging using the HERALDO method that can provide strong reference wave by adopting an extended reference. Using the HERALDO method, we obtained x-ray imaging with the resolution of 24 nm, close to the diffraction-limited resolution of 21 nm. Consequently, this compact x-ray microscope using the HERALDO coupled with the table-top x-ray laser can be applied to record transient phenomena of nanoscale materials with a temporal resolution of few ps.

Acknowledgments This work was supported by IBS (Institute for Basic Science) under IBS-R012-D1.

References

1. Suckewer, S., Jaeglé, P.: X-ray laser: past, present, and future. Laser. Phys. Lett. **6,** 411–436 (2009)
2. Brizuela, F., et al.: Microscopy of extreme ultraviolet lithography masks with 132 nm tabletop laser illumination. Opt. Lett. **34,** 271 (2009)
3. Kang, H.C., et al.: Single-pulse coherent diffraction imaging using soft x-ray laser. Opt. Lett. **37**, 1688–1690 (2012)
4. Kim, H.T., et al.: Single-shot nanometer-scale holographic imaging with laser-driven x-ray laser. Appl. Phys. Lett. **98,** 121105 (2011)
5. Kim, H.T., et al.: Demonstration of a saturated Ni-like Ag x-ray laser pumped by a single profiled laser pulse from a 10-Hz Ti:sapphire laser system. Phys. Rev. A. **77,** 023807 (2008)
6. Kim, H.T., et al.: Characteristics of a Ni-like silver x-ray laser pumped by a single profiled laser pulse. J. Opt. Soc. Am. B. **25,** 76–84 (2008)
7. Keenan, R., et al.: High-repetition-rate grazing-incidence pumped X-Ray laser operating at 18.9 nm. Phys. Rev. Lett. **94,** 103901 (2005)
8. Zhu, D., et al.: High-resolution x-ray lensless imaging by differential holographic encoding. Phys. Rev. Lett. **105,** 043901 (2010)
9. Janulewicz, K.A., Kim, C.M.: Role of the precursor in a triple-pulse pumping scheme of a nickel-like silver soft-x-ray laser in the grazing-incidence-pumping geometry. Phys. Rev. E. **82,** 056405 (2010)
10. Yu, T.J., et al.: Generation of high-contrast, 30 fs, 1.5 PW laser pulses from chirped-pulse amplification Ti:sapphire laser. Opt. Express. **20,** 10807 (2012)
11. Sung, J.H., et al.: 0.1 Hz 1.0 PW Ti:sapphire laser. Opt. Lett. **35,** 3021 (2010)
12. Hulin, S., et al.: Soft-x-ray laser scheme in a plasma created by optical-field-induced ionization of nitrogen. Phys. Rev. E. **61,** 5693–5700 (2000)
13. Kim, H.T., et al.: Development of x-ray sources using PW laser systems at APRI GIST. In: Klisnick, A., Menoni, C.S. (eds.) SPIE optical engineering + applications, pp. 88490T. International Society for Optics and Photonics, Bellingham (2013)
14. Phuoc, K.T., et al.: All-optical Compton gamma-ray source. Nat. Photon. **6,** 308–311 (2012)

Keyhole Reflection-Mode Coherent Diffractive Imaging of Nano-Patterned Surfaces Using a Tabletop EUV Source

Elisabeth Shanblatt, Matthew Seaberg, Bosheng Zhang, Dennis Gardner, Margaret Murnane, Henry Kapteyn and Daniel Adams

Abstract We demonstrate the first reflection-mode keyhole coherent diffractive imaging (CDI) of non-isolated samples from a single diffraction pattern. A table-top high harmonic generation (HHG) beam at 30 nm wavelength, with a highly curved wave-front, is used to illuminate Ti nano-patterns on a Si substrate at a 45° angle of incidence. The 30 nm illumination beam profile is first characterized using ptychograhic CDI. Keyhole CDI (kCDI) is then used to image the nano-sample. In contrast to ptychography CDI, keyhole CDI needs only one diffraction pattern, and therefore requires no scanning of the sample. This is a significant advantage for dynamic imaging, allowing a sequence of time-delayed images of the same region to be easily acquired. Our technique opens the door for imaging dynamics in nanostructures with sub-10 nm spatial resolution and fs temporal resolution.

1 Introduction

Imaging objects at the nano-scale is crucial for understanding and optimizing function for nanoscience and for next-generation nanotechnology. Recent developments in lensless coherent diffractive imaging (CDI) make possible high-resolution nanoscale imaging, particularly when combined with short wavelength illumination such as extreme ultraviolet (EUV) or X-ray high harmonic beams. In CDI, light is scattered by an object and then collected in the far field without any focusing optics after the object. The scatter pattern is processed by an iterative phase retrieval algorithm, which reconstructs the amplitude and phase of the object. CDI has been used to characterize objects in transmission [1, 2], but few experiments have been performed in a reflection geometry [3–7]. This type of imaging geometry is required when a thick, reflective sample needs to be imaged.

Ptychography CDI, a technique where the sample is scanned across the beam [8] has recently been demonstrated in reflection mode in a full-field microscope with no restriction on the angle of incidence [9]. Ptychography CDI reconstructs the object and illumination independently.

E. Shanblatt (✉) · M. Seaberg · B. Zhang · D. Gardner · M. Murnane · H. Kapteyn · D. Adams
JILA, University of Colorado at Boulder, Boulder, CO 80309, USA
e-mail: elisabeth.shanblatt@colorado.edu

© Springer International Publishing Switzerland 2016
J. Rocca et al. (eds.), *X-Ray Lasers 2014*, Springer Proceedings in Physics 169,
DOI 10.1007/978-3-319-19521-6_33

Keyhole CDI (kCDI) [10, 11] uses knowledge of the beam, which must have a highly curved wavefront and be isolated, to allow imaging of non-isolated samples. Previous keyhole results were obtained in a transmision geometry. Here we demonstrate kCDI in a reflection geometry. Unlike ptychography CDI, once the beam is known kCDI can reconstruct an image from a single diffraction pattern. This capability is very promising for dynamic pump-probe imaging, a technique that will prove crucial to understanding processes such as nanoscale transport phenomena.

2 High Harmonic Generation

In this demonstration, the coherent EUV light is obtained through High Harmonic Generation (HHG). HHG is an extreme nonlinear process that can produce coherent light from the ultraviolet to the soft X-ray range [12].

In HHG, an intense femtosecond laser is focused into a gas-filled hollow waveguide. The intense laser field ionizes the atoms the gas, and accelerates the electron first away from, and then back to the parent ion. Finally, the electron can recombine with the parent ion, converting it's oscillation energy into coherent HHG beams. By tuning the pressure of the gas, the phase velocity of the driving laser can be matched with that of the high harmonic wave, and a harmonic beam with full spatial coherence is produced. Although a large number of harmonics will be generated, the desired one can be selected using narrrow-bandwidth multilayer mirrors.

3 Experiment

Our HHG beam is produced using a Ti:sapphire laser centered at a wavelength of 785 nm and operating at 5 kHz repetition rate with 1.5 mJ pulse energy and 22 fs pulse duration. The beam is coupled into a 200 μm inner-diameter hollow waveguide filled with argon at 60 Torr. Harmonics around the 27th order (30 nm) are produced. The driving laser is filtered out using two silicon rejector mirrors set near Brewster's angle in conjuction with two 200-nm thick aluminum filters.

The beam is then passed through an adjustable pinhole (to achieve a hard edge) and reflected off of two multilayer Mg/SiC mirrors in a z-fold geometry. The first mirror is flat while the second one is a focusing curved mirror with a focal length of 5 cm. Due to the geometry, the curved mirror cannot be set at normal incidence, and is instead set at $\approx 2^\circ$. This non-normal incidence introudces some coma and astigmatism to the beam. The beam is directed onto the sample at a 45° angle of incidence, with the sample positioned 300 μm upstream of the circle of least confusion. The scattered light is collected by an Andor iKon CCD (2048×2048 pixels of size 13.5×13.5 μm) located 67 mm from the sample and at normal incidence.

The sample was fabricated using electron-beam lithography to pattern titanium onto a silicon wafer. The features on the sample are approximately 30 nm tall (Fig. 1).

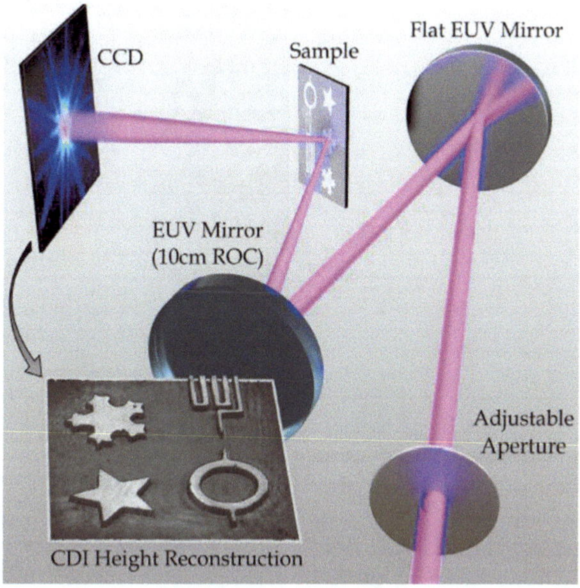

Fig. 1 Schematic of the experimental setup with an inset showing the ptychography CDI reconstruction. Figure adapted from Ref. [9]

3.1 Ptychography CDI

Keyhole CDI relies on knowledge of the beam, so in order to reconstuct the beam, ptychography CDI was performed. Ptychography CDI involves scanning the beam across the sample to obtain a number of diffraction patterns with significant overlap. A phase retrival algorithm then reconstructs the object and the beam independently. A ptychography scan was performed with a round beam approximatly 10 μm in diameter and 90 scan positions with a 2.5 μm step size.

3.2 Tilted Plane Correction

Due to the off-axis geometry and high NA of the system, the diffraction pattern is distored relative to the normal incidence case. This distortion can be corrected by interpolating the diffraction pattern onto a linear spatial frequency grid; a process known as tilited plane correction [7]. This allows fast fourier transforms to be used in the reconstruction, which significantly improves computational speed.

4 Results and Discussion

Using the technique described above, we achieved the first reflection-mode keyhole CDI result on a tabletop in conjunction with the first general full-field demonstration of reflection ptychography CDI.

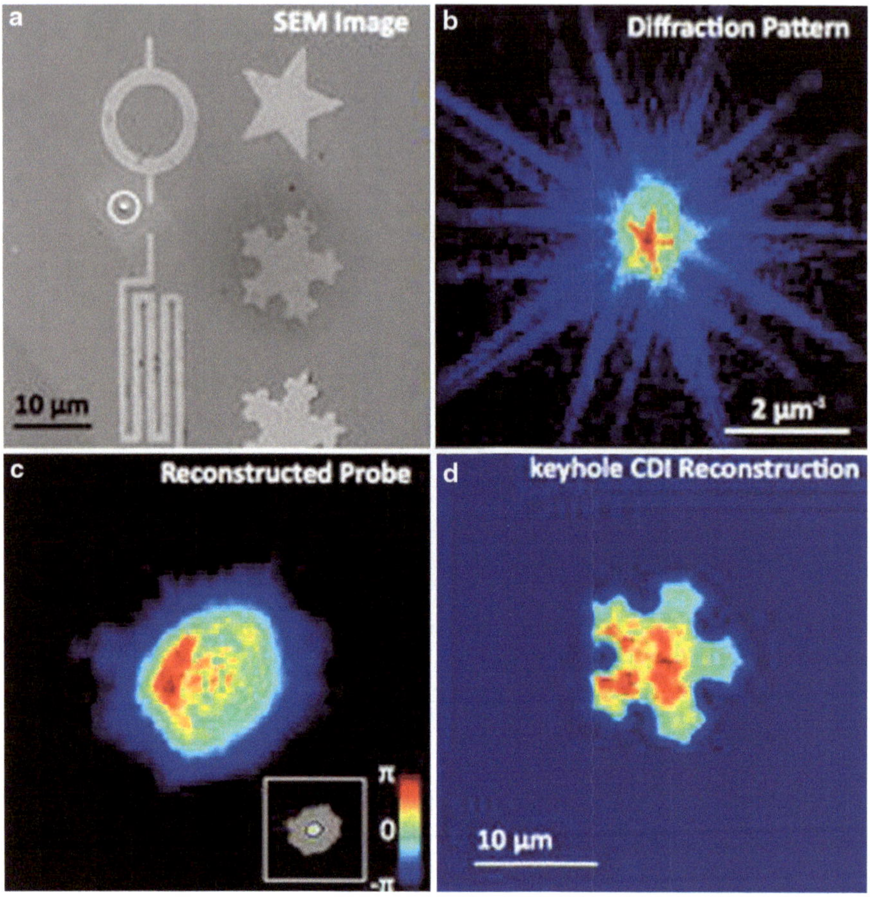

Fig. 2 **a** SEM image of the sample. **b** Diffraction pattern scaled to the 1/4 power. **c** Reconstructed beam. **d** Object reconsturcted using keyhole CDI. Figs. (**a**–**c**) adapted from Ref. [9]

To reconsturct the image using kCDI, tilted plane correciton was first performed on one of the diffraction patterns from the ptychography scan, and the reconstruction was then achieved using the method described in Ref. [11].

First, a random phase was generated at the detector, and then propagated back to the sample plane. The total reflectivity r was separated into two parts: a constant baseline reflectivity r_0, and the difference in reflectivites between the features and substrate, Δr. The magnitude of r_0 can be found either by trial-and-error, or from the ptychographic reconstruction. Next, the product of the beam and r_0 is subtracted from the total exit surface wave (ESW). This quantity is then multiplied by the conjugate of the probe phase, which gives the exit surface wave of just the object. We used the shrink-wrap approach [13] to apply the support constraint to the field at the sample plane.

Also applied was the phase constraint, which forces sample ESW's phase to be within a set range. Since the sample is approximately a binary valued object and Δr has a constant phase, the constraint is valid for this sample. The sample and detector plane constraints are applied iteratively until the algorithm converges, and the result is shown in Fig. 2d.

This technique readily generalizes for use with different light sources such as synchrotrons and xFELS, and can provide a variety of uses for both nano-science and in industry.

References

1. Miao, J., Charalambous, P., Kirz, J., Sayre, D.: Extending the methodology of X-ray crystallography to allow imaging of micrometre-sized non-crystalline specimens. Nature. **400,** 342–344 (1999)
2. Seaberg, M.D., Adams, D.E., Townsend, E.L., Raymondson, D.A., Schlotter, W.F., Liu, Y., Menoni, C.S., Rong, L., Chen, C.-C., Miao, J., Kapteyn, H.C., Murnane, M.M.: Ultrahigh 22 nm resolution coherent diffractive imaging using a desktop 13 nm high harmonic source. Opt. Express. **19,** 22470–22479 (2011)
3. Harada, T., Nakasuji, M., Nagata, Y., Watanabe, T., Kinoshita, H.: Phase imaging of extreme-ultraviolet mask using coherent extreme-ultraviolet scatterometry microscope. Jpn. J. Appl. Phys. **52,** 06GB02 (2013)
4. Roy, S., Parks, D., Seu, K.A., Su, R., Turner, J.J., Chao, W., Anderson, E.H., Cabrini, S., Kevan, S.D.: Lensless x-ray imaging in reflection geometry. Nat. Photonics. **5,** 243–245 (2011)
5. Zürch, M., Kern, C., Spielmann, C.: XUV coherent diffraction imaging in reflection geometry with low numerical aperture. Opt. Express. **21,** 21131–21147 (2013)
6. Sun, T., Jiang, Z., Strzalka, J., Ocola, L., Wang, J.: Three- dimensional coherent x-ray surface scattering imaging near total external reflection. Nat. Photonics. **6,** 586–590 (2012)
7. Gardner, D.F., Zhang, B., Seaberg, M.D., Martin, L.S., Adams, D.E., Salmassi, F., Gullikson, E., Kapteyn, H.C., Murnane, M.M.: High numerical aperture reflection mode coherent diffraction microscopy using off-axis apertured illumination. Opt. Express. **20,** 19050–19059 (2012)
8. Maiden, A., Rodenburg, J.: An improved ptychographical phase retrieval algorithm for diffractive imaging. Ultramicroscopy. **109,** 1256–1262 (2009)
9. Seaberg, M.D., Zhang, B., Gardner, D.F., Shanblatt, E.R., Murnane, M.M., Kapteyn, H.C., Adams, D.E.: Tabletop nanometer extreme ultraviolet imaging in an extended reflection mode using coherent Fresnel ptychography. Optica. **1,** 1–8 (2014)
10. Abbey, B., Nugent, K.A., Williams, G.J., Clark, J.N., Peele, A.G., Pfeifer, M.A., Jonge, M.D., McNulty, I.: Keyhole coherent diffractive imaging. Nat. Phys. **4,** 394–398 (2008)
11. Zhang, B., Seaberg, M.D., Adams, D.E., Gardner, D.F., Shanblatt, E.R., Shaw, J.M., Chao, W., Gullikson, E.M., Salmassi, F., Kapteyn, H.K., Murnane, M.M.: Full field tabletop EUV coherent diffractive imaging in a transmission geometry. Opt. Express. **21,** 21970–21980 (2013)
12. Popmintchev, T., Chen, M.-C., Arpin, P., Murnane, M.M., Kapteyn, H.C.: The attosecond nonlinear optics of bright coherent X-ray generation. Nat. Photonics. **4,** 822–832 (2010)
13. Marchesini, S., He, H., Chapman, H.N., Hau-Riege, S.P., Noy, A., Howells, M.R., Weierstall, U., Spence, J.C.H.: X-ray image reconstruction from a diffraction pattern alone. Phys. Rev. B. **68,** 140101 (2003)

Nanoscale Imaging with Soft X-Ray Lasers

C. S. Menoni, J. Nejdl, N. Monserud, I. D. Howlett, D. Carlton, E. H. Anderson, W. Chao, M. C. Marconi and J. J. Rocca

Abstract This paper describes recent advances in full field aerial soft x-ray (SXR) laser microscopes. It also shows these microscopes can be readily configured into a differential holographic microscopy (DHM) geometry to acquire image plane holograms from which amplitude and phase of the object wave are obtained. The ability to quantitatively determine phase is critical at SXR wavelengths when the absorption contrast is low.

1 Introduction

Microscopy is a pervasive technology impacting many areas of science and technology. Extreme ultraviolet (EUV) and soft x-ray (SXR) illumination with wavelengths of tens of nanometers brings to microscopy the possibility of significantly enhanced spatial resolution compared to optical illumination. Furthermore, it opens up new opportunities that are made possible by the distinct interaction of the SXR illumination and materials. EUV/SXR microscopes have been implemented using illumination from synchrotrons, incoherent sources, high harmonic generation sources and SXR lasers [1–5]. The highest spatial resolution, 10 nm, was obtained using illumination from the Advanced Light Source at Lawrence Berkeley Laboratory [4].

Table-top soft-ray laser (SXRL) beams with pulse energies of tens of microjoules combined with SXR optics have enabled the implementation of imaging systems that can capture space-time resolved absorption contrast images with nanoscale resolution in transmission and reflection configurations [5–8]. The high

C. S. Menoni (✉) · J. Nejdl · N. Monserud · I. D. Howlett · D. Carlton · E. H. Anderson · W. Chao · M. C. Marconi · J. J. Rocca
National Science Foundation Engineering Research Center for Extreme Ultraviolet Science and Technology, Colorado State University, Fort Collins, USA
e-mail: carmen.menoni@colostate.edu

C. S. Menoni · N. Monserud · I. D. Howlett
Electrical and Computer Engineering, Colorado State University, Fort Collins, USA

D. Carlton · E. H. Anderson · W. Chao · M. C. Marconi · J. J. Rocca
Center for X-Ray Optics, Lawrence Berkeley National Laboratory, Berkeley, CA, USA

J. Nejdl
ELI Beamlines project, Institute of Physics of AS CR, Prague, Czech Republic

© Springer International Publishing Switzerland 2016

J. Rocca et al. (eds.), *X-Ray Lasers 2014*, Springer Proceedings in Physics 169,
DOI 10.1007/978-3-319-19521-6_34

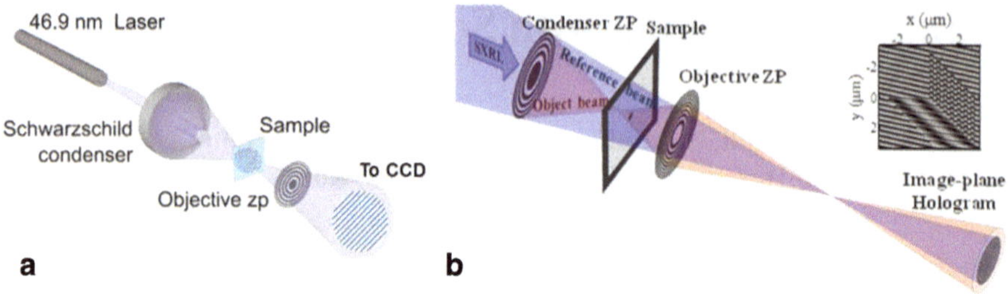

Fig. 1 Schematics of the geometry of: **a** the full field 46.9 nm aerial microscope and, **b** differential holographic microscopy setup

monochromaticity of SXRLs also contributes to the high throughput when employing reflective, diffractive or a combination of both as condensers and objectives in the microscope.

We describe a recent characterization of full field SXRL microscopes that have achieved 62 nm spatial resolution. We also describe results of the first demonstration of Differential Holographic Microscopy (DHM) from which two dimensional phase and amplitude images of a weakly absorbing test pattern are reconstructed from the image plane hologram. The ability to quantitatively determine phase with this microscope open up opportunities to determine variations in density or thickness of weakly absorbing samples at the nanoscale.

2 Experimental Details

The geometry of the 46.9 nm microscope used in this work previously described is shown in Fig. 1a [9]. Briefly, a Sc/Si reflective condenser illuminates the sample which is imaged by an objective zone plate objective (OZP) lens onto an array detector. The 46.9 nm capillary discharge SXRL produces pulses of energy between 0.01 and 0.1 mJ [10–12] which has been instrumental for capturing single shot aerial images and holograms [5, 13].

To implement DHM, a second objective zone plate was incorporated between the Schwarzschild condenser and the sample. This is shown schematically in Fig. 1b. When illuminated by the converging wavefront from the Schwarzschild the 0th order of this ZP generates the reference beam, while the 1st order illuminates the sample. The reference beam propagates practically unaltered through the sample and OZP towards the array detector where it interferes with the object wave created by the objective zone plate at the detector plane. Different OZP were used in the microscope. The absorption contrast images were acquired using a OZP of 0.47 NA and 860 $\times$ magnification. The image plane holograms were obtained using OZPs of 0.23 and 0.46 NA to achieve magnifications of 600 $\times$ and 900 $\times$ respectively.

The ZPs and samples were fabricated by electron-beam lithography and subsequent processing [14]. The 0.5 mm diameter ZPs were engineered as free standing

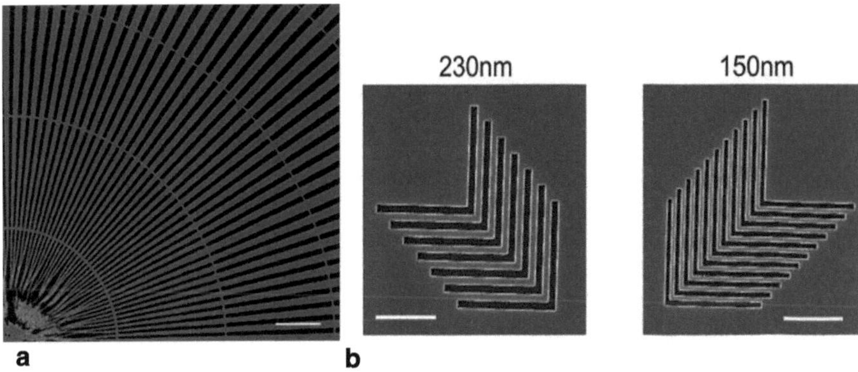

Fig. 2 **a** Scanning electron microscope image of the Siemens star with spokes' period in the inner ring of 200 nm, **b** Elbow gratings with periods of 230 and 150 nm. The white bar scale is 2 μm

structures onto silicon nitride membranes. A similar process was used to fabricate the Siemens star used for absorption contrast imaging (Fig. 2a). The samples used for DHM consisted of a set of Si elbow gratings with periods of 150 and 230 nm fabricated on a Si membrane (Fig. 2b). The absorption contrast of this sample is 30 % at $\lambda = 46.9$ nm. The patterns were located in a $20 \times 20\,\mu m^2$ region at the center of the $0.5 \times 0.5\,mm^2$ membrane window. This geometry allowed the reference beam to propagate through the sample with minimum distortion. Off axis holograms were obtained with the DHM microscope with typical exposures of 5–30 laser shots with the laser operating at 2 Hz.

3 Results

3.1 Absorption Contrast Imaging

A broad area absorption contrast image of the Siemens star obtained the SXRL microscope is shown in Fig. 3a. It took 100 laser shots to acquire this image. Intensity line-outs from different regions of the image were obtained to determine the spatial resolution of the microscope using the knife-edge test.

We have previously shown that angular illumination affects the resolution of the 46.9 nm microscope [9]. Moreover, it is known that when analyzing a broad area image as shown in Fig. 3a, the spatial resolution varies from $k\lambda/NA$ near the optical axis to $k\lambda/(NA_{ozp} - NA_{czp})$ on the periphery of the image. To account for these effects, the knife edge test was applied to four separate quadrants in the image as indicated in Fig. 3a. Within each quadrant, 5 different regions were analyzed to determine the 10–90 % intensity rise on the falling edge of periodic patterns as shown in Fig. 3b. An spatial resolution of 83, 85, 64 and 62 nm were obtained for quadrant 1–4 respectively.

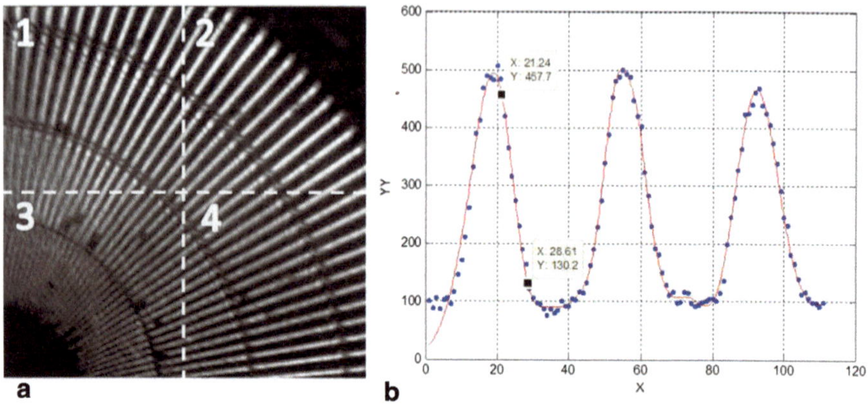

Fig. 3 **a** SXR image of a Siemens star pattern obtained with a 0.47 NA OZP with a magnification of 860×. **b** Line-out from a section of the image. The knife-edge spatial resolution was calculated to be 83, 85, 64 and 62 nm for quadrant 1–4

We have also previously analyzed the coherence properties of the illumination in the 46.9 nm SXRL microscope when using a 0.32NA OZP [9]. This analysis showed the illumination is fully coherent. For the knife-edge test, $k = 0.68$ for fully coherent illumination [15]. Therefore, the half period resolution for the 0.47NA OZP is calculated to be 68 nm near the optical axis and 109 nm on the periphery, which is in agreement with the measurements.

3.2 *Differential Holographic Microscopy*

The DHM microscope produced off-axis holograms as that shown in Fig. 4a for the 230 nm elbow grating. This image plane hologram was obtained with the 0.23NA OZP and a magnification of 600×. The off-axis hologram was analyzed using a Fourier transform method which consisted of applying a point-symmetrical band-pass filter centered at the carrier frequency to extract the amplitude and phase of the complex field of the object wave [16]. The amplitude map shows reduced contrast, as expected from an object designed to have only 30 % absorption contrast at $\lambda = 46.9$ nm. Instead the phase map has very high contrast. The histogram constructed from the analysis of the phase map enabled the determination of the phase shift from the centre frequency to be (2.3 ± 0.3) rad. This phase shift accurately predicts the height of the Si dense line to be 100 nm.

Images from a 150 nm half period elbow gratings with 900 × magnification were also obtained [17]. With increased magnification the fringe visibility in the hologram reduced. This resulted in larger spread of the phase shift compared to that obtained from the histogram of Fig. 3d.

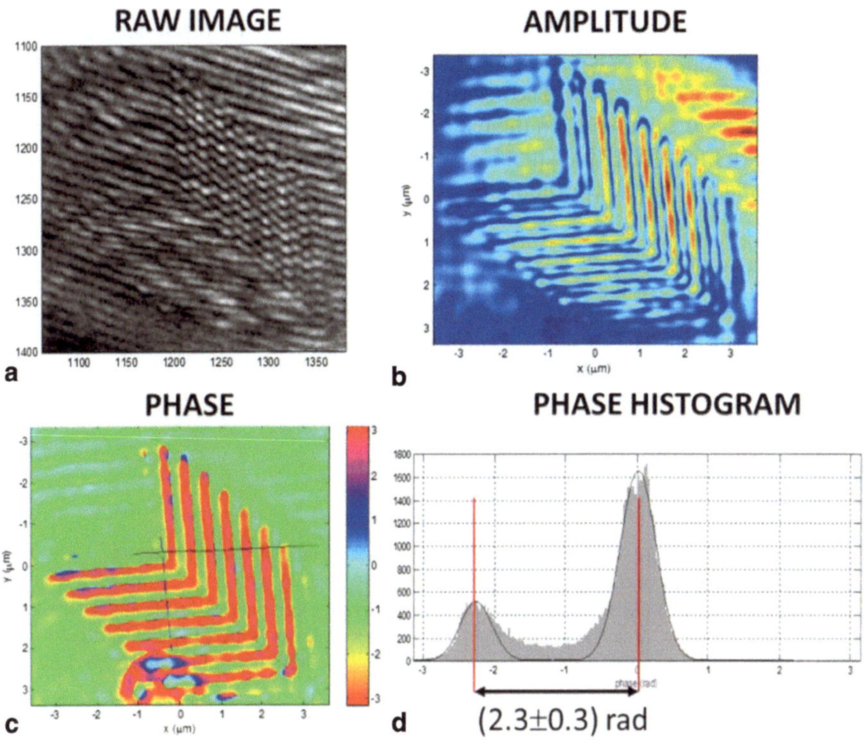

Fig. 4 **a** Hologram of the 230 nm half period elbow pattern. Two dimensional maps showing the variation in the amplitude **b** and phase, **c** of the object wave at the detector plane. The circular pattern at the *bottom* of **c** is caused by in homogeneity of the object beam arising from illumination of a condenser zone plate with a central opening, **d** The phase histogram from **c** shows the phase shift from the 0th order is (2.3 ± 0.3) rad

An in-depth analysis of the factors that affect the quality of the holograms obtained by the DHM set up of Fig. 1b has been presented [18]. A detailed ray tracing calculation showed that while the longitudinal coherence of the SXRL, ~700 μm [19], does not play significant demands on the alignment of the microscope, the constraints imposed by the transverse coherence do. Furthermore, it was shown that accurate determination of the amplitude and phase of the electric field phasor at the image plane requires that both centre-symmetric components of a given spatial frequency be transmitted by the OZP.

4 Conclusions

We have extended the spatial resolution of full field areal microscopes to 62 nm when using $\lambda = 46.9$ nm illumination and a 0.47 NA objective. We have also demonstrated SXR DHM with the addition of a ZP to the geometry of the full field aerial microscope that serves to illuminate the object and create the reference beam. The DHM microscope successfully obtained quantitative phase and amplitude maps of low absorbing Si dense lines. The common-path nature of the method which significantly reduces the demands on the temporal coherence of the illumination coupled with the high flux of SXRLs will allow for single-shot image plane holography with high spatial and temporal resolution.

Acknowledgements This work was supported by the NSF ERC for Extreme Ultraviolet Science and Technology under NSF Award EEC-0310717 using equipment developed under NSF Award MRI-ARRA 09-561. J.N. was partially supported by the Fulbright program and by the Ministry of Education, Youth and Sport of the Czech Republic (project ECOP Nos. CZ.1.07/2.3.00/20.0279, CZ.1.07/2.3.00/30.0057, and ELI-Beamlines CZ.1.05/1.1.00/02.0061).

References

1. Martz, D.H., et al.: High average brightness water window source for short-exposure cryomicroscopy. Opt. Lett. **37**, 4425 (2012)
2. Wachulak, P.W., et al.: Water-window microscopy using a compact, laser-plasma SXR source based on a double-stream gas-puff target. Appl. Phys. B-Lasers Opt. **111**, 239 (2013)
3. Lee, K.H., et al.: Ultrafast direct imaging using a single high harmonic burst. Opt. Lett. **38**, 1253 (2013)
4. Chao, W., et al.: Real space soft x-ray imaging at 10 nm spatial resolution. Opt. Express. **20**, 9777 (2012)
5. Brewer, C.A., et al.: Single-shot extreme ultraviolet laser imaging of nanostructures with wavelength resolution. Opt. Lett. **33**, 518 (2008)
6. Brizuela, F., et al.: Reflection mode imaging with nanoscale resolution using a compact extreme ultraviolet laser. Opt. Express. **13**, 3983 (2005)
7. Carbajo, S., et al.: Sequential single-shot imaging of nanoscale dynamic interactions with a table-top soft x-ray laser. Opt. Lett. **37**, 2994 (2012)
8. Brizuela, F., et al.: Extreme ultraviolet laser-based table-top aerial image metrology of lithographic masks. Opt. Express. **18**, 14467 (2010)
9. Brizuela, F., et al.: Imaging at the Nanoscale With Practical Table-Top EUV Laser-Based Full-Field Microscopes. IEEE J. S. Top. Quantum Electron. **18**, 434 (2012)
10. Rocca, J.J., et al.: Demonstration of a Discharge Pumped Table-Top Soft-X-Ray Laser. Phys. Rev. Lett. **73**, 2192 (1994)
11. Macchietto, C.D., et al.: Generation of millijoule-level soft-x-ray laser pulses at a 4-Hz repetition rate in a highly saturated tabletop capillary discharge amplifier. Opt. Lett. **24**, 1115 (1999)
12. Heinbuch, S., et al.: Demonstration of a desk-top size high repetition rate soft x-ray laser. Opt. Express. **13**, 4050 (2005)
13. Malm, E.B., et al.: Tabletop single-shot extreme ultraviolet Fourier transform holography of an extended object. Opt. Express. **21**, 9959 (2013)
14. Anderson, E.H.: Specialized electron beam nanolithography for EUV and X-ray diffractive optics. IEEE J. Quantum Electron. **42**, 27 (2006)
15. Heck, J.M., et al.: Resolution determination in X-ray microscopy: an analysis of the effects of partial coherence. J. X-Ray Sci. Technol. **8**, 95 (1998)
16. Takeda, M., et al.: Fourier-transform method of fringe-pattern analysis for computer-based topography and interferometry and illumination spectrum. J. Opt. Soc. Am. **72**, 156 (1982)

17. Nejdl, J., et al.: Image Plane Holographic Microscopy With a Table-Top Soft X-Ray Laser. IEEE Photonics J. **7**, (2015)
18. Nejdl, J., et al.: In: Klisnick, A., Menoni, C.S. (eds.) Proceedings of SPIE Volume 8849. X-Ray Lasers and Coherent X-Ray Sources: Development and Applications X, (2013)
19. Urbanski, L., et al.: Spectral linewidth of a Ne-like Ar capillary discharge soft-x-ray laser and its dependence on amplification beyond gain saturation. Phys. Rev. A. **85**, 033837 (2012)

Nanoscale Imaging Using Coherent and Incoherent Laboratory Based Soft X-Ray Sources

H. Stiel, A. Dehlinger, K.A. Janulewicz, R. Jung, H. Legall, C. Pratsch, C. Seim and J. Tümmler

Abstract Nanoscale imaging of biological samples in the lab as well as mask inspection in extreme ultraviolet lithography near the production line with sub 30 nm resolution require high spectral brightness soft x-ray sources. Laser produced plasma (LPP) sources and plasma based X-ray lasers (XRL) emit soft X-ray radiation in the wavelength region of interest between 2 and 20 nm. Whereas LPP sources easily can be tuned to the so called water window (2.2–4.4 nm) the output of an XRL is restricted to relatively few fixed wavelengths in the extreme ultraviolet range. However due to the relatively high degree of coherence the XRL is well suited also for nanoscale imaging using coherent techniques like coherent diffraction imaging or Fourier transform holography.

1 Introduction

Nanoscale imaging techniques with sub 100 nm resolution like X-ray microscopy (XRM) [1–3], coherent diffraction imaging (CDI) [4], [5] or Fourier transform holography (FTH) [6], [7] have emerged as with a great potential for applications in biology, biophysics, medicine, material and environmental sciences [8]. For a wide application of these techniques at a laboratory scale high spectral brightness laboratory soft x-ray sources as well as appropriate optical components are required.

From the application point of view two spectral ranges (cp. Fig. 1) are of special interest: the so called water window between 2.2 and 4.4 nm and the extreme

H. Stiel (✉) · R. Jung · J. Tümmler · C. Seim
Max-Born Institut für Nichtlineare Optik und Kurzzeitspektroskopie (MBI), Berlin, Germany
e-mail: stiel@mbi-berlin.de

H. Stiel · A. Dehlinger · R. Jung · H. Legall · C. Seim
Berlin Laboratory for innovative X-ray Technologies (BLiX), Berlin, Germany

A. Dehlinger · H. Legall
Technical University (TU), Berlin, Germany

K.A. Janulewicz
Department of Physics and Photon Science, Gwangju Institute of Science and Technology (GIST), Gwangju, Republic of Korea

C. Pratsch
Helmholtz-Zentrum Berlin (HZB), Berlin, Germany

© Springer International Publishing Switzerland 2016
J. Rocca et al. (eds.), *X-Ray Lasers 2014*, Springer Proceedings in Physics 169,
DOI 10.1007/978-3-319-19521-6_35

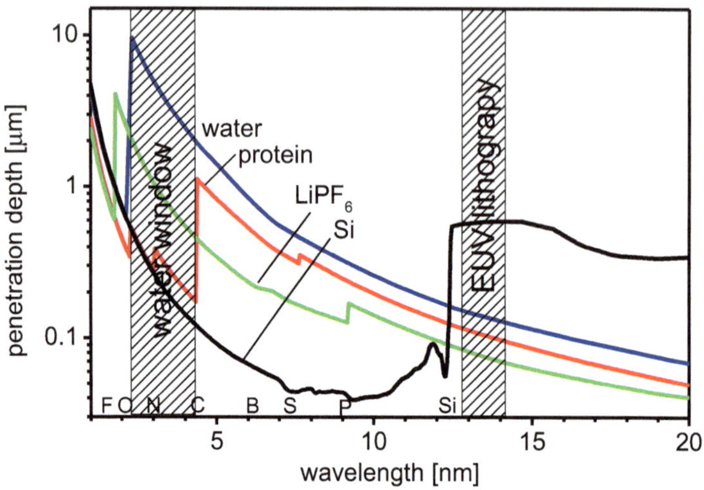

Fig. 1 Soft X-ray penetration depths for materials of biological and technological interest. Between 2.2 and 4.4 nm (*water window*) the penetration depth in water is about one order of magnitude higher than in protein providing a high contrast for x-ray imaging. In the region above 12.4 nm the penetration depth in Si is relatively high making this material ideally suited as a support for investigations of nanoparticles. The region around 13.5 nm is of great technological interest for *EUV lithography*. Due to the small penetration depth of soft X-rays EUV imaging often rely on techniques in reflection mode [10]

ultraviolet (EUV) range around 13 nm. Whereas the water window is of special interest for applications in life and environmental sciences the imaging techniques in the EUV range are of great technological importance e.g. for actinic mask inspection in lithography [9].

In this contribution we report about two highly brilliant X-ray sources, a laser produced plasma source for the water window and a plasma based X-ray laser (XRL) for the EUV region The capability of these sources for applications in XRM and FTH will be discussed.

2 Full Field Laboratory X-Ray Microscopy in the Water Window

Full field laboratory X-ray microscopy in the water window requires both a highly brilliant X-ray source and appropriate optics for illumination of the sample and imaging. Laser produced plasmas (LPP), due to its small source are ideally suited for sample illumination. LPP are scalable to high average photon fluxes if high repetition rate lasers are used as a pump source [1], [2].

Our laboratory X-ray transmission microscope (LTXM) is based on a LPP using a liquid nitrogen jet as target. An intense laser-pulse (slab laser system FhG-ILT, Aachen, Germany, 1064 nm, 100 mJ, 450 ps, 1.3 kHz) is focused on a liquid

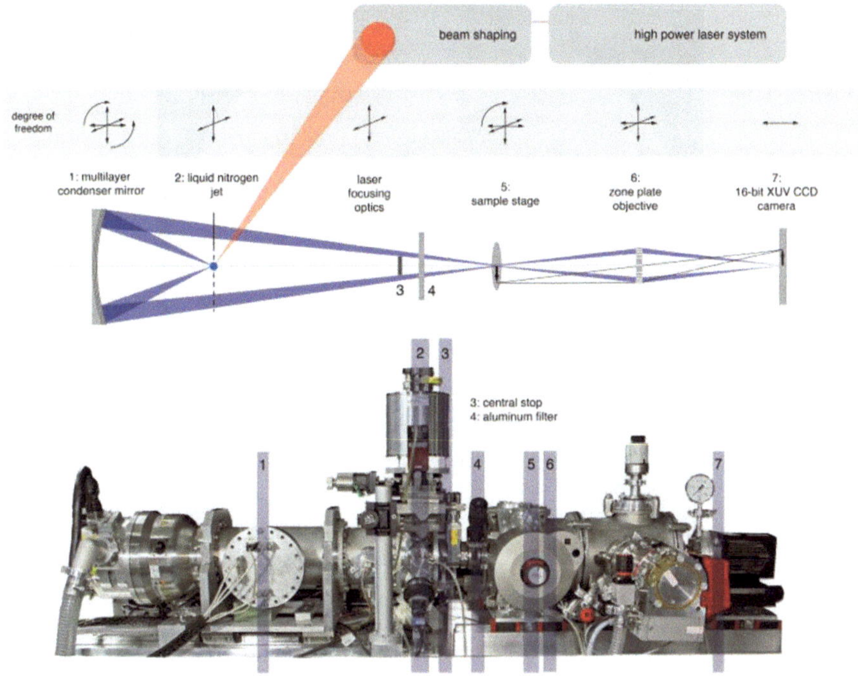

Fig. 2 Water window laboratory transmission X-ray microscope (*LTXM*). The light source is a laser produced nitrogen plasma (laser not shown, parameters see text). Other optical elements are explained in text

nitrogen jet, producing hot dense plasma. The most prominent lines in the spectrum emitted by the plasama originate from N-VII and N-VI ions at 2.478 nm (Ly_α line) and 2.879 nm (He_α line). The brightness of the source was determined by two independent experiments measuring the source size and the photon flux. It yields about 8×10^{11} ph/mm^2 mrad^2 @ 2.478 nm at 100 W average laser power. This value is comparable with the brightness of bending magnet beamlines of second generation synchrotron sources. The isotropic emission of the source is collected by a Cr/V multilayer mirror (FhG-IOF, Jena, Germany, D = 54 mm, ROC = 350 mm, $r \approx 0.6\,\%$ @ 2.478 nm). Due to its extremely small bandwidth of 8 pm the condenser mirror collects only the emission from the 2.478 nm line, thus monochromatizing the radiation. The flux on the sample yields to about 10^6 ph/μm^{2}s at 100 W average laser power mainly limited by the relatively low reflectance of the coating of the multilayer. The sample is imaged using an objective zone plate (HZB-BESSY, Berlin, Germany, d = 66.5 μm, f = 692 μm, NA = 0.05, dr = 25 nm) with an efficiency of about 5 %. Typically the photon flux on CCD (Roper Scientific PIXIS-XO: 2048B, back illuminated, cooled sensor, pixel size $13.5 \times 13.5\ \mu m^2$) amounts to 20 ph/s at a magnification of 1000. Taking this flux and assuming a S/N of about 30 (corresponding to 1000 ph on CCD) a recording time in the order of 1 min is required for a high quality image. The microscopy chamber (BRUKER ASC,

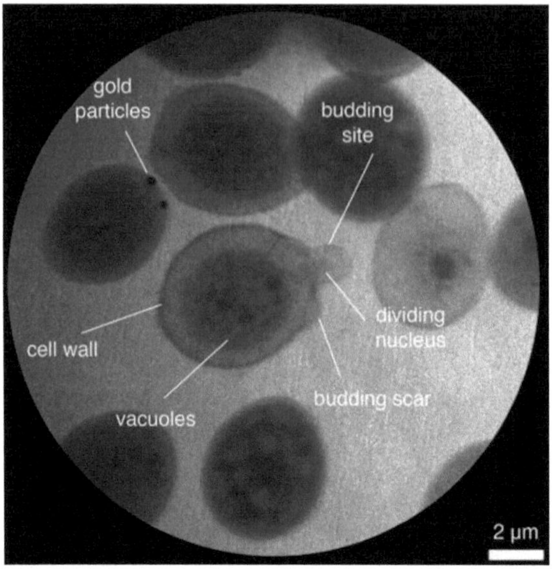

Fig. 3 Nanoscopic image of a vitrified yeast cell (*saccharomyces cerevisiae*)

cp. Fig. 2) contains the motorized sample and zone plate stages. It is equipped with a probe lock, hence specimen can be quickly changed during operation, enabling the investigation of series of plunge-frozen samples.

2.1 Results: Imaging of Cells

We applied the LTXM for imaging of cryo plunged frozen yeast cells. The micrograph (Fig. 3) shows a vitrified yeast cell (*saccharomyces cerevisiae*). Cell wall and the cell's budding cite on the right side are distinctly recognizable. When taking a closer look, a structure, what could be the dividing nucleus, just a glimpse darker than the cytoplasm, seems to reach into the newly forming cell. Just below the budding site, a budding scar, from a previous budding process, is visible. The darker circular shaped features in the center of the cell could be vacuolar granules.

3 High Repetition Rate X-Ray Laser

Figure 4 shows a schematic of the high repetition rate X-ray laser (XRL) operating in grazing incidence pump (GRIP) geometry. The two pump pulses are provided by a 100 Hz thin disk laser (TDL) chirped pulse amplification (CPA) system which has been described elsewhere [11]. Shortly, the TDL system consists of a front-end

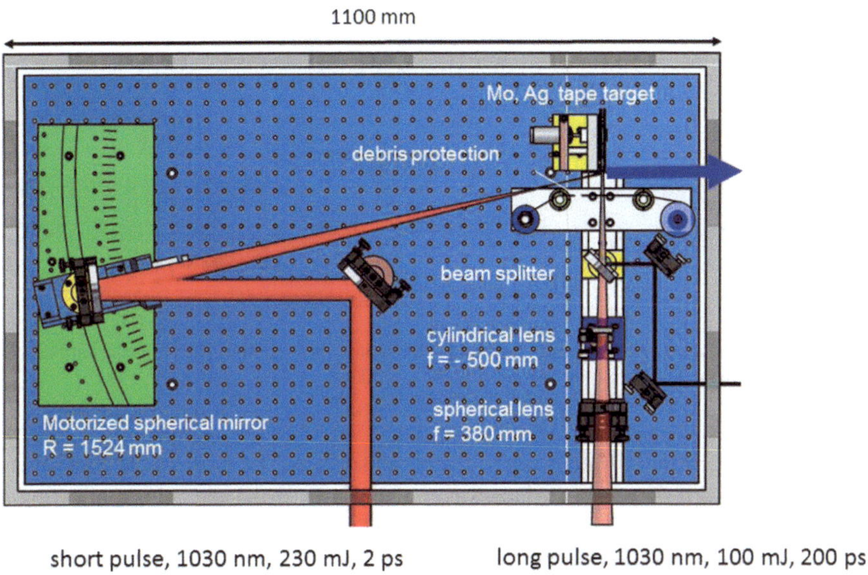

Fig. 4 Scheme of the 100 Hz XRL in GRIP geometry

with an Yb:KGW oscillator, stretcher and Yb:KGW regenerative amplifier followed by two regenerative amplifiers and one multipass amplifier. The front end delivers an output energy of 1 mJ, a pulse duration of about 1.5 ns at 1030 nm. The output is divided into two pulses each of them is amplified in a regenerative amplifier to a level of about 200 mJ. Whereas the pulse from the first regenerative amplifier is compressed to a duration of about 200 ps using a grating compressor the output of the second regenerative amplifier is given to a thin disk multipass amplifier which amplifies the pulses to an energy up to 400 mJ and compressed in the grating compressor to about 2 ps pulse duration. The long pulse (200 ps, $E \approx 100$ mJ) is focused by a cylindrical ($f = -500$ mm) and a spherical lens ($f = 380$ mm) onto the target at normal incidence giving a line focus of about 30 μm in width. The generated plasma column will then be heated by a short pulse (2 ps, $E > 230$ mJ) focused according to the GRIP method by a spherical mirror (Edmund Optics $f = 762$ mm) into the preformed plasma.

We have found for our Mo target an optimum GRIP angle of about 24°. The delay between the two pulses has appeared as a very critical parameter [12–14]. Therefore the delay between long and short pulse can be adjusted by adapting the round trip time of the two regenerative amplifiers as well as finely-tuned by an additional delay stage. Because of the high repetition rate all optical components are protected against debris by thin glass plates or foil, and the XRL output is guided through an aperture to reduce debris contamination on the following optical elements. In the experiments for output optimization of the XRL we used Mo slab targets with a length of 50 mm and a width between 1 and 6 mm. The target was attached to a motorized stage with four degrees of freedom allowing the adjustment in 3 axes

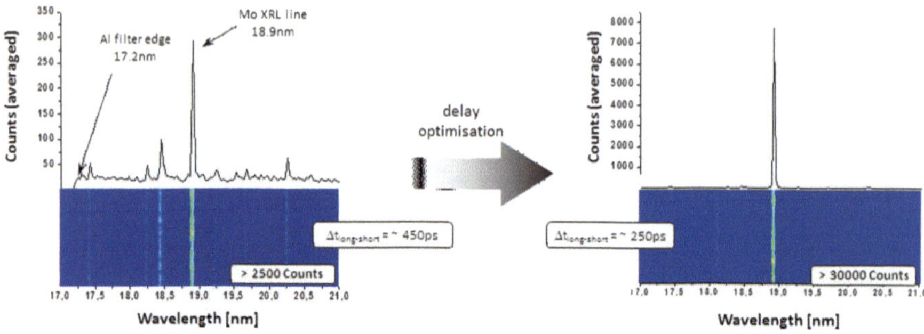

Fig. 5 Spectral output.of the XRL before and after delay optimization. The highest output was achieved for a delay between long and short pulse of 250 ps

as well as the continuous renewing of the target surface. The most stable XRL operation was found if the target surface was renewed after 5–10 laser shots, allowing a continuous operation of some minutes at 100 Hz repetition rate of the pump laser. For a long time operation of the XRL at 100 Hz a continuously moving metal tape has been developed which allows an operation time of some hours [15].

3.1 Results: XRL Output Optimization

The spectral output of the XRL has been measured using a flat field grating spectrometer. It consists of a filter wheel equipped with Al filters (thickness 0.2–1 μm), an entrance slit (100 μm), an aberration corrected concave grating (HITACHI #0437, 1200 l/mm) on a rotational stage and a back-illuminated CCD camera (ANDOR, DO420A-BN, 1024 × 256 pixel). Both CCD camera and grating were calibrated at the Metrology Light Source (MLS) of Physikalisch-Technische Bundesanstalt (PTB). We started our investigations with the optimization of the XRL output in dependence on delay between short and long pulse. For this purpose we measured the delay between the two pulses using a fast photodiode. For each delay adjustment the whole output spectrum has been measured. We have found that the signal peaks at 250 ps delay, very similar to results reported earlier [13] for XRL pumped by a Ti:Sa laser with similar pumping parameters. The spectral output of the XRL at optimum delay is shown on Fig. 5.

4 Conclusions and Outlook

We have shown that using appropriate pump laser concepts based on diode pumped slab or thin disk laser systems high repetition rate highly brilliant soft X-ray source are feasible. Based on an incoherent laser produced plasma source a laboratory full field x-ray microscope allows imaging of cryo plunged frozen cells with a resolution of about 30 nm limited only by the parameters of the objective zone plate.

In future it is planned to use the coherent radiation from our XRL in a FTH experiment using an advanced setup. In contrast to the experiment in [7], we use one diffractive optical element (DOE) to create the illumination of the sample as well as the holographic reference. Due to the DOE the sample and the reference are no longer spatially connected and hence this setup can be used to scan arbitrary large areas. Calculations also predict an increase in resolution and S/N ratio.

Acknowledgement Part of this work (LTXM) has been supported by BMBF program #13N8913. HS and KAJ gratefully acknowledge support by the german-korean collaboration program BMBF #KOR 10/016. HS, RJ, JT and CS gratefully acknowledge support by INREX (LaserLab Europe) program.

References

1. Martz, D.H., Selin, M., von Hofsten O., Fogelqvist, E., et al.: High average brightness water window source for short-exposure cryomicroscopy. Opt. Lett. **37**, 4425–4427 (2012)
2. Legall, H., Blobel, G., Stiel, H., Sandner, W., et al.: Compact x-ray microscope for the water window based on a high brightness laser plasma source. Opt. Express. **20**, 18362–18369 (2012)
3. Vaschenko, G., Brewer, C., Brizuela, F., Wang, Y., et al.: Sub-38 nm resolution tabletop microscopy with 13 nm wavelength laser light. Opt. Lett. **31**, 1214–1216 (2006)
4. Chapman, H.N., Nugent, K.A.: Coherent lensless x-ray imaging. Nat. Photonics. **4**, 833–839 (2010)
5. Sandberg, R.L., Song, C., Wachulak, P.W., Raymondson, D.A., et al.: High numerical aperture tabletop soft x-ray diffraction microscopy with 70-nm resolution. Proc. Natl. Acad. Sci. USA. **105**, 24–27 (2008)
6. Gorniak, T., Heine, R., Mancuso, A.P., Staier, F., et al.: X-ray holographic microscopy with zone plates applied to biological samples in the water window using 3rd harmonic radiation from the free-electron laser FLASH. Opt. Express. **19**, 11059–11070 (2011)
7. Malm, E.B., Monserud, N.C., Brown, C.G., Wachulak, P.W., et al.: Tabletop single-shot extreme ultraviolet fourier transform holography of an extended object. Opt. Express. **21**, 9959–9966 (2013)
8. Sakdinawat, A., Attwood, D.: Nanoscale x-ray imaging. Nat. Photonics. **4**, 840–848 (2010)
9. Brizuela, F., Howlett, I.D., Carbajo, S., Peterson, D., et al.: Imaging at the nanoscale with practical table-top EUV laser-based full-field microscopes. IEEE J. Sel. Top. Quantum Electron. **18**, 434–442 (2012)
10. Brizuela, F., Wang, Y., Brewer, C.A., Pedaci, F., et al.: Microscopy of extreme ultraviolet lithography masks with 13.2 nm tabletop laser illumination. Opt. Lett. **34**, 271–273 (2009)
11. Tümmler, J., Jung, R., Stiel, H., Nickles, P.V., Sandner, W.: High-repetition-rate chirped-pulse-amplification thin-disk laser system with joule-level pulse energy. Opt. Lett. **34**, 1378–1380 (2009)
12. Tuemmler, J., Janulewicz, K.A., Priebe, G., Nickles P.V.: 10-Hz grazing incidence pumped Ni-like Mo x-ray laser. Phys. Rev. E. **72**, 0374011–0374014 (2005)
13. Keenan, R., Dunn, H., Patel, P.K., Price, D.F., et al.: High-repetition-rate grazing-incidence pumped x-ray laser operating at 18.9 nm. Phys. Rev. Lett. **94**, 1039011–1039014 (2005)
14. Luther, B.M., Wang, Y., Larotonda, M.A., Alessi, D., et al.: Saturated high-repetition-rate 18.9-nm tabletop laser in nickellike molybdenum. Opt. Lett. **30**, 165–167 (2005)
15. Stiel, H., Tummler, J., Jung, R., Nickles, P.V., Sandner, W.: X-ray laser takes the 100 Hz barrier. SPIE. **7451**, 7451091–7451098 (2009)

Coherent Diffractive Imaging with a Laboratory-Scale, Gas-Discharge Plasma Extreme Ultraviolet Light Source

J. Bußmann, M. Odstrčil, R. Bresenitz, D. Rudolf, J. Miao, W. S. Brocklesby and L. Juschkin

Abstract Coherent diffractive imaging (CDI) and related techniques enable a new type of diffraction-limited high resolution microscopy and have been widely used in the extreme ultraviolet (EUV) and X-ray communities. In this experiment, we demonstrate CDI using a compact gas-discharge EUV light source with a wavelength of 17.3 nm (oxygen VI emission). Our image reconstruction method accounts for the partial spatial coherence of the radiation using a deconvolution technique. Our results are promising for future laboratory-scale CDI applications, including mask inspection for EUV lithography and EUV metrology.

1 Introduction

Microscopy using extreme ultraviolet (EUV) radiation and soft X-rays has found broad applications [1]. However, the spatial resolution of EUV and soft X-ray microscopy is currently limited by the quality of reflective or diffractive optics. An important approach to overcome this limitation is to measure the far field diffraction pattern from the sample and then directly phase it to obtain an image. This novel imaging technique, known as coherent diffractive imaging (CDI), was first demonstrated by Miao et al. in 1999 [2], in which the well-known phase problem was solved by combining the oversampling method with an iterative algorithm [3, 4].

Since then, various CDI methods have been developed and applied to image a wide range of materials and biological samples [5–8]. The resolution of CDI is

J. Bußmann (✉) · M. Odstrčil · R. Bresenitz · D. Rudolf · L. Juschkin
Experimental Physics of EUV, RWTH Aachen University, JARA-FIT, Steinbachstraße 15, 52074 Aachen, Germany
e-mail: j.bussmann@fz-juelich.de

J. Bußmann · R. Bresenitz · D. Rudolf · L. Juschkin
Peter Grünberg Institut (PGI-9), Forschungszentrum Jülich GmbH, JARA-FIT, 52425 Jülich, Germany

M. Odstrčil · W. S. Brocklesby
Optoelectronics Research Centre, University of Southampton, Southampton SO17 1BJ, UK

J. Miao
Department of Physics and Astronomy, and California Nano Systems Institute, University of California, Los Angeles, CA 90095, USA

© Springer International Publishing Switzerland 2016
J. Rocca et al. (eds.), *X-Ray Lasers 2014*, Springer Proceedings in Physics 169,
DOI 10.1007/978-3-319-19521-6_36

only limited by the highest spatial frequency diffracted from the sample and can thus achieve the diffraction limit. However, most of CDI experiments have been thus far conducted at the 3rd generation synchrotron and newer X-ray free electron lasers facilities. Since obtaining beamtime at these large-scale X-ray facilities is very competitive, it is highly desirable to perform CDI experiments with laboratory-scale light sources. In the EUV and soft X-ray spectral range, there are currently a few laboratory-based light sources with different coherence properties, which include laser-generated high harmonics, EUV lasers and plasma-based light sources. CDI has already been successfully demonstrated with EUV lasers and high harmonics [9, 10] and the first results with a plasma-based EUV light source were also reported by us [11].

Here, we present the first successful demonstration of CDI with a compact, partially coherent plasma EUV light source at 17.3 nm wavelength (O VI 3d-2p spectral line). From experimental data, we estimate the photon flux to be larger than $3 \cdot 10^{13}$ photons/(2π sr) per pulse at 1 kHz repetition rate and 2.3 J discharge energy. In our gas-discharge light source, EUV radiation is emitted by highly ionized atoms in a hot dense plasma [12, 13]. Similar EUV light sources have already been successfully used in a variety of applications such as EUV lithography [14, 15], reflectometry [16], EUV and soft X-ray microscopy [17–19], and magnetic polarization spectroscopy [20]. Tuning of the photon energy is possible by using emission lines from different working gases and ionization levels, e.g. O IV to O VI lines covering the 3p absorption edges of Fe, Co and Ni between 17 and 25 nm [20]. Therefore, our EUV light sources can potentially enable element-specific contrast, in particular in CDI experiments with intense spectral lines at atomic absorption edges.

2 Experimental Setup for Coherent Diffraction Imaging

The setup consists of the gas-discharge EUV light source, a pinhole to adjust the spatial coherence, a Bragg mirror to ensure temporal coherence, the sample and an EUV-sensitive CCD camera, which are installed in a high vacuum system (Fig. 1a). The distance between the light source and the pinhole (aperture with a diameter of 500 μm) was set to 40 cm, the pinhole-sample distance was 100 cm, and the sample-detector distance was 4.8 cm. From pinhole camera measurements, we obtained 700 μm for the light source size (FWHM of the Lorentz profile). The effective light source diameter seen by the sample through the pinhole was about 700 μm as well. However, the pinhole improved the spatial coherence of the light source by cropping the tails of the Lorentz profile. The transverse coherence length on the sample was estimated to be about 15 μm.

The pinch plasma emits multiple emission lines within a broad spectral range (EUV, VUV and visible light). To select a narrow-bandwidth part of the spectrum around the 17.3 nm line, and therefore to ensure a high degree of temporal coherence, we used a $[Si(8.04\,nm)/B_4C(5.36\,nm)]_{\times 50}$ multilayer Bragg mirror placed at 44° grazing incidence. The mirror tilt was optimized for high reflectivity using a

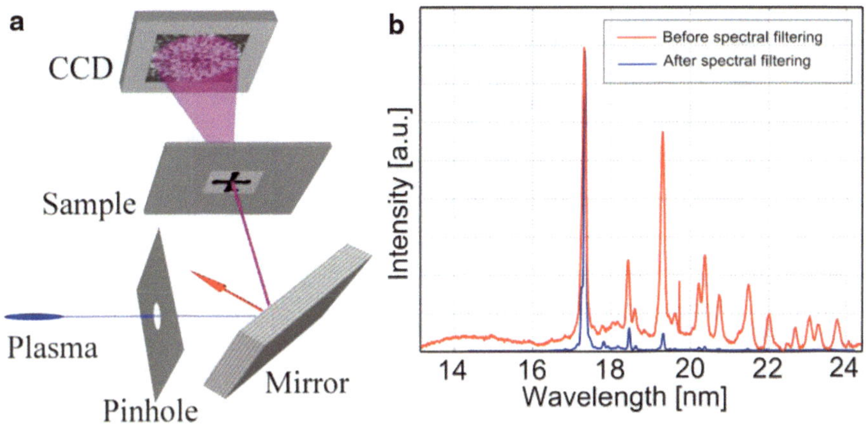

Fig. 1 **a** Experimental setup (not to scale). **b** Normalized spectrum of the oxygen plasma before and after reflection from the multilayer mirror with a strong O VI line at 17.3 nm

piezo-driven mirror holder (Smaract STT-25). The spectrum (Fig. 1b) was recorded by a spectrometer with a curved, blazed, gold-coated grating (1200 lines/mm, $\lambda/\Delta\lambda \approx 10^3$).

In our CDI measurements, the sample was a non-symmetric cross-like structure milled using a focused ion beam into an Au(500 nm)/Ti(10 nm) bilayer on Si_3N_4(50 nm) substrate (TedPella 21501), (Fig. 2a). Light diffracted from the sample was recorded by a back-illuminated CCD camera (Andor iKon-M, 1024 × 1024 pixels^2, 13 μm pixel size). We adjusted the camera settings to the slowest read-out rate (50 kHz), highest gain and a sensor temperature of $-50\,°C$ to reduce noise. Readout noise and dark current were measured prior to the data acquisition and subtracted directly from the recorded data. Hot pixels were replaced by their neighboring values and also cosmic rays have been removed. Multiple images were first acquired with different exposure times and then stitched together to extend the dynamic range. The maximum exposure time was 600 s ($\sim 6{\cdot}10^5$ pulses), with the final image containing about $6{\cdot}10^6$ photons (Fig. 2b). Only the central region of the CCD (256 × 256 pixels^2) was read out and used for the reconstruction.

3 Image Reconstruction Method

Reconstruction was performed using the Hybrid Input-Output algorithm [4] with the Shrinkwrap method [21]. The initial guess was estimated from the autocorrelation function support based on the triple-intersection method [22]. The measured data were binned to get pixel oversampling equal to 3.

Because the coherence was not sufficient for direct reconstruction, a modified modulus constraint [23].

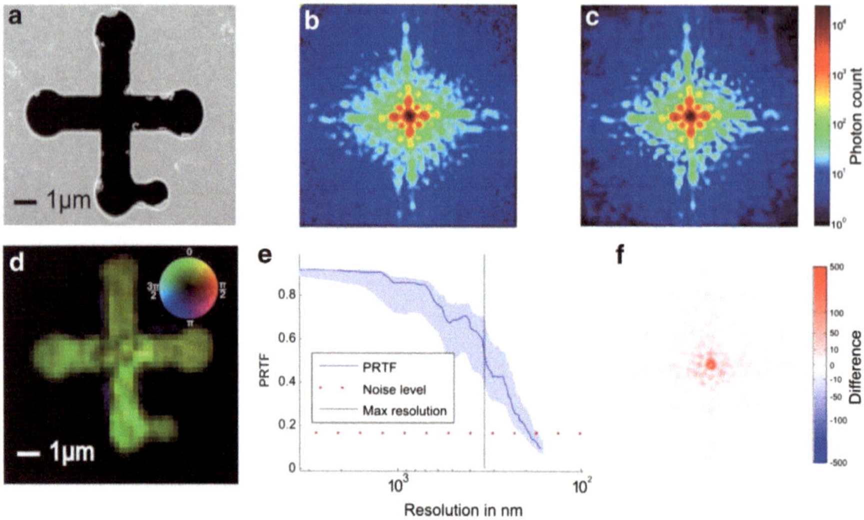

Fig. 2 **a** SEM image of the FIB-milled sample. Measured **b** and retrofitted **c** diffraction pattern on logarithmic scale. The color bar denotes the number of photons per pixel. **d** Reconstructed object. Inset: Color-coded phase and amplitude. **e** Phase Retrieval Transfer Function (PRTF) [22] used to determine the lateral resolution. The blue-shaded area denotes the variance of all successful reconstructions. **f** Square root of the difference between the diffraction patterns in **b** and **c**

$$\hat{\Psi}(k) = \sqrt{\frac{I(k)}{|\Psi(k)|^2 * g}}\Psi(k) \tag{1}$$

was used in every iteration step. Here, $\Psi(k)$denotes the complex-valued diffracted wave without the constraint, $I(k)$ denotes the measured diffraction intensity, g is the blurring kernel and $*$ is the convolution operator. We used a Gaussian kernel and dynamically estimated its standard deviation termed coherence factor during the reconstruction in order to minimize the misfit between the data and the model. This method significantly depends on the initial input. However, the initial support estimation from autocorrelation was precise enough to ensure stable reconstruction. Half of the reconstructions with lowest misfit were aligned in phase and value and the final reconstruction was obtained as a median of the selected reconstructions.

4 Results

The measured and retrofitted ($|\Psi(k)|^2 * g$ from Eq. 1) diffraction intensities are displayed in Fig. 2b, 2c.

The reconstructed image is shown in Fig. 2d, where the color represents the phase (inset) and the saturation represents the amplitude of the complex-valued sample transmission function. The image shows a spatial distribution of two phases, i.e. either no EUV light transmission (black) through the Au/Ti/Si_3N_4 layers or transmission through the cross-like structure milled through the layers (green). We determined the lateral resolution from the Phase Retrieval Transfer Function (PRTF) [24] to be (330 ± 50) nm.

The PRTF together with the variance of all successful reconstructions is presented in Fig. 2e. In Fig. 2f, the square root of the difference between the diffraction patterns in Fig. 2b and 2c indicates that the reconstruction works well for high spatial frequencies, whereas the center of the retrofit does not match well the measured pattern. Possible reasons explaining the deviation of the diffraction patterns in Fig. 2b, 2c can be the imperfect spatial coherence of the EUV light source and also imperfect spectral filtering by the multilayer mirror. In our experiment, the limiting factor for lateral resolution is the spatially coherent photon flux [11] as only a very small fraction of emitted photons was used for CDI. Possible ways to increase the coherent photon flux include increasing the pulse radiation energy, reducing the emitting source size, and increasing the repetition rate.

5 Summary

In conclusion, for the first time, we demonstrated a successful CDI experiment employing only partially coherent radiation from a laboratory-based gas-discharge EUV source. We imaged a simple cross structure at 17.3 nm wavelength (O VI 3d-2p line) with a spatial resolution of (330 ± 50) nm. To push the spatial resolution well below 100 nm, both the coherence and photon flux from the gas-discharge EUV light source have to be increased. In the next step, we plan to perform scanning CDI (Ptychography) of objects with a complex-valued transmission function. In the future, CDI and related methods with laboratory-based, partially coherent EUV light sources can be applied, e.g., for EUV metrology of lithography masks.

References

1. Sakdinawat, A., Attwood, D.: Nanoscale X-ray imaging. Nat. Photon. **4**, 840–848 (2010)
2. Miao, J., et al.: Extending the methodology of X-ray crystallography to allow imaging of micrometre-sized non-crystalline specimens. Nature. **400**, 342–344 (1999)
3. Miao, J., et al.: Phase Retrieval from the Magnitude of the Fourier transform of Non-periodic Objects. JOSA A. **15**, 1662–1669 (1998)
4. Fienup, J.R.: Reconstruction of an object from the modulus of its Fourier transform. Opt. Lett. **3**, 27–29 (1978)
5. Miao, J., et al.: Coherent X-ray diffraction imaging. IEEE J. Sel. Top. Quant. Electron. **18**, 399–410 (2012)
6. Spence, J.C.H.: Diffractive (lensless) imaging. In: Hawkes, P., Spence, J.C.H. (eds.) Science of Microscopy, pp. 1196–1227. Springer, New York (2007)
7. Nugent, K.: Coherent methods in the X-ray sciences. Adv. Phys. **59**, 1–99 (2010)

8. Chapman, H.N., Nugent, K.A.: Coherent lensless X-ray imaging. Nat. Photon. **4**, 833–839 (2010)
9. Sandberg, R.L., et al.: High numerical aperture tabletop soft x-ray diffraction microscopy with 70-nm resolution. PNAS. **105**, 24–27 (2008)
10. Seaberg, M.D., et al.: Ultrahigh 22 nm resolution coherent diffractive imaging using a desktop 13 nm high harmonic source. Opt. Expr. **19**, 22470–22479 (2011)
11. Juschkin, L., et al.: Tabletop coherent diffraction imaging with a discharge plasma EUV source. Proc. SPIE. **8849**, 88490Y (2013)
12. Bergmann, K., et al.: Highly repetitive, extreme-ultraviolet radiation source based on a gas-discharge plasma. Appl. Opt. **38**, 5413–5417 (1999)
13. Benk, M., Bergmann, K.: Brilliance scaling of discharge sources for extreme-ultraviolet and soft x-ray radiation for metrology applications. J. Micro/Nanolith. MEMS MOEMS. **11**, 021106 (2012)
14. Bergmann, K., et al.: Optimization of a gas discharge plasma source for extreme ultraviolet interference lithography at a wavelength of 11 nm. J. Appl. Phys. **106**, 073309 (2009)
15. Danylyuk, S., et al.: Diffraction-assisted extreme ultraviolet proximity lithography for fabrication of nanophotonic arrays. J. Vac. Sci. Technol. B. **31**, 021602 (2013)
16. Banyay, M., Juschkin, L.: Table-top reflectometer in the extreme ultraviolet for surface sensitive analysis. Appl. Phys. Lett. **94**, 063507 (2009)
17. Juschkin, L., et al.: EUV microscopy for defect inspection by dark-field mapping and zone plate zooming. J. Phys.: Conf. Ser. **186**, 012030 (2009)
18. Juschkin, L.: Imaging with plasma based extreme ultraviolet sources. Proc. SPIE. **8678**, 8678F (2012)
19. Benk, M., et al.: Compact soft x-ray microscope using a gas-discharge light source. Opt. Lett. **33**, 2359 (2008)
20. Wilson, D., et al.: Generation of circularly polarized radiation from a compact plasma-based extreme ultraviolet light source for tabletop X-ray magnetic circular dichroism studies. Rev. Sci. Inst. **85**, 103110 (2014)
21. Marchesini, S., et al.: X-ray image reconstruction from a diffraction pattern alone. Phys. Rev. B. **68**, 140101 (2003)
22. Crimmins, R., et al.: Improved bounds on object support from autocorrelation support and application to phase retrieval. JOSA A. **7**, 3–13 (1990)
23. Clark, J.N., Peele, A.G.: Simultaneous sample and spatial coherence characterisation using diffractive imaging. Appl. Phys. Lett. **99**, 154103 (2011)
24. Chapman, H.N., et al.: High-resolution ab initio three-dimensional X-ray diffraction microscopy. J. Opt. Soc. Am. A. **23**, 1179–1200 (2006)

On the Optical Transforms in the Fourier Space for Simulation of Coherent Imaging of Tilted Objects

I. A. Artyukov, N. L. Popov and A. V. Vinogradov

Abstract Coherent reflection imaging at slant and grazing angles attracted attention during past years (Fenter et al., J Synchrotron Rad 15:558–571 2008; Marathe et al., Opt Express 18:7253–7262 2010; Roy et al., Nat Photonics 5:243–245 2011; Sun et al., J Nat Photon 6:586–590 2012; Gardner et al., Opt Express 20:19050–19059 2012; Zürch, Opt Express 21(18):21131–21147, 2013; Zhang et al., Extended reflection coherent diffraction imaging of nanostructures on a tabletop, Bulletin of the American Physical Society, APS March Meeting: Vol. 59, number 1, Monday–Friday, March 3–7, 2014, Denver). It is a new imaging technology allowing X-ray study of near surface layers of thick and opaque samples or nanostructures and thin samples fixed on a reflecting substrate. This work develops diffraction theory and calculations relevant to coherent reflection imaging of tilted objects illuminated below critical angles of X-ray reflection.

1 The Concept

Coherent imaging implies that the optical field at the object surface (plane) is known and have to be recorded at a detector after propagation through an optical system or in the free space (lensless imaging). So simulation of image formation is a diffraction problem. Various approaches can be used to solve it: direct numerical solution of wave equation or application of approximations such as parabolic wave equation (PWE), Fresnel–Kirchhoff, Rayleigh–Sommerfeld and Fresnel integral (FI) formulas etc. Each approximate method has its own applicability area and these areas can partly intercept as it is studied in for example [8–11]. For simulation of image formation of tilted objects we use the approach of PWE:

$$2ik\frac{\partial u(x,z)}{\partial z} + \frac{\partial^2 u(x,z)}{\partial x^2} = 0 \quad (1)$$

here u(x, z) is a slowly varying factor in electromagnetic wave amplitude. There are three reasons for this choice. (1) PWE is valid for low divergent (paraxial) beams like those provided by modern coherent sources of X-rays: laboratory or

A. V. Vinogradov (✉) · I. A. Artyukov · N. L. Popov
Lebedev Physical Institute, Leninsky Prospekt 53, Moscow, 119991, Russia
e-mail: vinograd@sci.lebedev.ru

© Springer International Publishing Switzerland 2016

J. Rocca et al. (eds.), *X-Ray Lasers 2014*, Springer Proceedings in Physics 169,
DOI 10.1007/978-3-319-19521-6_37

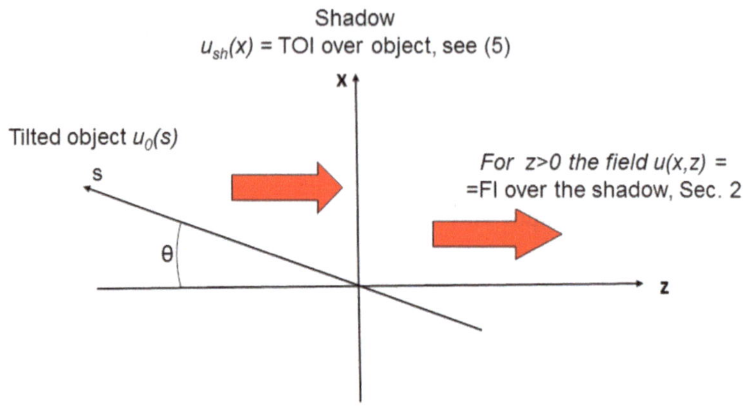

Fig. 1 To the theory of coherent imaging of a tilted object (Sect. 2)

free-electron lasers and high order harmonic generators of visible laser radiation. (2) Application of PWE basically guarantees that the energy conservation law is satisfied that is very attractive for simulations. (3) For objects located normally to the wave k-vector PWE is equivalent to Fresnel integral (FI) approximation which is widely used in imaging and optical sciences, confirmed in numerous experiments and is well elaborated for simulations and computing [15].

2 Analytical Approach, Cartesian Space

Initial (boundary) condition to (1) is posed on the plane of an object tilted to an angle θ (see Fig. 1):

$$u(x,z)\left|_{\mathrm{x}=-\mathrm{ztg}\theta}\right. = u_0(z) \tag{2}$$

For $\theta = \pi/2$ the solution to problem (1) and (2) is the Fresnel integral (FI):

$$u(x,z) = \sqrt{\frac{k}{2\pi iz}} \int\limits_{-\infty}^{\infty} u_0(s)ds \exp\left[i\frac{k(x-s)^2}{2z}\right]. \tag{3}$$

For arbitrary $\theta \neq \pi/2$ the solution to the same problem (1) and (2) is the tilted object integral (TOI) [12]:

$$u(x,z) = \sqrt{\frac{k}{2\pi i}}(x\cos\theta + z\sin\theta) \int\limits_{-z/\cos\theta}^{\infty} \frac{u_0(s)ds}{(z+s\cos\theta)^{3/2}} \exp\left[i\frac{k(x-s\sin\theta)^2}{2(z+s\cos\theta)}\right], \tag{4}$$

Where s is the coordinate on the object plane.

Formula (4) gives the field distribution in the halfspace $x > -z \cdot tg\theta$, see Fig. 1.

Suppose now that $s = 0$ is the right edge of the object, so that there is no object field at s < 0. Then we introduce "shadow field" $u_{sh}(x) = u(x, z = 0)$ and obtain from (4):

$$u_{sh}(x) = \sqrt{\frac{k}{2\pi i \cos\theta}} x \int_0^\infty \frac{u_0(s)ds}{s^{3/2}} \exp\left[i \frac{k(x - s\sin\theta)^2}{2s\cos\theta} \right], \ x > 0. \tag{5}$$

Shadow field (5) is an evident initial distribution for further Fresnel propagation in the free space towards a detector or an entrance aperture of optical system. Due to its evolution character PWE determines the shadow field only for $x > 0$. On the half axis $x < 0$ for calculation of FI we take $u_{sh}(x) = 0$ or use other *a priori* information if it is available.

The iterative lensless imaging algorithms based on Fresnel propagation theory use both direct and inverse Fresnel transforms [18]. Therefore it may be of interest to consider the inverse problem to Eq. (5). To our knowledge this problem is not solved yet. However as was shown in [19] the object field satisfies the integral equation:

$$u_0(s) - \frac{i\sqrt{s}}{\pi} \int_0^\infty \frac{u_0(t)\exp\left\{\frac{ik(t-s)\sin^2\theta}{2\cos\theta}\right\} dt}{\sqrt{t}(t-s)}$$
$$= \sqrt{\frac{2ik}{\pi s\cos\theta}} \int_0^\infty \exp\left\{ -\frac{ik(x - s\sin\theta)^2}{2s\cos\theta} \right\} u_1(x)dx, \ s > 0,$$

that can be used in iteration procedure.

Note that as was shown in [13] the field of a tilted object (4) can be found as a coordinate transformation of a well-known heat conductivity problem [14]. Introduce:

$$w(\xi, z) = \exp\left(ikxtg\theta + ikz\frac{tg^2\theta}{2} \right) u(x, z), \ x = \xi - ztg\theta. \tag{6}$$

It can be convinced that the new field w(ξ,z) instead of (1), (2) satisfies standard PWE but with the boundary condition posed on the horizontal axis:

$$2ik\frac{\partial w(\xi, z)}{\partial z} + \frac{\partial^2 w(\xi, z)}{\partial \xi^2} = 0, \ w(\xi, z)\,|_{\xi=0} = w_0(z) = e^{-ikztg^2\theta/2} u_0(z). \tag{7}$$

Hence at any tilt angle $\theta \neq \pi/2$ the field of an object (see (4)) can be alternatively by simple algebra found from the limiting case $\theta = 0$. The latter according to (4) (see also [14]) equals to:

$$w(\xi, z) = \sqrt{\frac{k}{2\pi i}} \xi \int_0^\infty \frac{w_0(s)ds}{s^{3/2}} \exp\left[i\frac{k\xi^2}{2(z+s)} \right]. \tag{8}$$

3 Analytical Approach, Fourier Space

The Fourier space is often used in optical computing [15–17]. Free space propagation from the shadow plane of the field Fourier transform $u_{sh}(p)$ is given by simple formula [18] that is equivalent to (3):

$$u(p,z) = u_{sh}(p)\exp\left\{-i\frac{p^2}{2k}z\right\}, \tag{9}$$

$$u_{sh}(p) = \frac{1}{2\pi}\int_{-\infty}^{\infty} dx e^{-ipx} u_{sh}(x). \tag{10}$$

If an object is normal to the wave k-vector, then $u_{sh}(x)$ equals the object field $u_0(x)$.

But if an object is tilted to the wave k-vector then $u_{sh}(p)$ has to be found using the above theory, i.e. by formula (5) and then by introducing (5) into (10). Following the note (6)–(8) it is sufficient to consider the limiting case $\theta = 0$ in (5):

$$u_{sh}(x) = \sqrt{\frac{k}{2\pi i}}x\int_0^{\infty}\frac{u_0(s)\exp\left\{\frac{ikx^2}{2s}\right\}ds}{s^{3/2}}, \quad x > 0. \tag{11}$$

Continue both $u_0(s)$ and $u_{sh}(x)$ to negative values:

$$u_0(s) = 0, \quad s < 0 \text{ and } u_{sh}(x) = 0, \quad x < 0 \tag{12}$$

and introduce Fourier transforms

$$u_0(q) = \frac{1}{2\pi}\int_0^{\infty} ds e^{-iqs}u_0(s),\ u_0(s) = \int_{-\infty}^{\infty} dq e^{iqs}u_0(q),$$

$$\text{with } u_0(q) = \frac{i}{\pi}P\int_{-\infty}^{\infty}\frac{dq' u_0(q')}{q'-q}, \tag{13}$$

and

$$u_{sh}(p) = \frac{1}{2\pi}\int_0^{\infty} dx e^{-ipx}u_{sh}(x),\ u_{sh}(x) = \int_{-\infty}^{\infty} dp e^{ipx}u_{sh}(p),$$

$$\text{with } u_{sh}(q) = \frac{i}{\pi}P\int_{-\infty}^{\infty}\frac{dq' u_{sh}(q')}{q'-q}. \tag{14}$$

Then two formulas equivalent to (11) can be obtained for $u_{sh}(x)$ in terms of the object field Fourier transform $u_0(q)$:

$$u_{sh}(x) = \int_{-\infty}^{\infty} dq u_0(q) e^{i\sqrt{2kq}x}, \tag{15}$$

plane wave method [13],

where $\surd q = +\surd q$ for positive and $\surd q = +i\surd(-q)$ for negative q, and

$$u_{sh}(x) = \int_{-\infty}^{\infty} dp e^{ipx} \frac{p}{k} u_0 \left(\frac{p^2}{2k}\right), \quad x > 0, \tag{16}$$

"stretched" Fourier transform.

All three formulas (11), (14) and (15) determine $u_{sh}(x)$ for $x > 0$. As was mentioned in Sect. 2 if no *a priori* information about $u_{sh}(x)$ at negative x is available then the lower limit of the integral (10) have to be replaced by 0.

4 Simulation Results

Formulas obtained in Sect. 3 were implemented in computer codes for simulation of diffraction patterns and coherent images produced by tilted objects. The most convenient and important for modeling is "stretched" Fourier transform (16). The generalization of (16) to arbitrary angle θ and 3D geometry has the form:

$$u_{sh}(x, y) = \int_{-\infty}^{\infty} dp_x dp_y e^{ip_x x + p_y y} \left(\sin\theta + \frac{p_x}{k}\cos\theta\right) u_0 \left(p_x \sin\theta + \frac{{p_x}^2}{2k}\cos\theta,\, p_y\right). \tag{17}$$

In order to use FFT, it is necessary to set a regular grid. Unfortunately, Eq. (17) eliminates the possibility of determination $u_0(p_s,\ p_y)$ and $u_{sh}(p_x,\ p_y)$ on regular grids. Therefore, the accuracy of the calculations $u_{sh}(x,\ y)$ and $u_{sh}(p_x,\ p_y)$ is greatly influenced by the accuracy of the interpolation function $u_0(p_s,\ p_y)$ used. In this paper, a Taylor series expansion in the neighborhood of the grid point is utilized for interpolation. Calculation of the derivatives $\partial_{m,n}(u_0(p_s, p_y))$, was reduced to finding the Fourier image (FFT) of the $(-i)^{m+n} s^m y^n\ u_0(s,\ y)$. The maximum order of the derivatives was 4 that consumes 15 times more memory. But it enables calculating $u_{sh}(p_x,\ p_y)$ only 15 times slower than a single FFT, with a relative error of $\sim 10^{-6}$. The relative error was determined by a comparison with a Laguerre–Gaussian beam, which is an analytical solution of PWE, see Fig. 2.

4.1 *Diffraction Patterns Produced by Tilted Objects*

Figure 3 shows the 3D layout of the object and the detector. The 2D amplitude of the object has the shape of a human profile. The phase of the object is taken to be zero. Under the scheme the field amplitude in a linear scale is shown, with white background corresponding to zero. The geometric parameters are: $\lambda = 10\,\text{nm}$,

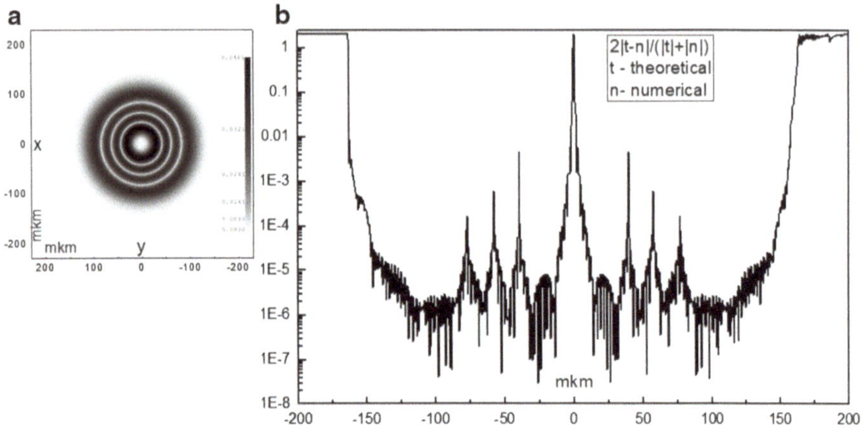

Fig. 2 The comparison of the theoretical and the numerical field $u_{sh}(x, y)$. **a**—amplitude of the Laguerre–Gaussian beam (l,p = 5,3). **b**—relative error. $\lambda = 1.06$ mkm, $\theta = 55°$

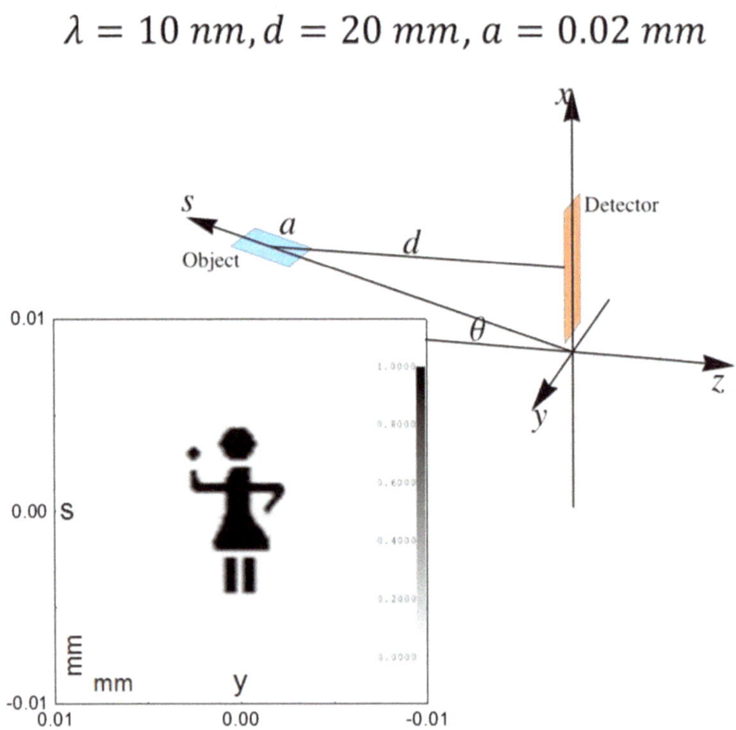

Fig. 3 The scheme for 3D diffraction modelling and the picture displayed in the tilted plane (the phase equals to zero)

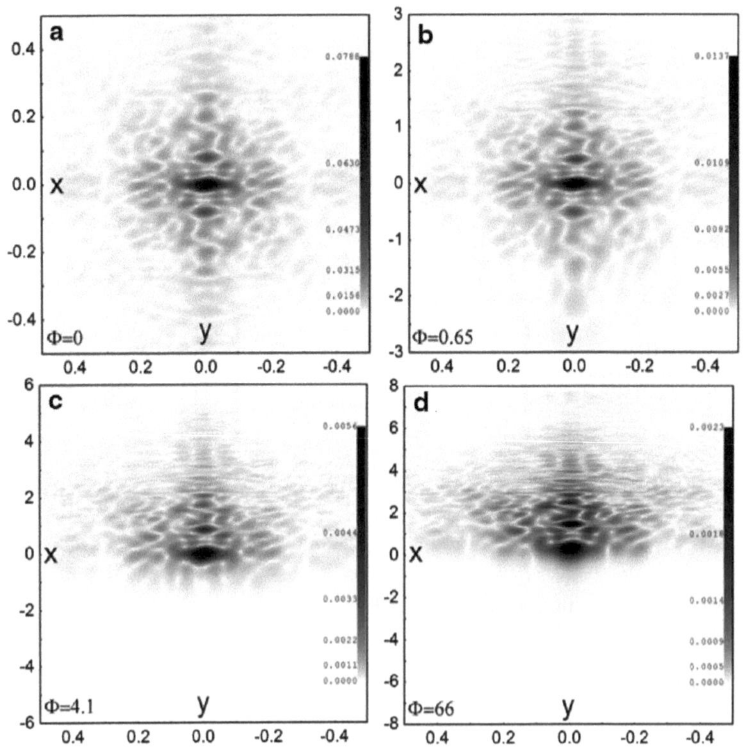

Fig. 4 Diffraction patterns produced by an object for different tilt angles: $\theta = 90°$—(**a**), $\theta = 10°$—(**b**), $\theta = 1°$—(**c**) and $\theta = 4°$—(**d**). The value of $\Phi = 2\lambda\cos\theta/(\delta\sin^2\theta)$ is responsible for the influence of tilting on the diffracted field structure, δ is the object feature size [19]

$a = 0.02$ mm, $d = 20$ mm, that roughly corresponds to far-field diffraction $d \approx 0.5$ (a^2/λ). Figure 4 shows the result of calculating the diffraction field for various tilt angles θ between object and beam. The 2D profiles of the field amplitude are shown in a logarithmic scale. The qualitative changes of diffraction picture appear at $\theta \leq 4°$. Diffraction pattern becomes blurred and asymmetric.

4.2 Coherent Image Behaviour as a Function of Tilt Angle

Location of an object (Siemens star), the lens and the detector is shown in Fig. 5. Wavelength is $\lambda = 10$ nm, and the numerical aperture NA = 0.2. Figure 6 shows the calculation result of the field amplitude in a logarithmic scale, depending on the angle θ. The image of the Siemens star remains recognizable even for the angle $\theta = 0.5°$.

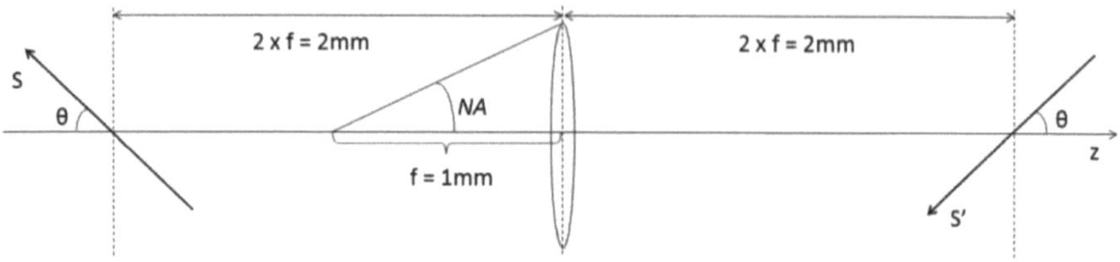

Fig. 5 The scheme for lens 3D coherent imaging

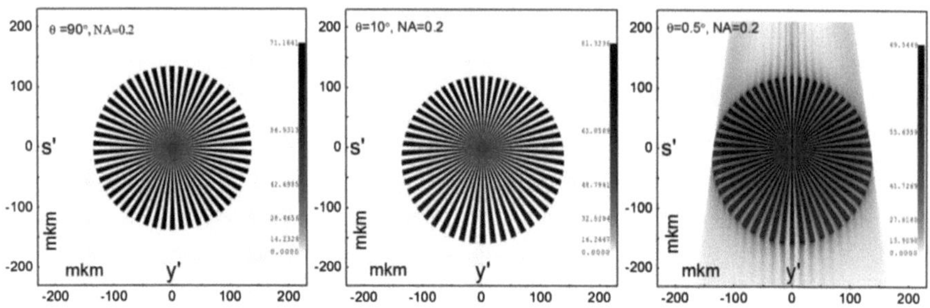

Fig. 6 Coherent image as function of tilt angle θ

5 Summary

1) Parabolic wave equation is used to derive TOI (tilted object integral)—the extension of Fresnel integral to diffraction from tilted objects.
2) TOI for arbitrary tilt angle $\theta \neq \pi/2$ is expressed by simple algebra through TOI at $\theta = 0$.
3) The concept of "shadow" field is suggested for simulation of lensless and lens coherent imaging of tilted objects.
4) High order interpolation provides the accuracy of shadow field computing $\sim 10^{-6}$. This serves as an input to Fresnel codes for further field propagation to the detector plane.
5) The results may be useful for simulation of optical systems designed for objects investigated under critical angles of X-ray reflection.

Acknowledgement The work was supported in part by RFBR Grant, no. 130201009 A. The authors are indebted to I N Bukreeva, A S Busarov, I V Epatko and R M Feshchenko for fruitful discussions.

References

1. Fenter, F., Park, C., Kohli, V., Zhang, Z.: Image contrast in x-ray reflection interface microscopy: comparison of data with model calculations and simulations. J. Synchrotron Rad. **15**, 558–571 (2008)
2. Marathe, S., Kim, S.S., Kim, S.N., Kim, C., Kang, H.C., Nickles, P.V., Noh, D.Y.: Coherent diffraction surface imaging in reflection geometry. Opt. Express. **18**, 7253–7262 (2010)

3. Roy, S., Parks, D., Seu, K.A., Su, R., Turner, J.J., Chao, W., Anderson, E.H., Cabrini, S., Kevan, S.D.: Lensless x-ray imaging in reflection geometry. Nat. Photonics. **5**, 243–245 (2011)
4. Sun, T., Jiang, Z., Strzalka, J., Wang, O.L.: Three-dimensional coherent x-ray surface scattering imaging near total external reflection. J. Nat. Photon. **6**, 586–590 (2012)
5. Gardner, D.F., Zhang, B., Seaberg, M.D., Martin, L.S., Adams, D.E., Salmassi, F., Gullikson, E., Kapteyn, H., Murnane, M.: High numerical aperture reflection mode coherent diffraction microscopy using off-axis apertured illumination. Opt. Express. **20**, 19050–19059 (2012)
6. Zürch, M.: Christian Kern and Christian Spielmann. Opt. Express. 21(18), 21131–21147, 2013
7. Zhang B., Seaberg, M., Gardner, D., Shanblatt, E., Murnane, M., Kapteyn, H., Adams, D.: Extended reflection coherent diffraction imaging of nanostructures on a tabletop, Bulletin of the American Physical Society, APS March Meeting: Vol. 59, number 1, Monday–Friday, March 3–7, 2014; Denver. http://meetings.aps.org/link/BAPS.2014.MAR.Y50.14 (2014)
8. John, C.H.: scalar rayleigh-sommerfeld and kirchhoff diffraction integrals: a comparison of exact evaluations for axial points. JOSA. **63**(8), 1003–1008 (1973)
9. Vinogradov, A.V., Popov, A.V., Kopylov, Yu.V., Kurokhtin, A.N.: Numerical Simulation of X-Ray Diffractive Optics. A & B Publishing House, Moscow (1999)
10. Luckt, R.L.: Rayleigh–Sommerfeld diffraction and Poisson's spot. Euro. J. Phys. **27**, 193–204 (2006)
11. Chang, H.-T.: Foundations of scalar diffraction theory, wave optics analysis of coherent optical systems, http://teacher.yuntech.edu.tw/htchang/Chap. 3.pdf
12. Artyukov, I.A., Popov, A.V., Vinogradov, A.V.: Wave field transformation at coherent imaging of a flat reflection mask, Proc. SPIE. **7451**, 745114-1-745114-3 (2009)
13. Artyukov, I.A., Mitrofanov, A.N., Popov A.V., Vinogradov A.V., Theory and Computation Towards Coherent Reflection Imaging of Tilted Objects, Proc. 12th Int. Conf. on X-ray Lasers, pp. 329–340. Springer, Berlin (2010)
14. Carslaw, H.S., Jaeger, J.C.: Conduction of Heat in Solids, 2nd edn. Clarendon, Oxford (1959)
15. Kelly, D.P.: Numerical calculation of the Fresnel transform. JOSA A. **31**, 755–764 (2014)
16. Delen, N., Hooker, B.: Free-space beam propagation between arbitrarily oriented planes based on full diffraction theory: a fast Fourier transform approach. J. Opt. Soc. Am. A. **15**(4), 857–867(1998)
17. Modregger, P., Lubbert, D., Schafer P.,Kohler R., Weitkamp, T., Hanke, M., Baumbach, T.: Fresnel diffraction in the case of an inclined image plane. Opt. Express. **16**(7), 5141–5149 (2008)
18. Paganin, D.M.: Coherent X-ray Optic. Clarendon Press, Oxford (2005)
19. Artyukov, I.A., Feshchenko, R.M., Popov, N.L., Vinogradov, A.V.: Coherent scattering from tilted objects. J. Opt. **16**, 035703 (2014)

Part VI
EUV Lithography

EUV Research at Berkeley Lab: Enabling Technologies and Applications

Patrick P. Naulleau, Christopher N. Anderson, Weilun Chao, Peter Fischer, Kenneth A. Goldberg, Eric M. Gullikson, Ryan Miyakawa, Seong-Sue Kim, Donggun Lee and Jongju Park

Abstract The tremendous progress in the development and deployment of lab scale extreme ultraviolet (EUV) sources over the past decade has opened up the door to a wide variety of new users beyond the traditional synchrotron community. The practical use of such sources, however, is heavily dependent on the availability of EUV optical components. In this manuscript, we describe recent advances at Berkeley Lab in the development of reflective and diffractive optical structures for imaging, wavefront encoding, metrology, spectral filtering, and more.

1 Introduction

Advances in high harmonic generation (HHG) and extreme ultraviolet (EUV) laser sources [1–4] have enabled the spread of coherent EUV applications beyond synchrotrons. Such applications often rely on the availability of specialized EUV optical components such as multilayer mirrors and diffractive optics. Although such components have been developed over the years for synchrotron use, specific application to lab-scale sources has driven new developments. Here we provide an overview of some of these developments including in the area of broadband optics, spectral filtering optics, and high efficiency zoneplates. We also describe an EUV microscopy application making use of such components with an HHG source.

2 Multilayer Optics

A key enabling technology for EUV optics is the multilayer coating [5], [6]. In addition to serving as high efficiency mirrors, multilayers can also act as spectral filters. For example when using HHG sources with diffractive optics, it may be necessary

P. P. Naulleau (✉) · C. N. Anderson · W. Chao · P. Fischer · K. A. Goldberg · E. M. Gullikson · R. Miyakawa
Center for X-Ray Optics, Berkeley Lab, Berkeley, 94720, CA, USA
e-mail: pnaulleau@lbl.gov

S.-S. Kim · D. Lee · J. Park
Samsung Electronics Co., Ltd., Hwasung, 445-701, Gyeonggi, Korea

© Springer International Publishing Switzerland 2016

J. Rocca et al. (eds.), *X-Ray Lasers 2014*, Springer Proceedings in Physics 169,
DOI 10.1007/978-3-319-19521-6_38

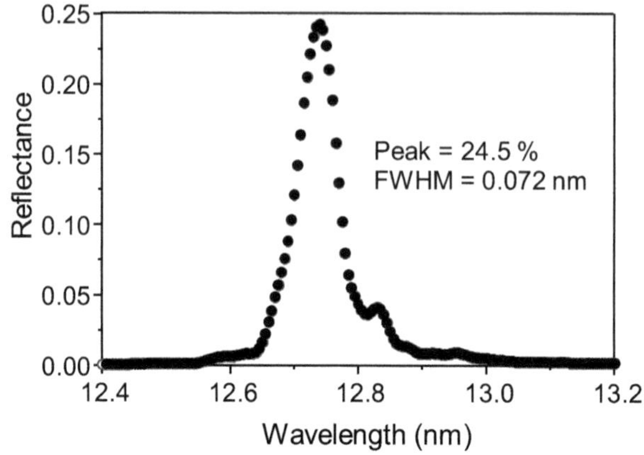

Fig. 1 Reflectance of line-narrowing multilayer achieving $\lambda/\Delta\lambda$ of 175

to narrow the intrinsic bandwidth of the source. The relative bandwidth reported in the literature [7] from high efficiency HHG sources is typically in the 50–100 range significantly restricting the number of zones one can use in the zoneplate. Figure 1 shows the results from the development of a line narrowing multilayer. The mirror is designed for a centroid wavelength of 12.7 nm and an angle of incidence of 6°. The mirror achieved a relative bandwidth ($\lambda/\Delta\lambda$)of 175 and peak reflectance of 24.5 % representing a bandwidth reduction of approximately 10 $\times$ compared to a conventional multilayer mirror with a loss in reflectivity of only about 2.6 $\times$. The Mo/Si mirror was fabricated using magnetron sputtering and is comprised of 300 bilayers and has a gamma (ratio of Mo thickness to bilayer thickness) of 0.125. The measurements in Fig. 1, as well as all subsequently reported multilayer measurements, were performed using the Center for X-ray Optics Calibrations and Standards beamline at the Advanced Light Source synchrotron facility.

Even when the intrinsic bandwidth of a single harmonic is suitable, the use of HHG source may require the extraction of a specific harmonic. Multilayers can also be designed for this purpose by achieving the optimal balance between reflecting the target harmonic and suppressing adjacent harmonics. To this end, we attempt to place harmonic at nulls in the multilayer wavelength-dependent reflectance. Figure 2 shows the reflectance curve for such a mirror which was also specified to be a approximately 45° turning mirror. The locations of the harmonics are shown as red dashed lines in the plot. The mirror has a peak reflectivity of 42 %.

When working with ultrashort (broadband) pulses, it may in fact be required to have extended bandwidth multilayers to preserve the pulse width. This however will come at the cost of reflectivity. Figure 3 shows a broadband Mo/Si mirror designed for near 14-nm operation and achieving a reflectivity of approximately 20 % with a bandwidth of 3 nm. The Mo/Si mirror is comprised of 100 layers total in an aperiodic configuration determined using a model based optimization process [8].

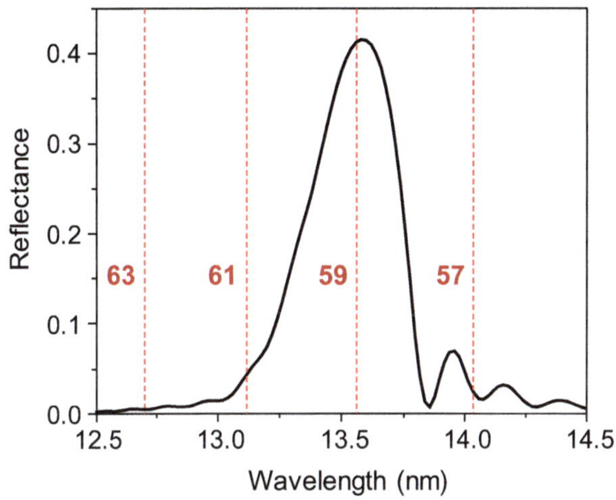

Fig. 2 Reflectance of harmonic selecting multilayer achieving peak reflectivity of 42 %

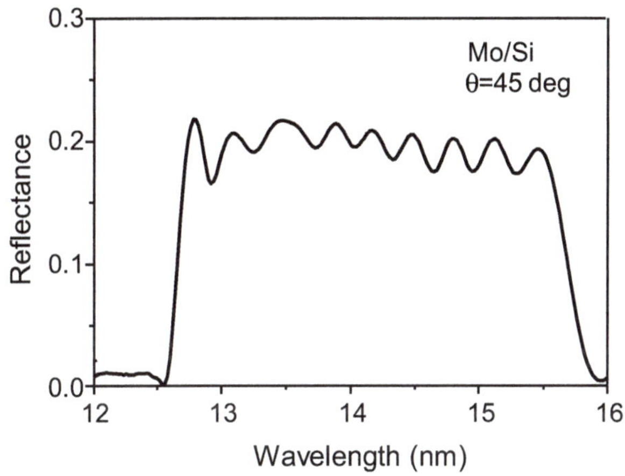

Fig. 3 Reflectance of broadband EUV optic

3 Diffractive Optics

Diffractive optics play a crucial role in EUV systems. Their relatively low cost, small size, and high resolution make them very attractive especially for use with low divergence sources. A drawback of these optics, however, is low efficiency. Below we describe a variety of methods to improve optical efficiency.

Diffractive optics [9–12] are normally patterned onto thin membrane substrates, however the limited transmission of these membranes in the EUV regime can significantly impact the total system efficiency. To avoid this problem, free-standing diffractive optics are required. This is especially important in the 40-nm wavelength

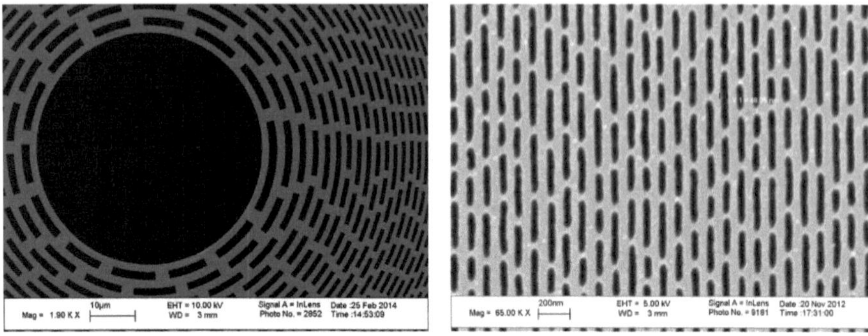

Fig. 4 Free-standing zoneplate for 47-nm operation. Bridges are placed between the zones to keep the whole free-standing structure together

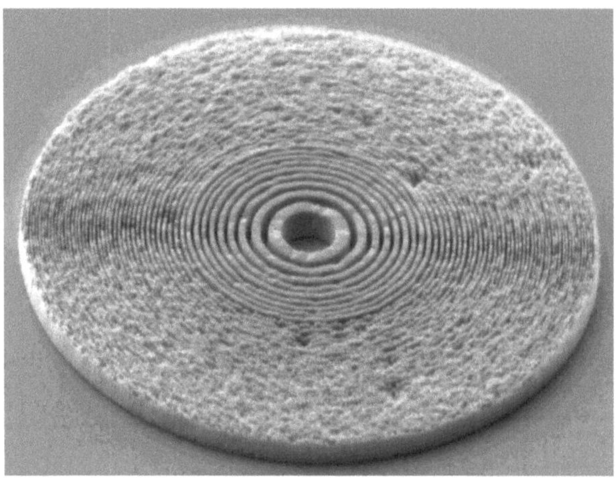

Fig. 5 Double patterned 760-nm thick zoneplate with 40-nm outer zone width

range where typical membrane transmission is extremely low. Figure 4 shows a free-standing zoneplate designed for use at 47 nm wavelength and having an outer zone width of 50 nm. Such zoneplates are used, for example, in soft-x-ray laser ablation mass spectroscopic imaging [13].

Another potentially important factor in zoneplate efficiency is zone thickness. This is especially true at shorter wavelengths. High resolution, high aspect ratio patterning, however, can be quite challenging. Multiple patterning provides a mechanism to mitigate the challenge. We have used double patterning of zoneplates [12] in the past to interlace lower density zones thereby surpassing the single exposure resolution limits of the patterning tool, but this same method can also be used to pattern multiple zoneplates directly on top of each other thereby increasing the total thickness again surpassing the single patterning limits. Figure 5 shows a 40-nm outer zone width zoneplate with a zone thickness of 760 nm (an aspect ratio of 19 to 1).

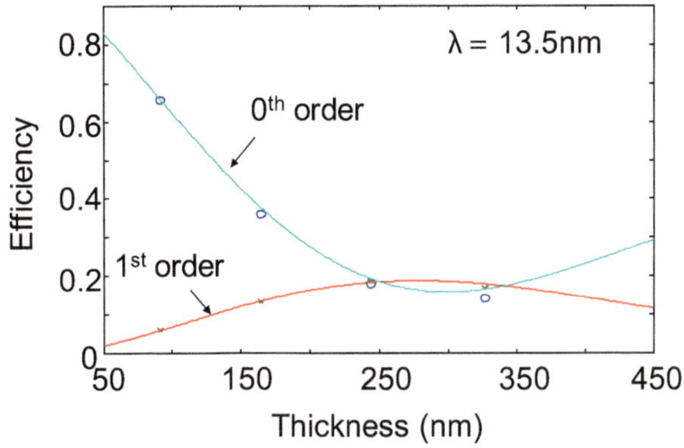

Fig. 6 Measured diffraction efficiency of etched silicon nitride test gratings as a function of groove thickness

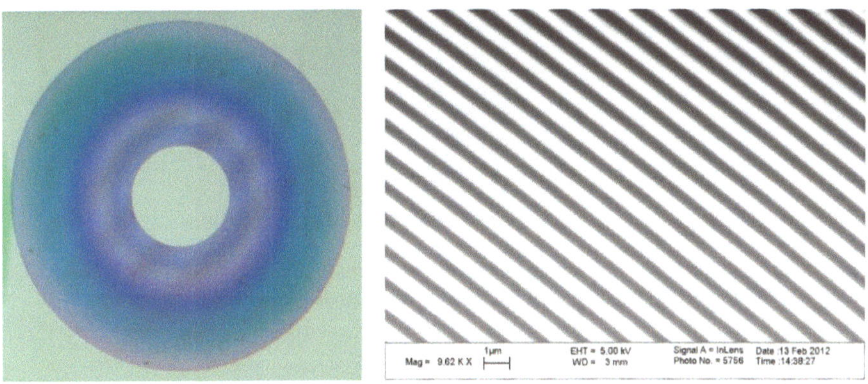

Fig. 7 Etched silicon nitride zoneplate used as HHG condenser optic

Another way to improve zoneplate efficiency is by way of phase shift materials. At 13.5 nm, silicon nitride, a common membrane material, exhibits good phase shift properties. This means that instead of patterning the zoneplate onto the membrane, we can also etch zoneplates into the membrane. Figure 6 shows diffraction efficiency measurements from silicon nitride test gratings where we have achieved an output diffraction efficiency of approximately 20 %. The experimental results are very well modelled based on the tabulated [14] optical properties of silicon nitride. The scatterometry measurements were performed using the Center for X-ray Optics Calibrations and Standards beamline. The nitride process has been successfully applied to operational zoneplates as shown in Fig. 7 which is used as an HHG condenser.

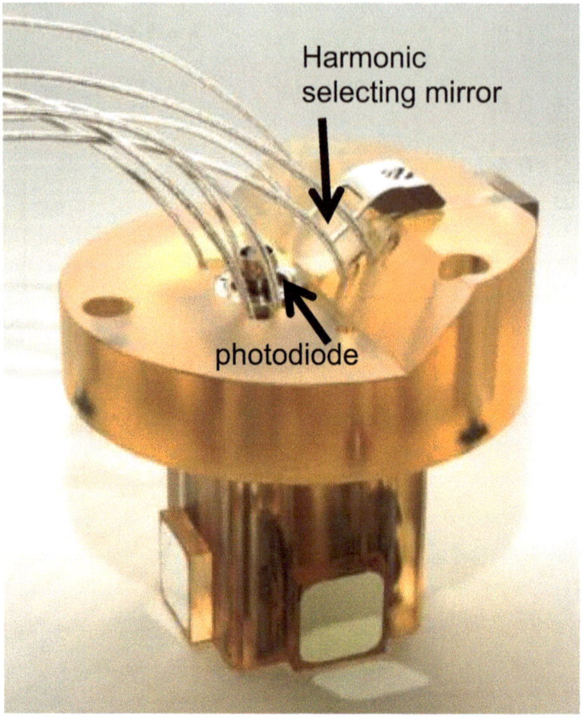

Fig. 8 Photograph of the scanning reflection EUV mask microscope optical housing. The zoneplate is mounted to the bottom of the housing and not visible in the photograph

4 Example System

Microscopy is a crucial scientific technique in the EUV regime. The optical components we have described above can be integrated with lab scale sources to make these capabilities available beyond the synchrotron community. An example of such a system is a scanning reflection EUV mask microscope [15] based on a commercial HHG source [16]. The system uses a harmonic selecting multilayer turning mirror to direct the light from the HHG source down at an angle of 6° from normal and towards a zoneplate placed 500-μm above the surface of a reflection EUV mask. The light is then focused to the mask and reflected back towards a photodiode integrated into the same optical housing holding the turning mirror and the zoneplate. The mask is then scanned under the focused beam to generate the two dimensional image. Figure 8 shows a photograph of the optical housing. The zoneplate is mounted to the bottom of the housing and not visible in the photograph. Figure 9 shows an image of 100-nm lines and spaces recorded from an EUV mask using the microscope described above [15].

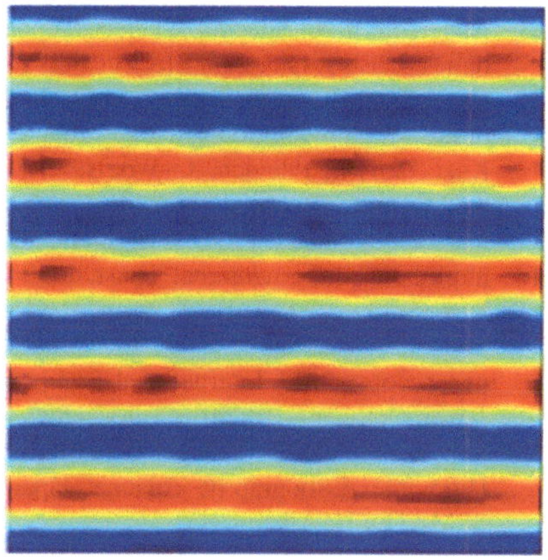

Fig. 9 Image of 100-nm lines and spaces recorded from an EUV mask using the optical system in Fig. 8

5 Summary

Good progress has been made recently in the area of compact coherent EUV sources. Using state of the art diffractive optics and reflection coatings, such sources can readily be integrated into a wide variety of optical systems including the scanning reflection EUV microscope described here.

This work was supported in part by Samsung Semiconductor through the U.S. Department of Energy under Contract No. DE-AC02-05CH11231.

References

1. Reagan, B.A., Wernsing, K.A., Curtis, A.H., Furch, F.J., Luther, B.M., Patel, D., Menoni, C.S., Rocca, J.J.: Demonstration of a 100-Hz repetition rate gain-saturated diode-pumped table-top soft x-ray laser. Opt. Lett. **37**, 3624 (2012)
2. Zhang, X., Libertun, A.R., Paul, A., Gagnon, E., Backus, S., Christov, I.P., Murnane, M.M., Kapteyn, H.C., Bartels, R.A.: Highly coherent light at 13 nm generated by use of quasi-phase-matched high-harmonic generation. Opt. Lett. **29**, 1357 (2004)
3. Kim, S., Jin, J., Kim, Y., Park, I., Kim, Y., Kim, S.: High-harmonic generation by resonant plasmon field enhancement. Nature. **453**, 757–60 (2008)
4. Midorikawa, K.: High-order harmonic generation and attosecond science. Jpn. J. Appl. Phys. **50**, 9001 (2011)
5. Spiller, E.: Reflective multilayer coatings for the far UV region. Appl. Opt. **15**, 2333 (1976)
6. Underwood, J.H., Barbee, T.W. Jr.: Layered synthetic microstructures as Bragg diffractors for X rays and extreme ultraviolet: theory and predicted performance. Appl. Opt. **20**, 3027 (1981)
7. Takahashi, E., Nabekawa, Y., Midorikawa, K.: Generation of 10-mJ coherent extreme-ultraviolet light by use of high-order harmonics. Opt. Lett. **27**, 1920 (2002)
8. Aquila, A., Salmassi, F., Dollar, F., Liu, Y., Gullikson, E.: Developments in realistic design for aperiodic Mo/Si multilayer mirrors. Opt. Exp. **14**, 10073 (2006)
9. Schmahl, G., Rudolph, D. (eds.): X-Ray Microscopy, Proceedings International Symposium, Goettingen, Germany, Sept. 14–16, 1983, Vol. 43. Springer, Berlin (1984)
10. Anderson, E.H., Kern, D.: Nanofabrication of zone plates for X-Ray microscopy. In: Michette, A., Morrison, G., Buckley, C. (eds.) X-Ray Microscopy III. Springer, Berlin (1992)

11. Anderson, E.H.: Specialized electron beam nano-lithography for EUV and X-Ray diffractive optics. IEEE J. Quantum Electron. **42**, 27–35 (2006)
12. Chao, W., Tyliszczak, T., Kim, J., Fischer, P., Rekawa, S., Anderson, E.H.: Latest development of ultra-high resolution soft x-ray zone plate microscopy. X-Ray Microscopy X. (2010)
13. Kuznetsov, I., Filevich, J., Woolston, M., Bernstein, E., Crick, D., Carlton, D., Chao, W., Anderson, E., Rocca, J., Menoni, C.: Composition depth profiling by soft x-ray laser-ablation mass spectrometry. Proc. SPIE. **8849**, 88490G (2013)
14. http://henke.lbl.gov/optical_constants/
15. Kim, S., Lee, D., Park, J., Kim, E., Jeon, C., Cho, H., Jeon, B., choi, C., Anderson, C., Myakawa, R., Naulleau, P.: EUV mask imaging system based on the scanning reflective microscopy. 2013 International Symposium on Extreme Ultraviolet Lithography, Toyama, Japan, October 6–10, 2013, (proceedings available from SEMATECH, Albany, NY)
16. Yoo, B.: EUVO: EUV Light Generation System. 2013 International Symposium on Extreme Ultraviolet Lithography, Toyama, Japan, October 6–10, 2013, proceedings available from SEMATECH, Albany, NY

Table Top Soft X-ray Laser Used for Fabrication of Periodic Nanostructures

W. Li, D. Patel, C.S. Menoni, W. Chao and M.C. Marconi

Abstract In this paper we report the fabrication of periodic structures utilizing a soft X-ray (SXR) table top laser and Talbot printing. Patterns are first recorded in photoresist, which are then used as sacrificial mask to transfer nanofeatures into noble metals. The laser source used in these experiments is a 46.9 nm table top capillary laser which enables a high spatial resolution to this lithography tool. The smallest critical dimension achieved was 40 nm.

1 Introduction

Nanolithography is a key technology enabling new opportunities in various fields of research ranging from biomedical to semiconductor industry [1–3]. The constant advancement of nano-sciences fuelled a constant effort to implement simple methods to fabricate nano-scale structures reliably. Consequently a robust and cost effective fabrication method will be an important contribution for the advancement of this field.

In this paper we describe a lithography approach that combines the classic Talbot image effect and a table top soft X-ray laser [4, 5]. With this lithography method we succeeded in duplicating and transferring nano-features from master mask into metal films deposited on Si substrates.

Talbot image is based on a Fresnell diffraction effect produced when periodic patterns are illuminated by coherent light. This one-to-one image is periodically replicated along the light propagation axis with a period defined by the Talbot distance (Z_t):

$$Z_t = \frac{2nP^2}{\lambda} \tag{1}$$

M.C. Marconi (✉) · W. Li · D. Patel · C.S. Menoni
NSF Engineering Research Center for Extreme Ultraviolet Science and Technology and Electrical and Computer Engineering Department, Colorado State University, Fort Collins, USA
e-mail: marconi@engr.colostate.edu

W. Chao
Center for X-Ray Optics, Lawrence Berkeley National Laboratory, Berkeley, USA

© Springer International Publishing Switzerland 2016
J. Rocca et al. (eds.), *X-Ray Lasers 2014*, Springer Proceedings in Physics 169,
DOI 10.1007/978-3-319-19521-6_39

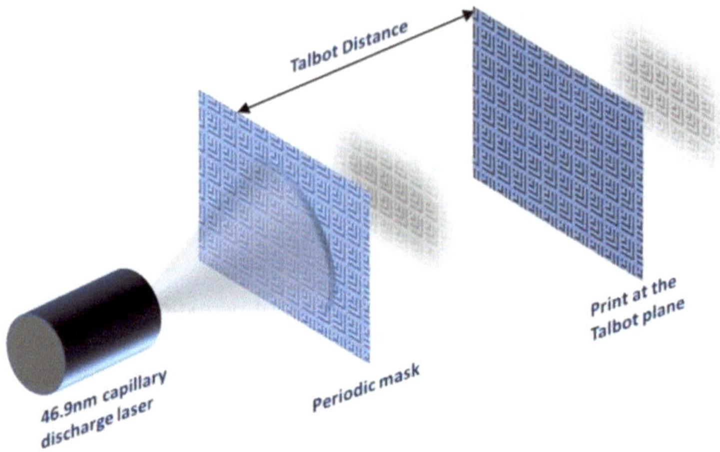

Fig. 1 Experiment set up. The SXR capillary laser illuminated a periodic semi-transparent mask that projects a Talbot image in a sample located at the first Talbot plane

Where n is an integer indicating the order of the Talbot image, P is the period of nanopatterns on the mask and λ is the wavelength of illumination. After the Talbot image is recorded in photoresist, an anisotropic ion beam etching process is used to transfer the pattern into a metal film.

2 Experiment Set Up

Figure 1 is a scheme of the experimental setup. It is composed of a semi-transparent mask with a periodic pattern, a coherent light source and a positioning system capable of positioning the mask and the sample at the Talbot distance accurately. The coherent light source used in this experiment is a 46.9 nm capillary discharge table-top soft X-ray laser. The positioning system is composed of piezo stages actuators and three eddy current sensors to control the distance with and accuracy of 100 nm.

The Talbot mask fabricated on an ultra thin membrane is designed specifically for the wavelength of illumination (46.9 nm). Two different types of masks were designed, fabricated and tested. The first one is free standing mask shown in Fig. 2a. It is based on a $2 \times 2\,\text{mm}^2$, 50 nm thick silicon nitride membrane. Using e-beam lithography the periodic pattern was defined on top of the membrane followed by CF_4 reactive ion etching that transferred the pattern into the silicon nitride. The self-standing structure allows for 100 % transmission through the openings while the 50 nm thick silicon nitride absorbs approximately 98 % of the SXR laser light. The second kind of mask is shown in Fig. 2b. In this case if is fabricated on a $2 \times 2\,\text{mm}^2$, 200 nm thick silicon membrane. After performing the e-beam lithography, nanopatterns in 4 % HSQ resist (60 nm in thickness) are used as the absorbing

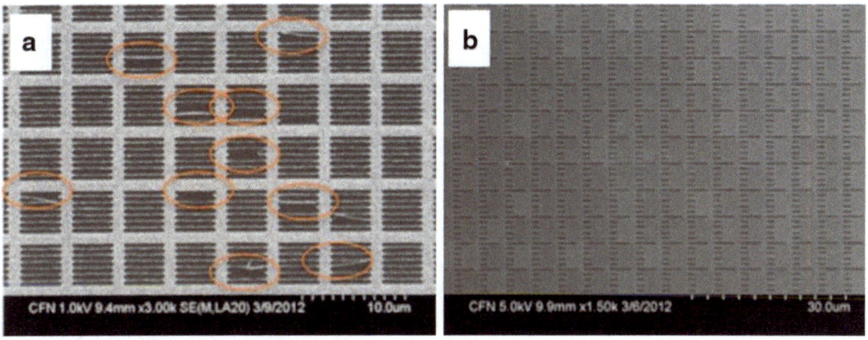

Fig. 2 **a** Free standing mask fabricated on a SiN membrane. The *red circles* indicate defects developed during the fabrication. **b** Talbot mask fabricated on a Si membrane with opaque regions defined in cured HSQ resist

regions for the SXR laser light. Around 37 % of the light is transmitted after passing through the 200 nm silicon membrane, while the 60 nm developed HSQ absorbs 96 % of the laser light. These designs make both masks effectively functional for this application.

While the self standing structure allows for a more efficient diffractive mask (higher transparency), it has the drawback that not all periodic designs are allowed with this approach and also that due to the thin membrane structure, usually defects are developed in the fabrication process. The red circles in Fig. 2a show an example of such defects. However, the unique defect-tolerant capability of the Talbot imaging corrects these defects producing a defect-free print in the photoresist.

3 Nanostructures Fabricated by Talbot Lithography

In this section the nanostructures fabricated are described. A comparison between the patterns and the mask clearly illustrated the defect tolerant characteristic of this lithography approach. In the end, an alternative exposure strategy is illustrated showing the ability to reduce critical dimension (CD).

3.1 Results for Pattern in Photoresist

Using the experimental setup depicted in Fig. 1, prints in photoresist of a periodic mask were produced as shown in Fig. 3. Figure 3a is the print in photoresist obtained with the self-standing mask shown in Fig. 2a. Despite of the numerous defects in the mask (broken bridges highlighted by the red circles), the print in the photoresist is defect-free. This is a unique characteristic of the Talbot lithography method. The

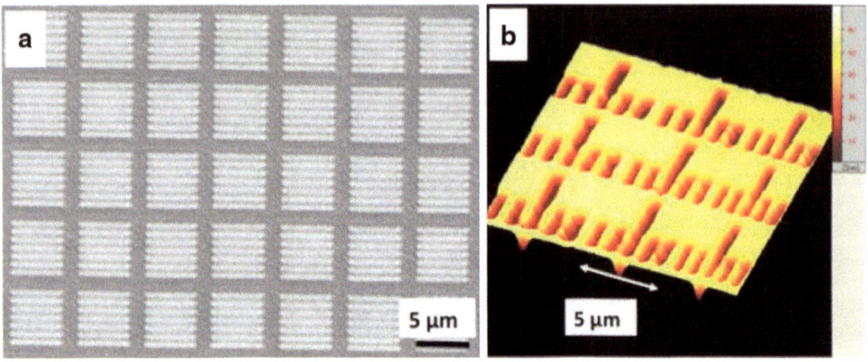

Fig. 3 **a** Print in photoresist obtained using the mask shown in Fig. 2a. The print comes out with no defects. **b** AFM scan of the print in photoresist produced by the mask shown in Fig. 2b

second kind of masks is also adequately replicated in photoresist. Figure 3b is an AFM scan of the print produced by the mask shown in Fig. 2b.

3.2 Pattern Transferred to Metal Surface

The photoresist prints shown in Fig. 3 were used as sacrificial layer to transfer the patterns to a metallic surface, to fabricate nanoscale structure. To complete the fabrication process an anisotropic Ar ion beam etching system was used. A Si substrate was coated with Au (or Ag) and subsequently coated with photoresist that was print using the Talbot lithography method described in the former section. The unprotected regions of the sample were etched down with the ion beam etching system in this way transferring the print onto the metallic layer. Figure 4a is a SEM scan of the structure defined by the mask in Fig. 2a transferred onto a Au layer. Figure 4b is the periodic structure of the mask shown in Fig. 2b transferred onto a Ag layer. In both cases the structure was successfully replicated in the metal-coated samples.

3.3 Double Exposure to Reduce the Feature Size

An alternative exposure strategy was used to further reduce the critical dimension (CD) of the fabricated arrays. The schematic description is of the method is shown in Fig. 5. Figure 5a is an SEM image of the mask used, composed of tiles with 300 nm slits. In a first step the sample was exposed with half the dose corresponding to a complete activation of the photoresist. After that, the sample was moved along a desired direction, in this case we moved vertically in the direction indicated by the arrow in Fig. 5a. Finally, the sample was exposed with the remaining half dose.

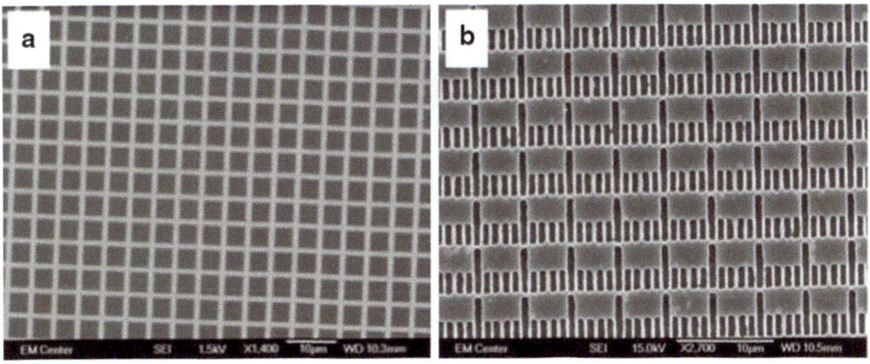

Fig. 4 **a** Array of slits fabricated on a Si substrate coated with Au. The structure was obtained by ion etching using the print shown in Fig. 3a as a sacrificial mask. **b** SEM image of the array of lines fabricated on Ag using the print in Fig. 3b as sacrificial mask

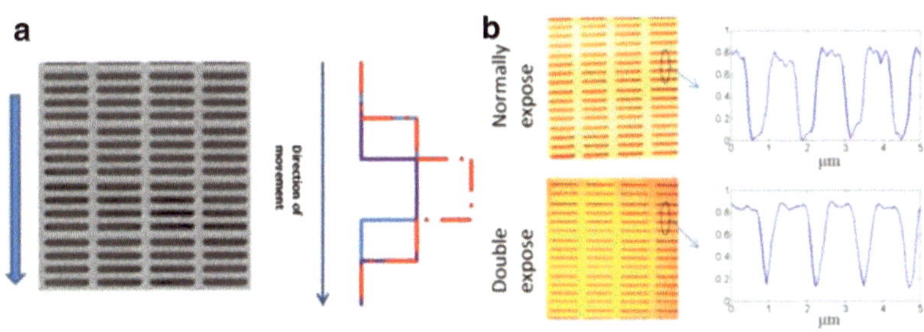

Fig. 5 **a** SEM image of the Talbot mask used. It is composed of an array of tiles each one formed by four 300 nm slits. **b** Prints in photoresist obtained by two consecutive exposures after displacing the sample in the direction of the *blue arrow*. The size of the printed line reduced from 320 to 200 nm

This simple procedure allows for a higher (complete activation) of the photoresist in narrower regions than the size of the mask. This method has been proved to be helpful regarding to the sharpness of edge, which is shown in Fig. 5b. The linewidth is reduced from 500 nm to 320 nm by moving 200 nm in between two consecutive exposures. This exposure strategy needs to be tuned for specific non-linear effect of the photoresist in order to optimize the results. With this method, a 40 nm FWHM lines were achieved using a mask with 75 nm linewidth, by displacing 50 nm in between of two exposures. These narrower lines are shown in Fig. 6.

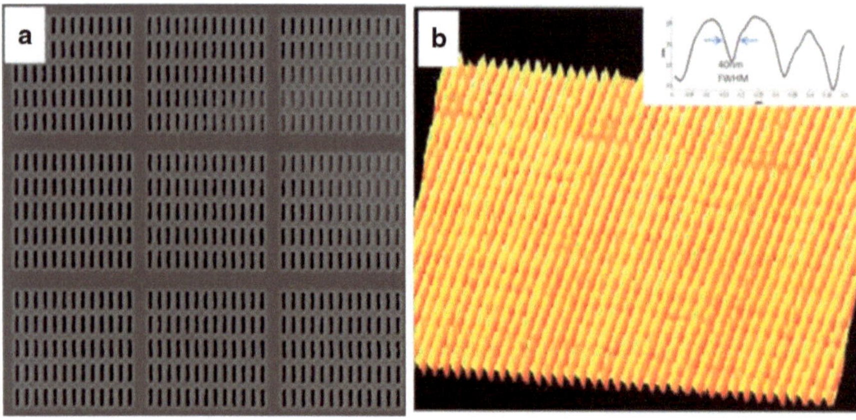

Fig. 6 **a** SEM image of a Talbot mask composed of an array of tiles each one with 75 nm slits. **b** Prints in photoresist obtained by two consecutive exposures after displacing the sample. The size of the printed line reduced to 40 nm

4 Conclusions

We have described a compact lithography system that uses a table top SXR laser capable to produce defect-free prints of nanoscale structures over large areas. The simple method is even potentiated recurring to a double exposure strategy that allowed printing of 40 nm lines. This simple system can be of great interest for fabricating prototypes or small volume production of nanoscale structures.

References

1. Pelletier, V., Asakawa, K., Wu, M., Adamson, D.H., Register, R.A., Chaikin, P.M.: Aluminum nanowire polarizing grids: fabrication and analysis. Appl. Phys. Lett. **88**, 211114 (2006)
2. Baker, G.A., Moore, D.S.: Progress in plasmonic engineering of surface-enhanced Raman-scattering substrates toward ultra-trace analysis. Anal. Bioanal. Chem. **382**, 1751 (2005)
3. Campbell, M., Sharp, D.N., Harrison, M.T., Denning, R.G., Turberfield, A.J.: Fabrication of photonic crystals for the visible spectrum by holographic lithography. Nature. **404**, 53 (2000)
4. Li, W., Martinez Esquiroz, V., Patel, D., Urbanski, L., Menoni, C.S., Rocca, J.J., Stein, A., Chao, W., Anderson, E., Marconi, M.C.: Coherent nanopatternig with Table Top Soft X-ray Lasers. Invited paper. Proceedings of the X-Ray Lasers and Coherent X-Ray Sources: Development and Applications X. SPIE Proceedings Vol. 8849. 8849OW-1, (2013)
5. Urbanski, L., Marconi, M.C., Isoyan, A., Stein, A., Menoni, C.S., Rocca, J.J.: Defect tolerant Talbot nano-patterning. Proceedings ICXRL: 2012, Chap. 42, pp. 263–267. Springer Proceedings in Physics Volume 147. S. Seban, J. Gautier, D. Ros, P. Zeitoun Editors, ISSN 0930–8989. Published (2013)

Applications for Coherent Narrow-Band Sources in EUV Lithography

R. Miyakawa and P. Naulleau

Abstract In this paper, we discuss the unique application of coherent, narrow-band sources in both wavefront metrology and mask defect inspection which play essential roles in the EUVL ecosystem. For metrology, coherent sources can efficiently supply power to wavefront sensors, whereas current systems using incoherent sources must rely on highly inefficient spatial filters. This advantage may be especially important as next-generation EUVL tools move to higher numerical apertures. For defect inspection, sources with high temporal coherence allow zone plate lenses to replace costly EUV multilayer optics. These zone plates can be programmed to operate with contrast-enhancing techniques like Zernike phase contrast microscopy. Additionally, the coherence allows inspection tools to operate in a scanning mode, similar to a STXM microscope, which can achieve a large field of view without field-dependent aberrations.

1 Introduction

Extreme ultraviolet lithography (EUVL) is a key technology for the semiconductor industry as it continues to shrink the critical dimensions of semiconductor devices. Using a wavelength of 13.5 nm, next-generation EUVL tools will be capable of supporting the 11-nm semiconductor device node and beyond. While production platforms like the ASML NXE:3300 have demonstrated full-field high-resolution printing, there are still several remaining challenges that EUVL faces before it can be commercialized [1]. In this paper, we discuss two such challenges: wavefront metrology and actinic mask defect inspection, and how the introduction of standalone, narrow-band, coherent EUV sources can play a key role in overcoming these challenges.

R. Miyakawa (✉) · P. Naulleau
Center for X-ray Optics, Lawrence Berkeley National Lab, 1 Cyclotron Rd., Berkeley, CA 94720, USA
e-mail: rhmiyakawa@lbl.gov

© Springer International Publishing Switzerland 2016

J. Rocca et al. (eds.), *X-Ray Lasers 2014*, Springer Proceedings in Physics 169,
DOI 10.1007/978-3-319-19521-6_40

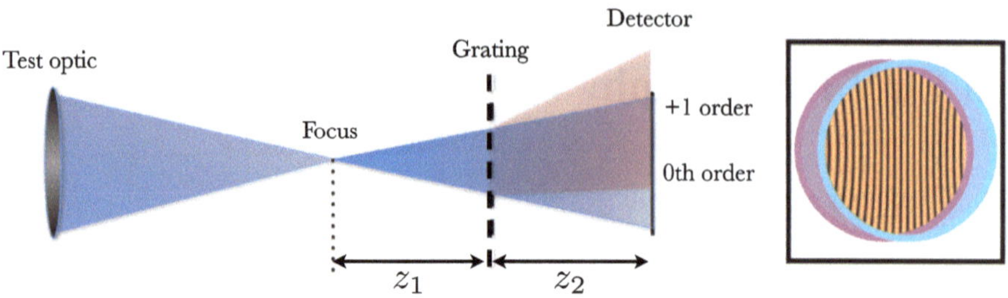

Fig. 1 LSI setup at EUV. A grating is placed after the focus of the test optic. The detector captures the interference between the diffracted orders of the grating

2 Wavefront Metrology

Characterizing and removing aberrations in the optics is a key part of maximizing the resolution capability of a lithography tool. However, most wavefront sensors suitable for measuring EUV optics require coherent illumination. Since lithography tools typically employ plasma-based incoherent sources, the incoming light must be heavily filtered to create enough coherent flux to perform the measurements. This problem is exacerbated at high-NA, where smaller and smaller filters are necessary to meet the coherence requirements. Although typical EUV laser and HHG sources have lower overall output power, a much larger fraction of this power can be used by the wavefront sensors making these sources an attractive alternative to plasma sources.

2.1 Lateral Shearing Interferometry

Lateral shearing interferometry (LSI) is a leading candidate for next-generation EUV optical testing. Unlike point-diffraction interferometers like PS/PDI [2], the shearing interferometer is a self-referencing test that does not require a diffraction-limited spatial filter aperture on the fast side of the optic which can approach sizes below 16 nm for high-NA optics. This is an important advantage, because these small apertures present several challenges; achieving adequate absorption requires aggressive aspect ratios which are difficult to fabricate, and small apertures are prone to contamination buildup, which renders filters unusable after a short period of time (Fig. 1).

LSI works by interfering the test wavefront, $W(x, y)$, with a shifted (sheared) copy of itself, $W(x + s, y)$. The interferogram is thus a representation of the derivative of the test wavefront in the direction of the shear:

$$U(x, y) = W(x + s, y) - W(x, y) \approx s \cdot \frac{\partial W(x, y)}{\partial x} \quad (1)$$

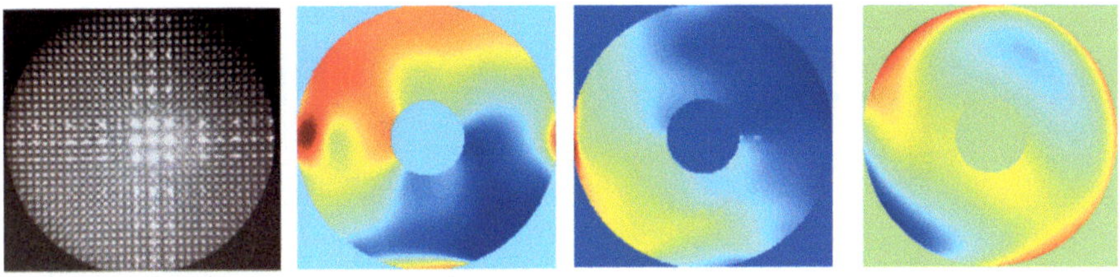

Fig. 2 Shearing results on the Samsung SERM microscope

The shearing measurement is taken in both the x and y directions and $W(x, y)$ can be subsequently reconstructed using numerical methods [3]. When performing LSI at EUV wavelengths, there are limited options for the optical setup to create sheared copies of the wave on the detector. Beamsplitters for instance are not a viable solution for short wavelength systems due to the lack of materials with suitable reflection and transmission properties. Consequently, LSI setups for EUV optical systems generally use diffractive elements such as gratings to create the sheared waves. Care must be taken when implementing grating-based shearing interferometers at high NA, because grating misalignment can cause systematic aberrations. However, because LSI is a common-path interferometer, it is resistant to many other systematic effects such as system vibration.

An EUV lateral shearing interferometer with a HHG source was recently used to characterize the Samsung SERM microscope, demonstrating an RMS wavefront error of $\lambda/17$, as shown in Fig. 2.

2.2 *AIS Wavefront Sensor*

The AIS wavefront sensor is an alternative wavefront measurement scheme that will be used to support the upgrade of the SEMATECH Albany and Berkeley METs to 0.5 NA. AIS scales well to high numerical apertures, and unlike LSI does not suffer from systematic aberrations caused by grating tilt. AIS uses a photodiode as a detector, and therefore has a much lower profile than sensors requiring a full vacuum-compatible CCD.

A schematic of a simple AIS experimental setup is shown in Fig. 3. Light from a coherent EUV source is incident on a scanning mirror that serves to control the range of spatial frequencies entering the system. Different illumination settings steer the diffracted light from the reticle to a specific part of the test optic pupil. This light picks up phase curvature caused by aberrations in the optic, which manifests as a small focus shift in the image plane. In the image plane, a focus sensor measures the plane of best focus for each grating orientation at each pupil probe location. Localized wavefront curvature is computed from the measured focus shifts and fed into an algorithm that reconstructs the aberrations in the system.

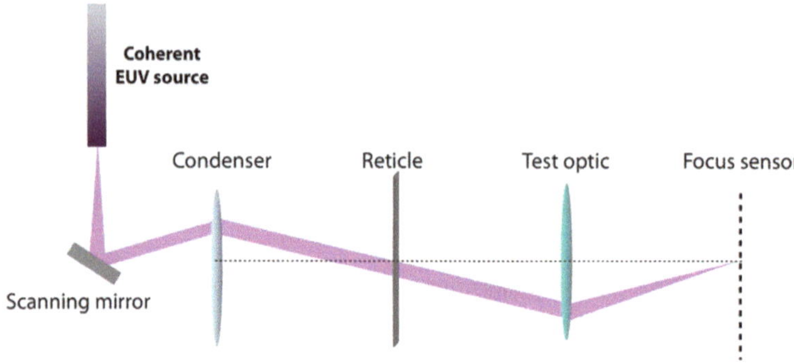

Fig. 3 A schematic of the AIS wavefront sensor operating with a coherent EUV source

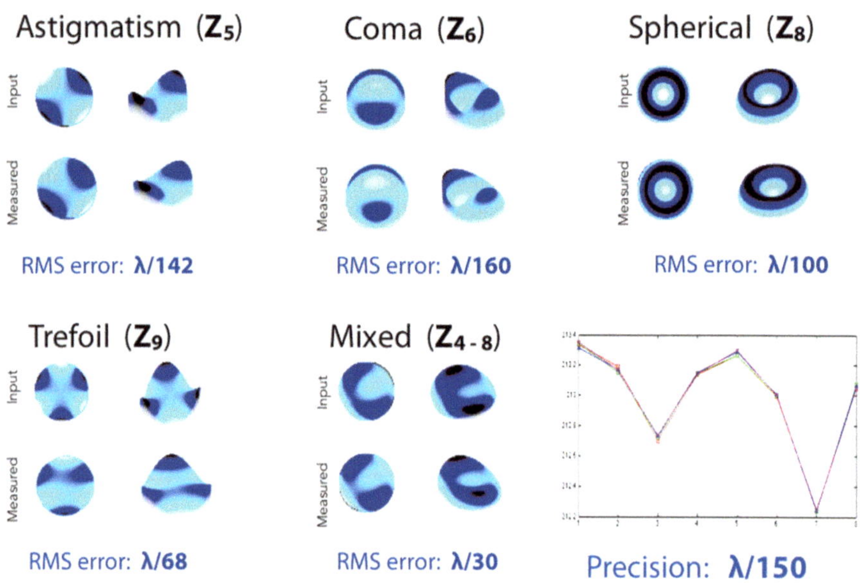

Fig. 4 Visible light prototype results for the AIS wavefront sensor that will be deployed on the SEMATECH Albany and Berkeley METs

A visible light prototype of the AIS was constructed to validate the working principle of the sensor. Figure 4 shows the results of this test demonstrating $\lambda/30$ accuracy with a precision of $\lambda/150$.

3 Mask Inspection Tools

Actinic EUVL mask metrology tools are a key part of EUV commercialization. Among the currently available mask inspection solutions, the majority operate on synchrotron sources. While plasma-based solutions are viable, they require expensive reflective optics, because poor temporal coherence prohibits use with zone

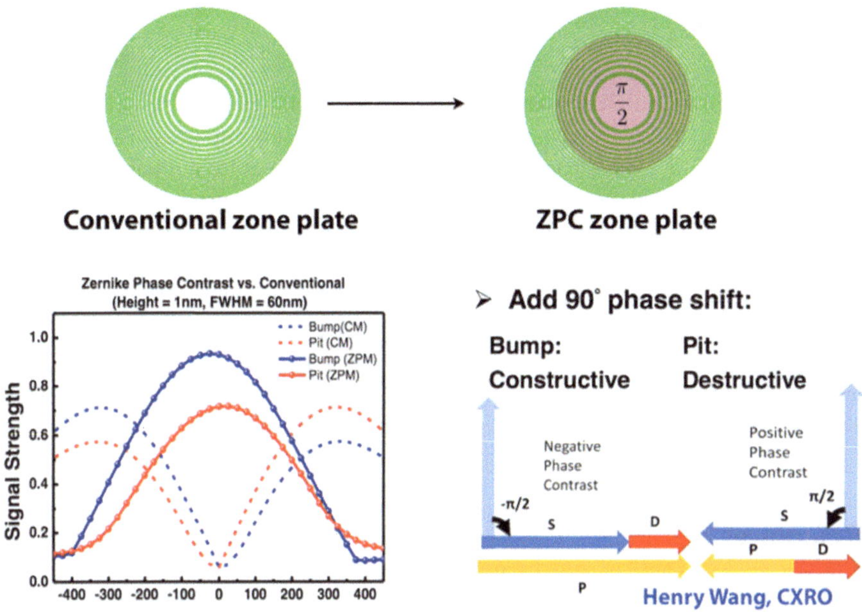

Fig. 5 Visible light prototype results for the AIS wavefront sensor that will be deployed on the SEMATECH Albany and Berkeley METs

plates. Narrowband solutions like HHG sources offer an attractive alternative, since their spectra are compatible with zone plate bandwidth constraints. While coherence is normally avoided in full-field imaging systems, it is ideal for scanning systems, where the coherence of the imaging is determined by the numerical aperture of the detector rather than the coherence of the source. The Samsung SERM tool is a recent example of an actinic, scanning, EUV microscope operating on a HHG source using zone plate optics.

The use of zone plates also opens up the possibility of implementing phase-enhancing and other advanced imaging techniques into the mask inspection tool. Figure 5 shows how a Zernike phase contrast microscope could be implemented by adding a 90° phase shift to the imaging zone plate. This serves to enhance phase-only defects such as embedded multilayer defects that would normally be invisible. Because the zone plate patterns are customizable, arbitrary phase functions can be encoded into the pupil function. In the SEMATECH Berkeley SHARP microscope, for example, special zone plates are encoded with a cubic phase function which is used to increase the depth of focus of the system.

4 Discussion

Coherent, narrowband EUV sources have unique properties that are well-suited to various parts of the EUV ecosystem, and could play an important role in commercializing EUVL. Wavefront metrology continues to require a reliable source of coherent light which is becoming more and more difficult to provide as the numerical apertures of the lithography tools increase. Mask inspection tools using these sources are compatible with zone plate optics that can create scanning systems that are much more affordable than full-field AIMS tools based on plasma sources.

Acknowledgements This work was performed at LBNL's Center for X-ray Optics and was supported by SEMATECH through the U.S. Department of Energy under Contract No. DE-AC02-05CH11231.

References

1. Wagner C., et al.: Proc. SPIE **7969**, 79691F (April, 2011)
2. Naulleau P., Goldberg, K., et al.: Extreme-ultraviolet phase-shifting point-diffraction interferometer: a wave-front metrology tool with subangstrom reference-wave accuracy. Appl. Op. **38**, 7252–7263 (1999)
3. Rimmer, M.: Method for evaluating lateral shearing interferograms. Appl. Op. **13**, 3 (1974)

Lensless Proximity EUV Lithography with a Xenon Gas Discharge Plasma Radiation

Hyun-su Kim, Serhiy Danylyuk, Sascha Brose, Peter Loosen, Klaus Bergmann, William S. Brocklesby and Larissa Juschkin

Abstract The possibilities and limitations of proximity and interference lithography under extreme ultraviolet (EUV) radiation are explored. Utilizing partially coherent EUV radiation at central wavelength of 10.9 nm from a Xe gas discharge plasma source, it is shown that proximity printing in the Fresnel diffraction regime can produce the high-resolution features even with low-resolution masks, and also yield sub-30 nm edge resolution in the resist. The scalability limit within non-paraxial case has been also studied. The effect of the diffraction behind the transmission mask is evaluated with respect to the achievable resolution. The finite-difference time-domain simulation of the diffraction patterns behind the Ni/Nb-based transmission mask is performed varying the pitch size. The results demonstrate that the method can be used to produce patterns with resolution down to 7.5 nm half-pitch with 2:1 mask demagnification utilizing achromatic Talbot effect with transverse electric (TE)-polarized light.

1 Introduction

At the end of the twentieth century one of the promising lithography technologies was the proximity lithography using hard X-rays. With this short wavelength, the influence of the diffraction limit on the resolution is negligible and high-resolution patterns are possible. Further reduction of the wavelength was found to be ineffective due to increased photoelectron blur in photoresist. Due to these challenges and rapid development of the immersion lithography, this technology has not progressed

Hyun-su Kim (✉) · L. Juschkin
Chair for the Experimental Physics of EUV, RWTH Aachen University and JARA-FIT, Steinbachstr. 15, 52074, Aachen, Germany
e-mail: lighttn@gmail.com

Hyun-su Kim · W. S. Brocklesby
Optoelectronics Research Center, University of Southampton, Southampton SO17 1BJ, UK

S. Danylyuk · S. Brose · P. Loosen
Chair for the Technology of Optical Systems, RWTH Aachen University and JARA-FIT, Steinbachstr. 15, 52074 Aachen, Germany

K. Bergmann
Fraunhofer Institute for Laser Technology, Steinbachstr. 15, 52074 Aachen, Germany

© Springer International Publishing Switzerland 2016

J. Rocca et al. (eds.), *X-Ray Lasers 2014*, Springer Proceedings in Physics 169,
DOI 10.1007/978-3-319-19521-6_41

past the pre-production state. The next step in history of the optical lithography is a radical reduction of the wavelength from current 193–13.5 nm, which should enable the possibility of scaling below 10 nm half-pitch levels. This radiation lies in the so-called extreme ultraviolet (EUV) wavelength range, where photon energy is significantly higher and, as a consequence, the interaction mechanisms between light and matter differ from well-studied ultraviolet radiation.

During the last decade, a number of interference schemes were investigated including Lloyd mirror, single grating and double grating schemes. However all of them have rather high requirements on spatial and/or temporal coherence, making them very inefficient for use with available high power laboratory sources. Recently, it was shown that it is possible to improve the resolution of the proximity printing by utilizing the Fresnel diffraction of the short wavelength radiation on the nanostructures. This allows achieving sub-30 nm resolution with mask-to-wafer gap reaching 30 μm, which is much more realistic to maintain in an industrial environment.

In this work we discuss the feasibility and resolution limits of such Fresnel diffraction-assisted proximity printing and also a rigorous simulation approach for numerical analysis of the scalability of Talbot interference lithography utilizing EUV radiation. In this work, the radiation at 10.9 nm with 3 % bandwidth, which is generated with a Xe gas discharge produced plasma (DPP) [1] has been used. This work is focused also on applications of the method with partially coherent high power laboratory sources.

2 Modelling and Methodology

Accurate simulation of the near-field diffraction is important for understanding limitations of EUV proximity lithography. The scalar simulation can be performed with the Fresnel-Kirchhoff or Rayleigh-Sommerfeld diffraction integrals, which shows essentially same results in our setup. This scalar approximation is valid when the wavelength is much smaller than both the proximity gap and the aperture size. EUV wavelength is short enough for this approximation dealing with sizes above several hundreds of nanometers and the proximity gaps of several micrometers. The concept of the EUV proximity lithography is illustrated in Fig. 1, which shows the intensity distribution propagated behind an aperture and profiles at several distances from the aperture.

As the Fresnel diffraction patterns produced by simple apertures can become rather complex, i.e. on a round aperture a set of concentric rings can be observed with a number of rings depending on the observation position. This is a potential method for creation of complex nanophotonic structures of novel designs. By tuning the exposure time and modifying the proximity gap the structures of different shapes with down to sub-100 nm resolution can be produced.

The question is whether this approach can be scaled down to sub-10 nm structuring. The approximations taken within the analytical theories are no longer valid

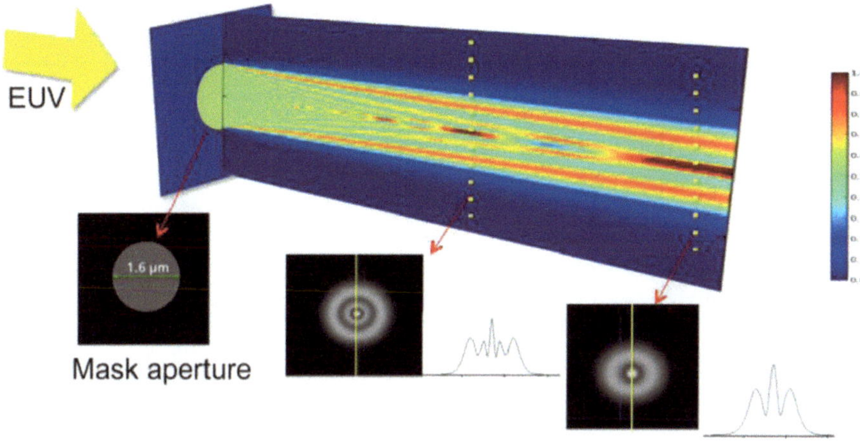

Fig. 1 Intensity distribution behind the aperture and the profiles as a function of distance under EUV radiation

when the size of the structures on the mask starts to be comparable with the wavelength of radiation. Therefore we employed Maxwell's equations solver utilising finite-difference time-domain (FDTD) approach to investigate the scalability limits of EUV Talbot lithography. In the simulation, the grid size was set to λ/15, which provides better than 2 % accuracy and the parameters of the lithography system were chosen according to the parameters of the recently developed transmission mask [2]. The thickness of the absorber was kept constant at 100 nm in the simulation. The space-to-pitch ratio was set to 0.5. The period of the pattern of the mask was varied from 10 to 100 nm.

3 Results

The demonstration has utilized a high power Xe gas DPP source developed at the Fraunhofer Institute for Laser Technology in Aachen, Germany. EUV radiation was optimized at 10.9 nm central wavelength with 3 % bandwidth at full-width half-maximum utilizing Xe-Argon gas mixture. The source spectrum is shown in Fig. 2. The diameter of the source emission zone varies from 150 to 350 μm depending on the gas mixture and discharge parameters.

The results of the numerical simulations and the corresponding experimental data of the proximity diffraction pattern of micro sized apertures are presented in Fig. 3. The intensity profiles produced by the triangular aperture of the length of 1.56 μm are shown in Fig. 3a: simulation and Fig. 3b: experiment. The same for the round aperture with the diameter of 1.6 μm and the proximity gap of 15 micron are schown in Fig. 3c: simulation and Fig. 3d: experiment. As expected from analytical modeling of the diffraction, the change in the Fresnel number manifests itself in the

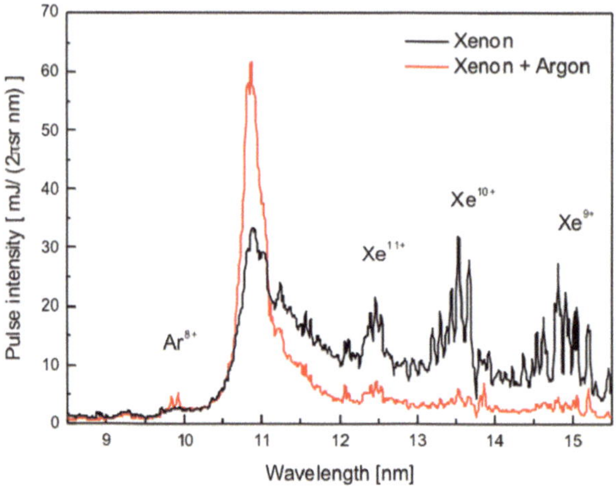

Fig. 2 EUV spectrum generated with a Xe gas or Xe/Ar gas mixture discharge plasma

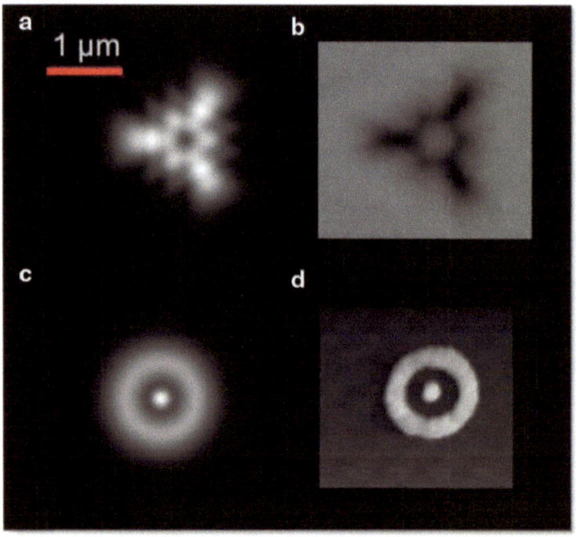

Fig. 3 Results of proximity lithography with a triagular aperture in **a** simulation and, **b** experiment, and with a circluar aperture in **c** simulation and **d** experiment

number of visible rings. The negative high-resolution HSQ resist of 40 nm thickness was exposed 15 µm away from the mask. The resulted structures after the EUV exposures and development were characterized using the atomic force microscope (AFM). A good agreement with the calculated aerial image is observed. Good correlation between scalar diffraction simulations and experiments confirms that the Fresnel diffraction is the main governing process in EUV proximity lithography of micrometer sized structures with 30 nm edge resolution.

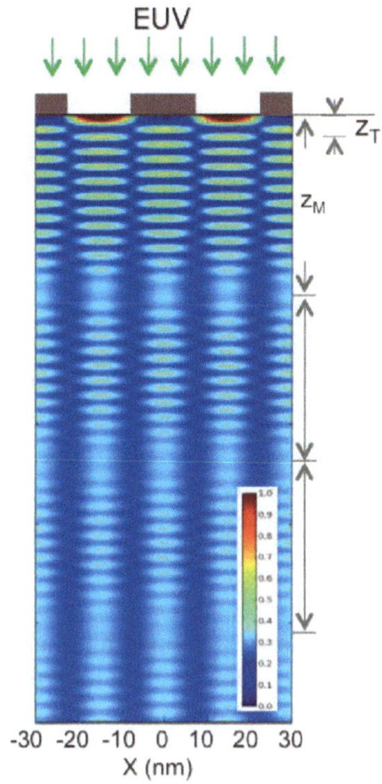

Fig. 4 Achromatic Talbot pattern behind the grating

The possibility to change the aerial image dramatically and controllably just by changing the proximity gap together with the ability to fine-tune the exposed resist area by variation of the exposure time provides a high degree of flexibility for the exposures. In the combination with the developed simulations, this should enable fast and efficient design of suitable transmission masks for wide range of nanophotonic structures including ring-disc cavities or other complex nano-resonant antennas [3].

The finite spectral bandwidth of the source radiation sets a limit of scaling the structuring down to sub-10 nanometers. Different wavelengths yield differnt Talbot distances (z_T) in the interference scheme of Talbot lithography. As a result of the wavelengths mix, the pattern get smeared along the propagation axis starting at a specific distance (Talbot mixture length, z_M). The simulated pattern propagation for 15 nm half pitch grating is shown in Fig. 4. Since lithography requires sufficient contrast for the image converting to the patterns in resist, the studies of the contrast and maximum intensity in dependence on the pitch size were performed. The results are presented in the Fig. 5.

The gradual decrease of the maximal intensity and contrast with half pitch reduction on the mask is clearly visible. Using TE polarized light the contrast

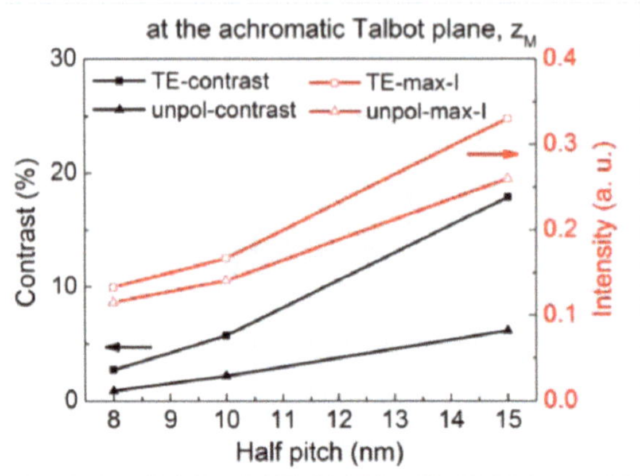

Fig. 5 The contrast and maximum intensity of achromatic Talbot patterns at the plane of z_M in dependence on half pitch on the mask. The resulting half pitch in the aerial image is a quarter of the mask pitch

above $\sim 16\,\%$ can be maintained down to 15 nm half pitch on the mask, which corresponds to 7.5 nm half pitch on the wafer. For the investigated mask design, this is a realistic resolution limit for the broadband Talbot lithography [4].

4 Summary

The laboratory setup for the EUV proximity lithography has demonstrated the ability to create complex nanoantenna arrays and can be used for cost-effective small and medium-scale manufacturing of nanophotonic structures. Utilizing a compact partially coherent DPP EUV source, the knife-edge resolution of 28 nm was achieved. Precise control over the mask-to-wafer gap increases flexibility of the diffraction-assisted EUV lithography, providing the possibility to produce different nanostructures without changing the design of the mask. In the non-paraxial case, we have investigated theoretically the scalability of Talbot EUV lithography. The FDTD simulation has demonstrated that the structuring with achromatic Talbot lithography can be effectively used with amplitude transmission masks. It is possible to utilize the achromatic Talbot effect to achieve 2:1 demagnification of masks period. The contrast and intensity of the resulting aerial image were shown to decrease with the reduction of the pitch size. The practical resolution limit in the achromatic Talbot regime was estimated to be at 7.5 nm half pitch for TE polarized illumination.

References

1. Bergmann, K., Danylyuk, S.V., Juschkin, L.: Optimization of a gas discharge plasma source for extreme ultraviolet interference lithography at a wavelength of 11 nm. J. Appl. Phys. **106**, 073309 (2009)
2. Brose, S., Danylyuk, S., Juschkin, L., Dittberner, C., Bergmann, K., Moers, J., Panaitov, G., Trellenkamp, St, Loosen, P., Grützmacher, D.: Broadband transmission masks, gratings and filters for extreme ultraviolet and soft X-ray lithography. Thin Solid Film. **520**(15), 5080–5085 (2012)
3. Danylyuk, S., Kim, H., Brose, S., Dittberner, C., Loosen, P., Taubner, T., Bergmann, K., Juschkin, L.: Diffraction-assisted extreme ultraviolet proximity lithography for fabrication of nanophotonic arrays. J. Vac. Sci. Technol. B. **31**, 021602 (2013)
4. Danylyuk, S., Loosen, P., Bergmann, K., Kim, H., Juschkin, L.: Scalability limits of Talbot lithography with plasma-based extreme ultraviolet sources. J. Micro/Nanolith. MEMS MOEMS. **12**(3), 033002 (2013)

Part VII
Characterization, Pump Lasers, Optics

Wave Front Study of Fully Coherent Soft X-Ray Laser Using Hartmann Sensor

Lu Li, Yong Wang, Shoujun Wang, Eduardo Oliva, Liang Yin, T. T. Thuy Le, David Ros, Gilles Maynard, Stephane Sebban, Jorge J. Rocca and Philippe Zeitoun

Abstract Fully coherent soft x-ray laser (SXRL) was achieved seeding high harmonics (HH) seed into a solid-plasma amplifier. Wave front of both HH seed and amplified seed were measured using Hartmann wave front sensor. Wave front root-mean-square (RMS) value and stability were observed with improvement of more than two times. In terms of wave front quality, optimized HH seeding parameters on injection time, injection angle and plasma length were discussed. With the benefit of the intensity and phase detection, electromagnetic field of HH seed and amplified seed beam were studied by back-propagation. The reconstructed SXRL source confirmed its high peak brightness.

1 Introduction

With the purpose of high resolution soft x-ray laser (SXRL) phase contrast and diffraction imaging, wave front of the SXRL beam must be excellent. Indeed, diffraction or holographic imaging of a sample with a coherent source having an aberrant wave front will reduce the resolution and may even distort the image. Wave front characterization is also related to the radiation intensity of the beam spot with micrometer dimension, as explained in the relation with Strehl ratio. Previous work on wave front of the HH [1] and gas-plasma-based seeded SXRL

S. Sebban · P. Zeitoun (✉)
Laboratoire d'Optique Appliquee, UMR7639, ENSTA ParisTech, CNRS, Ecole Polytechnique, Palaiseau 91120, France
e-mail: philippe.zeitoun@ensta-paristech.fr

Lu Li
Laboratoire d'Optique Appliquee, UMR7639, ENSTA ParisTech, CNRS, Ecole Polytechnique, Palaiseau 91120, France

School of Nuclear Science and Technology, Lanzhou University, Lanzhou 730000, China

Y. Wang · S. Wang · L. Yin · J. J. Rocca
National Science Foundation Engineering Research Center for Extreme Ultraviolet Science and Technology, Colorado State University, Fort Collins, CO 80523, USA

E. Oliva · T. T. Thuy Le · D. Ros · G. Maynard
Laboratoire de Physique des Gaz et des Plasmas, Universite Paris XI, Orsay cedex, France

© Springer International Publishing Switzerland 2016

J. Rocca et al. (eds.), *X-Ray Lasers 2014*, Springer Proceedings in Physics 169,
DOI 10.1007/978-3-319-19521-6_42

sources [2] showed near-diffraction limited characteristics. However, the wave front of HH-seeded SXRL from solid-based plasma amplifier had not been studied prior to this work. Despite its crucial role in the applications, we therefore performed this experiment [5] to measure the wave front of seeded SXRL based on solid-plasma amplifier.

2 Experiment on Wave Front Measurement

In this SXRL wave front measurement, we used the Hartmann wave front sensor developed at Laboratoire d'Optique Applique (LOA) in collaboration with Imagine Optics and SOLEIL [3]. After the sensor calibration at LOA [4], we conducted the experiment in Colorado State University (US) to measure the wave front of HH-seeded SXRL with the solid-plasma amplifier. The experiment set-up has been described in the article [5]. A direct approach of seeded SXRL wave front measurement using Hartmann wave front sensor was firstly proposed. As illustrated in Fig. 1, based on a table-top Ti:Sapphire CPA laser (central wavelength 810 nm), HH seed beam was generated focusing ($f = 1.1$ m) IR laser (13 mJ, 60 fs) into neon gas jet. Soft x-ray plasma amplifier based on a molybdenum slab target was created adapting multi-pulse grazing incident pumping scheme. Early precursor pulse (30 mJ, 200 ps) and the pre-pulse (250 mJ, 200 ps), separated by 5 ns time delay, were irradiated with line-focusing on the molybdenum slab target surface in normal incident angle. A combination of spherical lens and cylindrical lens were adapted to make the line foci into 40 μm width [6], which allows to create a plasma approaching the Ni-like ionization stage. The expanding plasma was subsequently heated by a short pump pulse (500 mJ, 5 ps) arriving at 600 ps after the pre-pulse. A similar focusing line as pre-pulse obtained by the same approach was delivered to the plasma with grazing incident angle of 23° where time compensation of the travelling wave was taken into account. By this rapid heating into the plasma region, electron density can achieve $\sim 2.6 \times 10^{20}$ cm^{-3} and electron temperature could reach ~500 eV [7], where a large transient population inversion was formed. Sequentially, the HH seed pulse was injected into the plasma amplifier assisted by a toroidal mirror. Due to the wavelength matching between the 43rd HH order (18.9 nm) and the lasing line ($4d^1 S_0 \rightarrow 4p^1 P_1$) in the plasma amplifier, the HH seed pulse (43rd order) can be significantly amplified at the optimized seeding time [8]. A 300 nm thick Aluminium filter was used to stop the straight IR lasers. The Hartmann wave front sensor was then placed at 1.6 m after the plasma amplifier.

Firstly, HH driven laser was set as *P* polarization. However, no diffraction pattern was observed. The illuminated pattern turned out as regular square shape as the apertures on the Hartmann plate. It means that there was no diffraction happened (or in negligible level) after passing through the aperture array. This is indicating that the amplification of the seed was not achieved. Owing to the high dense plasma, hard x-rays are considered as the main contribution to the observed light. Alternatively, we changed the polarization of HH driven laser to *S* polarization. Detected

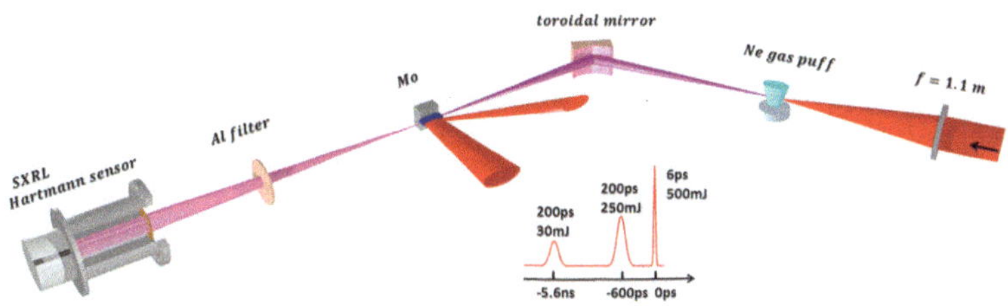

Fig. 1 Experimental setup of direct measurement of seeded SXRL wave front using SXRL Hartmann sensor

image of the output beam showed strong bright background and beam shape with obvious diffraction patterns associated with tilting lights at 22.5° deviated from the vertical. Amplified seed was considered contributing to the diffraction.

In order to get over the effect of the hard x-rays, a specific Mo-Si 45° multilayer flat mirror was employed. The flatness of this mirror is $\lambda/10$ tested at 632 nm for the full pupil of 25 mm. It was placed at 0.86 m after the plasma amplifier, where the estimated seeded SXRL beam size is $\sim$ 1 mm. Thus, the wave front aberration from this mirror could be neglected. Simultaneously, due to the narrow bandwidth of the efficient reflection of this multilayer mirror (specified for 18.9 nm), the effect of the hard x-rays was substantially reduced. In addition, the detection distance could be settled further with respect to the direct diagnostics, which allows a sufficient seeded SXRL beam size for the Hartmann wave front sensor detection. As it described in [5], the sensor was then positioned 3.5 m away from the plasma amplifier. One should beware of that the wave front of HH seed (without plasma amplifier) was measured without using 45° multilayer mirror (Fig. 1) for the sake of efficient single-shot detection.

3 Experiment Results and Discussion

A raw image of single-shot HH seed pulse was displayed in Fig. 2a. Considering the accuracy of the wave front analysing, a centre circular region was chosen for the data processing, which includes $\sim$80 % full beam energy [4]. Sequentially, the wave front analysing map was illustrated in Fig. 3a. It is interesting to note that the wave front distortions of tilt and focus are basically caused from the alignment precision with respect to the calibration beam, which is normally at micrometers level. However, the real wave front aberrations of the soft x-ray laser beam (HH seed and amplified seed) are normally in nano-meters level. We, therefore, ticked out the "Tilt 0°", "90°" and "Focus" during the analysing (see Fig. 3a). Wave front RMS value of the HH seed was detected as 0.53 λ, which consists of coma (0.34 λ), astigmatism at 0° (0.14 λ) and astigmatism at 45° (0.05 λ). Coma dominated the HH seed wave front aberration in this measurement.

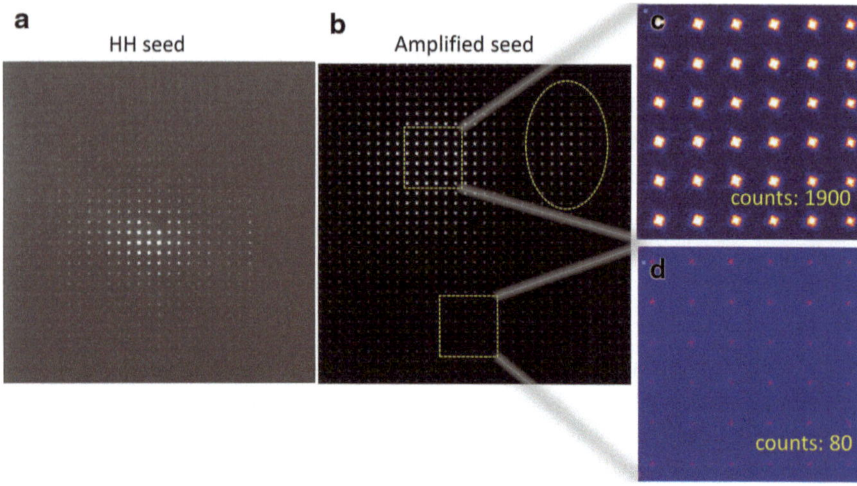

Fig. 2 Raw images of (**a**) HH seed, and (**b**) amplified seed beam detected by SXRL Hartmann sensor

When HH seed injected into the plasma amplifier (4 mm long in current case) at proper time delay between the seeding and the short pump pulse ($\Delta t = 2$ ps in this case), the energy can be significantly amplified. The amplified seed was then detected by the Hartmann wave front sensor. A sample raw image of the single-shot amplified seed beam recorded on the sensor was shown in Fig. 2b. There are three interesting regions in the full image. First of all, the bright beam primarily located in the second quadrant of the image is corresponding to the amplified seed with apparent diffraction patterns tilted by 22.5°, as displayed in the close-up (Fig. 2c). Secondly, the weak part as illustrated in Fig. 2d is principally contributed by the ASE, where the diffraction is obscure. It is important to note that ASE signal is over-spread the CCD chip. Phase analysing by the sensor is extremely dependent on the diffraction pattern, and the ASE signal may reduce the accuracy of wave front measurement. Finally, it is interesting to point out that there is a secondary bright beam observed in the first quadrant of the image as marked with dashed ellipse, which is smaller and less bright than the main amplified seed. It is, therefore, important to have a proper clipping approach to reduce the effect of ASE and secondary amplified beam during the wave front processing. Several different approaches have been discussed in the thesis work [4]. Within this context, the analysed wave front map of the amplified beam was then represented in Fig. 3b. Wave front RMS value of the HH seed was detected as 0.23 λ, which composed of coma (0.19 λ), astigmatism at 0° (0 λ) and astigmatism at 45° (0.04 λ). Wave front (RMS) of the amplified seed was improved more than two times during the amplification comparing to the input HH seed beam. Astigmatism was slightly optimized, while the coma was significantly improved.

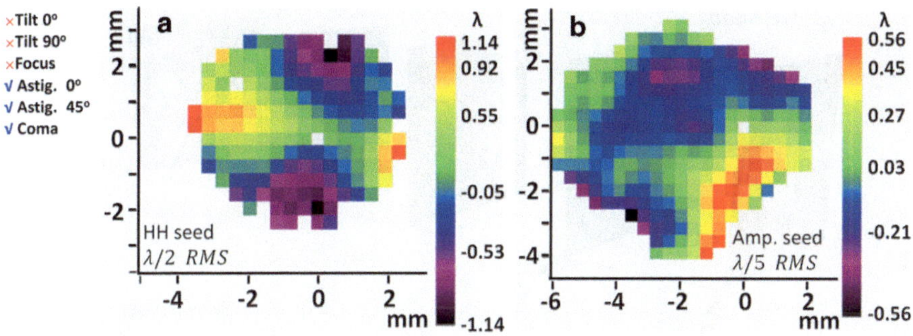

Fig. 3 Processed wave front map of **a** HH seed and **b** amplified seed beam by HASO software

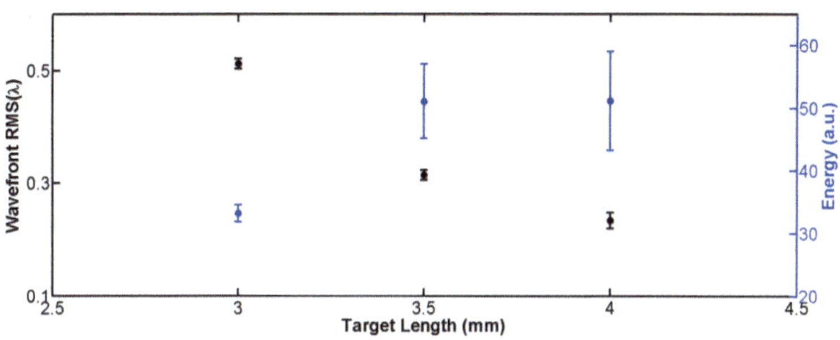

Fig. 4 Residual wave front RMS value of seeded soft x-ray laser from 3.0, 3.5 and 4.0 mm long plasma amplifiers

Furthermore, over 27 consecutive shots of both HH seed and amplified seed have been worked out [4, 5], same improvements of the wave front were observed over all the cases. Besides that, the stability of the wave front RMS value was upgraded over two times after HH seed passing through the plasma amplifier. Sequentially, we studied the effect of the plasma column length to the wave front. A series of measurements with different plasma column lengths (3.0, 3.5 and 4.0 mm) were carried out. As demonstrated in the Fig. 4, for 3 mm plasma, the wave front of amplified seed remains at $0.51\lambda \pm 0.09\lambda$, while it improves to $0.31\ \lambda \pm 0.09\lambda$ and $0.23\lambda \pm 0.01\lambda$ for 3.5 and 4 mm plasma, respectively. In another hand, the output energy of amplified is increasing with the plasma amplifier length and reaching saturation around 3.5 mm, which is in consistence with previous experiment results [6, 8]. Once again, an apparent improvement of wave front stability with respect of HH seed was achieved after plasma amplifier across the three plasma lengths. Moreover, the factors of HH seed injection time and injection angles were discussed in our recent work [4, 5, 9]. The optimizing injection time (with respect of the short pump pulse) and injection angle (in respect of the slab target plane) turned out as $\Delta t = 2$ ps and $\theta = 5$ mrad, respectively.

It is now interesting to mention that previous wave front measurement of amplified seed in the gas-based plasma amplifier [2] showed that the narrow gain zone in

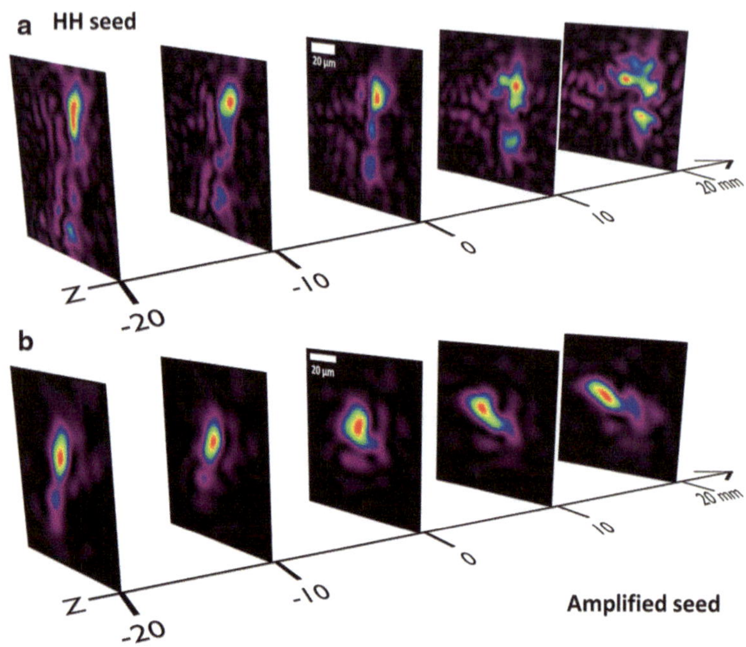

Fig. 5 The back-propagation of the (**a**) HH seed and (**b**) amplified seed source field along the Rayleigh length from −20 to 20 mm. $Z = 0$ mm corresponds to exit location of the plasma amplifier. Negative value indicates the direction deviated from the sensor

the acts as a spatial filter of the incoming beam. We are considering similar explanation in the current wave front measurement based on the solid-plasma amplifier. The coma normally generates a central spot surrounded by many rays extending far from it, which is selectively amplified when HH seed passing through the plasma amplifier. It can generate the observed drop of coma aberration. However, the gain density distribution of the gas-based [2] and solid-based [10] plasma amplifiers are different. Moreover, owing to the calibration limit of this Hartmann wave front sensor is 0.03λ RMS, the astigmatism at 45° did not show strong improving evidence while the astigmatism at 0° was drop down below the limit.

Taking the advantage of the Hartmann wave front sensor which recording both intensity and phase of the investigating beam, one could describe the complex amplitude of the electromagnetic field on the sensor plane [4]. In consequence, the electromagnetic field at any location can be reconstructed by an inverse Fresnel transformation. Point spread function in a free space was then performed for both detected HH seed and amplified seed, as illustrated in Fig. 5a and b, respectively. The "Z = 0 mm" position is corresponding to the exit location of the plasma amplifier. Negative value indicates the direction deviated from the sensor. Here, the HH seed was assumed as monochromatic beam (18.9 nm). The reconstructed HH source showed elongated profile in vertical, and ∼20 µm in horizontal. One can also observe that there are multi-spot in elongated vertical with a main spot at the top. This can be caused by the process of HH generation in the gas puff and the imaging from the toroidal mirror. Nevertheless, this multi-spot phenomena could give an explanation of the multi-amplification observed in the amplified seed beam, as described in Fig. 2b. For the back-propagated amplified beam (see Fig. 5b), the secondary amplified seed beam was not taken into account in the original amplified seed beam on the detection plane. The source at $Z = 0$ mm is relatively symmetrical shape with spot size of 20 μm$\times$20 μm FWHM. Significant beam clearness by the plasma amplifier can be declared when comparing the back propagation beam profiles of HH seed and amplified seed.

4 Summary

We observed that wave front RMS value and its stability of amplified seed improved $\sim$ 2.5 times and over 4 times, respectively, with respect to that of HH seed. As coma dominated the wave front distortions, which is caused by the toroidal, HH-seeded SXRL is capable to reach diffraction-limited ($\lambda/14$) beam by optimizing the HH seed beam quality and using aberration-free injection mirror. The reconstructed source confirms the high brightness of $\sim 1\times 10^{26}$ photons $\cdot$ s^{-1} $\cdot$ mm^{-2} $\cdot$ mrad^{-2}.

References

1. Gautier, J., et al.: Optimization of the wave front of high order harmonics. Eur. Phys. J. D. **48**, 459–463 (2008)
2. Goddet, J.P., et al.: Aberration-free laser beam in the soft x-ray range. Opt. Lett. **34**, 2438 (2009)
3. Mercere, P., et al.: Hartmann wave-front measurement at 13.4 nm with λ_{EUV}/120 accuracy. Opt. Lett. **28**, 1534 (2003)
4. Li, L.: Metrology of high harmonics seeded soft x-ray laser based on solid-plasma amplifier. Ecole Polytechnique, (2015, in press)
5. Li, L., et al.: Wavefront improvement in an injection-seeded soft x-ray laser based on a solid-target plasma amplifier. Opt. Lett. **38**, 4011–4014 (2013)
6. Wang, Y., et al.: High-brightness injection-seeded soft-x-ray-laser amplifier using a solid target. Phys. Rev. Lett. **97**, 123901 (2006)
7. Berrill, M., et al.: Improved beam characteristics of solid-target soft x-ray laser amplifiers by injection seeding with high harmonic pulses. Opt. Lett. **35**, 2317 (2010)
8. Wang, Y., et al.: Phase-coherent, injection-seeded, table-top soft-x-ray lasers at 18.9 nm and 13.9 nm. Nat. Photonics **2**, 94 (2008)
9. Li, L., et al.: Wavefront of seeded soft x-ray laser based on a solid-target plasma amplifier. Proceedings of SPIE, volume 8849 (2013)
10. Zeitoun, Ph., et al.: X-ray chirped pulse amplification: towards GW soft x-ray lasers. Appl. Sci. **3**, 581–592 (2013)

Recent Advances in Multilayer Reflective Optics for EUV/X-Ray Sources

R. Soufli, J. C. Robinson, M. Fernández-Perea, E. Spiller, N. F. Brejnholt, M.-A. Descalle, M. J. Pivovaroff and E. M. Gullikson

Abstract This paper discusses the development of (i) corrosion-resistant multilayers for the 25–80 nm region (ii) multilayer mirrors for the first 0.5-NA Micro-Exposure Tools at 13.5 nm and (iii) multilayer mirrors for the soft gamma-ray range.

1 Introduction

Multilayer mirrors are often essential optical elements in experiments and applications involving extreme ultraviolet (EUV) and x-ray sources. Multilayer interference coatings achieve high reflective performance due to constructive wave interference phenomena and thus enable imaging at near-normal incidence angles in the EUV/soft x-ray wavelength region. They also enable efficient operation at grazing incidence angles larger than the critical angle, thus greatly easing the fabrication, mounting and alignment of grazing-incidence x-ray optical systems. In the last few years, novel EUV/x-ray sources have emerged (4th generation synchrotrons, free-electron lasers, tabletop lasers, high-harmonic generation and attosecond sources) ushering a new era in the fields of materials science, chemistry, plasma physics, biology and life sciences. Increasingly efficient and sophisticated EUV/x-ray multilayer optics are also needed for space-borne telescopes for solar physics and astrophysics, radiation detection and medical imaging applications, high-energy physics and semiconductor photolithography.

This paper summarizes recent results and ongoing work from the development of reflective multilayer coatings for wavelengths ranging from the EUV to the soft gamma-ray regions of the spectrum. All multilayer coatings discussed below were developed and fabricated at Lawrence Livermore National Laboratory (LLNL)

R. Soufli (✉) · J. C. Robinson · M. Fernández-Perea · N. F. Brejnholt · M.-A. Descalle · M. J. Pivovaroff
Lawrence Livermore National Laboratory, Livermore, CA, USA
e-mail: regina.soufli@llnl.gov

E. Spiller
Spiller X-ray Optics, Livermore, CA, USA

E. M. Gullikson
Lawrence Berkeley National Laboratory, Berkeley, CA, USA

© Springer International Publishing Switzerland 2016

J. Rocca et al. (eds.), *X-Ray Lasers 2014*, Springer Proceedings in Physics 169,
DOI 10.1007/978-3-319-19521-6_43

using DC-magnetron sputtering techniques. More details on the deposition parameters related to the multilayer coatings discussed in this paper can be found in the References provided in the following sections.

2 Corrosion-Resistant Multilayer Mirrors for the 25–80 nm Wavelength Region

The wavelength region just below the Mg $L_{2,3}$ absorption edge (corresponding to wavelengths longer than 25 nm) contains several emission lines of highly ionized materials and coincides with the operational range of EUV sources such as table-top lasers, high-harmonic generation sources, synchrotrons and free-electron lasers as well as solar physics studies. Mg/SiC has been shown to be the best-performing multilayer coating in the 25–80 nm wavelength region, as it possesses a unique combination of consistently high reflectivity, good spectral selectivity, thermal stability to 350° C and near-zero stress [1]. However, Mg/SiC suffers from Mg-induced corrosion, an insidious problem which leads to degradation of the Mg/SiC multilayer film and its reflectivity, as shown in Fig. 1a, 1b. This problem has prevented Mg/SiC coatings from being used in applications that require long lifetime stability, which is a crucial requirement in applications such as space-borne telescopes for EUV solar physics [2]. It was established that corrosion starts from the top of the multilayer and is caused by environmental agents (reactive ions), which reach the Mg layers buried under the SiC capping (top) layer through pinholes and other defects typical in sputtered thin films. It was determined that Mg corrosion products initially appear as micron-size spots invisible to the eye, which may later grow and expand in volume, resulting in visible, eruptive effects which cause delamination of the top layers in the multilayer and exposure of the inner Mg layers.

To address this problem, Al-Mg corrosion barrier layers were developed [3]. The Al and Mg layers are deposited as two separate layers (underneath the SiC capping layer) and they spontaneously intermix to form a partially amorphous Al-Mg layer (Fig. 1c, 1d) which provides efficient corrosion resistance while maintaining the favorable reflective properties of the original, unprotected Mg/SiC multilayer. The efficacy of this corrosion resistance concept was verified experimentally on Mg/SiC films aged for 3 years. Mg/SiC multilayers with Al-Mg corrosion barriers were demonstrated, with high narrowband reflectivity in up to 3 bands [4]. The phenomenon of spontaneous intermixing and amorphization of sputtered Al and Mg layers with nanometer-scale thickness was observed for the first time during this work. Research is now in progress to determine in detail the physics of the Al-Mg layer formation, including the timescales of spontaneous intermixing and amorphization and any crystalline phases (Al, Mg or AlMg) possibly being present in the intermixed Al-Mg layer [5].

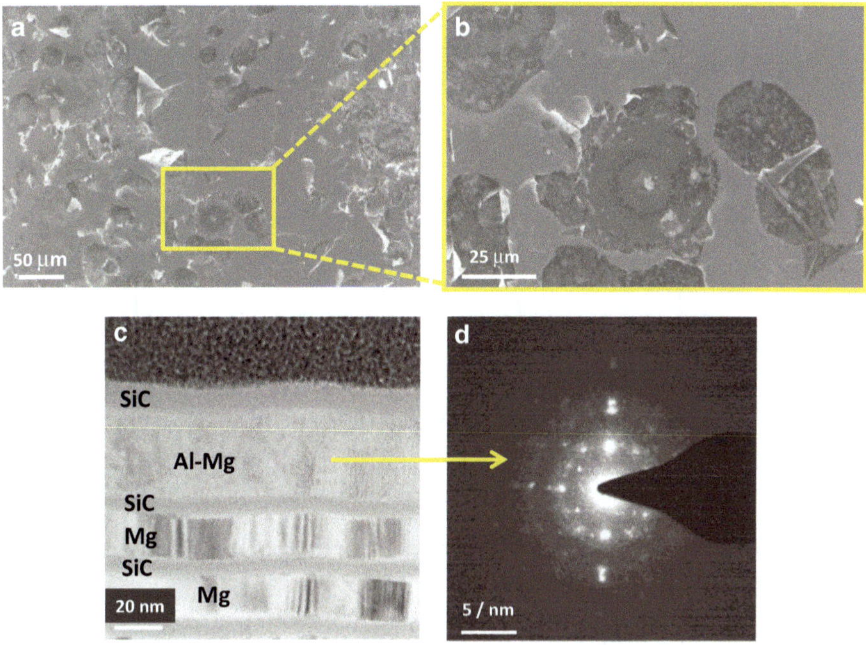

Fig. 1 **(a)**, **(b)** Scanning Electron Microscopy (*SEM*) images obtained at LLNL of the top surface of a severely corroded Mg/SiC multilayer film aged for 3 years. Several areas of the coating have erupted due to the formation of voluminous Mg corrosion products, with the top portions of the coating either still partially attached to the surface, or entirely missing (from Ref. [3]) **(c)** Cross-sectional Transmission Electron Microscopy (TEM) image of the top layers of a corrosion-resistant Mg/SiC multilayer designed for operation at wavelengths around 46 nm at normal incidence. The spontaneously intermixed, Al-Mg corrosion barrier is shown underneath the top SiC layer. (from Ref. [3]). **(d)** Electron diffraction image taken from the middle area of the intermixed Al-Mg layer. The appearance of diffuse rings and dispersed diffraction peaks indicates the presence of amorphous or nanocrystalline material. TEM and electron diffraction imaging were performed at Evans Analytical Group, (Sunnyvale, California)

3 Multilayer Mirrors for the First 0.5-NA Micro-Exposure Tools at 13.5 nm

Micro-Exposure Tools (METs) are small field (30 μm $\times$ 200 μm) tools developed to provide early learning towards the extendibility of EUV lithography using 13.5 nm wavelength of illumination, especially in the areas of photoresist and mask development. METs with a numerical aperture (NA) of 0.3 have been operational for over 10 years, using laboratory (e.g: laser plasma) or synchrotron EUV sources. Next-generation METs with a NA of 0.5 (MET5) have been recently proposed [6] to demonstrate EUV patterning of features with resolution of 11 nm (half-pitch) and below. A modified Schwarzschild design is used for the MET5 projection optics. The maximum aspherical sag and slope of the MET5 mirror substrates are a factor

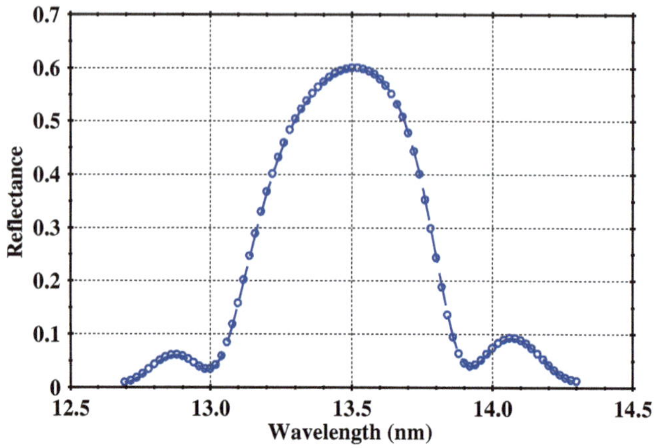

Fig. 2 Measured EUV reflectance of a Mo/Si multilayer with N = 30 bilayers deposited on top of a Cr under-layer. The coating has 60.1 % peak reflectivity and − 135 MPa stress. The substrate was a Si wafer with < 0.1 nm rms high-spatial frequency roughness. This coating was developed for the M1 mirror of the MET5 system. Measurement was performed at ALS beamline 6.3.2

of 7–12 higher than the previous MET (NA = 0.3) optics, thus presenting a significant leap in the state-of-the-art in substrate manufacturing [7, 8]. Both the convex primary (M1) and concave secondary (M2) projection mirrors are rotationally symmetric aspheres. The angles of incidence (AOI) of operation in the MET5 system, defined from the normal incidence direction, range from 4–14° across a clear aperture radius of 14.5–46.1 mm for M1 and from 1 to 4° across a clear aperture radius of 39.2–125.4 mm for M2. The maximum height difference (sag) between the center and the edge of the reflective surface is about 3 mm for the M1 mirror and 33 mm for the M2 mirror.

The MET5 multilayer coatings are subject to extremely stringent wavefront error and wavelength matching tolerances. To minimize the wavefront error contributions of the multilayer optics in the MET5 system, the MET5 multilayer coatings were especially optimized to achieve simultaneously the highest reflectivity, lowest stress and lowest figure error (the latter being proportional to the total coating thickness). They consist of N = 30 bi-layers of Mo/Si and contain a Cr under-layer that was optimized separately for each mirror. The resulting multilayer coating stress values are on the order of − 100 MPa (compressive) with peak reflectance around 60 %, as shown in Fig. 2. The multilayer coating profile in the lateral direction was designed for both M1 and M2 mirrors to produce a phase and a centroid wavelength (of the reflectance vs. wavelength curve) that remain constant at all locations within the mirror clear aperture, at the angles of incidence of the MET5 system. Multilayer thickness control across the curved surface of each mirror was achieved using a velocity modulation technique during deposition [9]. The non-compensable, multilayer-added figure error tolerance due to deviations from the

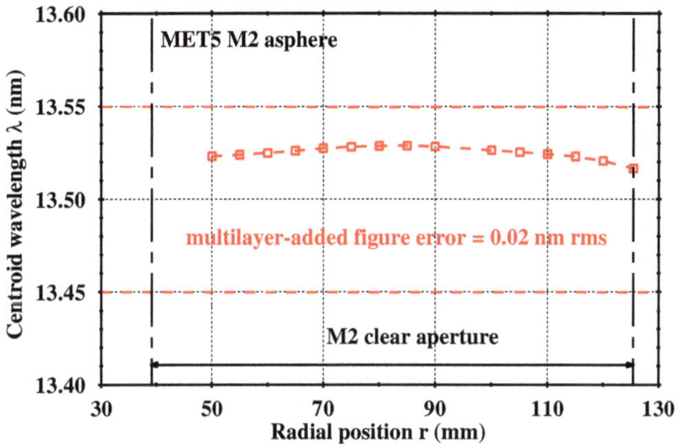

Fig. 3 EUV centroid wavelength measured across the clear aperture of the multilayer-coated M2 mirror at the angles of incidence (*AOI*) of operation in the MET system. A non-compensable multilayer-added figure error of 0.02 nm rms was determined from these results. The dash lines indicate the goal wavelength tolerance (13.5 ± 0.05 nm) at AOI. Measurements were performed at ALS beamline 6.3.2

as-designed multilayer thickness was set to 0.08 nm rms for each MET5 mirror. Figure 3 shows the centroid wavelength obtained from EUV reflectance measurements on a multilayer-coated aspherical M2 test mirror, at the M2 angles of incidence of operation. EUV reflectance measurements were obtained at beamline 6.3.2. of the Advanced Light Source (ALS) synchrotron at Lawrence Berkeley National Laboratory. The non-compensable, multilayer-added figure error of the M2 multilayer coating was determined to be 0.022 nm rms, after subtraction of a polynomial term $ax^2 + b$ which is entirely compensable via alignment shifts of the mirrors in the MET5 system. The M2 mirror centroid wavelength, calculated from the M2 profile at AOI shown in Fig. 3 and weighted by area within the clear aperture, was found to be 13.52 nm, which is within the centroid wavelength specification of 13.5 ± 0.05 nm. A more detailed discussion of the multilayer coating results from both MET5 M1 and M2 mirrors can be found in Ref. [8].

4 Multilayer Mirrors for the Soft Gamma-Ray Range

In the gamma-ray photon energy range, Laue lenses (crystals) and coded apertures are typically being used as focusing or collimating elements. Multilayer-coated mirrors, if used as focusing optics operating in the soft gamma-ray, could afford large improvements in sensitivity and resolution. Efficient multilayer mirrors have recently been demonstrated experimentally at the highest reported photon energies (384, 511 and 642 keV). WC/SiC multilayer coatings with ultra-short-periods (1.5 nm) deposited on flat, super-polished glass substrates [10] were used for this purpose and achieved 52.6 % peak reflectivity at 384 keV, at grazing incidence angle of 0.063°, measured at beamline ID15A of the European Synchrotron Radiation Facility (ESRF) [11, 12]. The effects of incoherent Compton scattering, which is negligible at lower photon energies but becomes significant in the soft gamma-ray, were also quantified for the first time as part of this work, using a Monte-Carlo particle simulation code. It was shown that considerable Compton scattering occurs for incidence angles in-between the Bragg resonances. Nevertheless, Compton scattering becomes insignificant for angles below the critical angle and around the Bragg reflectivity peaks, because most of the radiation is reflected by the multilayer thin film structure before it has the chance to reach the substrate and scatter [11]. These results foreshadow the use of multilayer reflective optics in soft gamma-ray applications such as nuclear [13] and medical physics and astrophysics. Of particular interest is the 511 keV photon energy corresponding to the electron-positron annihilation line, which is relevant in laser plasma diagnostics, nuclear medicine and astronomy. The ultra-short multilayer period thicknesses required for operation at gamma-ray photon energies, approach the limits of continuous layer formation and thus challenge the state-of-the-art in deposition technologies. Manipulating the deposition physics towards achieving the thinnest possible layers with the sharpest and most stable interfaces inside the multilayer structure, as needed to reach the highest reflective performance, remains an ongoing research endeavor.

Acknowledgements The contributions of the following colleagues are gratefully acknowledged: Jennifer Alameda, Jay Ayers, Shannon Ayers, Sherry Baker, Tom McCarville, Aaron Sperry, Chris Walton (LLNL), Luis Rodríguez-De Marcos, Jose Méndez, Juan Larruquert (Instituto de Óptica, Consejo Superior de Investigaciones Científicas), Klaus Ziock (Oak Ridge National Laboratory), Veijo Honkimäki, Eric Ziegler, Thomas Buslaps (ESRF), Finn Christensen and Anders Jakobsen (Danish Technical University (DTU)-Space). This work was performed under the auspices of the U.S. Department of Energy by Lawrence Livermore National Laboratory under Contract DE-AC52-07NA27344. Funding was provided in part by LLNL's Laboratory Directed Research and Development Program, by Zygo Corporation and by the National Nuclear Security Administration's Office of Nonproliferation and Verification Research and Development.

References

1. Takenaka, H., Ichimaru, S., Ohchi, T., Gullikson, E.M.: J. Electron Spectrosc. Relat. Phenom. Soft-X-ray reflectivity and heat resistance of SiC/Mg multilayer. **144–147**, 1047–1049 (2005)
2. Soufli, R., Fernández-Perea, M., Robinson, J.C., Baker, S.L., Alameda, J., Gullikson, E.M.: Corrosion-resistant, high-reflectance Mg/SiC multilayer coatings for solar physics in the 25–80 nm wavelength region. Proc. SPIE. **8443**, 84433R (2012)
3. Soufli, R., Fernández-Perea, M., Baker, S.L., Robinson, J.C., Alameda, J., Walton, C.C.: Spontaneously intermixed Al-Mg barriers enable corrosion-resistant Mg/SiC multilayer coatings. App. Phys. Lett. **101**, 04311 (2012)
4. Fernández-Perea, M., Soufli, R., Robinson, J.C., Rodríguez-de Marcos, L., Mendez, J.A., Larruquert, J.I., Gullikson, E.M.: Triple-wavelength, narrowband Mg/SiC multilayers with corrosion barriers and high peak reflectance in the 25–80 nm wavelength region. Opt. Express. **20**, 24018–24029 (2012)

5. Soufli, R., et al.: Corrosion-resistant Mg/SiC multilayer coatings for EUV laser sources in the 25–80 nm wavelength region. Proc. SPIE. **8849**, 88490D (2013)
6. Goldstein, M., Hudyma, R., Naulleau, P., Wurm, S.: Extreme-ultraviolet microexposure tool at 0.5 NA for sub-16 nm lithography. Opt. Lett. **33**, 2995–2997 (2008)
7. Cummings, K., et al.: Update on the Sematech 0.5 NA Extreme Ultraviolet Lithography (EUVL) Microfield Exposure Tool (MET). Proc. SPIE. **9048**, 90481M (2014)
8. Glatzel, H., et al.: Projection optics for EUVL micro-field exposure tools with 0.5 NA. Proc. SPIE. **9048**, 90481K (2014)
9. Soufli, R., Hudyma, R.M., Spiller, E., Gullikson, E.M., Schmidt, M.A., Robinson, J.C., Baker, S.L., Walton, C.C., Taylor, J.S.: Sub-diffraction-limited multilayer coatings for the 0.3 numerical aperture micro-exposure tool for extreme ultraviolet lithography. Appl. Opt. **46**, 3736–3746 (2007)
10. Fernández-Perea, M., et al.: Ultra-short-period WC/SiC multilayer coatings for x-ray applications. NIM. A. **710**, 114–119 (2013)
11. Fernández-Perea, M., et al.: Physics of reflective optics for the soft gamma-ray photon energy range. Phys. Rev. Lett. **111**, 027404 (2013)
12. Brejnholt, N.F., Soufli, R., Descalle, M.-A., Fernández-Perea, M., Pivovaroff, M.J., Christensen, F.E., Jakobsen, A.C., Honkimäki, V.: Demonstration of multilayer reflective optics at photon energies above 0.6 MeV. Opt. Express. **22**, 15364–15369 (2014)
13. Pivovaroff, M.J., Ziock, K.P., Fernandez-Perea, M., Harrison, M.J., Soufli, R.: Gamma-ray mirrors for direct measurement of spent nuclear fuel. NIM. A. **743**, 109–113 (2014)

Spectral and Coherence Properties of the Ne-like Zn X-Ray Laser at PALS

A. Le Marec, M. Kozlova, J. Nejdl, L. Meng, F. Tissandier, O. Guilbaud, A. Calisti and A. Klisnick

Abstract We present a refined measurement of the temporal coherence of the zinc X-ray laser generated at PALS, using a wavefront-division interferometer. The corresponding bandwidth of the lasing line is shown (i) to be broader than those of other types of X-ray lasers, previously characterized with the same instrument, and (ii) is compatible with the amplification of subpicosecond pulses. Similar measurements were also performed for shorter amplification lengths and the small-signal gain was carefully characterized.

1 Introduction

The pulse duration of plasma-based X-ray lasers (XRL) is currently limited to the picosecond timescale, even in the seeded mode where a high-order harmonic pulse of a few 10 fs is injected to be amplified in the inverted medium [1]. Over the last few years, efforts have been made to shorten the duration of XRLs to the femtosecond range, and this requires a better understanding of the processes that control the pulse duration of these sources. Among the parameters at stake, the spectral properties of the amplifier play an important role [2]. The latter are strongly related to the plasma parameters such as the electron density and the electron and ion temperatures, since

A. Le Marec (✉) · A. Klisnick
ISMO, CNRS, Université Paris-Sud, 91405 Orsay Cedex, France
e-mail: andrea.le-marec@u-psud.fr

M. Kozlova · J. Nejdl
IoP ASCR, ELI Beamlines Project, Prague, Czech Republic

IPP ASCR, Prague, Czech republic

L. Meng
Institute of Fluid Physics, China Academy of Engineering Physics, Mianyang, China

F. Tissandier
LOA, CNRS, ENSTA, Palaiseau, France

O. Guilbaud
LPGP, CNRS, Université Paris-Sud, Orsay, France

A. Calisti
Aix Marseille Université, CNRS, PIIM UMR 7345, Marseille, France

© Springer International Publishing Switzerland 2016
J. Rocca et al. (eds.), *X-Ray Lasers 2014*, Springer Proceedings in Physics 169,
DOI 10.1007/978-3-319-19521-6_44

they control the intrinsic broadening of the spectral lines. In particular the general qualitative behaviour predicted by the models is that the spectral bandwidth is enlarged when the plasma density and (ion) temperature are increased. On the other hand, collisional XRLs can be generated with different pumping techniques, leading to markedly different plasma conditions in the gain zone. A meticulous and comprehensive characterization of the spectral bandwidth of the different types of XRLs is thus necessary in order to assess their potential to amplify pulses with subpicosecond duration, and to provide benchmarking for numerical simulations.

The aim of the experiment described in this paper was to investigate the temporal coherence and related spectral width of the quasi-steady-state (QSS) XRL emitting at 21.2 nm (transition $(2p^5_{1/2}3p_{1/2})_{J=0} \rightarrow (2p^5_{1/2}3s_{1/2})_{J=1}$ in Ne-like zinc), generated at the PALS (Prague Asterix Laser System) facility in Prague, Czech Republic. This experimental campaign is a follow-up of a first experiment carried out in 2012, where the same measurement was attempted and encouraging preliminary measurements were obtained [3]. However, the specificities of the zinc XRL, in particular in terms of spatial coherence or repetition rate, made the measurement challenging. This led to several improvements in the experimental setup and method, which were implemented for the present work, and to a more reliable and extended spectral characterization of the zinc XRL, which will be presented below.

2 Experimental Method and Setup

The temporal coherence is measured using a specifically designed variable path-difference wavefront-division interferometer [4], which has been previously used with other types of XRLs [2]. The measurement is based on acquiring series of single-shot interferograms while increasing the path difference (i.e. the delay) between the arms of the interferometer, leading to a reduced fringe visibility. A 2D map of the visibility across each interferogram is obtained by processing a Fourier transform of the fringe pattern with a small sliding window, after background subtraction. The measured visibilities for each shot are then plotted and fitted with an appropriate function (Gaussian in the present case). The coherence time is defined as the delay at which the maximum visibility (i.e. visibility at zero delay) is reduced by a factor 1/e. Using the Wiener–Khintchine theorem, the spectral profile of the lasing line, hence the corresponding spectral FWHM, is inferred from the Fourier transform of the visibility curve. The maximum visibility at zero delay is related to the mutual coherence between the two half-beams at the entrance of the interferometer. It is generally less than one, due to the limited spatial coherence of the beam, but should be kept above 0.5, in order to ensure that the fringe modulation remains above noise when the delay is increased. This is achieved by adjusting the distance of the interferometer such as the transverse coherence length is at least equal to the size of the overlapping region over which fringes are produced.

The 21.2 nm XRL is generated by irradiating a 3 cm-long zinc slab target with the PALS iodine laser (1.315 μm) [9]. The pump-laser sequence consists of a prepulse ($\sim$ 2 J–300 ps) followed after 10 $\pm$ 0.5 ns by the main driving pulse

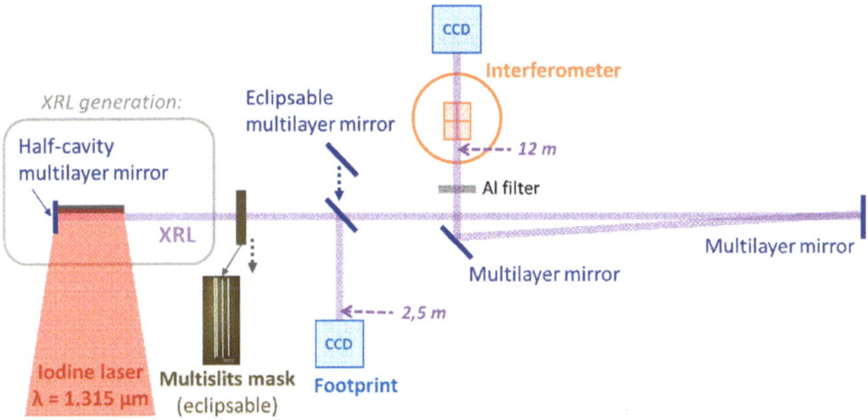

Fig. 1 Experimental setup showing the folded XRL beam path from the plasma source to the interferometer

(~600 J–300 ps). Saturated lasing is achieved through double-pass in a half-cavity, which consists of a flat Mo:Si multilayer mirror (reflectivity 30 %) positioned at ~ 5–7 mm from the plasma end. As mentioned above, several modifications were introduced to the measurement setup and procedure compared to the previous campaign. First of all, each measurement was performed in a single day, so as to avoid the effects of possible day-to-day variations of the pump laser. Owing to the relatively low repetition rate (two shots per hour), this led to a severe limitation of the number of data shots included in each visibility scan (~ 15–16, as compared to ~ 50 with other XRL types previously investigated). The distance from the source to the interferometer was increased to 12 m (instead of 7 m in 2012) in order to ensure a maximum visibility larger than 0.5 (see above). Finally a single-shot diagnostic of the spatial coherence of the XRL beam consisting of a multislits mask (see Fig. 1) was implemented. This diagnostic was used to check that the spatial coherence of the beam did not vary significantly from shot-to shot (see also [5]). Hence the observed loss of visibility when increasing the path difference in our measurement can be attributed to the effect of finite temporal coherence.

3 Results and Discussion

Figure 2 shows the measured visibilities for a 3-cm-long amplifier operated in double-pass, where the delay between the interfering beams was varied between −0.35 ps and +1.06 ps. As expected, the larger distance from the source (12 m) at which the interferometer was placed leads to a larger maximum visibility than in the previous experimental campaign: V_{max} ~0.55 (compared to ~ 0.3 in 2012). In spite of the limited number of shots (13) the data are well fitted with a Gaussian function, yielding a coherence time (as defined in Sect. 2) $\tau_c = 0.68 \pm 0.07$ ps. This is consistent with our previous measurement performed in 2012 with the interferometer

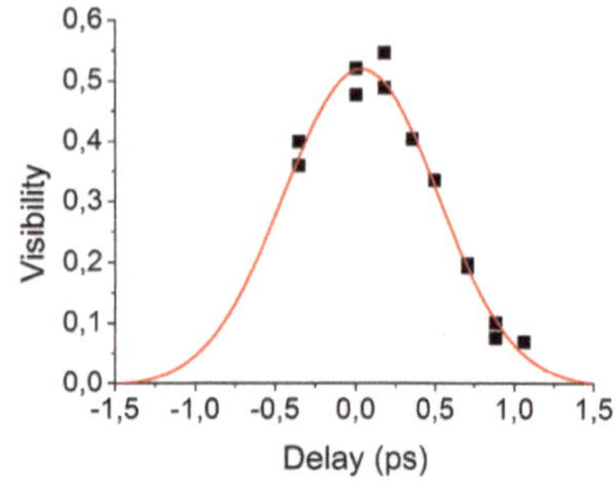

Fig. 2 (■) measured visibilities as a function of the delay, (–) best fit

placed at a distance of 7 m (see Table 1) [3]. This thus confirms that the coherence time of the QSS-pumped zinc XRL is shorter than the other types of XRLs previously characterized with the same instrument, and is actually shorter than 1 ps.

Following the Wiener–Khintchine theorem, the corresponding spectral profile of the source has a Gaussian shape with a FWHM $\Delta\nu = 2\sqrt{\ln 2}/\pi\,\tau_c$ and an associated Fourier limit duration given by $\tau_{FL} = \sqrt{\ln 2}\tau_c$. The bandwidth of the QSS XRL is thus $\Delta\lambda/\lambda \approx 5.5 \times 10^{-5}$, currently the broadest bandwidth ever measured for a saturated XRL. Such a large bandwidth is related to the specific plasma conditions at which QSS collisional pumping is achieved (see Sect. 1), namely both a high ion temperature ($\sim$ 150–200 eV) and a high electron density ($\sim$ 2–5 $\times 10^{20}$ cm^{-3}). The Fourier limit duration corresponding to the measured bandwidth is then $\tau_{FL} = 0.57 \pm 0.06$ ps. This is almost twice as short as the shortest duration measured so far for a seeded XRL pulse [1]. Table 1 summarizes the maximum visibility, coherence time, spectral bandwidth and Fourier limit duration inferred for each XRL-interferometer distance (5 m and 7 m in the first experimental campaign in 2012, 12 m in the second experimental campaign in 2013).

In order to investigate the spectral gain narrowing of the XRL line, we have performed two additional measurements with shorter amplifier lengths. This has been done by operating the XRL in single-pass (no half-cavity mirror) with target lengths of 3 and 2 cm. The measured bandwidths have been compared to 1D-radiative transfer calculations based on J. A. Koch's model [6, 7]. The calculated amplified linewidth is strongly dependent on the gain value g_0 used as an input. Therefore, a careful characterization of gain has been undertaken during our experimental campaign. The time-integrated intensity of the XRL beam, operated in single pass, was recorded as a function of the amplification length. Three series of measurements were performed, with the CCD detector placed at three different

Table 1 Summary of the coherence time measurements of the experimental campaigns of 2012 and 2013

XRL distance	V_{max}	τ_c (ps)	τ_{FL} (ps)	$\Delta\lambda/\lambda$ ($\times 10^{-5}$)
2012: 05 m	~ 0.3	1.03 ± 0.04	0.86 ± 0.03	3.6 ± 0.1
2012: 07 m	~ 0.3	0.71 ± 0.03	0.59 ± 0.02	5.3 ± 0.2
2013: 12 m	~ 0.55	0.68 ± 0.07	0.57 ± 0.06	5.5 ± 0.6

distances from the source: 2.5, 12.5 and 13.25 m. In the latter case the intensity was directly measured from the interferograms, while for the two other distances it was measured from the beam footprint. In all cases the contribution from the plasma thermal emission was subtracted from the measured intensity. The data were then fitted with a Linford-type function [8], which describes the dependence of the frequency-integrated intensity versus length, in a 1D approximation. The inferred gain values range from $2 \pm 0.2\ \text{cm}^{-1}$ to $3.1 \pm 0.2\ \text{cm}^{-1}$, i.e. significantly lower than $g_0 = 7\ \text{cm}^{-1}$, as reported in an earlier gain measurement of the PALS XRL [9]. Using these measured gain values we find a good agreement between the computed amplified linewidths and our experimental data. These results will be presented and discussed in a future paper.

4 Conclusions

We have successfully characterized the temporal coherence and the bandwidth of the zinc X-ray laser generated at PALS in the quasi-steady-state pumping regime. This completes the benchmarking of the different X-ray laser types, which has been performed using the same wavefront-division interferometer. The QSS zinc X-ray laser presents a broader bandwidth than the previously investigated X-ray lasers, due to its more favorable plasma conditions for spectral broadening. Gain narrowing of the lasing line was also observed. The measured data are consistent with line transfer calculations using the gain values measured during the same experimental campaign. Finally, the bandwidth of the zinc x-ray laser appears to be compatible with the amplification of subpicosecond pulses, which could be generated by seeding it with a femtosecond high-order harmonic pulse.

Acknowledgements This work benefits from LASERLAB-EUROPE—grant agreement no. 284464, EC's Seventh Framework Programme, ELI: EXTREME LIGHT INFRASTRUCTURE—project no. CZ.1.05/1.1.00/02.0061 and ECOP—project nos. CZ.1.07/2.3.00/30.0057 and CZ.1.07/2.3.00/20.0279.

References

1. Wang, Y., Berril, M., Pederaci, F., et al.: Phys. Rev. A. **79**, 023810 (2009)
2. Klisnick, A.: Spectral properties of collisional XUV lasers for the amplification of femtosecond pulses, these proceedings
3. Meng, L.M., Klisnick, A., Kozlová, M., et al.: Proceedings of 13th ICXRL (2013)
4. Klisnick, A., et al.: J. Quant. Spectrosc. Radiat. Transf. **99**, 370 (2006)
5. Kozlová, M.: Overview of development of laser driven secondary sources at PALS and ELI, these proceedings
6. Koch, J.A., MacGowan, B.J., et al.: Phys. Rev. A. **50**, 1877 (1994)
7. Guilbaud, O.: Phd thesis, Université Paris-Sud, Orsay, France (2004)
8. Linford, G.J., Peressini, E.R., Sooy, W.R., et al.: Appl. Opt. **13**, 379–390 (1994)
9. Rus, B., Mocek, T., Präg, A.R., Kozlová, M., et al.: Phys. Rev. A. **66**, 063806 (2002)

On Stability of Exact Transparent Boundary Condition for the Parabolic Equation in Rectangular Computational Domain

R. M. Feshchenko

Abstract Recently a new exact transparent boundary condition (TBC) for the 3D parabolic wave equation (PWE) in rectangular computational domain was derived. However in the obtained form it does not appear to be unconditionally stable when used with, for instance, the Crank–Nicolson finite-difference scheme. In this paper two new formulations of the TBC for the 3D PWE in rectangular computational domain are reported, which are likely to be unconditionally stable. They are based on an unconditionally stable fully discrete TBC for the Crank–Nicolson scheme for the 2D PWE. These new forms of the TBC can be used for numerical solution of the 3D PWE when a higher precision is required.

1 Introduction

In our previous paper we demonstrated a new transparent boundary condition for the 3D PWE in rectangular computational domain (RCD) [1]. The numerical implementation used was based on the standard Crank–Nicolson finite-difference (FD) scheme [2] together with a discretization of the continues Baskakov–Popov 1D TBC [3].

In this paper we consider numerical implementations of the same 3D TBC in RCD but based on different FD schemes and 1D boundary conditions. In particular we use the alternating-direction-implicit FD scheme (ADIS) [4] as well as Crank–Nicolson FD scheme both combined with a fully discrete unconditionally stable 1D TBC that we derive here. Furthermore, we access the precision, numerical complexity and stability of the mentioned combinations.

For a recent review of TBC for parabolic type equations see [5].

R. M. Feshchenko (✉)
P.N. Lebedev Physical Institute, Russian Academy of Sciences,
53 Leninski Prospect, Moscow 119991, Russia
e-mail: rusl@sci.lebedev.ru

© Springer International Publishing Switzerland 2016
J. Rocca et al. (eds.), *X-Ray Lasers 2014*, Springer Proceedings in Physics 169,
DOI 10.1007/978-3-319-19521-6_45

2 1D Fully Discrete Boundary Condition

Fully discrete boundary conditions are derived directly from the discretized PWE and, therefore, are FD scheme dependent. Their main advantage over discretized continues TBCs (like Baskakov–Popov TBC) is that they preserve unconditional stability of the respective FD scheme.

Let us derive an exact fully discrete TBC for the Crank–Nicolson discretization [2] of the 2D PWE in free space:

$$2ik\frac{\partial u}{\partial z}+\frac{\partial^2 u}{\partial x^2}=0, \tag{1}$$

where u is the field amplitude. The computational domain is defined as $|x|\leqslant a$, where a is a positive real number. The Eq. (1) holds outside this domain and at its boundaries at $x=\pm a$. The initial condition for Eq. (1) is $u(x,z=0)=u_0(x)$, where u_0 is assumed to be equal to $\exp(iqx)$ if $|x|\geqslant a$ (q is a parameter). The uniform Crank–Nicolson discretization of Eq. (1) is as follows

$$-u^{n+1}_{p+1}+Bu^{n+1}_{p}-u^{n+1}_{p-1}=u^{n}_{p+1}-\tilde{B}u^{n}_{p}+u^{n}_{p-1}, \tag{2}$$

where

$$B=2-4ikh^2/\tau,\quad \tilde{B}=2+4ikh^2/\tau,\quad 0\leqslant p\leqslant N,\quad 0\leqslant n\leqslant N_\tau,$$

$N+1$ is the total number of x axis nodes and $N_\tau+1$ is the total number of z axis nodes. Parameters $h=2a/N$ and τ are steps in x and z directions, respectively. The boundary condition will be imposed at x corresponding to either $p=1$ or $p=N$.

Applying the one-sided Z-transform to Eq. (2), then performing simple transformations and finally applying the inverse Z-transform one can obtain the following TBC for upper boundary (upper sign) and lower boundary (lower sign) of the computational domain:

$$u^{n+1}_{N_b+1}-u^{n+1}_{N_b-1}= \\ \mp 2\sum_{l=0}^{n+1}\beta^{n+1-l}u^{l}_{N_b}+2i\sin qh\, e^{iqhN_b}g^{n+1}\pm 2e^{iqhN_b}\sum_{l=0}^{n+1}\beta^{n+1-l}g^{n+1}, \tag{3}$$

where $N_b=N$ for the upper boundary and $N_b=1$ for the lower boundary. It can be shown that the coefficients β^n and g_n can be calculated using the following relations:

$$\beta^n=\frac{1}{n}\sum_{k=0}^{n-1}\varphi^{n-k}\beta^k,\quad g^n=\left(\frac{\alpha+2-2\cos qh}{\alpha-2+2\cos qh}\right)^n,\ \alpha=\frac{2ikh^2}{\tau} \tag{4}$$

where

$$\varphi^k=-\frac{1}{2}-(-1)^{k-1}\left[1-\frac{1}{2}\left(\frac{1+\alpha/4}{1-\alpha/4}\right)^k\right],\quad \beta^0=\frac{\sqrt{\alpha^2-4\alpha}}{2}. \tag{5}$$

The discrete TBCs (3) are analogous to the fully discrete boundary condition derived in [6]. The main differences are that we used the second order approximation to the derivative by transversal coordinate x and we considered the case where $u(x, 0) = \exp(iqx) \neq 0$.

3 3D Finite-Difference Scheme

We use two FD schemes for RCD in combination with the discrete unconditionally stable 1D TBC derived above: a conventional Crank–Nicolson scheme [2] and the ADIS [4]. The 3D PWE is as follows

$$2ik\frac{\partial u}{\partial z} + \frac{\partial^2 u}{\partial x^2} + \frac{\partial^2 u}{\partial y^2} = 0. \tag{6}$$

In the case of Crank–Nicolson FD scheme we use generally the same numerical implementation as outlined in [1]. The only difference is that we replace the discretized Basakov–Popov TBC with that from (3) (see (17)–(20) and (25)–(32) in [1]).

In the case of ADIS we use the conventional 1D Crank–Nicolson scheme with the discrete TBC from (3) to propagate the field amplitude in both x and y coordinates. This FD implementation is by definition unconditionally stable.

The numerical complexity of both Crank–Nicolson and ADIS numerical implementations is $\sim P \times N^4 + Q \times N \times N_\tau + R \times N_\tau^2$, where P, Q and R are coefficients – the same as that of the old Basakov–Popov based implementation (see [1]).

4 Numerical Experiments

In order to quantitatively assess the accuracy of the proposed TBCs we carried out numerical experiments involving free space propagation of a Gaussian beam of hard X-ray radiation incident at some angle ψ. For the hard X-ray radiation dielectric constants of all materials are close to unity and the PWE is an established method for the simulation of X-ray radiation propagation [7]. In our simulations the wavelength was 0.1 nm ($E = 12.4$ kev).

Since the Gaussian beam is an exact solution of the PWE we can calculate the difference between it and the results of any FD implementation. The Gaussian beam is as follows

$$\begin{aligned} u_g(x, y, z) = \frac{w^2}{w^2 + i2z/k} &\exp\left[ik \sin\psi((x - z\sin\psi\cos\alpha/2)\cos\alpha + \right. \\ &\left.(y - z\sin\psi\sin\alpha/2)\sin\alpha)\right]\times \\ &\exp\left[-((x - z\sin\psi\cos\alpha)^2 + (y - z\sin\psi\sin\alpha)^2)/(w^2 + i2z/k)\right], \end{aligned} \tag{7}$$

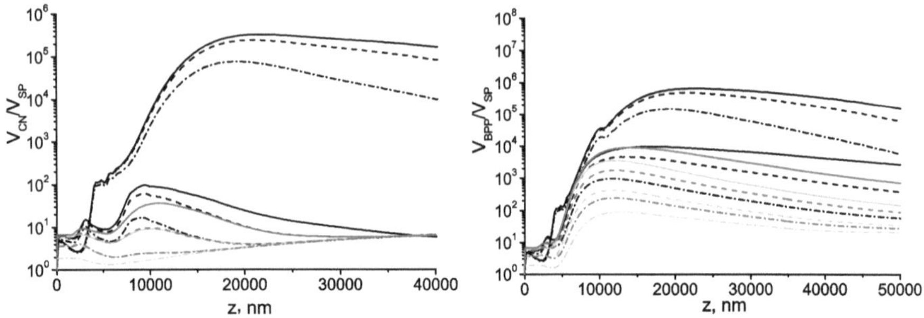

Fig. 1 The ratio of differences V (see (8)) of the Crank–Nicolson FD scheme with fully discrete 1D TBC to that of the ADIS with fully discrete 1D TBC (*left*) and of the Crank–Nicolson FD scheme with discretized 1D Basakov–Popov TBC to that of the ADIS with fully discrete 1D TBC (*right*) for a Gaussian beam in free space. The steps τ and h were: —— $h = 0.25$ nm, - - - - $h = 0.5$ nm, — · — $h = 1$ nm; *black* – $\tau = 200$ nm, *dark gray* – $\tau = 100$ nm, *gray* – $\tau = 50$ nm, *light gray* – $\tau = 25$ nm

where α is the azimuthal angle assumed to be $\pi/4$, so that the beam propagates diagonally through a corner of the RCD. The waist radius w is 18 nm. The size of the computational domain is 200×200 nm with the grid step equal to 1, 0.5 or 0.25 nm. The longitudinal step is 200, 100, 50 or 25 nm. The incidence angle ψ is 0.04, so that the beam reaches the corner after propagation over the distance of around 3500 nm.

For each numerical experiment we calculate the relative difference V between a FD implementation and the exact solution (7). The function V is defined as

$$V(\tau n) = \sum_{m,s=0}^{N} |u_{m,s}^{n} - u_g(\tau n, hm, hs)|^2 / \sum_{m,s=0}^{N} |u_g(\tau n, hm, hs)|^2. \qquad (8)$$

The ratio of functions V of the Crank–Nicolson FD scheme with discretized 1D Basakov–Popov TBC to that of the ADIS with fully discrete 1D TBC is shown in Fig. 1, left. The same ratio for the Crank–Nicolson FD scheme with fully discrete 1D TBC to that of the ADIS with fully discrete 1D TBC is shown in Fig. 1, right.

From Fig. 1 it is clear that both new implementations significantly (by several order of magnitude) better that old one. Their precision also improves faster with the decrease in h and τ than the precision of the old Basakov–Popov based implementation.

5 Conclusion

In this paper we developed two new numerical implementations of the 3D TBC in RCD. We showed that the 3D TBC based on ADIS combined with the fully discrete 1D TBC is the most precise one and is unconditionally stable. On the other hand the

3D TBC based on the Crank–Nicolson scheme with the same fully discrete 1D TBC while not exactly unconditionally stable is still significantly better than the old TBC based on the discretized Basakov–Popov 1D TBC. The computational complexity of all three TBCs appears to be the same.

The results presented here are fully applicable to the 2D and 3D time-dependent Schrodinger equations [8].

Acknowledgements The authors would like to thank A. V. Vinogradov for the fruitful discussions about the 3D parabolic wave equation and its applications. This work was supported in part by the basic research program of Russian Academy of Sciences "Femtosecond optics and new optical materials" (subprogram "New optical materials") and by the Russian Foundation for Basic Research (RFBR) grant No. 13-02-01009-a.

References

1. Feshchenko, R.M., Popov, A.V.: J. Opt. Soc. Am. A. **28**, 373–380 (2011)
2. Crank, J., Nicolson, P.: Proc. Cambridge Phil. Soc. **43**, 50–67 (1947)
3. Popov, A.V.: Radio Sci. **31**, 1781–1790 (1996)
4. Peaceman, D.W., Rachford, H.H.: J. Soc. Indust. Appl. Math. **3**, 28–42 (1955)
5. Antoine, X., Arnold, A., Besse, C., Ehrhardt, M., Schädle, A.: Comm. Comput. Phys. **4**, 729–796 (2008)
6. Arnold, A., Ehrhardt, M., Sofronov, I.: Comm. Math. Sci. **1**, 501–556 (2003)
7. Kopylov, Yu. V., Popov, A.V., Vinogradov, A.V.: Opt. Comm. **118**, 619–636 (1995)
8. Feshchenko, R.M., Popov, A.V.: Phys. Rev. E. **88**, 053308 (2013)

Gain Dynamics in Injection-Seeded Soft X-Ray Laser Plasma Amplifiers

S. Wang, Y. Wang, E. Oliva, L. Li, M. Berrill, L. Yin, J. Nejdl, B. Luther, C. Proux, T. T. Thuy Le, J. Dunn, D. Ros, Ph. Zeitoun and J. J. Rocca

Abstract We present the first measurement of the gain dynamics in an injection-seeded soft x-ray plasma amplifier (Wang et al. Nat. Photon. **8**, 381, (2014)). A sequence of two time-delayed spatially-overlapping high harmonic pulses was injected into a $\lambda = 18.9$ nm Ni-like Mo plasma amplifier to measure the regeneration of the population inversion that follows the gain depletion caused by the amplification of the first seed pulse. Collisional excitation is measured to re-establish in about ~ 1.75 ps the population inversion depleted during the amplification of the seed pulse. The measured gain-recovery time is compared to model simulations to gain insight on the population inversion mechanisms that create the transient gain in these amplifiers. The result supports the concept of a soft x-ray laser amplification scheme to generate ultra-intense fully phase-coherent ultrashort soft x-ray laser pulses based on the continuous extraction of energy from a plasma- based amplifier by a stretched seed pulse.

1 Introduction

There is great interest in the generation of intense, coherent soft x-ray beams for scientific and technical applications. Great efforts have been made in the development of compact plasma-based soft x-ray lasers. High energy soft x-ray laser pulses of several milliJoules can be generated in a dense plasma column based on the amplification of spontaneous emission (ASE) [2]. Lower pulse energies can be obtained by grazing incidence pumping using pump energies near one order of magnitude

S. Wang (✉) · Y. Wang · L. Yin · J. Nejdl · B. Luther · J. J. Rocca
National Science Foundation ERC for Extreme Ultraviolet Science and Technology and Colorado State University, Fort Collins, CO, 80523, USA
e-mail: sjwang@engr.colostate.edu

E. Oliva · L. Li · C. Proux · T. T. Thuy Le · D. Ros · Ph. Zeitoun
Loratoire d'Optique Appliquée (LOA), ENSTA-École Polytechnique, Chemin de la Hunière, 91761, Palaiseau, France

M. Berrill
Oak Ridge National Laboratory, OAK Ridge, 37831, USA

J. Dunn
Lawrence Livermore National Laboratory, Livermore, CA, 94550, USA

© Springer International Publishing Switzerland 2016
J. Rocca et al. (eds.), *X-Ray Lasers 2014*, Springer Proceedings in Physics 169,
DOI 10.1007/978-3-319-19521-6_46

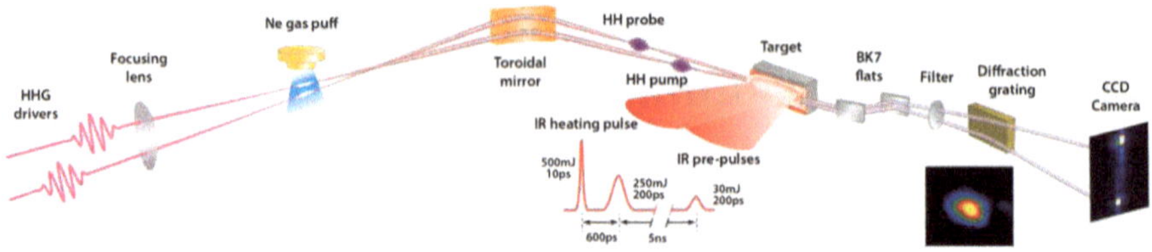

Fig. 1 Schematic of the set-up used to measure the gain dynamics in an injection-seeded soft-X-ray laser plasma amplifier

lower [3, 4]. Gain saturated operation was obtained at wavelengths down to 8.8 nm, and amplification was observed at wavelengths down to 7.3 nm is Ni-like Sm. These self-seeded amplified spontaneous emission lasers can generate bright beams, however their degree of coherence is limited. Full spatial and temporal coherence can be generated by seeding the plasma amplifier with a high harmonic (HH) seed. The wavelength of HH is tuned to match that of the amplifier. Besides improving coherence this seeding scheme has the additional advantage of producing laser pulses of shorter pulse width, decreased divergence, defined polarization, and higher brightness. Injection seeded soft x-ray lasers operating at several wavelengths between 32.6 and 13.2 nm were demonstrated from the Ne-like and Ni-like ions [5–7].

The output energy of the seeded soft x-ray laser has the potential to be greatly boosted by extending the chirped pulse amplification (CPA) [8] technique to this spectral region [9]. Understanding of the gain dynamics of injection-seed plasma amplifier is crucial to the efficient implementation of x-ray CPA. When an ultrashort seed pulse of sufficient energy is injected in the amplifier its intensity grows to saturate the gain and extract most of the energy stored in the population inversion, burning a hole in the gain temporal profile. However, since the upper laser level continues to be pumped by electron collisions, the gain will rapidly recover. We have conducted a detailed pump and probe experiment to understand this gain dynamics.

2 Experimental Setup

To conduct the experiment, a Ti:sapphire laser beam was separated into two. In turn, each beam was further split to create the population inversion in the plasma amplifier, and two independently controllable HH seed pulses, the HH pump and the HH probe (Fig. 1). Use of the same laser to generate the infrared plasma heating pulses and HH seed pulses ensures jitter-free synchronization. The first beam was shaped into a sequence of two pre-pulses, 200 ps in duration with the energy of ~ 30 mJ and ~ 250 mJ separately and delayed by 5 ns, impinging at normal incidence onto the target, followed after 600 ps by a short heating pulse of 10 ps duration and ~ 500 mJ energy impinging at a grazing incidence angle of 23°. All the pulses were focused into a 30 μm wide, 3.5 mm FWHM line focus on a 3 mm wide Mo slab target. This pumping geometry takes advantage of pump beam refraction to increase its energy deposition in the plasma region with optimum density for amplification [3,4]. The 23° grazing incidence angle couples the pump beam into the region where the plasma density is about 2×10^{20} cm^{-3}.

The two Ti:sapphire pulses in the second beam were compressed to about 60 fs using a separate grating compressor, and were focused on a Ne gas jet with a vertical separation of ~90 μm to allow for the generation of two independent HH seed pulses. We observed that the two HH seed pulses could be operated independently without interference as long as they are separated in time by at least 250 fs. To ensure their overlap on the plasma amplifier, the beams generating the HH seed pulses were spatially superimposed 16 mm after the gas jet, and this overlap point was relay-imaged with a magnification of one onto the middle point of the plasma amplifier column using a grazing incidence toroidal mirror. The output of the soft x-ray amplifier was separated from the straight 800 nm laser light by a set of filters and was spectrally dispersed with a variable-line-spacing diffraction grating. The angular separation between the pump and probe pulses vertically separates the two pulses on the CCD camera, thus allowing for their simultaneous detection. This vertical offset allows us to record any shot-to-shot variation due to either the HH seed pulses or the plasma amplifier itself.

3 Experiment Results

First we measured the energy of a single injected HH pulse as a function of time respect to the peak of the Ti:Sapphire heating pulse. The results are shown in Fig. 2a by the blue curve. The amplification period is observed to last ~10 ps, with a maximum occurring ~5 ps after the peak of the short infrared heating pulse.

Subsequently we proceeded to conduct the pump/probe experiment by injecting the HH Pump pulse at the delay of $T_1 = 3$ ps respected to the heating pulse and scanning the probe delay Δt from 250 fs to 7 ps respected to the HH pump, as shown in Fig. 2b by black curve. At 250 fs, the output energy of the probe pulse was practically at the detection threshold which indicates the practically complete depletion of the gain by the pump pulse. Indeed, our modeling shows that the HH pump enters the plasma with an intensity of several 10^8 Wcm^{-2} and exits with an intensity of ~10^{11} Wcm^{-2}, nearly ten times greater than that necessary to saturate the gain. At a HH pump–HH probe delay of 0.5 ps, amplification is already observable, but not yet a full recovery of the gain. At $\Delta t = 1.5$ ps, the energy of the amplified HH probe equals its value without the HH pump and continues to increase, reaching a maximum for a delay of 1.75 ps. We repeated the measurement by injecting the pump pulse latter into the gain curve at 6 ps, red curve in Fig. 2b. We again observed near full depletion of the gain by the pump pulse and gain recovery. However by this late time the gain has already decreased significantly respect to its peak value and it never fully recovers.

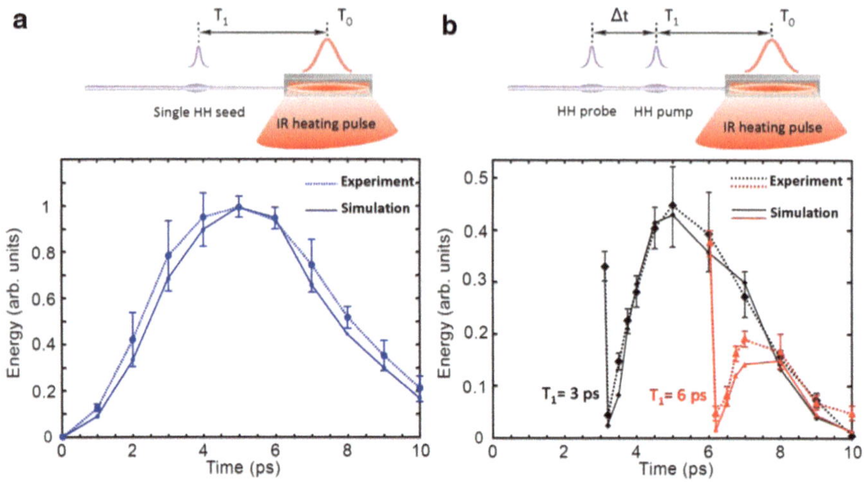

Fig. 2 Experimental results compared with Maxwell-Bloch simulations. **a** Measured (*dotted line*) and simulated (*solid line*) variation of the energy of the amplified seed pulse as a function of time delay respect to the peak of the heating short pulse. **b** Measured (*dotted line*) and simulated (*solid line*) variations of the energy of the HH probe pulse as a function of time for the HH pump injected at either 3 ps (*black*) or 6 ps (*red*) after the peak of the plasma heating short pulse. The HH probe pulse energy is normalized to the HH pump energy. Each point is the average value over ten independent measurements

4 Modelling and Discussion

The measurements compared well with simulations conducted with a time-dependent Maxwell-Bloch model [10] that uses as input the results from hydrodynamic simulations. As shown in Fig. 2, modeling reproduces very well the full experimental data for both delays, confirming the gain rapidly recovery in 1.5–1.7 ps and the late drop of the gain due to plasma over-ionization.

This measured rapid recovery of the gain indicates that a HH pulse injected more than 1.7 ps after the previous seed pulse can extracts a similar amount of energy, effectively doubling the emitted coherent soft x-ray laser energy**.** This is in good agreement with the Maxwell-Bloch calculation, as shown in Fig. 3. Thus, in a medium in which the duration of the gain exceeds that of the gain recovery the energy extracted can be greatly increased by either injecting a sequence of multiple seed pulses, or by increasing the duration of a single seed pulse, as proposed in the x-ray CPA scheme [9].

The experimental results could not be precisely reproduced unless the spatio-temporal dynamics of the lasing ion is fully taken into account. Figure 4 illustrates the result of the hydrodynamic simulation of the lasing ion abundance. These results are inputted into the Maxwell-Bloch code to simulate the pump-probe experiment.

As the HH beams cover regions of different ionization and density of the lasing ion, if only the local plasma parameters for amplification are considered, there is

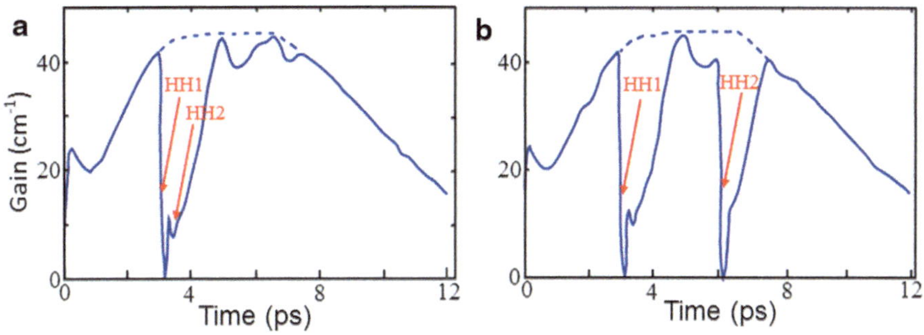

Fig. 3 Simulation of the temporal variation of the gain perturbed by two HH seeds. The *dotted lines* in both (***a***) and (***b***) presents the unperturbed gain, the solid lines indicate the bleaching of the population inversion after amplification of the HH pulses. **a** The separation of two HH pulses is 0.5 ps. **b** The second HH pulse is injected 2.7 ps after the first HH pulse

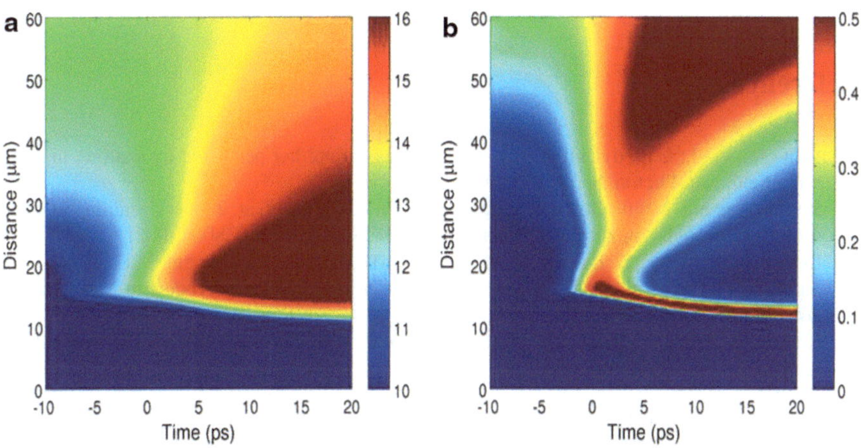

Fig. 4 Hydrodynamic simulation of the lasing ion abundance. **a** mean plasma ionization. **b** The abundance of Ni-like ions. The abrupt decrease of the density is caused by the plasma over-ionization, that is the cause of the termination of the gain observed in the experiment. For both figures, t = 0 ps is the *time* when the short heating pulse hits the plasma. The target surface lies at "0 μm" position with positive distances corresponding to vacuum

always disagreement with experimental data. The modelling has to consider the spatio-temporal variation of the plasma parameters to fit experimental data. Thus, this pump-probe experiment not only uncovers the gain dynamics but also serves as a diagnostic of the ionization in the hot-dense plasma with 100 s femtosecond resolution.

5 Conclusions

This x-ray pump–x-ray probe experiment uncovered the gain recovery dynamics in a soft x-ray laser amplifier and also serves as a diagnostic of the plasma ionization with a resolution of hundreds of femtoseconds. The results show that in a soft x-ray amplifier in which the duration of the gain exceeds that of the gain recovery the energy extracted can be greatly increased by injecting an stretched seed pulse that subsequently could be re-compressed.

Acknowledgements Work supported by the AMOS programme of the Office of Basic Energy Sciences, US Department of Energy, using equipment developed at the NSF ERC for Extreme Ultraviolet Science and Technology and NSF award MRI-ARRA 09-561, and by the LASERLAB3-INREX European project and SHYLAX plus CIBOR RTRA 'Triangle de la Physique' programmes.

References

1. Wang, Y., et al.: Gain dynamics in a soft x-ray laser amplifier perturbed by a strong injected x-ray field. Nat. Photon. **8**, 381 (2014)
2. Rus, B., et al.: Multimillijoule, highly coherent x-ray laser at 21 nm operating in deep saturation through double-pass amplification. Phys. Rev. A. **66**, 063806 (2002)
3. Luther, B.M., Wang, Y., Larotonda, M.A., Alessi, D., Berrill, M., Marconi, M.C., Rocca, J.J.: Saturated high-repetition-rate 18.9-nm tabletop laser in nickellike molybdenum. Opt. Lett. **30**, 165 (2005)
4. Keenan, R., Dunn, J., Patel, P.K., Price, D.F., Smith, R.F., Shlyaptsev, V.N.: High-repetition-rate grazing-incidence pumped x-ray laser operating at 18.9 nm. Phys. Rev. Lett. **94**, 103901 (2005)
5. Wang, Y., Granados, E., Larotonda, M.A., Berrill, M., Luther, B.M., Patel, D., Menoni, C.S., Rocca, J.J.: High-brightness injection-seeded soft-x-ray-laser amplifier using a solid target. Phys. Rev. Lett. **97**, 123901 (2006)
6. Wang, Y., Granados, E., Pedaci, F., Alessi, D., Luther, B., Berrill, M., Rocca, J.J.: Phase-coherent, injection-seeded, table-top soft-x-ray lasers at 18.9 nm and 13.9 nm. Nat. Photon. **2**, 94 (2008)
7. Pedaci, F., Wang, Y., Berrill, M., Luther, B., Granados, E., Rocca, J.J.: Highly coherent injection-seeded 13.2 nm tabletop soft x-ray laser. Opt. Lett. **33**, 491 (2008)
8. Strickland, D., Mourou, G.: Compression of amplified chirped optical pulses. Opt. Commun. **56**, 447 (1985)
9. Oliva, E., et al.: A proposal for multi-tens of GW fully coherent femtosecond soft x-ray lasers. Nat. Photon. **6**, 764 (2012)
10. Oliva, E., Zeitoun, P., Fajardo, M., Lambert, G., Ros, D., Sebban, S., Velarde, P.: Comparison of natural and forced amplification regimes in plasma-based soft-x-ray lasers seeded by high-order harmonics. Phys. Rev. A. **84**, 013811 (2011)

Part VIII
X-Ray Laser Material Interaction

Application of Soft X-Ray Lasers to Neutral Cluster Chemistry

E. R. Bernstein, F. Dong, Y. Q. Guo, J.-W. Shin, S. Heinbuch and J. J. Rocca

Abstract This presentation provides a review and overview of the chemical studies that are possible with the aid of a soft X-ray (EUV) laser. The laser operates from 1 to 10 Hz with an output energy/pulse of $\sim 10\,\mu J$ at a wavelength of 46.9 nm (26.44 eV/photon). This cluster single photon ionization source is coupled to a time of flight mass spectrometer and is employed to study a large variety of neutral clusters to learn their distribution, reactions, and catalytic chemistry behavior. The presentation demonstrates the advantages of this apparatus through comparisons of mass spectra obtained with other laser ionization sources (6.4 and 10.5 eV/photon) for the same or comparable samples. The future directions for such studies are presented and a new photoelectron/mass spectrometer constructed around a soft X-ray/EUV photon source is planned and discussed.

1 Introduction

Clusters are small collections of atoms or molecules, typically present or synthesized in the gas phase. Examples of such collections are M_m (M = metal atom), M_mX_n (X = B, C, N, O, P, S, ...), $(H_2O)_n$, $(NH_3)_n$, $(H_2S)_n$, $(SO_2)_n$, $Al_xC_yH_z$, Ti_mO_n $(NH)_z$, $(C_6H_6)(H_2O)_n$, $(RCOOH)_n$, $(RCOH)_n$, etc. These species can have catalytic, biological, and other values, but always appear as inhomogeneous samples as they are synthesized through laser ablation [1–8], condensation [9–12], or plasma means. The importance of such cluster or nano particle systems is that they can be employed to model, study, and derive mechanisms for intermolecular interactions, local phenomena in condensed phases, solvation, catalysis, chemical reactions because they are simple and small enough that one can access, both experimentally and theoretically, their structures, spectroscopic properties, and dynamics.

In the experimental study of neutral clusters, one must deal with the eventual issue of identification of the neutral species. This is most effectively accomplished

E. R. Bernstein (✉) · F. Dong · Y. Q. Guo · J.-W. Shin
Department of Chemistry and NSF ERCEUVST, Colorado State University, Fort Collins, CO 80523, USA
e-mail: erb@lamar.colostate.edu

S. Heinbuch · J. J. Rocca
Department of Electrical and Computer Engineering, NSF ERCEUVST, Colorado State University, Fort Collins, CO 80523, USA

© Springer International Publishing Switzerland 2016
J. Rocca et al. (eds.), *X-Ray Lasers 2014*, Springer Proceedings in Physics 169,
DOI 10.1007/978-3-319-19521-6_47

by ionization followed by typically time of flight mass spectrometry (TOFMS): clearly this ionization process must not cause unknown or unidentified cluster fragmentation, as this would not lead to the identification of the correct cluster system, M_mX_n, that is responsible for the chemistry, spectra, catalysis, etc. studied. Thus neutral clusters must be size selected or identified (TOFMS) in order to be useful models and mechanistic and dynamical probes for reactions and cluster properties. This identification is especially critical for catalytic studies, in which the cluster can be employed as a direct model for the local active, defect, surface site at which the catalytic chemistry occurs.

We have determined in these studies that for both neutral van der Waals (molecular) and covalent clusters, this structure mandates single photon ionization (SPI) with $>$ 16 eV/photon energy to ensure that all species will be detected with minimum cluster fragmentation. In the implementation of TOFMS as the neutral cluster identifier, one must demonstrate that SPI does not cause fragmentation of the neutral cluster.

Nonetheless, neutral clusters, especially molecular, van der Waals clusters, such as $(H_2O)_n$, $(NH_3)_n$, $(RCOOH)_n$, and $(ROH)_n$, can fragment upon SPI, even though the excess ionization energy (E(photon)—E(VI) = E_e (KE), VI = vertical ionization, E_e(KE) = kinetic energy of the ionized electron $>$ 10 eV) is typically removed by the electron. This fragmentation is not caused by the SPI process, but by an ion/molecule reaction in the newly formed, ionized cluster. Reactions such as $(H_2O)_n + h\upsilon \rightarrow (H_2O)_{n-1}H^+ + OH^-$ are quite typical, but easily identified in the SPI experiment, as will be discussed below. Thus, neutral cluster fragmentation following SPI can be identified and studied for reactions, mechanisms, and chemical and physical processes occurring upon cluster threshold ionization. We will give examples of how our home built soft X-ray(SXR)/EUV (26.44 eV/photon, 46.9 nm) laser [13] has been, and will be employed to study neutral cluster synthesis, structure, size, chemistry, and electronic and vibrational states.

2 Experimental Procedures

The apparatus for SXR/EUV SPI TOFMS technique is presented in Fig. 1. The design details of the components are covered in the references [1] and [13]. The SXR/EUV light is focused and directed by a gold-coated mirror system consisting of toroidal and plane elements. We estimate that the laser energy/pulse at the sample can be roughly 5 μJ/pulse. The neutral cluster beam is generated by a He expansion with a backing pressure of ca. 100 psi with a laser ablated sample (covalent clusters) or a 1–5 % backing gas mixture for a molecular sample. The sample generation chamber and the ionization region are separated by a 1–4 mm aperture skimmer. The supersonic nozzle and laser are operated at a 3–10 Hz rate. Much more detail, and the implementation of experiments at other photon energies can be found in the references provided.

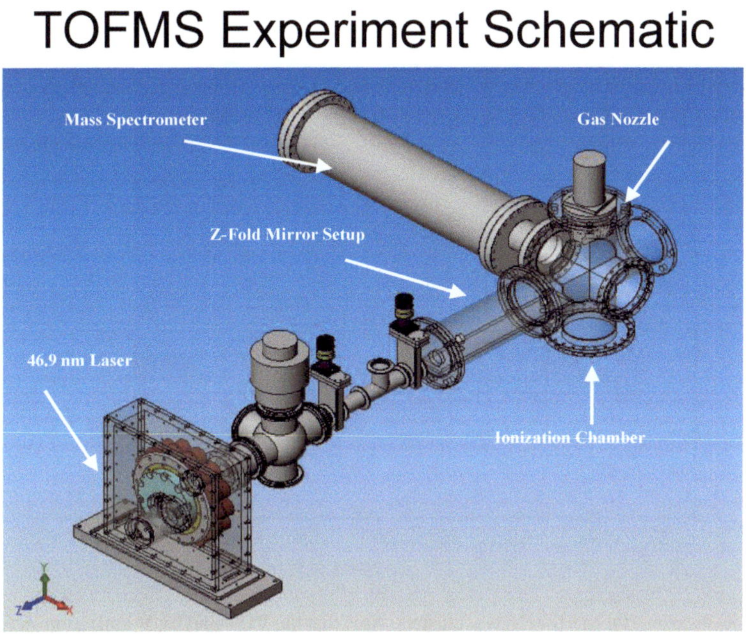

Fig. 1 Three-dimensional layout of the experimental apparatus with mass spectrometer (linear and reflectron)

3 Results and Discussions

We will now demonstrate by specific examples from our published works just how central and fundamental these studies can be for the elucidation of the behavior and properties of neutral covalent and molecular clusters.

3.1 Small Molecular Cluster Detection and Ion Chemistry [9–12]

This first task for the application of SPI EUV TOFMS to cluster studies is to demonstrate that a 26.44 eV photon will not destroy the neutral cluster distribution. We discuss the ammonia cluster behavior followed by the water cluster behavior as these two systems nicely represent one regime for the behavior of molecular systems.

Figure 2 shows the SPI TOFMS of ammonia clusters. These spectra compare the neutral cluster distribution detected at 10.5 eV/photon ionization with that detected at 26.44 eV/photon SPI. The VIE of ammonia clusters is ca. 9.5 eV. We conclude from these two nearly identical spectra that the photo generated electron effectively removes the excess ionization energy as photoelectron kinetic energy for both the

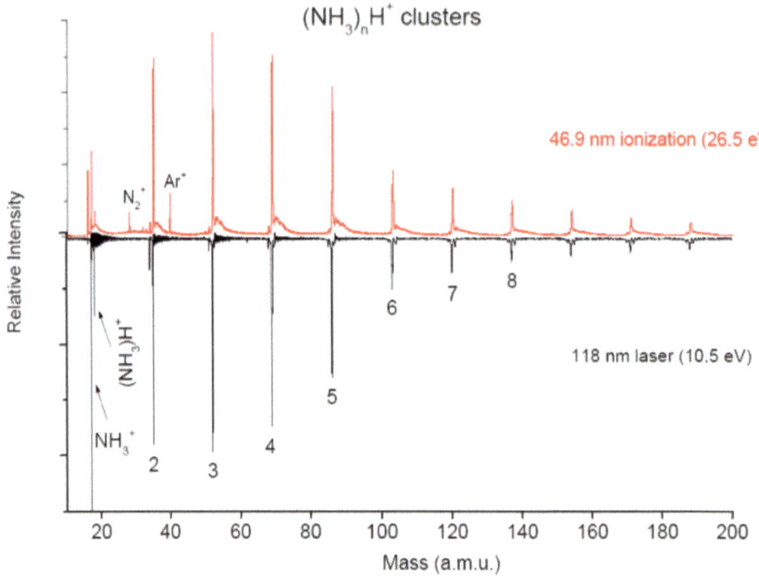

Fig. 2 Reflectron TOFMS for $(NH_3)_n$ clusters. Note that the VUV and EUV ionization yield nearly identical mass spectra

~ 1 and ~ 17 eV excess situations. The energy/probe for the two ionization situations is typically ca. 1–5 μJ/pulse, which effectively excludes multiphoton processes on the ns time scale (~ 2 ns for 26.44 and ~ 10 ns for 10.5 eV photons). Note that the mass features detected are not $(NH_3)_n$ but $(NH_3)_nH^+$: this implies that both SPIs cause the proton transfer reaction $(NH_3)_n + h\upsilon \rightarrow (NH_3)_{n-1}H^+ + e^- + NH_2$. Two types of post ionization reactions can be characterized in these experiments, in which 1 or 2 NH_3 molecules are dissociated from the cluster due to, not the excess ionization energy, but due to the proton transfer reaction ΔE following ionization. These results and their implications are further discussed in the original refs. [9–12]. Nonetheless, they represent here a very useful demonstration of the principle of SPI TOFMS studies of post ionization chemistry for van der Waals molecular clusters.

Figure 3 presents the SXR/EUV SPI mass spectra of water clusters. This mass spectrum is unique in the sense that the VIE of $(H_2O)_n$ is ca. 12.5 eV: of the 3 lasers employed for SPI of clusters, only the 26.44 eV/photon laser can be employed for this experiment. Multiphoton ionization of water clusters completely obscures the detail present in the $(H_2O)_n$ spectra. Figure 3 shows that the proton transfer reaction energy is again sufficient to release sequentially two water molecules. One can determine this from comparison of the linear and reflectron TOFMS. Again, more details of these results can be found in the original study [9–12]. Prior to these studies, $(H_2O)_{21}$ was viewed as a special cluster structure due to its greater intensity in mass spectra: the above described results for $(NH_3)_n$ and $(H_2O)_n$ have enabled the rate constants for post ionization proton transfer reaction generated fragmentation to

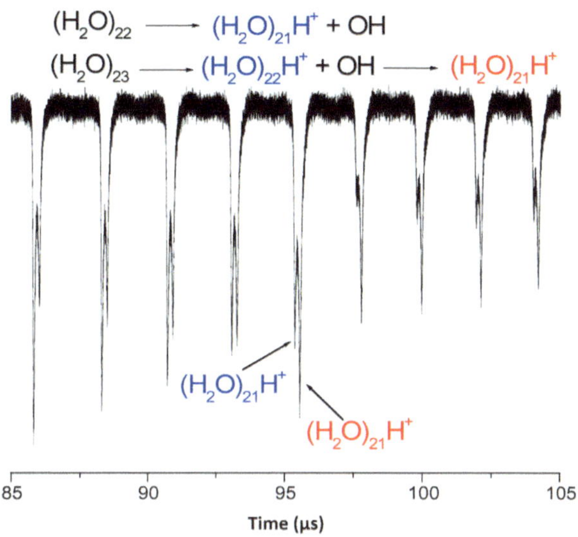

Fig. 3 Water cluster reflectron mass spectra around the region of water $(H_2O)_{20}$ showing the loss of one molecule following the proton transfer reaction for the ionized water H_2O^+

be determined. These values show that the proton transfer reactions are the intensity controlling features, as the water loss mechanism is consistent over the range of cluster sizes.

3.2 *Chemical Reactions, Catalysis, New Species: Neutral Covalent Cluster [1–8]*

Herein, we present studies of inhomogeneous neutral covalent systems: specifically, metal, metal oxide, and metal carbide clusters. Again, one first needs to demonstrate that SPI with EUV radiation does not fragment the target neutral clusters: Fig. 4 shows this well for Ta_n species. Metal clusters have low VIEs, so 6.4 eV/photon SPI is sufficient to ionize the clusters, but not cause fragmentation. This spectrum in Fig. 4 is nearly identical to that obtained at 26.44 eV/photon, except for the Ta atomic feature and a few relative intensities probably caused by small cross section issues at the two photon energies.

Al_n clusters (show some surprising features (at 6.4 eV/photons) possibly attributable to growth problems for Al_{4-6}, $Al_{8,9}$, and especially Al_{13}. A comparison of the two spectra (acquired at ca. 3 μJ/cm^2 for EUV and VUV lasers) shows that the problem of missing features is not one of neutral cluster growth or instability, but due to a variation in the ionization energy based on cluster electronic state energies.

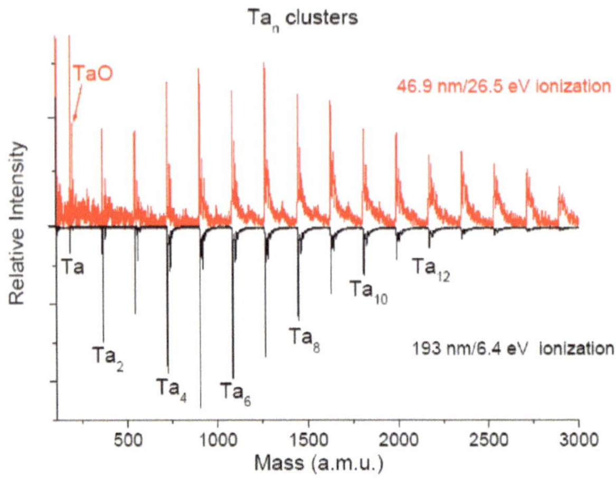

Fig. 4 Ta_n clusters detected at 6.4 and 26.44 eV ionization showing nearly identical spectra

The importance of SPI with different wavelength lasers (193, 118, 46.9 nm) can be appreciated from the 3 spectra of Ta_mO_n. At 193 nm (6.4 eV/photon) only Ta^+, TaO^+, TaO_2^+ ions are observed and these arise mostly from fragmentation due to multiphoton processes for higher mass neutrals. At 118 nm SPI, TaO and a few low mass clusters are observed below the Ta_4O_9 cluster mass. These spectra give no sense of the rich neutral cluster distribution for this metal oxide system. At 46.9 nm ionization a systematic distribution of neutral clusters at last to Ta_mO_n, $m = 9$, $n = 22$ is determined, with distributions (1,1) to (1,5), (2,4) to (2,7), (3,6) to (3,10), etc. One can readily understand how such a neutral cluster distribution can be studied for chemical reactions and catalytic processes.

Vanadium oxide is a well-known catalyst for hydrocarbon generation from petroleum products and for the manufacture of sulfuric acid. In this latter process SO_2 is converted to SO_3 in the presence of the catalyst. The actual surface or bulk site that does the conversion is not known, as is the case in almost all catalytic reactions. We employ neutral clusters to learn what possible sites could be active in the SO_2 to SO_3 conversion (e.g., VO, VO_3, V_2O_6, V_2O_3, V_4O_{12}, ...) (See Fig. 5). With 25.44 eV/photon SPI, oxygen rich clusters $VO_3(H_x)$, $x = 0,1,2$, $V_2O_6(H_x)$, $V_3O_8(H_x)$ $V_5O_{13}(H_x)$, and $V_5O_{15}(H_x)$ are uniquely detected and these clusters take SO_2 to SO_3 in a catalytic cycle with oxygen present. These clusters are so reactive they attach H atoms removed from metal and other sources the vacuum system. Additionally, the cluster VO forms VO_2 and SO in the presence of SO_2. Thereby, two paths exist for the catalytic generation of SO_3: (1) $SO_2 + 1/2O_2 \rightarrow SO_3$ (oxygen rich clusters); and (2) $SO_2 \rightarrow SO$ (oxygen poor clusters). In the latter instance, $SO + O_2 \rightarrow SO_3$ without a barrier at 300 K. The essence of this mechanism is the bond strength OS–O vs. OV–O and O_2V–O vs. O_2S–O. A full cycle regenerates the original vanadium oxide species either through both active sites reacting with each other or with O_2 and other V_mO_n sites, as appropriate.

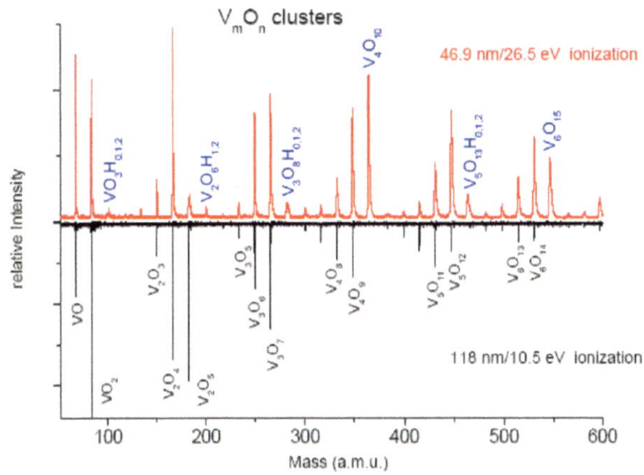

Fig. 5 V_nO_n clusters detected at 10.5 and 26.44 eV ionization showing the all important high oxygen content, especially catalytic, clusters at the EUV ionization

The same oxygen rich neutral clusters are catalytically active for the oxidation of hydrocarbons: VO_3, etc. clusters can sever the C–C bond of ethylene and generate CH_2 and H_2CO. This is a very important process for the formation of new alkanes and alkenes, as well as partially oxidized hydrocarbons like formaldehyde.

Finally, we discuss the discovery of a new series of neutral clusters or molecules, $Al_mC_nH_x$, (x = 0, 1, 2, ...) [6–8]. The initial attempt to determine the extent of the Al_mC_n series derived from an Al/C pressed disk sample under laser ablation. $Al_mC_nH_x$ species are readily generated by ablating Al into hydrocarbons Al/C into H_2. The observed series, Al_3C_x, x = 2,4,6,..., Al_5C_y, y = 1,3, and 5,..., Al_7C_z, z = 2,4, ..., seem strange and many "missing" clusters ($Al_{2,4,6,...}$) are apparent. This distribution of products seems to be an unusual synthetic issue, but at 26.44 eV/photon ionization we find all Al_mC_n species present in the synthetic mix. The problem of the "missing" clusters is not a synthetic issue, but one of high VIE for different parts of the cluster distribution. This phenomenon can be attributed to high ionization energy for clusters with closed electronic shells and fully occupied orbitals. Again, SPI with a photon of enough energy to ionization reliably all molecules generated, but not to fragment them, enables one to be confident that all species are detected in their neutral form.

A final, obvious, and practical use for an EUV laser is for the determination of optical damage and optical coatings for EUV optics [3]. In this experiment, laser ablation creates metal and metal compound materials that could be or have been suggested as good potential protective optical coatings for EUV optics. Examples of such coating films are Si_mO_n, Ti_mO_n, Hf_mO_n, Zy_mO_n, Ru_m, Ru_mO_n, and perhaps others. The experiments are similar to those described above only in these instances we are not concerned with fragmentation or learning the neutral cluster distribution, but simply with the issue of reaction of the neutral clusters with species such as, H_2O, ROH, NH_3, C_nH_{2n+2}, C_nH_{2n}, O_2, etc., under EUV radiation. To summarize the many experiments in this study, only Zr_mO_n and Hf_mO_n neutrals do not react with the above compounds. In general, oxygen rich clusters are more reactive than oxygen poor or stoichiometric clusters. Thus, capping layers of Hf_mO_n or Zn_mO_n that are oxygen deficient could be a good choice for hydrocarbon resistant EUV optics.

4 Conclusions

Concerning the EUV laboratory size laser, that generates photons that will ionize all possible molecules and radicals, we have shown that single photon ionization is a very valuable tool for the study of neutral cluster chemistry. In particular, access to a laboratory located, table-top, EUV laser enables use of it as a standard feature of the research rather than a major commitment once a year at national facility. We have demonstrated that neutral cluster distributions, structures, chemistry, and reaction mechanisms can be thereby determined, because theory and experiment can be accomplished for these simple model systems. Clearly new and improved EUV sources will expand the scope of these applications.

5 Future Directions

Given what we now have learned with these SPI EUV TOFMS studies of neutral clusters, an immediate next advance would be to expand these techniques to study the electronic and vibrational energy levels of the active and inactive neutral species to learn what features of these systems are important for their chemistry and general activity. These data could also be compared with theoretical studies of cluster structure and electronic and vibrational states so that both theory and experiments can be verified. Such studies can be implemented through a combination of negative ion mass spectrometry followed by photoelectron spectroscopy.

Finally, future work would not be complete without suggesting further development of EUV lasers with more available wavelength outputs, greater timing accuracy, higher repetition rates, longer times between services, and of course, more energy/pulse.

Acknowledgments The work was supported by the NSF Center for Extreme Ultraviolet Science and Technology and grants from the AFOSR and ARO.

References

1. Dong, F., Heinbuch, S., He, S.G., Xie, Y., Rocca, J.J., Bernstein, E.R.: Formation and distribution of neutral vanadium, niobium, and tantalum oxide clusters: Single photon ionization at 26.5 eV. J. Chem. Phys. **125**, 164318 (2006)
2. Dong, F., Heinbuch, S., Xie, Y., Rocca, J.J., Bernstein, E.R.: Experimental and theoretical study of the reactions between neutral vanadium oxide clusters and ethane, ethylene, and acetylene. J. Am. Chem. Soc. **130**, 193 (2008)
3. Heinbuch, S., Dong, F., Rocca, J.J., Bernstein, E.R.: Gas phase study of the reactivity of optical coating materials with hydrocarbons using a desk-size EUV laser†. J. Opt. Soc. Am. B. **25**, B85 (2008)
4. He, S.-G., Xie, Y., Dong, F., Heinbuch, S., Jakubikova, E., Rocca, J.J., Bernstein, E.R.: Reactions of sulfur dioxide with neutral vanadium oxide clusters in the gas phase. II. Experimental study employing single-photon ionization. J. Phys. Chem. A. **112**, 11067 (2008)
5. Dong, F., Heinbuch, S., Xie, Y., Rocca, J.J., Bernstein, E.R.: C=C bond cleavage on $VO_3(V_2O_5)_n$ clusters. J. Am. Chem. Soc. **131**, 1057 (2009)
6. Dong, F., Heinbuch, S., Xie, Y., Rocca, J.J., Bernstein, E.R.: Experimental and theoretical studies of neutral Al_mC_n and $Al_mC_nH_x$ clusters. Phys. Chem. Chem. Phys. **12**, 2569 (2010)
7. Heinbuch, S., Dong, F., Rocca, J.J., Bernstein, E.R.: Experimental and theoretical studies of reactions of neutral vanadium and tantalum oxide clusters with NO and NH_3. J. Chem. Phys. **133**, 174314 (2010)
8. Dong, F., Xie, Y., Bernstein, E.R.: Experimental and theoretical studies of neutral $Mg_mC_nH_x$ and $Be_mC_nH_x$ clusters. J. Phys. Chem. A. **135**, 054307 (2011)
9. Heinbuch, S., Dong, F., Rocca, J.J., Bernstein, E.R.: Single photon ionization of van der Waals clusters with a soft x-ray laser: $(CO_2)_n$ and $(CO_2)_n(H_2O)_m$,. J. Chem. Phys. **125**, 154316 (2006)
10. Dong, F., Heinbuch, S., Rocca, J.J., Bernstein, E.R.: Single photon ionization of van der Waals clusters with a soft x-ray laser: $(SO_2)_n$ and $(SO_2)_n(H_2O)_m$. J. Chem. Phys. **125**, 154317 (2006)
11. Dong, F., Heinbuch, S., Rocca, J.J., Bernstein, E.R.: Dynamics and fragmentation of van der Waal clusters: $(H_2O)_n$, $(CH_3OH)_n$, and $(NH_3)_n$ upon ionization by a 26.5 eV soft x-ray laser. J. Chem. Phys. **124**, 224319 (2006)
12. Heinbuch, S., Dong, F., Rocca, J.J., Bernstein, E.R.: Single photon ionization of hydrogen bonded clusters with a soft x-ray laser: $(HCOOH)_x$ and $(HCOOH)_y(H_2O)_z$. J. Chem. Phys. **126**, 244301 (2007)
13. Heinbuch, S., Grisham, M., Martz, D.: Demonstration of a desk-top size high repetition rate soft x-ray laser. Opt. Express. **13**, 4050 (2005)

Application of Laser Plasma Sources of Soft X-rays and Extreme Ultraviolet (EUV) in Imaging, Processing Materials and Photoionization Studies

H. Fiedorowicz, A. Bartnik, P. W. Wachulak, R. Jarocki, J. Kostecki, M. Szczurek, I. U. Ahad, T. Fok, A. Szczurek and Ł. Węgrzyński

Abstract In the paper we present new applications of laser plasma sources of soft X-rays and extreme ultraviolet (EUV) in various areas of plasma physics, nanotechnology and biomedical engineering. The sources are based on a gas puff target irradiated with nanosecond laser pulses from commercial Nd: YAG lasers, generating pulses with time duration from 1 to 10 ns and energies from 0.5 to 10 J at a 10 Hz repetition rate. The targets are produced with the use of a double valve system equipped with a special nozzle to form a double-stream gas puff target which allows for high conversion efficiency of laser energy into soft X-rays and EUV without degradation of the nozzle. The sources are equipped with various optical systems to collect soft X-ray and EUV radiation and form the radiation beam. New applications of these sources in imaging, including EUV tomography and soft X-ray microscopy, processing of materials and photoionization studies are presented.

1 Introduction

Soft X-ray and extreme ultraviolet (EUV) radiation wavelength ranging from about 1–50 nm can be produced in a high-temperature plasma generated by interaction of high power laser pulses with matter. It was demonstrated that laser plasma soft X-ray and EUV sources could be useful in various applications in physics, material science, biomedicine, and technology [1–4]. However, conventional laser plasma sources based on a solid target have debris production problem. We have demonstrated that using a double-stream puff target instead of a solid target it is possible to develop highly efficient and debris-free laser plasma soft X-ray and EUV sources [5–7]. The target is formed by injection of high-Z gas (xenon, krypton, argon, etc.) into a hollow stream from low-Z gas (hydrogen and helium) using a double nozzle.

H. Fiedorowicz (✉) · A. Bartnik · P. W. Wachulak · R. Jarocki · J. Kostecki · M. Szczurek · I. U. Ahad · T. Fok · A. Szczurek · Ł. Węgrzyński
Institute of Optoelectronics, Military University of Technology, 2, Kaliskiego Street, 00-908, Warsaw, Poland
e-mail: hfiedorowicz@wat.edu.pl

© Springer International Publishing Switzerland 2016
J. Rocca et al. (eds.), *X-Ray Lasers 2014*, Springer Proceedings in Physics 169,
DOI 10.1007/978-3-319-19521-6_48

The nozzle setup consists of a central nozzle in a form of a circular orifice surrounded by an outer nozzle in the form of a ring. The nozzle is supplied with gases from two electromagnetic valves mounted in a common body.

A detailed description of the double-stream gas puff target system, the target characterization measurements and the results of investigations on generation of soft X-ray and EUV radiation are presented in our previous publications [8, 9]. Strong soft X-ray and EUV emissions from the double-stream gas puff targets, exceeding the emissions from solid targets, have been demonstrated.

The gas puff target approach was used for developing a compact laser plasma EUV source for metrology applications [9]. The xenon target was irradiated with 4 ns/0.5 J pulses produced with repetition of 10 Hz from a commercial Nd:YAG laser. Conversion efficiency of the laser energy into EUV energy at 13.5 nm wavelength of about 2 % was measured in 7 % wavelength band, corresponding to about 0.5 % in 2 % band [10]. The source has been used in the measurements of optical characteristics of Mo/Si multilayer mirrors [11].

High-brightness soft X-ray source based on the gas puff target driven with the PALS laser facility [7, 12] has been used for the first time for processing materials and in experiments on radiobiology. Direct photo-etching of inorganic (silicon) and organic (polymers) materials with nanosecond pulses of soft X-ray and EUV radiation was demonstrated [13, 14]. The first experiments on radiobiology have shown the possibility for the use of the laser plasma X-ray source in research on radiation damage in plasmid DNA at very high doses in a single-shot exposure [15]. This allows using a laser plasma sources in prospective pulse-probe experiments.

Efficient processing of organic polymers (PMMA and PTFE) has been demonstrated with the compact EUV source operating at 10 Hz [16] and strong temperature effect on soft X-ray photo-etching of PTFE was shown [17]. The same source equipped with a multi-foil optic collector [18] has been used to study EUV emission from solids irradiated with intense EUV pulses [19]. A new technique for detection of surface changes of materials utilizing scattered or luminescent EUV radiation was proposed [20].

The use of a grazing incidence axisymmetrical ellipsoidal mirror as a collector strongly increased the EUV fluence on irradiated samples up to 100 mJ/cm^2 [21]. It made possible to increase dramatically the EUV ablation rates and improve micromachining of polymers. Efficient processing of non-organic materials (SI, Ge, NaCl, and CaF_2) has been also demonstrated [22]. Modification of polymer surfaces by creation of characteristic micro- and nanostructures was observed in a case of irradiation with EUV pulses at relatively low fluence ($<$ 10 mJ/cm^2) [21, 23–27]. It was found that such EUV patterning of surfaces can be useful for biocompatibility control of polymers [28]. Contrary to the traditional patterning techniques, based on UV sources, the use of EUV radiation, due to its very low penetration depths, makes possible to modify polymer surfaces without degradation of bulk material caused by interaction of highly penetrating UV photons [29]. Laser plasma source dedicated for EUV processing of materials has been developed [30].

The compact laser plasma EUV source for metrology applications, additionally equipped with an ellipsoidal mirror with the Mo/Si coating and operating with the nitrogen gas puff target, allowed producing quasi-monochromatic EUV radiation at 13.8 nm wavelength [31]. This source has been successfully used for EUV nanoimaging with the spatial resolution down to 50 nm, proving possibility to develop a compact, desk-top EUV imaging tool [32–34]. The tool was also used for EUV imaging of nanostructures [35] and crystalline thin films and nanofibers [36]. Comprehensive review of these studies has been presented at the previous ICXRL conference [37].

In this paper a short review of the research on application of our laser plasma soft X-ray and EUV sources performed during the last 2 years is presented. New applications in imaging, including EUV radiography tomography and soft X-ray microscopy, EUV processing materials, including micro-machining and modification polymer surfaces, and EUV photoionization are presented.

2 Imaging

2.1 EUV Radiography and Tomography

The compact laser plasma EUV source has been used for pulsed radiography and tomography of gas puff targets. The beam of EUV radiation for backlighting of the gas puff targets under study was produced by spectral selection of emission from xenon plasma using a Mo/Si multilayer mirror. To eliminate the visible light from the plasma a 200 nm thick Zr filter deposited on a 200 nm thick Si_3N_4 membrane was used. In this way quasi-monochromatic radiation at 13.5 nm wavelength with the bandwidth of about 1 nm was obtained. The EUV shadowgrams of the gas puff targets were registered with the use of the back-illuminated CCD camera (Reflex) equipped with a 512×512 pixels CCD chip [38].

The multi-jet gas puff targets with modulated gas density from high-order harmonic generation (HHG) experiments [39] produced using the nozzles in a form of linear array of 5, 7 or 9 orifices have been studied [38]. 2-D gas density map for the multi-jet gas puff target produced using the nozzle with 7 orifices at the backing pressure of 4 bar and corresponding gas density profiles for various distances are presented in [38]. The radiography setup with the laser plasma EUV source has been also used for characterization of the dual-gas multi-jet gas puff target we developed for HHG experiments [40] and elongated plasma channels [41].

The same setup was used for EUV tomography of the multi-jet gas puff targets [42, 43]. In this case the gas puff valve was mounted on top of a rotation stage to ensure 2π rotation of the nozzles while acquiring projections. The set of 900 EUV projections was used for tomographic reconstruction. The 3-D image of the multi-jet gas puff target has been obtained [42, 43]. A new technique for 3-D tomographic reconstruction of low density objects with the use of a compact laser plasma EUV source was demonstrated.

2.2 Soft X-Ray Microscopy

Laser plasma soft X-ray sources operating in the 'water window' wavelength range between 2.3 and 4.4 nm are used for microscopy of live biological objects [44–51]. It was shown that our laser plasma source with argon or nitrogen gas puff targets is an efficient source of radiation in this range [52]. We have used this source for development of a compact, desk-top soft X-ray microscope, based on a laser plasma source, operating in the 'water window' range. The microscope was equipped with an ellipsoidal grazing incidence mirror coated with nickel as a condenser to focus soft X-ray radiation from the plasma onto a sample [53]. A Wolter type grazing incidence hyperboloid/ellipsoid axisymmetrical mirror was used as an objective to form a magnified image onto a CCD camera. The first soft X-ray microscopy images in the 'water window' spectral range of biological samples (onion skin cells) with sub-micron spatial resolution have been obtained [54].

3 EUV Processing of Materials

Processing polymers is an important field that has many applications in microelectromechanical systems, biomedicine and photonics. Lasers have been proven as effective tools in micromachining of polymers [55]. Because organic polymers show high absorption in the ultraviolet region the laser ablation with UV lasers has become the most popular microprocessing technique and excimer lasers (157–351 nm) are extensively used for photoablation, chemical etching, lithography, and surface cleaning.

It was demonstrated, that processing polymers can be realized by direct photo-etching with the use of EUV radiation generated from laser plasma sources [13, 14, 56–58]. In the EUV photo-etching process a single photon carries enough energy to break any chemical bond, thus the mechanism is similar to the UV laser ablation. However, much shorter wavelength of EUV allows obtaining smaller features as compared to UV. Moreover, because of very high absorption of EUV (the penetration depth is less than 100 nm for $\lambda \cong 10$nm) the material is removed from the very thin near-surface layer and the material inside is not affected by the photons.

3.1 Laser Plasma EUV Source for Processing Polymers

The laser plasma EUV source dedicated for processing polymers has been designed and built. The source is composed of a vacuum chamber in a form of a vertical column placed onto a cubical base, housing a compact commercial Nd:YAG laser system (EKSPLA) generating 4 ns laser pulses with energy up to 800 mJ and vacuum pumping system. The source chamber is composed of three sections. Each section is pumped separately by oil-free vacuum pumps (differential pumping). In the first upmost section of the chamber the electromagnetic valve to produce a gas puff target and the laser beam focusing system are placed. The valve is mounted using the x-y-z translation stages allowing placing the gas puff target in the required position with accuracy of about 10 μm. The gas puff targets are formed by pulsed injection of working gas (krypton, xenon or krypton/xenon mixture) into a stream of helium using the electromagnetic valve system with a double-nozzle setup. The repetition rate of the system is determined by the repetition rate of the laser (10 Hz). The source is equipped with a grazing incidence axisymmetrical ellipsoidal mirror (Reflex) to focus EUV radiation. The mirror is mounted in the second central section of the vacuum chamber. It makes possible to focus EUV radiation onto a polymer sample mounted in the third section of the chamber, evacuated to high-vacuum. EUV radiation is focused to a spot of about 1 mm in diameter with fluency up to 100 mJ/cm^2 for the xenon gas puff target. Description of the source is given in [59].

3.2 *EUV Micromachining of Poly(Vinylidene Fluoride) (PVDF)*

Poly(vinylidene fluoride) (PVDF) is an important fluoropolymer because of its piezoelectric, pyroelectric and ferroelectric properties. It is also known to have an extremely high chemical stability and electrical resistivity. Micro- or even nanopatterning of PVDF is highly desirable for applications in multifunctional and integrated devices. Many works have been performed on surface processing of PVDF using ion beams, synchrotron X-ray and UV laser radiation, however, irradiation of PVDF with these sources resulted in strong modification of the molecular structure in a near-surface layer of the polymer. Using the laser plasma EUV source dedicated for processing polymers we have demonstrated for the first time efficient micromachining of PVDF without changing the chemical structure of the unprocessed material [60]. PVDF foils of 50 μm thickness (Goodfellow) were irradiated with EUV through a contact metal mask with square orifices $60 \times 60\ \mu m^2$. Micro holes etched through the foils have been obtained as a result of 1 min irradiation at 10 Hz repetition rate.

Investigation of the ablation products with QMS demonstrated a good agreement between the stoichiometric composition of PVDF molecules in the ablated and bulk polymer. XPS spectra acquired for the polymer after ablation are almost identical to the spectrum of pristine PVDF, indicating preservation of the chemical structure of the remaining material. However, XPS measurements performed on the polymer irradiated with low fluence ($< 10\ mJ/cm^2$) indicated strong chemical modification in the near-surface layer. In this case defluorination and thus carbon enrichment in the surface material was revealed [60].

3.3 EUV Modification of Polymer Surfaces for Biocompatibility Control

We have demonstrated in our previous works that EUV radiation can be used for surface modification of polymers for biocompatibility control [28,61]. Surface modification of PTFE, PVDF and PC polymers for biocompatibility control has been studied using the laser-plasma EUV source. Modified surfaces were characterized by SEM and AFM. Up to several hundred nanometers high wall-type micro- and nanostructures were formed [62]. Simultaneous treatment of polymer surface by EUV radiation and ionized nitrogen injected in the interaction region has been studied [63]. Chemical analysis by XPS revealed decreased oxygen contents in PC samples and nitrogen enrichment in PTFE [62,64]. Exclusion of oxygen from polar groups leads to a polymer with increased hydrophobicity that was confirmed by contact angle measurements. Biocompatibility tests of PTFE and PVF samples modified with EUV photons and seeded with fibroblasts have shown strong cells adhesion to polymer surfaces [65].

4 EUV Photoionization

The laser plasma EUV source dedicated for processing materials have been also used for the first time in the experiments on EUV photoionization of atomic and molecular gases. Gases could be injected into the focus of the EUV beam using an additional gas nozzle mounted in the third section of the EUV source chamber. Formation of low-temperature photo-ionized neon plasmas induced by nanosecond EUV pulses from the laser plasma source and by femtosecond EUV pulses from the FLASH free electron laser was studied [66]. Luminescence of helium and neon gases induced by EUV pulses was measured [67] and significant differences between absorption spectra of neutral helium and low temperature photoionized helium plasmas have been detected [68]. Spectral investigations in the EUV/VUV region of photoionized plasmas induced in atomic and molecular gases using nanosecond EUV pulses were performed [69].

EUV photoionization of gases and formation of low-temperature plasmas have been also studied using more energetic EUV pulses generated as a result of irradiation of the gas puff target with laser pulses with time duration from 1 to 10 ns and energies from up to 10 J at 10 Hz repetition rate produced from the Nd:YAG laser system (EKPLA) [70]. These preliminary investigations have shown applicability of compact laser plasma EUV sources for research in a new field of EUV-induced plasmas.

The research was supported by the Foundation for Polish Science under the HOMING 2009 Programme (grant number HOM2009/14B), the EC's 7. Framework Program (LASERLAB-EUROPE II—grant agreement n° 228334 and LASERLAB-EUROPE III—grant agreement n° 284464, COST Actions MP0601 and MP1203, and the Erasmus Mundus programme EXTATIC), the National Centre for Science (award number DEC-2011/03/D/ST2/00296), and the National Centre for Research and Development Lider Programme, award number LIDER/004/410/L-4/12/NCBR/2013.

References

1. Michette, A.G.: Optical Systems For Soft X-Rays. Plenum Press, New York (1986)
2. Turcu, I.C.E., Dance, J.B.: X-Rays From Laser Plasmas: Generation and Applications, Wiley, London (1998)
3. Attwood, D.T.: Soft X-Rays and Extreme Ultraviolet Radiation: Principles and Applications, Cambridge University Press, Cambridge (1999)
4. Bakshi, V.: EUV Sources for Lithography, SPIE Press Book, Bellingham (2006)
5. Fiedorowicz, H., et al.: Enhanced X-ray emission in the 1-keV range from a laser-irradiated gas puff target produced using the double nozzle setup. Appl. Phys.B **70,** 305–308, (2000)
6. Fiedorowicz, H., et al.: Strong EUV emission from a double-stream xenon/helium gas puff target irradiated with a Nd:YAG laser. Opt. Commun. **184,** 161–167 (2000)
7. Fiedorowicz, H., et al.: High-brightness laser plasma soft X-ray source using a double-stream gas puff target irradiated with the Prague Asterix Laser System (PALS). J. Alloys & Comp., **362,** 67–70, 2004
8. Fiedorowicz, H., et al.: Generation of soft X-rays and extreme ultraviolet (EUV) using a laser-irradiated gas puff target. Laser & Particle Beams. **23,** 365–373 (2005)
9. Fiedorowicz, H., et al.: Compact laser plasma EUV source based on a gas puff target for metrology applications. J. Alloys Comp. **401,** 99–103 (2005)
10. Rakowski, R., et al.: Metrology of Mo/Si multilayer mirrors at 13.5 nm with the use of a laser-produced plasma extreme ultraviolet (EUV) source based on a gas puff target. Opt. Appl., XXXVI, 593–600, 2006
11. Rakowski, R., et al.: Characterization and optimization of the laser-produced plasma EUV source at 13.5 nm based on a double-stream Xe/He gas puff target. Appl. Phys. **B 101,** 773 (2010)
12. Jungwirth, K., et al.: The Prague Asterix Laser System. Phys. Plasmas. **8,** 2495–2501 (2001)
13. Juha, L., et al.: Ablation of various materials with intense XUV radiation. Nucl. Inst. Meth. Phys. Res. A. **507,** 577–581 (2003)
14. Fiedorowicz, H., et al.: Micromachining of organic polymers by direct photo-etching using a laser-plasma x-ray source. Microel. Eng. **73–74,** 336–339 (2004)
15. Davídková, M., et al.: A high-power laser-driven source of sub-nanosecond soft X-ray pulses for single-shot radiobiology experiments. Rad. Res. **168,** 382–387 (2007)
16. Bartnik, A., et al.: Micromachining of organic polymers by X-ray photo-etching using a 10 Hz laser-plasma radiation source. Microel. Eng. **78–79,** 452–456 (2005)
17. Bartnik, A., et al.: Ablation and surface modifications of PMMA using a laser plasma EUV source. Appl. Phys. **B 96,** 727–730 (2009)
18. Sveda, L., et al.: Multi-foil optic condenser for a laser plasma EUV source. Phys. Scr. **T123,** 131–134 (2006)
19. Bartnik, A., et al.: EUV emission from solids illuminated with a laser-plasma EUV source. Appl. Phys. **B 93,** 737–741 (2008)
20. Bartnik, A., et al.: Detection of surface changes of materials caused by intense irradiation with laser-plasma EUV source utilizing scattered or luminescent radiation excited with the EUV pulses. Appl. Phys. **B, 91,** 21–24 (2008)
21. Bartnik, A., et al.: Ablation and surface modifications of PMMA using a laser-plasma EUV source. Appl. Phys. **B 96,** 727–730 (2009)

22. Bartnik, A., et al.: Micro-and nanoprocessing of polymers using a laser plasma extreme ultraviolet source. Acta Phys. Pol. A. **117,** 384–390 (2010)
23. Bartnik, A., et al.: Creation of nanostructures on polymer surfaces irradiated with extreme ultraviolet pulses.Acta Phys. Pol. A. **118,** 109–110 (2009)
24. Bartnik, A., et al.: PMMA and FEP surface modifications induced with EUV pulses in two selected wavelength ranges. Appl. Phys. A. **98,** 61–65 (2010)
25. Bartnik, A., et al.: Physical and chemical modifications of PET surface using a laser-plasma EUV source. Appl. Phys. A. **99,** 831–836 (2010)
26. Bartnik, A., et al.: EUV-induced physico-chemical changes in near-surface layers of polymers. J. Electr. Spectr. & Rel. Phenom. **184,** 270–275 (2011)
27. Bartnik, A., et al.: Combined effect of EUV irradiation and acetone treatment on PET surface. Appl. Phys. A. **103,** 173–178 (2011)
28. Reisinger, B., et al.: EUV micropatterning for biocompatibility control of PET. Appl. Phys. A. **100,** 511–516 (2010)
29. Bauerle, D.: Laser Processing and Chemistry, Springer, Berlin (2000)
30. Bartnik, A., et al.: Laser-plasma EUV source dedicated for surface processing of polymers. Nucl. Instr. Methods Phys. Res. A. **647,** 125–131 (2011)
31. Wachulak, P.W., et al.: A compact, quasi-monochromatic laser-plasma EUV source based on a double stream gas puff target at 13.8 nm wavelength. Appl. Phys. B. **100,** 461–469 (2010)
32. Wachulak, P.W., et al.: Sub-70 nm resolution tabletop microscopy at 13.8 nm using a compact laser-plasma EUV source. Opt. Lett. **35,** 2337–2339 (2010)
33. Wachulak, P.W., et al.: A 50 nm spatial resolution EUV imaging-resolution dependence on object thickness and illumination bandwidth. Opt. Expr. **19,** 9541–9550 (2011)
34. Wachulak, P.W., et al.: Nanometer-scale incoherent imaging using laser-plasma EUV source. Acta Phys. Pol. A. **121,** 450–453 (2012)
35. Wachulak, P.W., et al.: Imaging of nanostructures with sub-100 nm spatial resolution using a desktop EUV microscope. Appl. Phys. B. **109,** 105–111 (2012)
36. Wachulak, P.W., et al.: Study of crystalline thin films and nanofibers by means of the laser-plasma EUV-source based microscopy. Rad. Phys. Chem. **93,** 54–58 (2013)
37. Wachulak, P.W., et al.: Nanoscale imaging using a compact laser plasma EUV source. AIP Confer. Proc. **1437,** 79–82 (2012)
38. Wachulak, P.W., et al.: Characterization of multi-jet gas puff targets for high-order harmonic generation using EUV shadowgraphy. Nucl. Instrum. Methods Phys. Res. B. **285,** 102–106 (2012)
39. Fok, T., et al.: High-order harmonic generation using a multi-jet gas puff target. Phot. Lett. Pol. **6,** 14–16 (2014)
40. Wachulak, P.W., et al.: Characterization of a dual-gas multi-jet gas puff target for high-order harmonic generation using extreme ultraviolet shadowgraphy. Laser & Part. Beams. **31,** 195–201 (2013)
41. Wachulak, P.W., et al.: Extreme ultraviolet tomography of multi-jet gas puff target for high-order harmonic generation. Phys. Plasmas. **21,** 103106 (2014)
42. Wachulak, P.W., et al.: Extreme ultraviolet tomography using a compact laser-plasma source for 3D reconstruction of low density objects.Opt. Lett. **39,** 523–535 (2014)
43. Wachulak, P.W., et al.: Extreme ultraviolet tomography of multi-jet gas puff target for high-order harmonic generation. Appl. Phys. B. **117,** 253–263 (2014)
44. Tomie, T., et al.: Three-dimensional readout of flash x-ray images of living sperm in water by atomic-force microscopy. Science. **252,** 691–693 (1991)
45. Aritome, H., et al.: An imaging x-ray microscope using a laser plasma source. Springer Ser. Opt. Sci. **67,** 199–201 (1992)
46. Nakayama, S., et al.: Zone-plate x-ray microscope using a laser plasma source. Jpn. J. Appl. Phys. **33,** L1280–L1282 (1994)
47. Ogata, T., et al.: Development of a grazing incodence soft x-ray microscope with a laser-produced plasma source. J. Electr. Spectr. & Rel. Phenom. **80,** 357–360 (1996)

48. Berglund, M.: Compact water-window transmission microscopy. J. Microsc. **197,** 268–273 (2000)
49. Johansson, G.A., et al.: Design and performance of a laser-plasma-based compact soft x-ray microscope. Rev. Sci. Instr. **73,** 1193–1197 (2002)
50. Kim, K.W., et al.: Compact soft x-ray transmission microscopy with sub-50nm spatial resolution. Phys. Med. Biol. 51**,** N99–N107 (2006)
51. Legall, H., et al.: Compact x-ray microscope for the water window based on a high brightness laser plasma source. Opt. Express. **20,** 18362–18369 (2012)
52. Wachulak, P.W., et al.: "Water window" compact, table-top laser plasma soft X-ray sources based on a gas puff target. Nucl. Instrum. Methods Phys. Res. B. **268,** 1692–1700 (2010)
53. Wachulak, P.W., et al.: Water-window microscopy using a compact, laser-plasma SXR source based on a double-stream gas-puff target. Appl. Phys. B. **111,** 239–247 (2013)
54. Wachulak, P.W., et al.: Sub 1-μm resolution "water-window" microscopy using a compact, laser-plasma SXR source based on a double stream gas-puff target. Nucl. Instrum. Methods Phys. Res. B. **311,** 42–46 (2013)
55. Lippert, T.: Laser application of polymers. Adv. Pol. Sci. **168,** 51–246 (2004)
56. Juha, L., et al.: Ablation of poly(methyl methacrylate) by a single pulse of soft X-rays emitted from Z-pinch and laser-produced plasmas. Surf. Rev.Lett. **9,** 347–352 (2002)
57. Makimura, T., et al.: Direct micromachining of quartz glass plates using pulsed laser plasma soft x-rays. Appl. Phys. Let. **86,** 103111 (2005)
58. Barkusky, F., et al.: Direct photo-etching of poly(methyl methacrylate) using focused extreme ultraviolet radiation from a table-top laser-induced plasma source. J. Appl. Phys. **101,** 124908 (2007)
59. Bartnik, A., et al.: Short wavelength laboratory sources: Principles and practices. Royal Society of Chemistry, Cambridge, pp. 366–372 (2015)
60. Bartnik, A., et al.: Efficient micromachining of poly(vinylidene fluoride) using a laser-plasma EUV source. Appl. Phys. A. **106**, 551–555 (2012)
61. Ahad, I.U., et al.: Surface modification of polymers for biocompatibility via exposure to extreme ultraviolet radiation. J. Biomed. Mater. Res. A 102 A, 3298–3310 (2014)
62. Ahad, I.U., et al.: Polycarbonate polymer surface modification by extreme ultraviolet (EUV) radiation. Acta Phys. Pol. A. **125**, 924–928 (2014)
63. Bartnik, A., et al.: Simultaneous treatment of polymer surface by EUV radiation and ionized nitrogen. Appl. Phys. A. **109**, 39–43 (2012)
64. Ahad, I.U., et al.: Nitrogen doping in biomaterials by extreme ultraviolet (EUV) surface modification for biocompatibility control. Eur. Cells Mater. **26** (Supp. 6), 145 (2013)
65. Ahad, I.U., et al.: Extreme Ultraviolet (EUV) surface modification of Polytetrafluoroethylene (PTFE) for control of biocompatibility. Nucl. Instrum. Methods Phys. Res. B—submitted
66. Bartnik, A., et al.: Photo-ionized neon plasmas induced by radiation pulses of a laser-plasma EUV source and a free electron laser FLASH. Laser Part. Beams. **31**, 195–201 (2013)
67. Bartnik, A., et al.: Luminescence of He and Ne gases induced by EUV pulses from a laser plasma source. Rad. Phys. Chem. **93**, 9–13 (2013)
68. Bartnik, A., et al.: Detection of significant differences between absorption spectra of neutral helium and low temperature phot-oionized helium plasmas. Phys. Plasmas. **20**, 113302 (2013)
69. Bartnik, A., et al.: Extreme ultraviolet-induced photoionized plasmas. Phys. Plasmas. **21**, 073303 (2014)
70. Bartnik, A., et al.: Spectral investigations of photo-ionized plasmas induced in atomic and molecular gases using nanosecond extreme ultraviolet (EUV) pulses. Phys. Scr. **T161**, 01461 (2014)

Rapid Calculation of Scattering Factors for Partially Ionized Plasmas in the EUV

A. K. Rossall and G. J. Tallents

Abstract The work presented demonstrates a newly developed, rapid algorithm for the calculation of atomic scattering factors in the EUV for partially ionized plasmas. Using element specific input data, the code REFRAC, calculates plasma refractive index, atomic scattering factors and scattering cross sections with a runtime short enough to be used in-line with fluid codes to model ablative flow from a planar target. The algorithm demonstrates good agreement with data from the Centre for X-ray Optics (CXRO) at Lawrence Berkeley National Laboratory for carbon. This work aims to accurately model moderate irradiance ($> 10^9$ W cm^{-2}) EUV laser absorption within solid material and expanding plasma, to characterise the ablation properties and highlight the potential applications of such lasers.

1 Introduction

Ablation by conventional optical and infra-red lasers has been comprehensively studied over the past few decades [1], giving rise to a range of applications such as deposition, micro-analysis and micro-structuring. The advent of the excimer UV laser, opened up further applications in laser ablation lithography and micro hole drilling in polymers [2, 3] and metals [4]. The ablation mechanism for optical and IR irradiation is dominated by photo-thermal effects, i.e. the photo-excitation of a material producing thermal heat; where the ablation mechanism for UV lasers is a combination of photo-thermal and photo-chemical, i.e. where UV photons directly break molecular bonds due to the increase in photon energy. Both methods result in energy deposition into the target, ionisation and subsequent expansion of the produced plasma, leaving an ablation 'crater' in the target material. By carefully controlling the laser pulse characteristics, e.g. fluence, temporal duration and wavelength, the ablation characteristics can be optimised for the desired application.

More recently, considerable advances have been made in higher fluence EUV and x-ray laser technology as demonstrated by free-electron lasers [5, 6] and capillary discharge lasers [7, 8]. With higher fluences now available, EUV and x-ray lasers

A. K. Rossall (✉) · G. J. Tallents
York Plasma Institute, Department of Physics, University of York,
Heslington, York YO10 5DD, UK
e-mail: andrew.rossall@york.ac.uk

© Springer International Publishing Switzerland 2016

J. Rocca et al. (eds.), *X-Ray Lasers 2014*, Springer Proceedings in Physics 169,
DOI 10.1007/978-3-319-19521-6_49

can be used to directly generate plasma with unique properties. These plasmas are produced by irradiating solid targets with EUV/x-ray lasers, heating the sample via direct photo-ionization, resulting in typically lower temperatures and higher densities than the traditional laser produced plasma, generated using infra-red, optical and UV pulses. For example, in a simple hydrocarbon the average energy required to break the C-H bond is 4.3 eV [9], achievable by excimer lasers (KrF - 5.0 eV, ArF - 6.3 eV), whereas the first ionisation energy of carbon is $E_i = 11.26$ eV [10]. By reducing the wavelength into the EUV/x-ray region, the photon energy, E_p, becomes sufficient to directly photo-ionise elemental components, transferring a set amount of energy ($E_p - E_i$) to the ejected electron. Reducing the lasing wavelength to the EUV/soft x-ray region allows for a tighter focus (due to a reduction in the diffraction limit) and enables the laser to penetrate into the solid material, as the critical density at these wavelengths are typically higher than solid. The tighter focus and the increased target penetration are potentially desirable properties for a number of applications, for example, micro-machining, mass spectrometry and the coating of refractory material onto substrates. Due to the unique properties of plasmas generated by EUV/x-ray lasers, these plasmas can contribute to our understanding of warm dense matter, essential in order to achieve inertial confinement fusion as the pre-ignition fuel is in the warm dense matter regime.

To promote research and to accelerate the development of industrial applications, there has been significant motivation to produce compact and affordable EUV/x-ray sources for use in parallel with large scale facilities such as Linac Coherent Light Source (LCLS) in the US and the European XFEL. The research group at Colorado State University (CSU) have developed a table-top size soft x-ray laser system [11–13] based upon capillary discharge excitation of an Ar gas producing a large soft x-ray amplification.

The work presented here utilises a novel combination of 2D fluid code modelling (POLLUX) with a rapid atomic physics algorithm (REFRAC) to simulate the EUV/x-ray interaction with solid material, the expected energy deposition within the target, the produced plasma parameters and the subsequent ablative flow away from the target. The REFRAC algorithm uses a super-configuration model for atomic structure and calculates ionisation and excited level populations, absorption cross-sections, atomic scattering factors and plasma refractive index. Population distributions and free electron density is updated in POLLUX each time step to accurately calculate laser absorption via photoionisation and inverse bremsstrahlung and the refarctive index as a function of position in the plasma enables more accurate ray-tracing within the code.

2 REFRAC

The algorithm REFRAC utilises a superconfiguration approach to calculate atomic scattering factors in the EUV and X-ray for partially ionised plasmas. Detailed level structure is initially obtained by using the Flexible Atomic Code (FAC) [14] to solve the radial wave equation. This structure is post-processed to group energetically similar levels into 'supershells', the average energy of which ($\langle E \rangle_{SS}$) is weighted by the degeneracy (g_m) as shown in Eq. 1, where E_m is the energy of a single detailed level.

$$\langle E \rangle_{SS} = \frac{\sum_m g_m E_m}{\sum_m g_m} \tag{1}$$

Ionisation and excited level populations are determined assuming local thermodynamic equilibrium, an assumption which is justified on hydrodynamic timescales of > ps, as the highly non-equilibrium plasma initially created equilibrates on the order of 10 s of fs [15]. Photoionisation cross-sections for each ionisation stage and superconfiguration level are calculated assuming a E^{-3} photon energy dependance, and a configuration specific constant ($A\,(n, Z)$) determined using data from FAC. This approximation allows for an analytical solution to the Kramers-Kronig relation to calculate the atomic scattering factor f_1 as shown by Eq. 2. In Eq. 2, $Z^{\star}$ is the number of bound electrons, Z is the ionisation stage, n is the supershell number, E is the photon energy and $E_{i,n}$ is the binding energy of an electron in supershell n of an ion with charge Z. This model has shown reasonable agreement with data from the CXRO for cold, solid carbon, as is shown in Fig. 1.

$$f_1^0(E) = Z^{\star} + \frac{1}{\pi r_e hc}\left(\sum_Z \sum_n \frac{A(n, Z)}{2E^2} \log\left(|1 - \frac{E^2}{E_{i,n}^2}|\right)\right) \tag{2}$$

The algorithm REFRAC is used to calculate the populations, free electron density and refractive index for each cell within the fluid code POLLUX and is updated for each time step. The populations and free electron density are used in the calculation

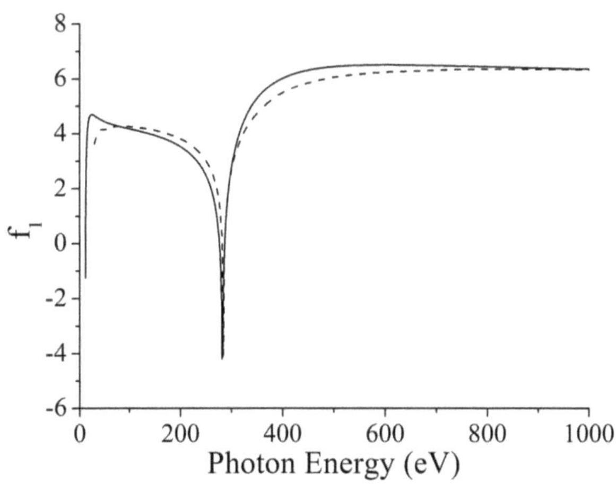

Fig. 1 Scattering factor, f_1, for cold, solid carbon as a function of photon energy. A comparison is made between REFRAC (*solid line*) and data from the Centre for X-ray Optics (CXRO) at Lawrence Berkeley National Laboratory (*dashed line*)

of the laser absorption as outlined previously [16] and the refractive index is used in the ray-tracing routine.

3 POLLUX

POLLUX is a 2D Eulerian radiative-hydrodynamic code, written at the University of York, and was originally developed to simulate the interaction of optical and infrared high power laser irradiation of a solid target and the subsequently produced strongly ionised plasma [17, 18]. The code solves the three first-order quasi-linear partial differential equations of hydrodynamic flow using the flux corrected transport model of Boris and Book [19] with an upwind algorithm [20] for the first term. Energy is absorbed by the plasma and distributed through electron-ion collisions, the equilibration of which is determined by the Spitzer plasma collision rate [21]. For calculation of the equation-of-state (EOS) variables, POLLUX utilizes in-line hydrodynamic EOS subroutines from the Chart-D equation-of-state package developed at Sandia National Laboratories [22].

POLLUX uses an explicit solver and therefore requires a maximum Courant number of $C \sim 1$, accounting for both spatial directions as shown below in equation 3,

$$C = \frac{u_x \Delta t}{\Delta x} + \frac{u_y \Delta t}{\Delta y} \sim 1 \tag{3}$$

where u_x and u_y are magnitudes of the particle velocities in the respective directions, Δt is the time step and Δx,Δy are the cell spatial dimensions. The Courant-Friedrichs-Lewy condition [23] is not the only constraint on simulation parameters, however, it is the most restrictive. To increase the resolution of the Eulerian mesh, the time step must also be reduced and as a result the runtime of the code increases exponentially as the resolution increases. Not adhering to the condition outlined by Eq. 1 results in numerical instability within the code.

4 Results

This simulation study is for the ablation of a planar parylene-N target using an Ar based capillary discharge laser, operating at a photon energy of 26.4 eV (46.9 nm), pulse length of 800 ps, a fluence of 8 J cm^{-2} and a focal diameter of 650 nm.

Figure 2 illustrates the mass density as a function of position within the target at the end of the EUV laser pulse, $t = 800$ ps. Taking the ablated depth as the position of the melt front at the end of the pulse on the laser axis, the simulation predicts an ablated depth of 395 nm with a mesh resolution of 10 nm in the x direction. The ablated depth as a function of laser fluence is shown in Fig. 3. It is found that over 70 % of the ablation in the x direction occurs during the initial 200 ps of the laser

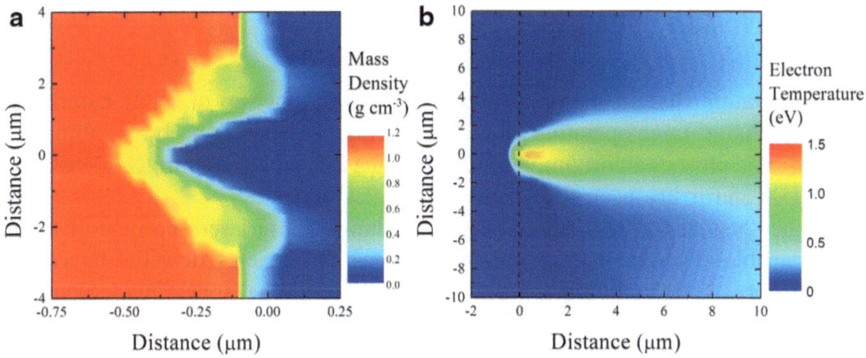

Fig. 2 Mass density (**a**) and electron temperature (**b**) as a function of position at the end of the EUV laser pulse, $t = 800$ ps

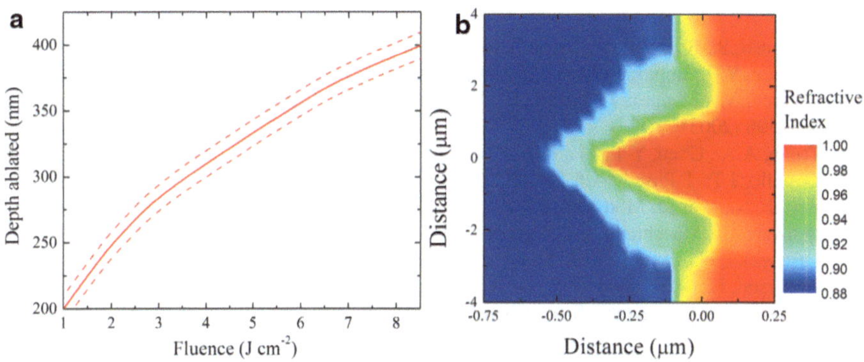

Fig. 3 Ablated depth as a function of laser fluence (**a**) and refractive index map (**b**) at the end of the EUV laser pulse, $t = 800$ ps

pulse, after which a large portion of the energy goes to increasing the size in the direction perpendicular to the laser pulse. This indicates shorter pulses would be more advantageous for high aspect ratio drilling or applications requiring a small feature size.

Figure 2 shows the electron temperature as a function of position within the plasma at the end of the EUV laser pulse, $t = 800$ ps. Typical electron and ion temperatures are between 1 and 2 eV at densities ranging up to $\sim 8 \times 10^{22}$ cm^{-3}. The peak energy density is observed around $t = 200$ ps and reaches over 30, 000 J cm^{-3}.

Plasma refractive index as a function of position within the plasma is shown in Fig. 3. The inclusion of ray tracing in the code causes a small increase in the peak electron temperature ($\sim$ 0.2 eV) on the laser axis with no observable difference in the ablated depth at this irradiance. Focussing and de-focussing effects caused by refractive index gradients within the plasma are expected to have a considerably more significant role at higher irradiance, above 10^{11} W cm^{-2} (compared to $\sim 9 \times 10^9$ W cm^{-2} used in these simulations).

5 Conclusion

This simulation study has investigated the ablative properties of an Ar based capillary discharge laser operating at 26.4 eV. This has been done by combining a rapid atomic physics algorithm, REFRAC, with a 2D radiative-hydrodynamic code, POLLUX and it has been found that shorter EUV pulses ($\sim$ 200 ps) would be of greater benefit to high aspect ratio machining. These lasers have been found to be capable of producing warm dense plasmas, with energy densities exceeding 30,000 J cm^{-3}. Refractive index gradients have a minimal focussing/defocussing effect at the fluence of 8 J cm^{-2}.

References

1. Amoruso, S., Bruzzese, R., Spinelli, N., Velotta, R.: J. Phys. B. **32**, R131–R172 (1999)
2. Tokarev, V.N., Lopez, J., Lazare, S.: Appl. Surf. Sci. **168**, 75–78 (2000)
3. Tokarev, V.N., Lopez, J., Lazare, S., Weisbuch, F.: Appl. Phys. A: Mater. Sci. Process. **76**, 385–396 (2003)
4. Tseng, A.A., Chen, Y-T., Ma, K-J.: Opt. Laser. Eng. **41**, 827–847 (2004)
5. Costello, J.T.: J. Phys. Conf. Ser. **88**, 012057 (2007)
6. Berrah, N., et al.: J. Mod. Opt. **57**, 1015–1040 (2010)
7. Benware, B., Macchietto, C., Moreno, C., Rocca, J.: Phys. Rev. Lett. **81**, 5804–5807 (1998)
8. Vaschenko, G., Etxarri, A.G., Menoni, C.S., Rocca, J.J., Hemberg, O., Bloom, S., Chao, W., Anderson, E.H., Attwood, D.T., Lu, Y., Parkinson, B.: Opt. Lett. **31**, 3615 (2006)
9. Zumdahl, S.S., Zumdahl, S.A.: Chemistry [Hardcover], 6th ed. Houghton Mifflin, Boston (2002)
10. Gallagher, J., Moore, C.E.: Tables of Spectra of Hydrogen, Carbon, Nitrogen, and Oxygen Atoms and Ions. CRC, Boca Raton (1993)
11. Benware, B.R., Moreno, C.H., Burd, D.J., Rocca, J.J.: Opt. Lett. **22**, 796 (1997)
12. Rocca, J.J., Shlyaptsev, V., Tomasel, F., Cortázar, O., Hartshorn, D., Chilla, J.: Phys. Rev. Lett. **73**, 2192–2195 (1994)
13. Rocca, J.J., Tomasel, F.G., Marconi, M.C., Shlyaptsev, V.N., Chilla, J.L.A., Szapiro, B.T., Giudice, G.: Phys. Plasmas. **2**, 2547 (1995)
14. Gu, M.F.: Can. J. Phys. **86**, 675–689 (2008)
15. Aslanyan, V., Tallents, G.J.: Phys. Plasmas. **21**, 062702 (2014)
16. Rossall, A.K., Aslanyan, V., Tallents, G.J.: SPIE Optical Engineering + Applications. Klisnick, A., Menoni, C.S. (eds.) 12–19 pp. 884912–888849 (2013)
17. Pert, G.J.: J. Comput. Phys. **49**, 1–43 (1983)
18. Pert, G.J.: J. Plasma Phys. **41**, 263–280 (1989)
19. Boris, J., Book, D.: J. Comput. Phys. **20**, 397–431 (1976)
20. Courant, R., Isaacson, E., Rees, M.: Commun. Pure Appl. Math. **5**, 243–255 (1952)
21. Spitzer, L., Härm, R.: Phys. Rev. **89**, 977–981 (1953)
22. Thompson, S.L.: Sandia Natl. Lab. Rep. **SC-RR-70-2** (1970)
23. Courant, R., Friedrichs, K., Lewy, H.: IBM J. Res. Dev. **11**, 215–234 (1967)

Transmission Measurement for Highly Transparent Metallic Sodium in the Extreme Ultraviolet Spectral Range: New Application of an Intense EUV Sources

Hiroyuki Daido, Yoji Suzuki, Tetsuya Kawachi and Alexander Sergeevich Pirozhkov

Abstract In order to perform transmission measurement of metallic sodium in the mm thickness range, we made the sample using a glove box which is quite different technique compared with the previous works. Using a CW source, we have made direct measurement of actual transmittance of a sodium sample with a thickness of $1 \sim 8$ mm in the spectral range > 115 nm which is restricted by magnesium fluoride (MgF_2) windows, resulting in, for example, $\sim 50\,\%$ transmittance of a 3 mm thick solid sodium sample including the transmittance of the windows and the reflection losses at each interface at the wavelength of ~ 120 nm, with a very weak temperature dependence up to 150 °C which is beyond the melting point of 97 °C. To confirm the high transmittance, we set up a simple transmission imaging configuration with a 8-mm thick sodium sample, resulting in clear images composed of 100 μm diameter tungsten bars recorded on a Charge Coupled Device detector. The results show that a short pulse intense source such as high order harmonics as well as short pulse EUV plasma source opens a new technique for investigating dynamics of phase transitions as well as hydrodynamics of molten metal.

1 Introduction

Metallic sodium is used as a media of large scale battery such as Sodium Sulfur (NaS) battery for wind generators, as well as coolant for fast breeder type reactors. In this report, we describe the optical property of sodium in the extreme ultraviolet (EUV) spectral range (~ 120 nm) using newly developed technique for preparing a sodium sample. We perform measurement of spectral transmittance and imaging of fine structures with an EUV source. We find that a several mm-thick sodium sample is highly transparent in the EUV spectral range. We try to make physical

H. Daido (✉) · Y. Suzuki
Tsuruga Head Office, Applied Laser Technology Institute, Japan Atomic Energy Agency, Tsuruga, Japan
e-mail: daido.hiroyuki@jaea.go.jp

T. Kawachi · A. S. Pirozhkov
Quantum Beam Science Directorate, Japan Atomic Energy Agency, Kyoto, Japan

© Springer International Publishing Switzerland 2016
J. Rocca et al. (eds.), *X-Ray Lasers 2014*, Springer Proceedings in Physics 169,
DOI 10.1007/978-3-319-19521-6_50

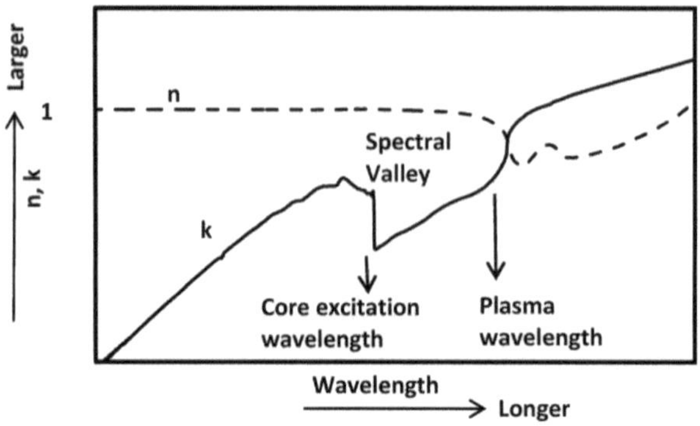

Fig. 1 Typical spectral features of refractive index *n* and the extinction coefficient *k* of an alkali metal

interpretation of high transmission in EUV range. Finally, we discuss prospects of pulsed intense EUV sources for this study.

A brief history of the study on EUV light transmission through sodium is listed as follows. In 1930s, Wood proposed anomalous optical property of alkali metals based on his spectroscopic experiments [1]. In 1960s, people at Oak Ridge National Laboratory performed series of experiments on optical constants of alkali metals with thin foil techniques [2–5]. They also showed anomalous transmission shorter than the plasma wavelength of each material. In 2010, Fukuda et al. proposed a UV visualization technique for liquid sodium dynamics [6]. Kawachi et al. performed a preliminary experiment on EUV transmission through a sodium sample using a copper laser plasma source in 2010. Suzuki, Daido et al. have performed precise transmission measurements and imaging experiments since 2012 [7, 8].

Alkali metals are potentially partially transparent between the plasma wavelength and the wavelength corresponding to the core excitation levels. Beyond the plasma wavelength, each material has its own space for free electron absorption in the EUV spectral range. Typical extinction coefficient as a function of wavelength of the incident light of alkali metals is schematically shown in Fig. 1. At the wavelength of μm range, reflectivity is high and the extinction coefficient is also very high. Below the plasma wavelength, the extinction coefficient falls down. At the wavelength of core excitation energy, the coefficient increases rapidly. Below the core excitation wavelength, the coefficient decreases. In this range, the light—matter interaction is characterized by atomic scattering coefficients given by Henke et al. [9]. Thus, between the plasma wavelength and core excitation one, the spectral valley corresponding to relatively high transmission is visible. The extinction coefficient of sodium which is the simplest alkali metal shows similar spectral feature as the materials having adjacent atomic numbers [10, 11]. If we concentrate on the spectral valley of the coefficient, a metallic aluminum shows rather steeper gradient and deeper valley [10]. In the same series of handbook we can also find a lot of empty regions corresponding to the valley [11]. These facts push us to perform a new experiment on sodium transmission around the spectral valley.

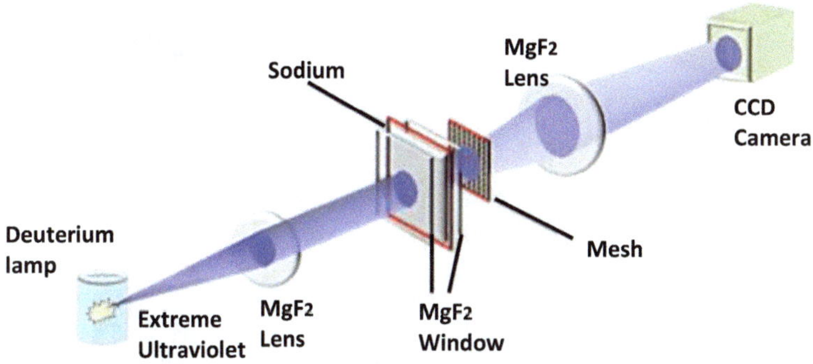

Fig. 2 Experimental setup of an imaging experiment

2 Measurement of Spectral Transmittance

In our work, we use a glove box dedicated for training of sodium handling to make sodium samples with thickness of greater than a few mm. On the other hand, in the previous experiments at Oak Ridge National Laboratory, sub-μm-thick sodium samples made with an evaporation technique under the ultra-high vacuum conditions of 10^{-8} Torr were used [4, 5]. In general, accurate extinction coefficients can be obtained when the attenuation is within a dynamic range of a measurement system. If the absorption depth is expected to be orders of magnitude different, we have to use a sample with suitable thickness. In our case, we have to employ a completely deferent technique to make a much thicker sample.

We measure the spectral transmittance of these samples. The experimental setup and the details of the measurement can be found in our previous paper [7]. A deuterium lamp emits extreme ultra-violet (EUV) light. The EUV light is passing through MgF_2 windows and a sodium sample. The transmitted light enters into the Seya-Namioka type spectrometer. Finally the spectral component of the transmitted light is detected by the photomultiplier. As a result, the spectral transmittance of the 3-mm sodium sample at room temperature is around a half which includes reflection losses of the MgF_2 windows and the sodium sample. We also perform temperature dependent transmission measurement, resulting in very weak dependency between 23 and 153 °C. However, we have to try to perform detailed and extensive experiments. Based on the measured transmittance, we can calculate the extinction coefficient of sodium resulting in several orders of magnitude smaller values than the previous results [5, 12].

In order to confirm the high spectral transmittance, we demonstrate an imaging of a fine structure with the transmitted EUV light. An experimental setup is shown in Fig. 2. As air is highly absorbing in the spectral region of interest, we pump down an imaging chamber from an atmospheric pressure down to 0.01 Torr and further down to 0.003 Torr. The EUV light passes through a MgF_2 lens for collimation and the light enters into a sodium sample. Significant portion of the EUV light passes

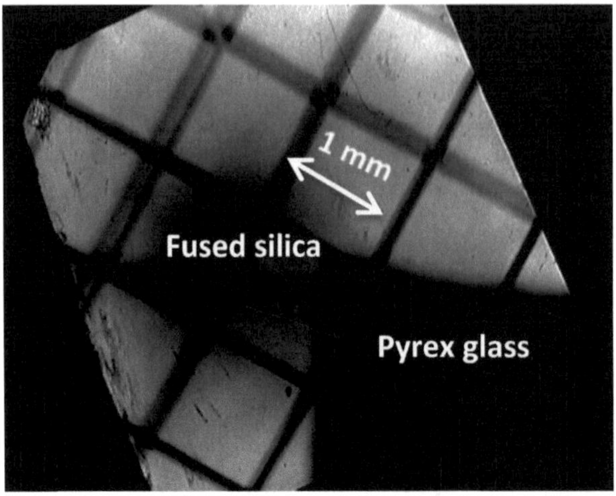

Fig. 3 An image of the metal mesh. A blackened area indicated as "*Pyrex glass*" corresponds to the area that is covered with a 1 mm-thick pyrex glass plate which is opaque in the vacuum ultraviolet spectral range. The notation "*Fused silica*" corresponds to the area covered with a 0.5 mm-thick fused silica plate where one can see slightly attenuated light

through the sodium sample and the light illuminates the metal mesh. The mesh is imaged on the CCD Camera as shown in Fig. 3. The mesh image becomes clearly visible after pumping down, which means the EUV light plays a dominant role in the image formation. We set a 1 mm-thick pyrex glass plate, whose shortest cut off wavelength is slightly longer than 300 nm, close to the mesh. This part is completely opaque, as expected. We obtained EUV images through a 8 mm-thick sodium sample, confirming its transmittance of at least a few % which is roughly consistent with the measured spectral transmittance.

Based on our new result, the extinction coefficient k is five orders of magnitude lower than the previous results. It means that the EUV light is transmitted through sodium samples with thickness of up to a few cm. The result also shows that a sodium sample acts as an EUV band-pass filter.

3 Physical Interpretation of High Transmission in EUV Range

The relaxation time τ given by electric conductivity is one of the important parameters for transmittance given by Drude model [10,11]. For example, if the temperature is 20 °C, the electric conductivity σ and the relaxation time τ, respectively are given as follows,

$$\sigma = \mathrm{ne}^2\tau/\mathrm{m}^*,\ \tau = 3.1 \times 10^{-14}\mathrm{sec},\ 1/\tau = 3 \times 10^{13}\mathrm{sec}^{-1},$$

where n, e and m* are the electrondensity, the electronic charge and the electron effective mass, respectively.

The present transmittance indicates that the relaxation time is a few orders of magnitude longer than that given by the conductivity. The theoretical aspect of this study should be promoted to show the reasonable physical interpretation.

4 Prospects of Pulsed Intense EUV Sources for this Study

A high speed and continuous imaging system for observing dynamic behavior of liquid sodium sample can be proposed with this technique. If we use MgF_2 windows for sodium sample, the wavelength range should be between 115 nm and $\sim$ 200 nm based on our present measurement. We note that utilization of EUV technique is advantageous even though other techniques can be used to observe inside or through a sodium medium. For example, sodium is transparent for hard x-rays and ultrasound. However, both of these techniques have their own difficulties. For the x-ray technique, the photon energy should be rather high in order to see through at least a few mm thicknesses of sodium and its container. A possible solution for Particle Image Velocimetry (PIV), for example, might be utilization of 5 μm diameter tantalum particles combined with quasi-monochromatic radiation just above the K edge of tantalum at 67.4 keV. Further, a high brightness quasi-monochromatic 67 keV x-ray source is not readily available except at SPring-8. As for the ultrasound, although it can be transmitted through longer distance in principle, the achievable resolution is typically sub-mm. Further, an image reconstruction for the ultrasound takes rather long time such as a few hours at this moment. In this case we cannot make a movie with this technique.

5 Summary

We measured spectral transmittance of sodium to obtain the extinction coefficient and performed imaging of fine structures with an EUV source. The measured transmittance is several orders of magnitude lower than the previous result. We demonstrated the images of a fine mesh illuminated by EUV light through a 8 mm-thick sodium sample. The relaxation time given by electric resistivity cannot account for the present high transmittance. Physical interpretation of high transmission is now under consideration. We proposed to use pulsed intense EUV sources for visualization of fluid dynamics of liquid sodium [8]. High speed and continuous imaging of a sodium sample is realized with the EUV imaging techniques.

Acknowledgements We are deeply indebted to Prof. T. Fukuda of Osaka University as well as Profs. S. Kubodera and M. Kaku of University of Miyazaki for variable contribution to the study. We also indebted to the members of International Nuclear Information and Training Center and Applied Laser Technology Institute, Japan Atomic Energy Agency. H. D. thanks Dr. R. Soufli of LLNL for providing us information.

References

1. Wood, R.W.: Remarkable optical properties of the alkali metals. Phys. Rev. **44**, 353–360 (1933)
2. Sutherland, J.C., Hamm, R.N., Stevenson, J.R., Arakawa, E.T., May 1967 "Optical Properties of Sodium in the Vacuum Ultraviolet," Oak Ridge National Laboratory Report ORNL-TM-1776 (1967)
3. Inagaki, T., Emerson, E.T., Arakawa, E.T., Williams, M.W.: Optical properties of solid Na and Li between 0.6 and 3.8 eV. Phys. Rev. **B13**, 2305–2313 (1976)
4. Sutherland, J.C., Arakawa, E.T., Hamm, R.N.: Optical properties of sodium in the vacuum ultraviolet. J. Opt. Soc. Am. **57**, 645–650 (1967)
5. Sutherland, J.C., Hamm, R.N., Arakawa E.T.: Extinction coefficient and imaginary part of the dielectric constant for sodium and potassium above the plasma energy. J. Opt. Soc. Am. **59**, 1581–1583 (1969)
6. Fukuda, T., Takata, T., Horiike, H., Kimura, N., Kamide, H.: May 17–21, 2010 "Direct Observation and Control of Liquid Sodium Flow Dynamics Using VUV-LIF-PIV Technique Under EXB Lorentz Force," Proc. 18th Int. Conf. Nucl. Eng. ICONE18, paper number ICONE18-29671, Xi'an, China (2010)
7. Daido, H., Suzuki, Y., Kawachi, T., Fukuda, T., Kaku, M., Kubodera, S.: Demonstration of partially transparent thick metallic sodium in the vacuum ultraviolet spectral range. Opt. Express. **21**, 28182, (7 pages) (2013)
8. Daido, H., Suzuki, Y., Kawachi, T., Fukuda, T., Kaku, M., Kubodera, S., Pirozhkov, A.S.: Transmission imaging of sodium in the vacuum ultra-violet spectral range: new application for an intense VUV source. Proc. SPIE. **8849**, 884908-1-11 (2013)
9. Henke, B.L., Gullikson, E.M., Davis, J.C.: X-Ray interactions: photoabsorption, scattering, transmission, and reflection at $E = 50 - 30{,}000\,\mathrm{eV}$, $Z = 1 - 92$. At. Data. Nucl. Data. Tables. **54**, 181–342 (1993)
10. Palik, E.D. (ed.): Handbook of Optical Constants of Solids. Academic, San Diego (1985)
11. Palik, E.D. (ed.): Handbook of Optical Constants of Solids II. Academic, San Diego (1991)
12. Sato, S., Miyahara, T., Hanyu, T., Yamaguchi, S., Ishi T.: Extreme ultraviolet absorption spectra of alkali metals. J. Phys. Soc. Jpn. **47**, 836–843 (1979)

Opacity Bleaching by Extreme Ultraviolet Radiation Incident on Solid Density Targets

V. Aslanyan and G. J. Tallents

Abstract The wavelength range of Extreme Ultraviolet (EUV) and x-ray lasers corresponds to strong absorption of photons by direct photoionization. Irradiation by high intensities of photons in this region leads to a rapid drop in the opacity due to a rise in the average degree of ionization of the plasma. A one-dimensional radiation transport model, which considers non-equilibrium conditions, has been used to model this "bleaching" of opacity by a short wavelength laser pulse as it propagates through a target. The effect of rapidly changing plasma conditions on inverse bremsstrahlung is explored.

1 Introduction

Recent technological progress has made lasers operating in the Extreme Ultraviolet (EUV) wavelength range (below $\sim$ 100 nm) available for industry and research. Small table-top capillary discharge lasers operating at 46.9 nm [1] and laser-pumped EUV lasers [2] have been developed, while free electron lasers operating at wavelengths from the EUV to x-ray have been constructed [3]. These lasers enable a unique light-matter interaction regime as the energy of a single photon is sufficient to directly photoionize atoms and the short wavelength produces small diameter focal spots. Photoionization rapidly creates a plasma, which can expand outward along a narrow channel [4]. Such plasma behaviour may be useful in, for example, laser cutting of micro-features where narrow laser focii enabled by the small diffraction limit of short wavelength lasers allow small-feature microstructures to be produced. The plasma is also produced at solid state densities allowing fundamental studies of high density plasma material [5].

The transport of radiation through a plasma is governed by its opacity. The opacity of a plasma is dependent on the density and distribution of ions and electrons which are often treated through appropriate equilibrium relations, such as Local Thermodynamic Equilibrium (LTE). We have simulated the propagation of a drop in opacity using a non-LTE model with a self consistent electron temperature and

V. Aslanyan (✉) · G. J. Tallents
York Plasma Institute, Department of Physics, University of York,
Heslington, YO YO10 5DD, UK
e-mail: va567@york.ac.uk

© Springer International Publishing Switzerland 2016

J. Rocca et al. (eds.), *X-Ray Lasers 2014,* Springer Proceedings in Physics 169,
DOI 10.1007/978-3-319-19521-6_51

constant density. A constant density is assumed as the timescales for ion motion (> 1 ps) are much greater than the assumed heating times (< 100 fs).

2 Model

We have used a stand-alone collisional-radiative model [6], previously used to study the validity of Local Thermodynamic Equilibrium (LTE), to simulate the propagation of EUV photons into bulk matter. The model calculates the time dependent local electron temperature, populations of ions occupying quantum states and the resulting electron density, starting with solid state initial conditions. This model is extended to a single spatial dimension by solving for these conditions in a series of cells at regularly spaced intervals, with radiation transport between them.

The atomic model considers, for every ionization stage of an element, a ground state and a single excited state formed by a superconfiguration approach [7], typically from three leading principal quantum levels. Cross sections used to calculate the rates of collisional processes between levels are taken from work by Van Regemorter citeregemorter [8], while between ionization stages they are given by Lotz [9]. Detailed balance [10] is used to calculate inverse rates. Cross sections for photoionization $\sigma_{i,j}$ and spontaneous emission coefficients A are taken from the NIST database [11]. At high densities, dielectronic processes are assumed to be insignificant compared to other collisional processes and are ignored.

Our model considers a plasma of a single element of atomic number Z, at a constant total ion density N_T. The populations of ions of a given ionization stage i and level j are denoted by $N_{i,j}$. Electron density is calculated assuming quasineutrality, given by $n_e = \sum_{i=0}^{Z} i N_{i,j}$. The electrons are assumed to rapidly come into equilibrium and therefore always form a Maxwellian distribution with a temperature T_e. The temperature of electrons is related to their energy density self consistently through $\varepsilon = \frac{3}{2} n_e T_e$, the equation of state of an ideal gas.

We take the temporal profile of the radiation intensity to be a simple step function, which begins at time $t = 0$, incident on matter at $x = 0$. Radiation is transported along a single axis and its intensity I is modelled by

$$\frac{dI}{dx} = (\kappa_{bf} + \kappa_{ff}) I, \tag{1}$$

where the distance of propagation is x and the two absorption coefficients κ correspond to photoionization (bound-free) and inverse bremsstrahlung (free-free) and are defined as the product of the plasma's opacity and its mass density. The photoionization cross section is approximately constant for ions of different ionization stages, provided that the photon energy is greater than the energy gap; $\sigma_{i,j} = \sigma \Theta(E_{photon} - E_{i,j})$, where Θ is the Heaviside step function. The bound-free absorption coefficient simplifies to

$$\kappa_{bf} = \sigma \sum_{E_{i,j} < E_{photon}} N_{i,j}. \tag{2}$$

For EUV photons, we take the free-free inverse bremsstrahulung absorption coefficient to be given by [10]

$$\kappa_{ff} = \sum_{i=0}^{Z} \frac{32\pi^3}{3\sqrt{3}} \frac{e^6(\hbar c)^2}{m_e c^2} \frac{1}{\sqrt{2\pi m_e c^2 T_e}} \frac{N_{i,j} n_e i^2}{E^3_{photon}} \left[1 - \exp\left(-\frac{E_{photon}}{T_e}\right)\right]. \quad (3)$$

3 Bleaching wave

As photoionization transfers ions to higher ionization stages, we can see from Eq. (2) that ions which energetically contribute to the absorption coefficient become depleted, in a so-called "bleaching" process. This in turn allows radiation to propagate further into the material and so the bleaching of absorption propagates.

The change in absorption coefficient for carbon with time and distance is illustrated in Fig. 1a and c, for EUV radiation of photons of energy 50 eV at an intensity of 10^{15} Wcm^{-2}. In order to track the progress of the wave, we shift the absorption profile at a given timestep along the x-axis by a displacement D until it overlaps with the profile at the previous timestep. The value of D is determined as that which minimizes $\sum_k |\kappa_t(x_k) - \kappa_{t+1}(x_k - D)|$. For simulations where inverse bremsstrahlung has been switched off and on, profiles shifted in this way are shown in Fig. 1b and d, respectively.

These calculations allow the displacement D to be plotted as a function of time and hence the wave velocity dD/dt to be determined. This method can be used to compare the cases with inverse bremsstrahlung switched off and on, as shown in Fig. 2a and b, respectively. The wave velocity tends to a constant in the former case, while inverse bremsstrahlung tends to rapidly decelerate the bleaching wave in the latter.

Free electrons in dense plasmas affect atoms through Ionization Potential Depression [12]. One leading model for this effect, from Stewart and Pyatt, has a change in ionization energy given by

$$\Delta E = \frac{T_e}{2(Z^* + 1)} \left[\left(\frac{(Z^* + 1) Z e^3 n_e^{1/2}}{\frac{4}{3}\pi \varepsilon_0^{3/2} T_e^{3/2}} + 1 \right)^{\frac{2}{3}} -1 \right], \quad (4)$$

where physical constants have their usual meanings and Z^* is the ionization fraction. When ΔE exceeds an ionization energy, that ionization stage disappears into the continuum. This expression implies that at high densities, there is a non-zero free electron population even as $T_e \to 0$. From Eq. (3), we see that this situation leads to extremely high inverse bremsstrahlung rates and therefore slow bleaching wave velocities.

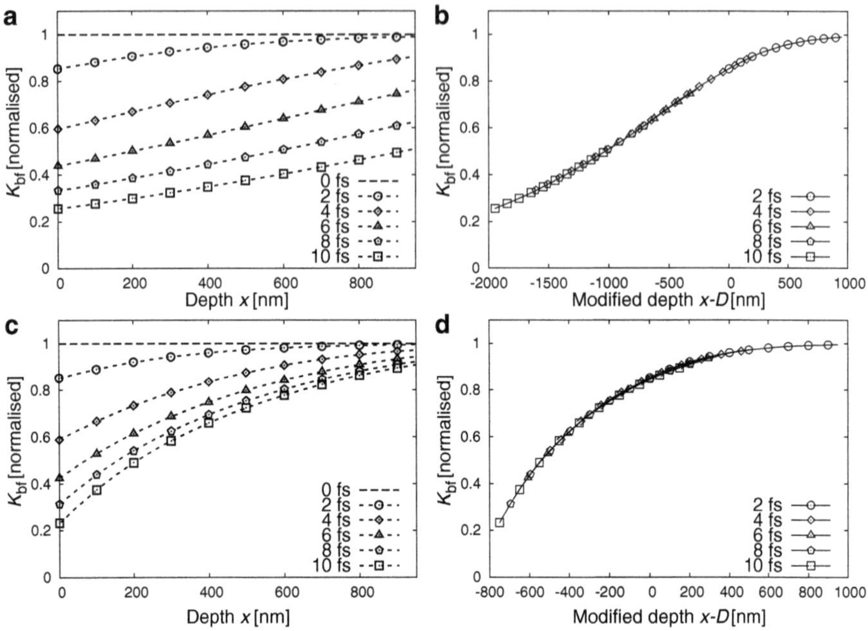

Fig. 1 Simulations with inverse bremsstrahlung switched (**a**, **b**) off and (**c**, **d**) on. The drop in absorption coefficient for carbon at solid density (2.3 g cm^{-3}) is shown after irradiation by photons of energy 50 eV at an intensity of 10^{15} Wcm^{-2} as a function of (a,c) depth x and (b,d) adjusted displacement $x - D$, so that the absorption coefficient profiles at different times (as indicated) overlap. Results are for different times as labelled

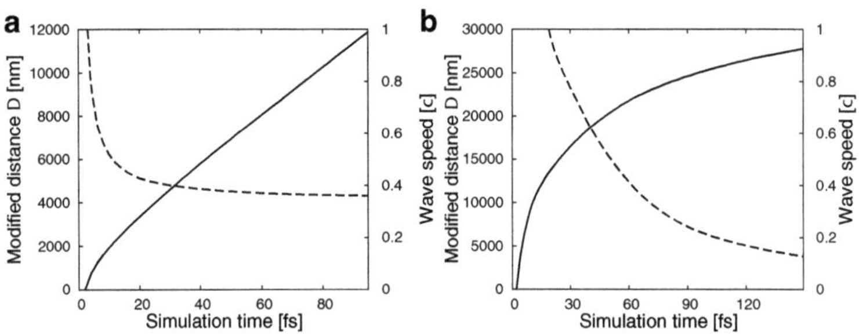

Fig. 2 Distance of propagation D (*solid*) and wave velocity $\frac{1}{c}\frac{dD}{dt}$ (*dashed*) for simulations with inverse bremsstrahlung **a** off and **b** on. Simulation parameters as for Fig. 1

4 Conclusion

We have discussed the concept of a drop in the bound-free absorption, the leading absorption mechanism for absorption in the EUV regime. We have shown that a bleaching wave propagates through a solid-density plasma driven by an intense

EUV laser pulse. Inverse bremsstrahlung, in particular when including ionization potential depression acts to slow the progress of this bleaching wave. However, the effects of both inverse bremsstrahlung and ionization potential depression are reduced at lower densities, such as those in an expanding ablation plume.

Currently, changes in intensity propagate forward instantaneously, equivalent to modelling the speed of light as infinite, as a consequence of Eq. (1). This approximation is only valid when the bleaching wave velocity is significantly lower than the speed of light. We plan to modify the model so that light propagates at the standard value of c rather than instantaneously, as is relevant for much lower densities.

Acknowledgements The authors would like to acknowledge EPSRC, Laserlab Europe and the Fusion Doctoral Training Network for funding this work. The authors are also grateful to L. J. Easy for computational aid.

References

1. Heinbuch, S., Grisham, M., Martz, D., Rocca, J.J.: Opt. Express **13**, 4050–4055 (2005)
2. Tallents, G.J.: J. Phys. D: Appl. Phys. **36**, R259–R276 (2003)
3. Milton, S.V., Gluskin, E., Arnold, N.D., Benson, C., Berg, W., Biedron, S.G., Borland, M., Chae, Y.C., Dejus, R.J., Den Hartog, P.K., Deriy, B., Erdmann, M., Eidelman, Y.I., Hahne, M.W., Huang, Z., Kim, K.J., Lewellen, J.W., Li, Y., Lumpkin, A.H., Makarov, O., Moog, E.R., Nassiri, A., Sajaev, V., Soliday, R., Tieman, B.J., Trakhtenberg, E.M., Travish, G., Vasserman, I.B., Vinokurov, N.A., Wang, X.J., Wiemerslage, G., Yang, B.X.: Science **292**, 2037–2041 (2001)
4. Rossall, A.K., Aslanyan, V., Tallents, G.J.: Proc. SPIE **8849**, 884912 (2013)
5. Vinko, S.M., Ciricosta, O., Cho, B.I., Engelhorn, K., Chung, H.K., Brown, C.R.D., Burian, T., Chalupsky, J., Falcone, R.W., Graves, C., Hajkova, V., Higginbotham, A., Juha, L., Krzywinski, J., Lee, H.J., Messerschmidt, M., Murphy, C.D., Ping, Y., Scherz, A., Schlotter, W., Toleikis, S., Turner, J.J., Vysin, L., Wang, T., Wu, B., Zastrau, U., Zhu, D., Lee, R.W., Heimann, P.A., Nagler, B., Wark, J.S.: Nature **482**, 59–62 (2012)
6. Aslanyan, V., Tallents, G.J.: Phys. Plasmas **21**, 062702 (2014)
7. Peyrusse, O.: J. Phys. B: At. Mol. Opt. Phys. **33**, 4303–4321 (2000)
8. Van Regemorter, H.: ApJ **136**, 906–915 (1962)
9. Lotz, W.: Z. Physik **232**, 101–107 (1970)
10. Salzmann, D.: Atomic Physics in Hot Plasmas—International Series of Monographs on Physics, vol. 97. Oxford University Press, New York (1998)
11. NIST Atomic Spectra Database. http://www.nist.gov/pml/data/asd.cfm. Accessed 10 May 2014.
12. Stewart, J.C., Pyatt, K.D.: ApJ **144**, 1203–1211 (1965)

Ablation Plume Induced by Laser Euv Radiation

O. Frolov, K. Kolacek, J. Schmidt and J. Straus

Abstract In this paper we report results of study laser plasma plumes created by nanosecond EUV laser pulses at wavelength of 46.9 nm in different materials (Au, Al, Si, Cu). Experiment with attenuation of Al filters for verification of energy of laser pulses was performed. In the ablation plume experiment low energy of laser pulse in the most cases was not sufficient for creation of ablation plumes in Cu, Al and Si. In contrast, higher energy has a greater effect in all the materials.

1 Introduction

Capillary discharge EUV lasers, first demonstrated in 1994 [1], remain to date as the highest average power table-top source of coherent EUV radiation [2, 3]. The capillary discharge Ne-like Ar laser operating at $\lambda = 46.9$ nm is also the table-top EUV laser most broadly utilized in applications. It has been used in interferometric studies of high-density plasmas [4], high-resolution microscopy [5, 6] holographic imaging [7], nanoscale patterning and machining [8–11], material ablation [12–15], single-photon ionization mass spectrometry studies of nanoclusters [16], the measurement of optical constants of materials [17], and other applications.

Laser ablation have been studied extensively for more than 50 years since the discovery of lasers in the 1960s. Plasma formation occurs when a high powered pulsed laser is focused onto a target such that the energy density exceeds the ablation threshold of the material. The laser-target interaction involves many processes, including heating, melting, vaporization, ejection of particles, and plasma creation and expansion [18]. The depth of laser ablation crater over which the laser energy is absorbed, and thus the amount of material removed by a single laser pulse, depends on the material's optical properties and the laser wavelength and pulse length. Usually, plasma cloud induced by laser irradiation consists of excited or ground-state neutrals, electrons, and ions. In the early works about laser ablation, visible plumes produced by fluorescence from excited atoms in the cloud were registered [19–21]. The properties of the plume, e.g. the mass distribution, ion and atom velocity, and

O. Frolov (✉) · K. Kolacek · J. Schmidt · J. Straus
Institute of Plasma Physics, v.v.i., Academy of Sciences of the Czech Republic, 182 00 Prague, Czech Republic
e-mail: frolov@ipp.cas.cz

© Springer International Publishing Switzerland 2016
J. Rocca et al. (eds.), *X-Ray Lasers 2014*, Springer Proceedings in Physics 169,
DOI 10.1007/978-3-319-19521-6_52

the angular distribution of the plume species, play an important role for applications of laser ablation in mass analysis of laser-induced plasma [22] and in the production of thin films by pulsed laser deposition [23–25]. In particular, the thickness distribution in film deposition on a substrate is determined by the plume shape that has evolved during the expansion from the target surface to the substrate [26–28]. Focused visible/IR laser pulses are absorbed in the expanding plasma plume at densities 100–1000 times smaller than the solid density. But, EUV laser pulses penetrate into the solid and create plasma directly at the solid density.

In this paper, we investigate an interaction of nanosecond EUV laser pulses with solid materials (Au, Al, Si, Cu). Energy of laser radiation plays an important role in the ablation of materials. Therefore, estimation of laser energy based on attenuation of Al filters was performed for two capillary lengths (232 and 400 mm). Higher energy of laser radiation ($\sim$40 mJ) is sufficient for creation visible ablation plumes in all of the studied materials. In the case of low energy of laser pulse ($\sim$1.5 mJ) visible ablation plume was observed only in experiment with ablation of thin layer of gold.

2 Experimental Apparatus

The apparatus CAPEX (CAPillary EXperiment) consists of a Marx generator, a coupling section, a pulse forming line (fast cylindrical capacitor), a main spark gap and a capillary (Fig. 1). In early experiments with surface modification of materials [11, 29], typical capillary current was about 30 kA with pre-ionization current of 10 A (5 μs prior to the spark gap breakdown). It was found that at these conditions capillary lifetime was limited on about only 100 shots [30]. At present, due to optimization of ground electrode geometry, increasing of amplitude and duration of pre-ionization current (100 A and 20 μs) and decreasing of the main capillary current (typically 15–20 kA) lifetime of a capillary is significantly longer (up to 1200 shots). Our driver is capable to operate in a repetitive regime with frequency up to a few Hz [31]. A detailed description of our capillary discharge driver CAPEX

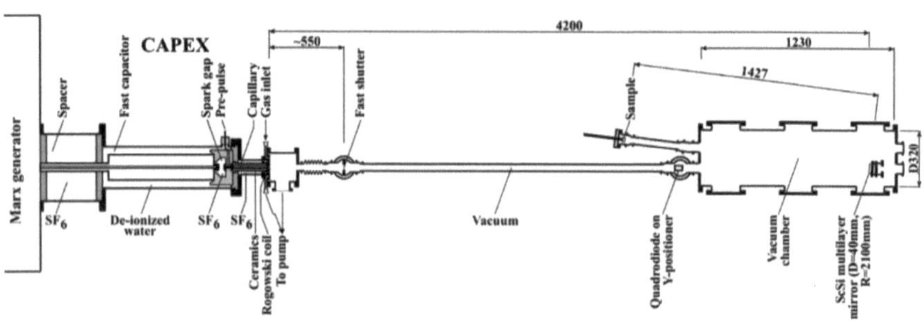

Fig. 1 Experimental apparatus *CAPEX* with extension for ablation experiments

can be found in the early-published papers [31–33]. For ablation experiments the apparatus was extended for vacuum chamber with a spherical multilayer mirror, interaction tube with samples and for registered detectors (Canon EOS 550D or PCO SensiCam).

3 Results

3.1 Energy Measurements

First measurement of the EUV pulse energy was made by commercially available NIST-calibrated aluminum oxide photodiode. Despite the fact that photodiode was calibrated by cw illumination, it could be used to measure pulse energy (by NIST recommendation). The cw responsivity expressed in A/W could be used as the pulse responsivity expressed in C/J. Obtained energy in experiment with 232 mm long capillary was 12.85 nJ. It seemed this energy was too low for the results that we have obtained from the ablation experiments. In the next step we tried to calibrate our photodiode with golden cathode at NIST. Unfortunately, this calibration was not successful, relative uncertainty at wavelength of $\sim$ 47 nm was more than 700 %. Thereafter, we have tried to estimate energy of laser pulse based on quantum efficiency of the golden photocathode [34]. It was found that laser pulse energy was approximately 3 times higher for the 400 mm long capillary ($\sim$ 12.3 μJ) in comparison with the 232 mm long one ($\sim$ 4.1 μJ). And finally laser pulse energy have defined from measurements of intensity of EUV radiation with Al filters of different thickness. These experiments were made for two capillary currents (20 kA and 30 kA) with capillary length of 400 mm. Dependence of the intensity of EUV radiation detected by vacuum photodiode with golden cathode on thickness of Al filters is shown in Fig. 2. After correcting attenuation of filters pulse energy was $\sim$ 1.52 mJ (transmission $T_{0.4\mu m\,Al} = 0.0867$) for capillary current of 20 kA and $\sim$ 40 mJ (transmission $T_{0.4\mu m\,Al} = 0.0878$) for capillary current of 30 kA.

3.2 Ablation Plume Experiment

As the first, visible ablation plumes were studied for different materials and energy. With laser pulse energy of $\sim$ 1 mJ the ablation plumes were observed only in thin layer of gold (thickness of 50 nm), in other materials only interaction spots were visible. In the case of higher energy ($\sim$ 40 mJ) in the all materials the ablation plumes were registered at least during the first shot (Fig. 3). On the next step, influence of charged electrode was studied. Rectangular charged electrode was placed 2 mm from surface of golden layer (which was a ground electrode) and 1.5 mm above axis of incidence of laser beam. No visible effect of positively or negatively charged electrodes on behavior of ablation plume was registered. Nevertheless, breakdown in the vacuum between electrodes have occurred at + 1.7 kV and − 11 kV. Photoelectrons

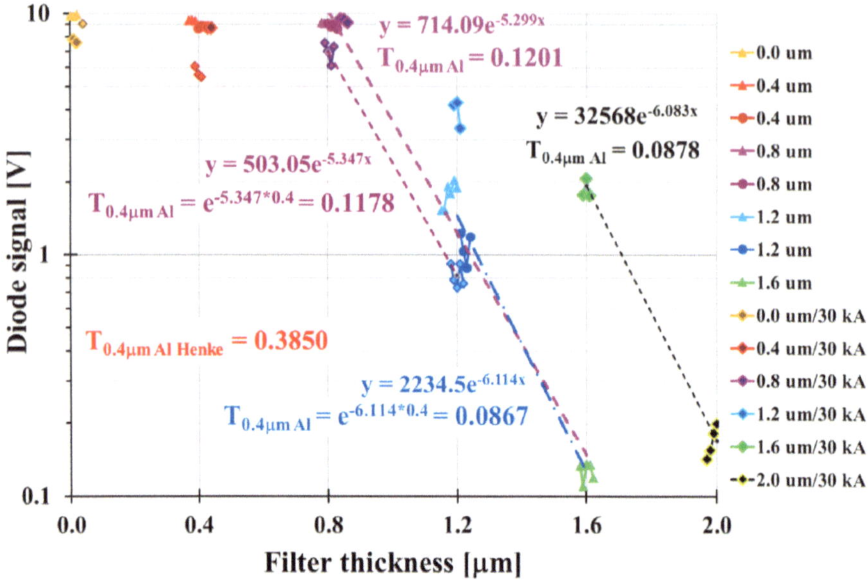

Fig. 2 Intensity of EUV signals in dependence of thickness of Al filters

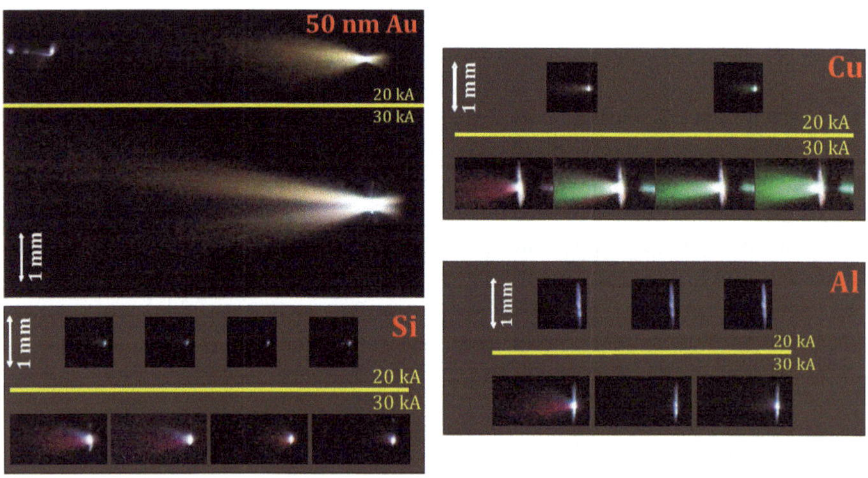

Fig. 3 Visible ablation plumes in the different materials for two capillary currents: 20 and 30 kA

have played major role in breakdown at positive voltage, breakdown was observed between corner of charged electrode and interaction spot on the surface of golden layer. On the other hand, breakdown at negative voltage was occurred between the surfaces of electrodes (by the shortest distance). Time evolution of the visible ablation plume was also studied (by changing delay between incidence of laser beam and trigger of camera). It was found that whole process from moment of interaction of laser beam with material surface to plume disappearance have lasted ~ 3 μs.

4 Conclusions

Energy of the EUV laser pulses was estimated from attenuation of Al filters: ~1.5 mJ for capillary current of 20 kA and ~ 40 mJ for capillary current of 30 kA. Laser pulse energy of ~ 1 mJ is not sufficient for creation of visible ablation plume on the surface Si, Cu and Al. With the laser energy of ~ 40 mJ situation is a little bit different: at least during the first shot ablation plume on the surface of materials are observed. In the case of layer of Au, weaker and stronger laser ablation plumes were registered.

Acknowledgment This work supported by the Czech Science Foundation under Contract GA1429772S and by the Grant Agency of the Ministry of Education, Youth and Sports of the Czech Republic under Contract LG13029.

References

1. Rocca, J.J., Shlyaptsev, V.N., Tomasel, F.G., Cortazar, O.D., Hartshorn, D., Chilla, J.L.A.: Demonstration of a discharge pumped table-top soft-x-ray laser. Phys. Rev. Lett. **73**, 2192 (1994)
2. Benware, B.R., Macchietto, C.D., Moreno, C.H., Rocca, J.J.: Demonstration of a high average power tabletop soft x-ray laser. Phys. Rev. Lett. **81**, 5804 (1998)
3. Macchietto, C.D., Benware, B.R., Rocca, J.J.: Generation of millijoule-level soft-x-ray laser pulses at a 4-Hz repetition rate in a highly saturated tabletop capillary discharge amplifier. Opt. Lett. **24**, 1115 (1999)
4. Filevich, J., Kanizay, K., Marconi, M.C., Chilla, J.L.A., Rocca, J.J.: Dense plasma diagnostics with an amplitude-division soft-x-ray laser interferometer based on diffraction gratings. Opt. Lett. **25**, 356 (2000)
5. Brewer, C.A., Brizuela, F., Wachulak, P., Martz, D.H., Chao, W., Anderson, E.H., Attwood, D.T., Vinogradov, A.V., Artyukov, I.A., Ponomareko, A.G., Kondratenko, V.V., Marconi, M.C., Rocca, J.J., Menoni, C.S.: Single shot extreme ultraviolet laser imaging of nanostructures with wavelength resolution. Opt. Lett. **33**, 518 (2008)
6. Brizuela, F., Vaschenko, G., Brewer, C., Grisham, M., Menoni, C.S., Marconi, M.C., Rocca, J.J., Chao, W., Liddle, J.A., Anderson, E.H., Attwood, D.T., Vinogradov, A.V., Artioukov, I.A., Pershyn, Y.P., Kondratenko, V.V.: Reflection mode imaging with nanoscale resolution using a compact extreme ultraviolet laser. Opt. Express. **13**, 3983 (2005)
7. Wachulak, P.W., Marconi, M.C., Bartels, R.A., Menoni, C.S., Rocca, J.J.: Analysis of extreme ultraviolet microscopy images of patterned nanostructures based on a correlation method. J. Opt. Soc. Am. **B25**, 1811 (2008)
8. Wachulak, P.W., Capeluto, M.G., Marconi, M.C., Menoni, C.S., Rocca, J.J.: Patterning of nano-scale arrays by table-top extreme ultraviolet laser interferometric lithography. Opt. Express. **15**, 3465 (2007)
9. Ottaviano, L., Bussolotti, F., Piperno, S., Rinaldi, M., Santucci, S., Flora, F., Mezi, L., Dunne, P., Kaiser, J., Reale, A., Ritucci, A., Zuppella, P.: Fabrication of metallic micropatterns using table top extreme ultraviolet laser interferometric lithography. Plasma Sources Sci. Technol. **17**, 024019 (2008)
10. Bravo H., Szapiro, B.T., Wachulak, P.W., Marconi, M.C., Chao, W., Anderson, E.H., Menoni, C.S., Rocca, J.J.: Demonstration of nanomachining with focuses extreme ultraviolet laser beams. IEEE J. Sel. Topics. Quantum Elec. **18**, 443 (2012)

11. Kolacek, K., Straus, J., Schmidt, J., Frolov, O., Prukner, V., Shukurov, A., Holy, V., Sobota, J., Fort, T.: Nano-structuring of solid surface by extreme ultraviolet Ar8+ laser. Laser. Part. Beams. **30**, 57 (2012)
12. Vaschenko, G., Etxarri, A.G., Menoni, C.S., Rocca, J.J., Hemberg, O., Bloom, S., Chao, W., Anderson, E.H., Attwood, D.T., Lu, Y., Parkinson, B.: Nanometer scale ablation with a table-top soft x-ray laser. Opt. Lett. **31**, 3615 (2006)
13. Juha, L., Bittner, M., Chvostova, D., Krasa, J., Otcenasek, Z., Prag, A.R., Ullschmied, J., Pientka, Z., Krzywinski, J., Pelka, J.B., Wawro, A., Grisham, M.E., Vaschenko, G., Menoni, C.S., Rocca, J.J.: Ablation of organic polymers by 46.9-nm-laser radiation. Appl. Phys. Lett. **86**, 034109 (2005)
14. Frolov, O., Kolacek, K., Straus, J., Schmidt, J., Prukner, V., Choukourov, A.: Application of EUV optics to surface modification of materials. Proc. SPIE 8777, Damage to VUV, EUV, and X-ray Optics IV; and EUV and X-ray Optics: Synergy between Laboratory and Space III, 877707 (2013)
15. Kolacek, K., Schmidt, J., Straus, J., Frolov, O., Prukner, V., Melich, R., Choukourov, A.: A new method of determination of ablation threshold contour in the spot of focused XUV laser beam of nanosecond duration. Proc. SPIE 8777, Damage to VUV, EUV, and X-ray Optics IV; and EUV and X-ray Optics: Synergy between Laboratory and Space III, 87770 N (2013)
16. Heinbuch, S., Dong, F., Rocca, J.J., Bernstein, E.R.: Single photon ionization of hydrogen bonded clusters with a soft x-ray laser: (HCOOH)x and (HCOOH)y(H2o)z. J. Chem. Phys. **126**, 244301 (2007)
17. Artioukov, I.A., Benware, B.R., Rocca, J.J., Forsythe, M., Uspenskii, Yu. A., Vinogradov, A.V.: Determination of XUV optical constants by reflectometry using a high-repetition rate 46.9-nm laser. IEEE J. Sel. Top. Quantum Electron. **5**, 1495 (1999)
18. Zeng, X., Mao, X.L., Greif, R., Russo, R.E.: Experimental investigation of ablation efficiency and plasma expansion during femtosecond and nanosecond laser ablation of silicon. Appl. Phys. A. **80**, 237 (2005)
19. Harris, T.J.: High-speed photographs of laser-induced heating. IBM J. Res. Develop. **7**, 342–345 (1963)
20. Geohegan, D.B.: Physics and diagnostics of laser ablation plume propagation for high-Tc superconductor film growth. Thin Solid Films. **220**, 138–145 (1992)
21. Kelly, R., Miotello, A., Mele, A., Giardini Giudoni, A.: Plume formation and characterization in laser-surface interactions. Laser Ablation Desorption. **30**, 225–289 (1998)
22. Vertes, A., Gijbels, R., Adams, F.: Solid sampling for analysis by laser ablation. Laser Ionisation Mass Analysis, Chap. 4. Wiley, USA **124**, 437–455 (1993)
23. Hubler, G.K.: Pulsed laser deposition. MRS Bull. vol. XVII, issue 2, 26–29 (1992)
24. Chrisey, D.B., Hubler, G.K.: Pulsed Laser Deposition of Thin Films, Chap. 3. Wiley, USA, 29–48 (1994)
25. Gorbunoff, A.: Laser-Assisted Synthesis of Nanostructured Materials, Fortschritt-Berichte VDI 357, VDI Verlag, Germany, 1–229 (2002)
26. Saenger, K.L.: Angular distribution of ablated material. Pulsed Laser Deposition of Thin Films. Chrisey, D.B., Hubler, G.K. (eds.), Wiley, USA, 199–227 (1994)
27. Amoruso, S., Bruzzese, R., Spinelli, N., Velotta, R.: Characterization of laser-ablation plasmas. J. Phys. B. **32**, 131–172 (1999)
28. Schou, J.: Laser beam-solid interactions: fundamental aspects. Material Surf Processing by Directed Energy Techniques. 33–62 (2006)
29. Frolov, O., Kolacek, K., Straus, J., Schmidt, J., Prukner, V., Shukurov, A.: Generation and application of the soft x-ray laser beam based on capillary discharge. J. Phys. Conf. Ser. **511**, 012035 (2014)
30. Straus, J., Kolacek, K., Neufuss, K., Kolman, B., Dubsky, J., Frolov, O., Schmidt, J., Prukner, V.: Modification of alumina-capillary inner-surface by pulse high-current discharge. Czech. J. Phys. **56**, B564–B570 (2006)

31. Schmidt, J., Kolacek, K., Frolov, O., Prukner, V., Straus, J.: Repetitive XUV laser based on the fast capillary discharge: Proc. SPIE 8140, x-ray lasers and coherent x-ray sources: Development and applications IX, SPIE, San Diego, 814015. (2011)
32. Schmidt, J., Kolacek, K., Straus, J., Prukner, V., Frolov, O., Bohacek, V.: Soft x-ray emission of a fast-capillary-discharge device. Plasma Devices Oper. **13**, 105–109 (2005)
33. Kolacek, K., Schmidt, J., Prukner, V., Frolov, O., Straus, J.: Recent progress in discharge-based soft X-ray lasers at IPP ASci CR: Proc. SPIE 6938, Atomic and Molecular Pulsed Lasers VII, SPIE, Washington, 93805–93805, (2008)
34. Schmidt, J., Kolacek, K., Frolov, O., Prukner, V., Straus, J., Kaufman, J.: Beam characteristics of CAPEX XUV argon laser: Proc. SPIE 8849, x-ray lasers and coherent x-ray sources: Development and applications X, SPIE, San Diego, 884917 (2013)

Study of Optical Emission During Spallative Ablation Induced by Soft X-Ray Laser Pulses

M. Ishino, M. Nishikino, N. Hasegawa, T. Pikuz, I. Skobelev, A. Faenov, M. Yamagiwa and T. Kawachi

Abstract Soft x-ray laser (SXRL) pulses were irradiated to the surfaces of LiF, Al, and Cu plate targets. On all irradiated target surfaces, the ablation and/or surface modification structures caused by the SXRL pulse irradiations were confirmed. However, the optical emission (visible spectral range) from the surfaces during the SXRL pulse irradiations were not observed, in spite of high sensitive registration technique. The theoretical model calculations for the interaction between SXRL pulse and material surfaces has revealed the spallative ablation of target surface with a low electron temperature. Our experimental results is consistent with the theoretical model calculations, qualitatively.

1 Introduction

A soft x-ray laser (SXRL) is an interest laser source, because of the features of short wavelength (high photon energy), small pulse width, and highly spatial coherence. High photon energy laser pulse with short duration has ability to interact directly with matters. The SXRL pulses having a wavelength of 13.9 nm and a pulse width of 7 ps can make the nanometer scale ablation and/or modification structures on material surfaces [1, 2].

The mechanisms of surface modifications induced by SXRL pulses were also investigated theoretically [1, 3, 4]. The atomistic model for the interaction between SXRL pulse and material reveals that the tensile stress created in heated layers by SXRL pulse can produce spallative ablation of target surface. The mechanisms of the surface modifications caused by SXRL pulses are essentially different from those by optical laser pulses.

M. Ishino (✉) · M. Nishikino · N. Hasegawa · T. Pikuz · M. Yamagiwa · T. Kawachi
Quantum Beam Science Center, Japan Atomic Energy Agency, Kyoto, Japan
e-mail: ishino.masahiko@jaea.go.jp

T. Pikuz · I. Skobelev · A. Faenov
Joint Institute for High Temperatures, Russian Academy of Science, Moscow, Russia

A. Faenov
Institute for Academic Initiatives, Osaka University, Osaka Japan

© Springer International Publishing Switzerland 2016
J. Rocca et al. (eds.), *X-Ray Lasers 2014*, Springer Proceedings in Physics 169,
DOI 10.1007/978-3-319-19521-6_53

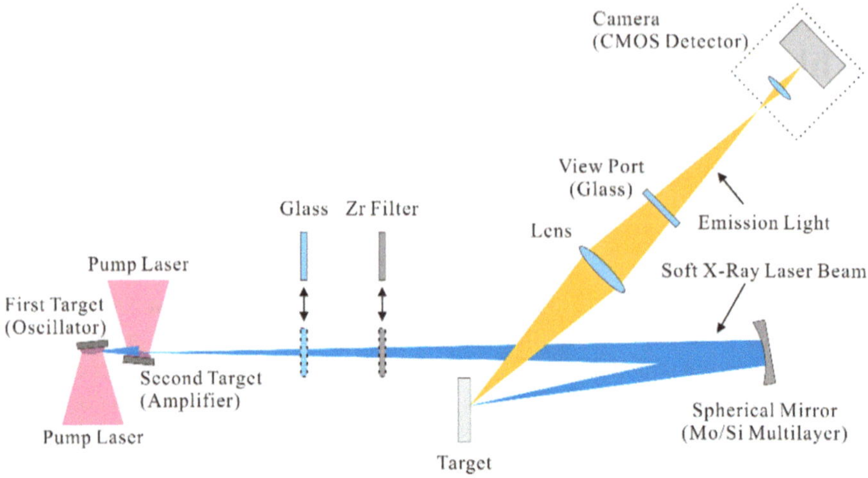

Fig. 1 Experimental setup for the SXRL irradiation and the emission observation

Understanding the ablation and surface modification mechanisms induced by the short pulse lasers is important not only in an application for the nanometer scale surface machining but also in describing or providing a fundamental physical process of the laser induced interaction with condensed matters. However, we can scarcely find the articles dealing with the ablation process occurred by the relatively low fluence region around the ablation threshold. Some reports concerned with the time-resolved observations can be found [5–7]. The development of the theory of ablation mechanism is important to understand the accurate behaviour of laser produced interaction on the material surface. Therefore an experimental information, such as the electron temperature, is one of the important values for theoretical model.

Now, we have initiated an experimental study for purchasing the ablation process, including the surface modification process, by the short duration pulse, especially by SXRL. In this article, we will report the observation results of the optical emission (visible spectral range) from the irradiated target surfaces.

2 Experimental Setup

The SXRL irradiation experiment was conducted at the SXRL facility of Japan Atomic Energy Agency. A schematic diagram of the experimental setup is shown in Fig. 1. The SXRL pulse was generated from a silver (Ag) plasma mediums using an oscillator-amplifier configuration with double Ag targets. The generated SXRL pulse had a wavelength of 13.9 nm (photon energy of 89 eV) and a pulse width of approximately 7 ps.

The SXRL pulse was focused on the plate target surface using a Mo/Si multilayer coated spherical mirror, which was optimized for soft x-rays of 13.9 nm at a normal incidence. The focused SXRL pulses were irradiated to the surfaces of LiF, Al, and Cu plate targets. A glass plate with 100 μm in thickness and a 0.2 μm thick Zr filter were placed in front of the spherical mirror, respectively. A glass plate was transparent for visible light and opaque for SXRL beam. The transmittance of 0.2 μm thick Zr filter for 13.9 nm was approximately 48 %. The Zr filter reduced the almost of scattered optical radiation from the laser produced Ag plasmas. These glass plate and Zr filter could be removed from the SXRL penetration axis independently. When a 0.2 μm thick Zr filter was adopted, the total energy of the SXRL beam reached on the sample surface was approximately 48 nJ/pulse. The typical fluences on target surfaces, which were measured as an averaged value in modification zone, were estimated to be 10–30 mJ/cm^2 [1,2].

The emission light of visible spectral range from target surface during an interaction with SXRL pulse was observed by the optical camera (visible range). The optical image of an emission light was transferred to the camera by the optical lens (f = 100 mm) through a glass view port. An image sensor of the camera was CMOS (complementary metal oxide semiconductor) device. The detectable energy threshold was estimated to be approximately 0.016 nJ (1.6×10^{-11} J).

The SXRL irradiation process was performed in a vacuum, and the camera was in air. The details of SXRL source and the focal properties of SXRL pulses on the target surfaces can be found in Ref. 2.

3 Results and Discussion

Figures 2a and 2b show the observed visible signals from the LiF target surfaces during the irradiation of SXRL single pulse, as an example of the result. As shown in Fig. 2a, a visible signal was clearly observed, when the SXRL pulse and also the optical radiation from Ag plasmas reached on the target surface through no filter. On the other hand, when a Zr filter was inserted within the SXRL trajectory, we could not confirm the any optical signal from irradiated surface [see Fig. 2b]. The observation results in other targets (Al and Cu) were the same as those in LiF targets.

On the irradiated LiF surfaces both with and without Zr filter, we could see the formation of the ablation structures (not shown). From the additional measurement, observed visible light signal was estimated to be emission from Ag plasma, which had a wavelength range of 400–650 nm. It can be concluded that the ablation and/or surface modification process induced by the SXRL pulse will occur without a plasma production.

We could irradiate a focused optical portion of an emission from the Ag plasmas on the target by inserting the glass plate in the trajectory, without changing an experimental geometry. The total energy of the optical light reached on the sample surface was measured to be 30 nJ/pulse, which was comparable to SXRL pulse energy (~48 nJ/pulse). In this case, the optical emission such as shown in Fig. 2a was observed (not shown). However, the irradiated LiF surface was observed by the optical microscope, and no modification structure was confirmed on the surface.

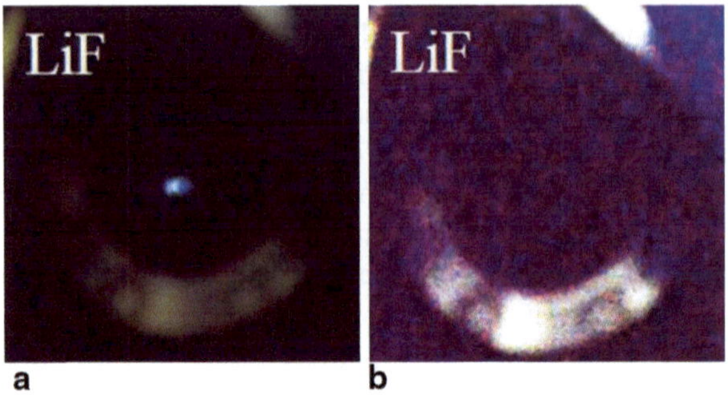

Fig. 2 Observed emission signals from the *LiF* surfaces during the SXRL irradiations **a** without or, **b** with Zr filter in the SXRL beam trajectory

The observed visible signals from the target surfaces were the scattered or reflected visible light of the focused Ag plasmas from the target surface. The optical irradiation did not influence the target surface.

The atomistic model calculation predicts that the absorbed energy of SXRL pulse having a fluence of 65 mJ/cm^2 for a surface modification creates an electron temperature of around 1.0 eV in a material [8]. In this experiment, the fluences of the focused SXRL pulses were lower than 30 mJ/cm^2, so the electron temperatures would be smaller than 1 eV. If the electron temperature is high and/or duration time of plasma emission induced by the SXRL pulse irradiation is long, the optical emission signal will be observed by the CMOS detector [9]. However, in this experiment, no emission signal could be observed, hence, we will be able to estimate an upper limit of the electron temperature and duration of plasma. The details of a relation between an emission intensity from the ablation plasma and a detectable intensity of the CMOS detector will be reported elsewhere [9].

The observation results indicate that the spallative ablations occurred by the irradiations of SXRL pulses without plasma productions having high electron temperature. This means that our experimental results are consistent with the theoretical calculations.

4 Summary

We tried to observe the emission signals from LiF, Al, and Cu surfaces, which were irradiated with the SXRL pulses with low fluence region around the threshold. In spite of the use of sensitive detector, the emission signal in visible spectral range accompanying the ablation and/or surface modification caused by SXRL pulse irradiation was not observed.

An atomistic theoretical model simulation predicts the surface modification, at low fluence region, occurs with an electron temperature quite lower than 1 eV. We can conclude that our experimental results of an absence of visible emission signal is qualitatively consistent with the theoretical prediction.

Acknowledgment This work was partly supported by Grant-in-Aid for Scientific Research (B), No. 25,289,244 (2013), from MEXT, Japan, and also supported by RFBR-JSPS collaboration program (RFBR Grant 14-02-92107).

References

1. Ishino, M., Faenov, A. Ya., Tanaka, M., Hasegawa, N., Nishikino, M., Tamotsu, S., Pikuz, T.A., Inogamov, N.A., Zhakhovsky, V.V., Skobelev, I. Yu., Fortov, V.E., Khohlov, V.A., Shepelev, V.V., Ohba, T., Kaihori, T., Ochi, Y., Imazono, T., Kawachi, T.: Nanoscale surface modifications and formation of conical structures at aluminium surface induced by single short exposure of soft x-ray laser pulse. J. Appl. Phys. **109**, 013504 (2011)
2. Ishino, M., Faenov, A.Y., Tanaka, M., Tamotsu, S., Hasegawa, N., Nishikino, M., Pikuz, T.A., Kaihoori, T., Kawachi, T.: Observations of surface modifications induced by the multiple pulse irradiation using a soft picosecond x-ray laser beam. Appl. Phys. A. **101**, 179–188 (2013)
3. Inogamov, N.A., Faenov, A. Ya., Khokhlov, V.A., Zhakhovskii, V.V., Petrov, Yu. V., Skobelev, I. Yu., Nishihara, K., Kato, Y., Tanaka, M., Pikuz, T.A., Kishimoto, M., Ishino, M., Nishikino, M., Fukuda, Y., Bulanov, S.V., Kawachi, T., Anisimov, S.I., Fortov, V.E.: Spallative ablation of metals and dielectrics. Contrib. Plasma Phys. **49**, 455–466 (2009)
4. Norman, G., Starikov, S., Stegailov, V., Fortov, V., Skobelev, I., Pikuz, T., Faenov, A., Tamotsu, S., Kato, Y., Ishino, M., Tanaka, M., Hasegawa, N., Nishikino, M., Ohba, T., Kaihori, T., Ochi, Y., Imazono, T., Fukuda, Y., Kando, M., Kawachi, T.: Nanomodification of gold surface by picosecond soft x-ray laser pulse. J. Appl. Phys. **112**, 013104 (2012)
5. Srojanovic, N., von der Linde D., Sokolowcki-Tinten, K., Zastrau, U., Perner, F., Förster, E., Sobierajski, R., Nietubyc, R., Jurek, M., Klinger, D., Pelka, J., Krzywinski, J., Juha, L., Cihelka, J., Velyhan, A., Koptyaev, S., Hajkova, V., Chalupsky, J., Kuba, J., Tschentscher, T., Toleikis, S., Düsterer, S., Redlin, H.: Ablation of solids using a femtosecond extreme ultraviolet free electron laser. Appl. Phys. Lett. **89**, 241909 (2006)
6. Suemoto, T., Terakawa, K., Ochi, Y., Tomita, T., Yamamoto, M., Hasegawa, N., Deki, M., Minami, Y., Kawachi, T.: Single-shot picosecond interferometry with one-nanometer resolution for dynamical surface morphology using a soft x-ray laser. Opt. Express. **18**, 14114–14122 (2010)
7. Tomota, T., Yamamoto, M., Hasegawa, N., Terakawa, K., Minami, Y., Nishikino, M., Ishino, M., Kaihori, T., Ochi, Y., Kawachi, T., Yamagiwa, M., Suemoto, T.: Experimental verification of femtosecond laser ablation schemes by time-resolved soft x-ray reflective imaging. Opt. Express. **20**, 29329–29337 (2012)
8. Starikov, S.V., Faenov, A. Ya., Pikuz, T.A., Skobelev, I. Yu., Fortov, V.E., Tamotsu, S., Kato, Y., Ishino, M., Tanaka, M., Hasegawa, N., Nishikino, M., Ohba, T., Kaihori, T., Ochi, Y., Imazono, T., Kawachi, T.: Soft, picosecond x-ray laser nanomodification of gold and aluminum surfaces. Appl. Phys. B. **116**, 1005–1016 (2014) (published online)
9. Ishino, M., Hasegawa, N., Nishikino, M., Pikuz, T., Skobelev, I., Faenov, A., Inogamov, N., Kawachi, T., Yamagiwa, M.: Very low electron temperature in warm dense matter formed by focused picosecond soft x-ray laser pulses. J. Appl. Phys. **116**, 183302 (2014)

Diagnosis of Radiation Heating in Iron Buried Layer Targets

M. Shahzad, G. J. Tallents, O. Culfa, A. K. Rossall, L. A. Wilson, S. J. Rose, O. Guilbaud, S. Kazamias, M. Pittman, K. Cassou, J. Demailly, O. Delmas, A. Mestrallain, M. Farjardo and D. Ros

Abstract Extreme ultra-violet (EUV) laboratory lasers can be used to probe energy transport in laser irradiated solid targets. We report on a recent experiment undertaken at LASERIX whereby the heating of laser-irradiated targets containing a thin layer of iron (50 nm) encased in plastic (CH) was diagnosed using EUV laser (13.9 nm) back-lighter probing. The heating laser pulse duration was 35 fs with focal irradiances of 3×10^{16} Wcm^{-2} and a deliberate prepulse 20 ps before the main pulse at irradiances of 3×10^{15} Wcm^{-2}. A one dimensional hydrodynamic fluid code HYADES has been used to simulate the temporal variation in EUV transmission using IMP opacity values for the iron layer and the simulated transmissions compared to measured transmission values. When a deliberate prepulse is used to preform an expanding plastic plasma, it is found that radiation heating is dominant in the heating of the iron layer giving rise to a rapid decrease in EUV opacity and an increase in the transmission of the 13.9 nm laser radiation as the iron ionizes to Fe^{5+} and above.

1 Introduction

Energy transport in high density plasma is important in inertial fusion and other applications of laser-produced plasmas. The transport of energy generally occurs

M. Shahzad (✉) · G. J. Tallents · O. Culfa· A. K. Rossall
York Plasma Institute, The University of York, York YO10 5DQ, UK
e-mail: ms1108@york.ac.uk

L. A. Wilson
Central Laser Facility, STFC Rutherford Appleton Laboratory, Oxfordshire OX11 0QX, UK

S. J. Rose
South Kensington Campus, Imperial College London, London SW7 2AZ, UK

O. Guilbaud · S. Kazamias · M. Pittman · K. Cassou · J. Demailly · O. Delmas · A. Mestrallain · D. Ros
Campus de l'ENSTA, Chemin de la Huniere, LASERIX, Universite Paris-Sud, Palaiseau Cedex, 91761, France

M. Farjardo
Centro de Fisica de Plasmas, Instituto Superior Técnico, Av. Rovisco pais 1, 1049-001 Lisbon, Portugal

© Springer International Publishing Switzerland 2016
J. Rocca et al. (eds.), *X-Ray Lasers 2014*, Springer Proceedings in Physics 169,
DOI 10.1007/978-3-319-19521-6_54

via radiation transfer, hot electron transport or thermal energy transport, with non-linear interactions between these processes if more than one is important. Energy transport can be diagnosed using buried layer targets where the emission or absorption properties of a thin layer of material in a solid target acts as a signature of heating [1]. The properties of the buried layer can also be used to diagnose equation of state and opacity information at high density [2].

Thermal electron transport in laser-produced plasmas cannot be modelled using Spitzer-Harm thermal conductivity [3] as the gradients of temperature and density are steep and the assumptions of classical plasma thermal conductivity break-down. An extension model assumes that the energy transport Q arises due to the movement of energy at some fraction f of the thermal velocity, where f is known as the flux limiter. We can write for the flux limited energy transport

$$Q_{FS} = f n_e k_B T_e v_{th} \tag{1}$$

where v_{th} is the thermal sound speed.

Hot electrons are produced by a variety of process in laser-produced plasma, but at moderate intensities (10^{14}–10^{18} Wcm^{-2}) resonance absorption dominates. Here the p-polarised laser light generates large fluctuations in charge density and resonantly drives a plasma wave. Hot electrons can heat a solid target uniformly as at energies $>$ 10 keV, they have typical penetration in solid $>$ 500 nm. However, due to their large penetration depths, energy deposition in the solid target can be low. Radiation from the expanding plume of a laser-produced plasma can heat the solid target as the penetration depth of typical radiation emission (photon energies 100 eV–1 keV) is 20–140 nm (see Fig. 4).

In this paper, we report energy transport measurements obtained when a 35 fs pulse interacts with solid plastic targets with a buried layer of iron. The laser irradiance employed is 3×10^{15} Wcm^{-2} with a 35 fs pre-pulse irradiance impinging 20 ps before the peak of the main pulse. Using a short laser pulse duration ensures that profile modification during resonance absorption is not important. We probe the heating of the buried iron layer using the LASERIX extreme ultra- violet (EUV) laser produced from a Ni-like silver transition at 13.9 nm wavelength. The iron becomes transparent to 13.9 nm radiation when the iron ionization energy exceeds the photon energy (89 eV), This occurs once Fe^{5+} with ionization energy 99.1 eV is produced. $Fe^{4+}$with ionization energy 75 eV is significantly more opaque.

2 Experiment

The targets comprised 39 nm frontside and 207 nm rearside parylene-N layers encasing a buried 50 nm iron layer. The targets were irradiated by p-polarized light of 35 fs duration at 3×10^{16} Wcm^{-2} and a pre-pulse of 3×10^{15} Wcm^{-2} 20 ps before the main pulse. The laser of wavelength 0.8 μm was focussed to a focal spot diameter of $\sim$ 40 μm at a range of angles of incidence 18–33°. The EUV laser

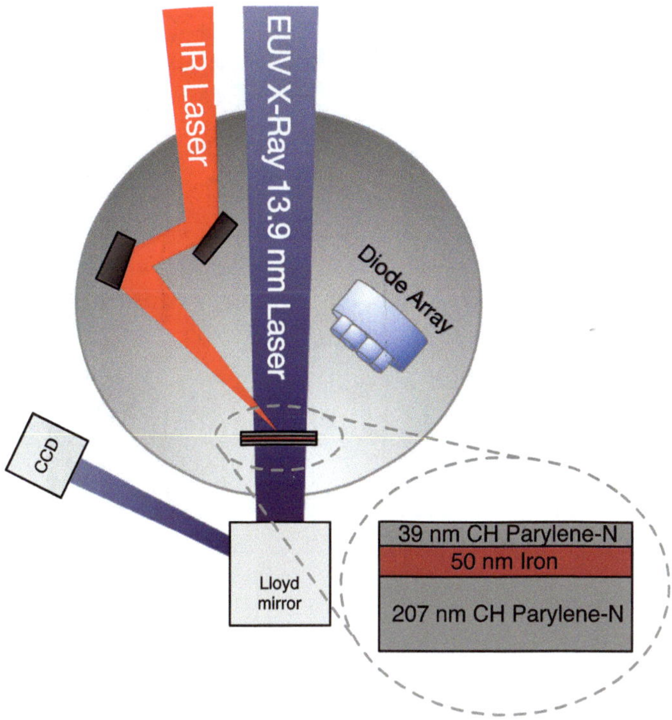

Fig. 1 Experiment layout with a 3×10^{16} Wcm^{-2} IR laser and 10 % prepulse incident upon a buried 50 nm iron layer encased in 39–207 nm parylene-N layers. Transmission through the iron layer was measured using an EUV laser back-lighter. A diode array operating in photovoltaic mode was introduced to determine the extent of hot electron heating. The buried iron target was rotated to vary the angle of incidence of the laser

pulse of estimated duration 3 ps was timed to probe the irradiated target at different times before and after the 35 fs main pulse was incident. An EUV imaging system (see Fig. 1) showed the spatial variation of the transmitted 13.9 nm laser radiation, enabling the extra transmission occurring due to the iron heating in the 35 fs laser focal region to be measured (see Fig. 2). Examples of the experimental transmission of the target at 13.9 nm as a function of time are shown in Fig. 3.

3 Analysis and Results

The one dimensional code HYADES [4] was used to simulate the infra-red laser interaction with the buried layer target. Flux limited thermal conduction and resonance absorption heating are modelled in the code. The effect of hot electron heating of the target was taken into account by depositing a 10 % fraction of the laser energy into the target, distributed according to the mass density of the target.

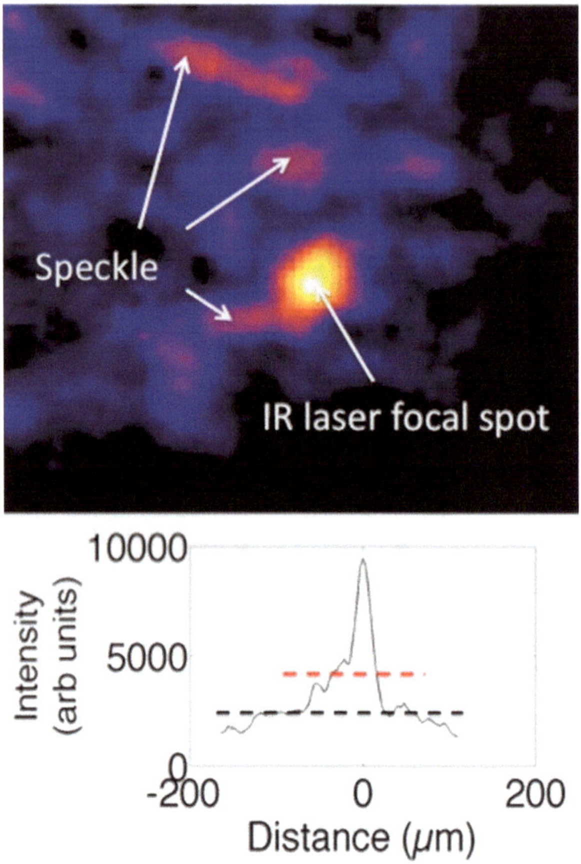

Fig. 2 Transmission CCD image with the infra-red laser incident. The EUV laser coherence caused speckle on the focussed EUV beam at the target. The red dashed line shows the level of speckle and the black line indicates the level of iron self emission

Radiation transport is modelled by multigroup diffusion. The HYADES predictions for density and temperature were post-processed using IMP opacities to deduce the transmission of the 13.9 nm radiation through the target [1]. The simulations best fit the experimentally measured 13.9 nm transmission when radiation transport is switched on, but hot electron heating with energy fraction (10 %) as envisaged to be feasible, has a negligible effect on the simulated transmission (see Fig. 3). The radiation absorption can be calculated by using the TOPS opacity code [5] and the predicted HYADES density and temperature conditions to determine the absorption for a range of photon energies. A comparison between the two competing heating processes show that the radiation absorption is relatively higher (see Fig. 4).

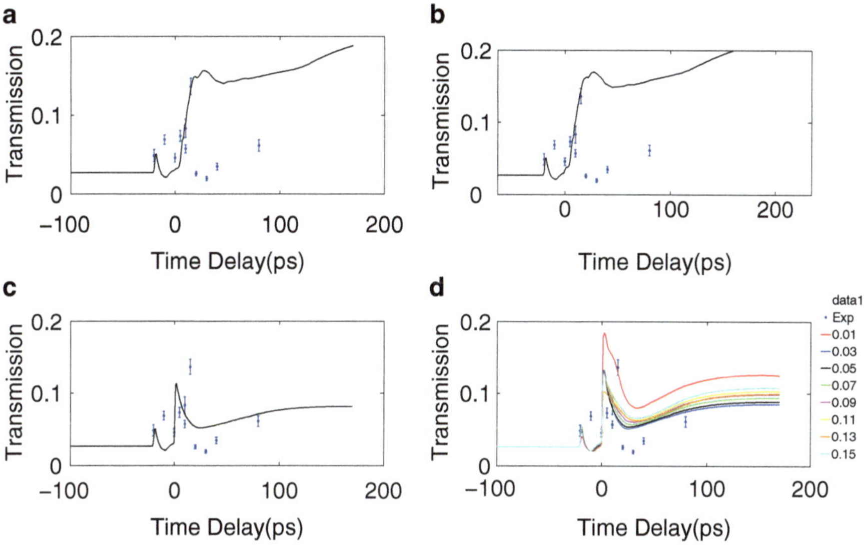

Fig. 3 Transmitted EUV laser light overlaid with HYADES transmission simulations such that **a** radiation and hot electron heating are turned off, **b** radiation heating is turned off but hot electron heating is turned on, **c** radiation heating is turned on, hot electron heating is turned off, and **d** radiation and hot electron heating are turned on. The heating laser irradiation 3×10^{16} Wcm^{-2} was incident at 18° with respect to the target normal. The flux limiter was set at 0.05

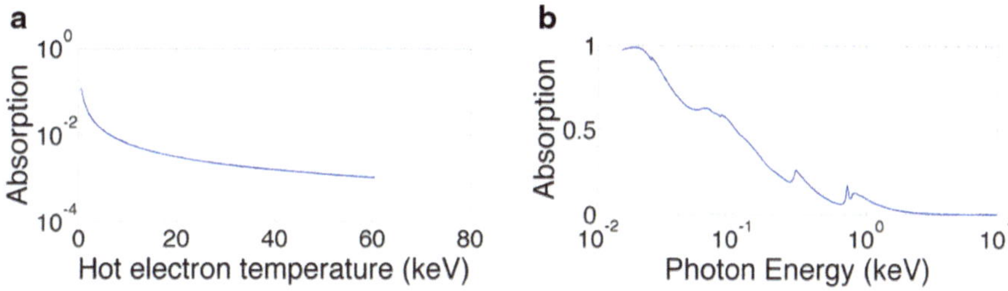

Fig. 4 **a** Hot electron absorption as a function of hot electron temperature. **b** Radiation absorption as a function of photon energy showing relatively higher absorption between 0.1 and 1 keV

4 Conclusion

We have shown that introducing a 35 fs prepulse can manipulate the energy transport mechanisms in the iron buried layer target, such that radiation heating dominates. Subsequent probing by the 13.9 nm Ni-like silver EUV laser allows us to measure the switch in opacity as the iron is heated to Fe^{5+} and above. The HYADES and IMP opacity codes have simulated this signature switch in transmission and have shown that hot electron heating is negligible within the iron layer.

References

1. Wilson, L.A., Tallents, G., Pasley, J., Whittaker, D.S., Rose, S.J., Guilbaud, O., Cassou, K., Kazamias, S., Daboussi, S., Pittman, M., Delmas, O., Demailly, J., Neveu, O., Ros, D.: Energy transport in short-pulse-laser-heated targets measured using extreme ultraviolet laser backlighting. Phys. Rev. E. **86**, 026406 (2012)
2. Edwards, M.H., Whittaker, D., Mistry, P., Booth, N., Pert, G.J., Tallents, G.J., Rus, B., Mocek, T., Koslová, M., McKenna, C., Delserieys, A., Lewis, C.L.S., Notley, M., Neely, D.: Opacity measurements of a hot iron plasma using an x-ray laser. Phys. Rev. Lett. **97**(3), 035001 (2006)
3. Malone, R.C., McCrory, R.L., Morse, R.L.: Indications of strongly flux-limited electron thermal conduction in laser-target experiments. Phys. Rev. Lett. **34**(12), 721–724 (1975)
4. Larsen, J.T., Lane, S.M.: HYADES—a plasma hydrodynamics code for dense plasma studies. J. Quant. Spectroscop. Radiat. Transf. **51**, 1–2 (1994)
5. Abdallah Jr, J., Clark, R.E.H.: Tops: a multigroup opacity code (No. LA-10454). Los Alamos National Lab, NM (USA) (1985)

MIX
Papier aus verantwortungsvollen Quellen
Paper from responsible sources
FSC® C105338

If you have any concerns about our products,
you can contact us on
ProductSafety@springernature.com

In case Publisher is established outside the EU,
the EU authorized representative is:
Springer Nature Customer Service Center GmbH
Europaplatz 3, 69115 Heidelberg, Germany

Printed by Libri Plureos GmbH
in Hamburg, Germany